A Life of Sir Francis Galton

A LIFE OF SIR FRANCIS GALTON

From African Exploration to the Birth of Eugenics

Nicholas Wright Gillham

2001

OXFORD
UNIVERSITY PRESS

Oxford New York
Athens Auckland Bangkok Bogotá
Buenos Aires Cape Town Chennai Dar es Salaam
Delhi Florence Hong Kong Istanbul Karachi Kolkata
Kuala Lumpur Madrid Melbourne
Mexico City Mumbai Nairobi Paris São Paulo Shanghai
Singapore Taipei Tokyo Toronto Warsaw
and associated companies in
Berlin Ibadan

Copyright © 2001 by Oxford University Press, Inc.

Published by Oxford University Press, Inc.
198 Madison Avenue, New York, New York 10016

Oxford is a registered trademark of Oxford University Press

All rights reserved. No part of this publication may be reproduced,
stored in a retrieval system, or transmitted, in any form or by any means,
electronic, mechanical, photocopying, recording, or otherwise,
without the prior permission of Oxford University Press.

Library of Congress Cataloging-in-Publication Data
Gillham, Nicholas W.
A life of Sir Francis Galton : from African exploration to the birth of Eugenics
p. cm.
Includes bibliographical references and index.
ISBN 0-19-514365-5 (acid-free paper)
1. Galton, Francis, Sir, 1822–1911.
2. Geneticists—England—Biography.
3. Eugenics—History.
4. Genetics—History.
I. Title.
QH429.2.G35 G53 2001
576.5'092—dc21
[B]   2001021612

1 3 5 7 9 8 6 4 2

Printed in the United States of America
on acid-free paper

To Carol

CONTENTS

	Preface	ix
	Prologue	1
	Francis Galton in Perspective	

I. ANTECEDENTS AND BEGINNINGS

1	An Enviable Pedigree	13
2	Metamorphosis	23
	From Birth to Medical School	
3	A Poll Degree from Cambridge	37
4	Drifting	47

II. GEOGRAPHY AND EXPLORATION

5	South Africa	61
6	Making Peace with Jonker Afrikaner	67
7	Expedition to Ovampoland	79
8	Fame and Marriage	93
9	Riding High with the Royal Geographical Society	107
	I. The Great Lakes of Africa	
10	Riding High with the Royal Geographical Society	123
	II. Stanley Faces Off with the Geographers	
11	Weather Maps and the Anticyclone	140

III. THE TRIUMPH OF THE PEDIGREE

12	Hereditary Talent and Character	155
13	Gemmules, Rabbits, Germs, and Stirps	173
14	Nature and Nurture	187
15	Sweet Peas and Anthropometrics	195

16	Probing the Mind	215
17	Fingerprints	231
18	The Birth of Biometrics	250
19	Galton's Disciples	269
20	Evolution by Jumps	286
21	The Mendelians Trump the Biometricians	303
22	The Triumph of the Pedigree *Eugenics*	324

Epilogue   345
Out of Pandora's Box:
The First International Congress of Eugenics

Notes	359
Bibliography	390
Index	398

PREFACE

With the rapid advances in modern human genetics, ethicists and scientists have begun to worry about whether a new eugenics, like some prehistoric monster, is emerging from its slumber and that eugenics will increasingly take control of mankind's hereditary destiny. Because such eugenic considerations impinge upon us once again it seemed the right time for a new biography of Francis Galton. After all, Galton invented the term and advocated the idea of applying the principle of selection to human beings to breed a better race.

Galton, a talented Victorian scientist, made lasting contributions in fields as diverse as African exploration, geography, meteorology, statistics, psychology, personal identification, and human heredity. Given that he was an important figure in the development of such a diversity of fields it is surprising that Galton has been the subject of only two biographies. The first is the four-volume labor of love published from 1914 to 1930 by his devoted disciple, Karl Pearson.[1] It is full of letters, photographs, and Pearson's own trenchant remarks. Nevertheless, one must go beyond the voluminous correspondence quoted in Pearson's biography to that preserved in the Galton Archive, held by the Manuscripts Room at University College, London, since Pearson tended to omit letters from certain people he did not respect, or disagreed with, or of whose importance he was unaware. Thus the Archive contains numerous letters from the accomplished statistician Francis Ysidro Edgeworth, whom Pearson did not hold in high regard. Not one is quoted in his biography. Pearson was involved in a long-running controversy with William Bateson, Mendel's great British champion, and Bateson's letters are reproduced sparingly although his alleged capacity for troublemaking and belittling of Pearson are referred to often.

Galton's second biographer, D. W. Forrest, published his study in 1974.[2] It brought Galton into sharper focus, filling many of the chinks that are empty in Pearson's monumental, but somewhat disjointed study. However, Forrest's biography failed to develop certain themes. Perhaps the most important of these involves the conflict that arose early in the twentieth century between Galton's three main disciples. On one side was William Bateson, representing the Mendelian view of the world, and on the other Karl Pearson and W. F. R. Weldon, the great advocates of biometrics. Their quarrel had important consequences for the way in which eugenics developed.

Lastly I felt that neither the Pearson nor the Forrest biographies satisfactorily placed Galton in context as the best biographies of eminent Victorians do. As Adrian Desmond wrote in his fine study of Thomas Henry Huxley, "the old history of ideas . . . displaced the person, made him or her a *dis*embodied ghost, a flash of transcendent genius. Only by embedding Huxley can we appreciate his role in the vast transformation that staggered our great-grandfathers."[3] Following Desmond's dictum, I have tried to avoid creating "a disembodied ghost" in favor of a creature of flesh and blood. After all, Galton meant well in his efforts to improve mankind, but he viewed the world through the lens of class, privilege, and the predominant role played by men in virtually all affairs in Victorian England. It is in this sense that I have tried to develop the man and his achievements for I do not believe he can be properly appreciated by applying modern or revisionist standards to his career. In this respect I should note that I have tried my best to set specific scenes or accounts of travel in the book as accurately as I was able based on existing factual material. In a few instances I have imagined a few details, for instance when Galton spoke before a large audience. These cases are clearly indicated by phrases like "Galton may have" or "one can imagine Galton."

The prologue, *Francis Galton in Perspective*, is meant to serve as a road map to Galton's career and to preview the significance of his many accomplishments in today's context. The first section of the book, *Antecedents and Beginnings*, investigates his ancestry, upbringing, training as a medical apprentice, and experience as a Cambridge undergraduate. The second section, *Geography and Exploration*, opens with Galton's post-Cambridge journey up the Nile to Khartoum and the story of his expedition to Namibia in 1851–52. Afterwards, he returned to England, married, and turned to travel writing. He also commenced on a long and close association with the Royal Geographical Society. Of particular interest is his relationship to the great British explorers who discovered the Central African lake system and the source of the Nile. I have tried to tell this part of the story from Galton's vantage point in London as a Fellow of the Royal Geographical Society. This section ends with a chapter on his foray into meteorology. The book, up to the middle of this section, follows Galton's career roughly chronologically. After that, I found, as did Galton's other biographers, that a more topical approach made sense.

The third section of the book, *The Triumph of the Pedigree*, considers Galton's investigations of hereditary mechanisms, psychology, fingerprinting, anthropometric measurements, and his role in the development of statistical methods. By now Galton was also a highly influential figure whose ideas concerning eugenic improvement of the human race received widespread approval both in Great Britain and abroad. The epilogue tells the story of the First International Congress of Eugenics, held the year after Galton's death. It concludes with a brief sketch of the consequences of the implementation of eugenic methods in Europe and the United States in the first half of the twentieth century.

Various people have been most helpful to me during the preparation of this manuscript. I am indebted to Ms. Gillian Furlong of the Manuscripts Room, University College, London, where the papers and correspondence of Sir Francis Galton are held. She and her efficient and helpful staff made my several visits there both pleasant and highly productive. I also wish to thank Dr. June Rathbone of the Galton Laboratory of University College for allowing me to examine Galton's book collection, which is under her care. Dr. Andrew Tatham, Keeper, the Royal Geographical Society, was most helpful in providing manuscripts and correspondence related to Galton's service with the Society. Miss Elizabeth Stratton kindly provided me with a print of a drawing of William Bateson that belongs to the John Innes Foundation Historical Collection. I also acknowledge with thanks the helpful staff of Perkins Library, Duke University, and my dear friend the late Professor Clyde deL. Ryals of the English Department at Duke, who made certain that the great collection of Victorian periodicals at Perkins was kept intact and easily available to users like myself. I am also most indebted to the Duke University Research Council for providing me with a travel grant that partially offset the expenses of my research trips to London. I am most grateful to my colleague Professor John Staddon of the Psychology Department at Duke for reading the entire manuscript. Professors Janis Antonovics, University of Virginia, and Seymour Mauskopf, Duke University, were kind enough to read early drafts of the first few chapters in the book while Professor Michael Wallis and Dr. Caryl Wallis, University of Sussex, read chapter 3 describing Galton's Cambridge education. My old friend Professor Irving Diamond of the Duke University Psychology Department was also most encouraging about this project in its early stages. I also thank the many Duke undergraduates I have been privileged to have in my seminar course "The Social Implications of Genetics." They are a bright, inquisitive lot and it is because of that course that I got the idea of writing this book. My Oxford editor, Mr. Kirk Jensen, has provided me with helpful guidance throughout. He has patiently read draft after draft of the manuscript and has constrained my prose as required while allowing me to paint with a broad brush. He also made excellent suggestions for condensing and tightening up the book. Finally, my dear wife Carol has patiently lived through Galton with me, reading a number of the earlier chapters and making many astute suggestions on how they could be improved. I dedicate this book to her with my greatest thanks for everything she has done over our many years together to make my life as rich and rewarding as it has been.

PROLOGUE

Francis Galton in Perspective

The January 11, 1999, issue of *Time* magazine ran a series of articles entitled "The Future of Medicine," which was devoted to the effects of the genetic revolution on mankind. One by Paul Gray was called "Cursed by Eugenics." He wrote that the "rise and fall of the theory known as eugenics is in every respect a cautionary tale. The early eugenicists were usually well-meaning and progressive types. They had imbibed their Darwin and decided that the process of natural selection would improve if it were guided by human intelligence. They did not know they were shaping a rationale for atrocities."[1]

The term "eugenics," which derives from the Greek stem meaning "good in birth," was coined by Francis Galton, a cousin of Charles Darwin's. After reading the *Origin of Species*[2] Galton concluded that it might be possible to improve mankind by selective breeding. To underpin his theory of evolution under natural selection Darwin began his book with a chapter entitled "Variation under Domestication." There he discussed the methods of artificial selection used to develop cultivated plants and domesticated animals. An obvious deduction at the time would have been that mankind might be improved similarly. Although Galton was far more interested in positive eugenics, the selective breeding of those perceived to be genetically superior, he recognized the complementary importance of negative eugenics, the prevention of those deemed genetically inferior from reproducing. In many ways Galton's view of a genetic utopia is captured best in an unpublished novel written shortly before his death titled *Kantsaywhere*. Inhabitants of Kantsaywhere were required to take an examination that vetted them genetically. Failures had inferior genetic material and were segregated in labor colonies where conditions were not onerous, but celibacy was enforced. Those passing the examination with a "second-class certificate" could propagate "with reservations." Those who did well took the honors examination at the Eugenics College of Kantsaywhere and were granted "diplomas for heritable gifts, physical and mental." These elite individuals were encouraged to intermarry.

What eugenics wrought in the first half of the twentieth century was much worse than anything Galton would have envisioned. More than 60,000 court-ordered sterilizations for eugenic reasons were carried out in the United States alone.[3] Furthermore, involuntary sterilizations for eugenic purposes took place in countries like Sweden, Norway, Switzerland, and the Canadian Province of Alberta. In some cases sterilizations continued into the 1970s. Worst of all was Nazi Germany.[4] Nearly 400,000 people were sterilized on the recommendation of Genetic Health Courts for supposed genetic diseases like alcoholism, feeblemindedness, and schizophrenia. A flurry of other negative eugenic legislation followed that was capped by the three Nuremberg Laws of 1935 designed to "cleanse" the German population of unwanted elements. The process was so horrific that for nearly two decades after World War II eugenics was little discussed and became almost forgotten. The Nazis also encouraged positive eugenics, at least among those of "Aryan" stock.[5] Its most striking manifestation was the Lebensborn (Well of Life) program in which the black-uniformed warriors of the SS coupled with unmarried women of suitable background for the good of the Fatherland. These women and their supposedly genetically superior children were cared for in maternity homes and child care institutions run by the program. The offspring could be adopted and were considered as future material for the SS.

Although Galton's name is linked inextricably to eugenics, he was a man of diverse interests and many achievements. To those who study the history of Africa, he is a nineteenth-century explorer and geographer. He was also a well-known travel writer. To meteorologists he is remembered as the discoverer of the anticyclone. Those who plumb the history of statistics will find Galton's name associated with regression, correlation, and the founding of biometrics. Psychologists, especially those interested in mental imagery, claim him as one of their own. Forensic experts recognize Galton as playing a central role in putting fingerprints as evidence on a firm scientific footing. And last, but certainly not least, Galton's name will always be linked with the founding of human genetics, the analysis of pedigrees, and twin studies. On his death Galton established a monument of lasting significance, the Galton Laboratory and Galton Professorship at University College, London. That chair has been occupied by great biometricians like Karl Pearson and R. A.

Fisher and renowned human geneticists such as Lionel Penrose and Henry Harris. These scientists made lasting contributions to our understanding of human genetic disease and to the development of important statistical tools used in genetics and elsewhere.

Because of his seemingly endless array of interests, Francis Galton is sometimes called a dilettante. This reveals a misunderstanding of both the man's achievements and the nature of Victorian science. Galton's research and published work revolved around two distinct sets of problems. During the first part of his career he was engaged in exploration and geography. Travel writing related to this interest as did meteorology, for the explorer is forever having to take account of the vicissitudes of the weather. The second part of his career was devoted to human heredity. To investigate the heritability of what Galton referred to as "talent and character," in the age before the rediscovery of Mendel's principles and the development of IQ tests, Galton used pedigrees, twin studies, and anthropometric measurements. He invented new statistical tools to analyze the masses of data he accumulated. He believed favorable physical characteristics correlated with superior mental qualities, but he had no way to measure the latter directly. Consequently, he ventured into psychology and personal identification, eventually lighting upon fingerprinting as a foolproof way to distinguish individuals.

The diversity of Galton's interests was not atypical for a Victorian scientist. His grandfather, Erasmus Darwin, was a highly successful physician, a serious student of botany and zoology, an inventor, and a talented poet. William Whewell, the Master of Trinity College, Cambridge, while Galton was a student there, had studied and written about philosophy, mathematics, mechanics, theology, and moral philosophy. He published a book about his theory of Gothic design and authored a treatise on the classification of minerals. Galton, like Charles Darwin, was independently wealthy and could spend full time on whatever interested him, but most scientists were not so lucky. T. H. Huxley, lacking a fortune, had to work hard to support himself and his family as a scientist and teacher. Many other scientists combined their investigations with an independent business that put bread on the table. Galton's friend, the mathematician William Spottiswoode, was printer to the Queen. Charles Booth, whose 17-volume work *Life and Labour of the People of London* (1891–1903) is a classic in early sociology, founded and chaired a successful steamship company with his brother Alfred.

Galton was born in 1822 and like Darwin, 13 years his senior, his initial trajectory was medicine. While studying medicine at King's College in London, Galton became friendly with his cousin who was living nearby. Under Darwin's influence Galton convinced his father to send him to Cambridge to study mathematics. Athough Galton failed to graduate with honors in mathematics, his enthusiasm for analyzing scientific problems quantitatively was unaffected. This is the unifying thread that runs throughout his career. Following Cambridge Galton, like many a modern undergraduate, spent a few years travelling and finding himself. Afterwards, under the auspices of the Royal Geographical Society, he mounted and financed his own expedition to Africa. This resulted in the first European exploration of the northern half of Namibia. His penchant for quantification was apparent in his detailed planning for the expedition and in the careful measurements he took of longitude, latitude, temperature, etc., often under trying circumstances. He was awarded a gold medal by the Society for this work largely in recognition of his quantitative contributions to geography. Galton then returned to London, married, and became a travel writer. Probably his best-known book, just republished once again, is *Art of Travel*, a practical guide for explorers and travellers in the bush.[6] In that book his pleasure in making calculations is often apparent as he advises the reader how to plan adequate supplies for an expedition or how to find one's way when lost.

Galton long served the Royal Geographical Society in an official capacity and his passion for quantitative geography was apparent in his papers and in his ceaseless prodding of explorers for precise data. He was at the center of an acrimonious dispute with Henry Morton Stanley following Stanley's relief of Livingstone. This was fueled partly because Galton regarded Stanley merely as an ambitious reporter and not a serious geographer. His interest in weather systems in connection with geography led him to discover the anticyclone and he played an important role in meteorological science in Great Britain for many years.

So Galton had already acquired fame as an explorer, geographer, travel writer, and meteorologist by midlife when he became interested in human heredity. His initial contribution to the subject, a two-part article "Hereditary Talent and Character" in *Macmillan's Magazine* in 1865, was a defining event for two reasons.[7] First, he used a new technique, pedigree analysis, to examine the inheritance of "talent and character." Today pedigree analysis is the essential analytical tool that human geneticists use in localizing the genes responsible for different human maladies both physical and mental. Second, as Galton's first biographer Karl Pearson wrote, "Hereditary Talent and Character" " . . . is really an epitome of the great bulk of Galton's work for the rest of his life; in fact all his labours on heredity, anthropometry, psychology and statistical method seem to take their roots in the ideas of this paper."[8]

Galton followed the article with his book *Hereditary Genius*.[9] This expanded greatly on his original theme and added an important new dimension, the normal distribution. He was enamoured by the possibilities the bell curve presented for the statistical analysis of quantitative data. He had seen how it could be applied to height or chest width, but what about human intelligence? With the IQ test still far in the future he hit upon the idea of using scores for the mathematics honours examination at Cambridge and the entrance examination to Sandhurst as proxies for measuring intelligence. He found that the scores for the latter examination began to approximate the normal distribution, probably the first time this distribution was applied to a numerical measure of intelligence. But most of *Hereditary Genius* was devoted to the pedigrees of eminent men. If eminence is heritable, he reasoned, those most closely related to the eminent man (e.g., fathers and sons) were most likely to be eminent. Galton then employed a technique we use commonly today. He converted qualitative information to numerical data by making the assumption that a man (women were excluded) was either eminent or he was not. He tabulated his results and believed he had supported his hypothesis. Darwin wrote Galton praising his new book highly and cited it approvingly in *The Descent of Man and Selection in Relation to Sex* (1871).[10]

But Galton had not satisfactorily ruled out the role of environment. Could the eminent man's father have helped him to garner a desirable position? Although he made various weak arguments to refute this possibility, his hereditarian thesis was seriously challenged by the Swiss botanist, Alphonse de Candolle. He pressed the environmentalist case in his book *Histoire des Sciences et des Savants depuis Deux Siècles* (1872).[11] Consequently, Galton set about gathering data on eminent English scientists to support his hypothesis. To obtain the information required, he designed a detailed questionnaire and, having tabulated the results, published his findings in a slim little volume, *English Men of Science: Their Nature and Nurture* (1875).[12] The phrase "nature and nurture" is probably as well known as natural selection, but while everyone associates the latter phrase with Charles Darwin, almost no one realizes that the former phrase was coined by Francis Galton. Raymond Fancher in his *Pioneers of Psychology* remarks that while Galton's analysis of his data was "naive" the real virtue of his book was "its demonstration that the statistical analysis of questionnaire data was a potentially valuable approach to psychological questions."[13]

Soon thereafter Galton had the clever idea of using twin studies to investigate the relative contributions of nature and nurture to human ability. He realized there were two kinds of twins, what we would call identical or monozygotic and nonidentical (fraternal) or dizygotic. Identical twins were usually reared in the same environment and Galton wanted to know what happened after they flew the nest. Did they retain similar habits once they were apart? He collected information on 35 sets of identical twins and in a magazine article recounted a series of anecdotes about uniquely similar behavior characteristics he found among certain pairs of identical twins.[14] Psychologists have done the same thing ever since. While they usually use IQ tests to measure intelligence and try to compare identical twins reared apart to assess the roles of nature and nurture in determining mental ability, they also tell anecdotes like Galton's. Some fine examples are to be found in Lawrence Wright's article "Double Mystery" published in the *New Yorker* in August, 1995.[15] These anecdotes have the effect of making identical twins seem even more similar, reinforcing the hereditarian argument. Dissimilarities, in contrast, are usually not mentioned probably because they are hard to compare, but the upshot is surely a distorted picture of heritability. Galton, like modern investigators, found that fraternal twins were quite dissimilar.

As Galton was completing *Hereditary Genius*, Darwin published a major new work, *Variation of Animals and Plants under Domestication* (1868),[16] where he laid out his "Provisional Hypothesis of Pangenesis." In contrast to the prevailing blending or "paint pot" theories of heredity, Darwin envisioned particles he called gemmules as the hereditary elements. These were transmitted from all parts of the body to the sexual organs. It was essential that his hypothesis explain the origin of the variations upon which natural selection acted, and he posited two mechanisms. The first resulted when injury to the reproductive organs occurred. This might lead to the improper aggregation of gemmules so that some were in excess and others in deficit resulting in variation and modification. Darwin's second mechanism imagined that heritable modifications could be induced in gemmules by the direct action of changed environmental conditions. These acquired modifications were transmissible and resulted in selectively advantageous alterations among the progeny. Galton rejected Darwin's notion of acquired characteristics on theoretical grounds and on the basis of negative results from experiments he carried out with rabbits in collaboration with Darwin. These tested the hypothesis that the gemmules were carried in the bloodstream.

To explain the phenomenon of "reversion," the occasional appearance of an ancestral characteristic among progeny, Darwin had postulated the existence of dormant or latent gemmules that remained hidden most of the time. Galton liked this idea and gradually developed a theory of particulate inheritance that also assumed the existence of two sorts of elements. One set (latent) was transmitted between generations while the other (patent) was responsible for determining the characteristics of an organism. Latent elements could give rise to patent elements, but not the reverse. The modern analogy would be the unidirectional transfer of information from genes to proteins. Galton's hypothesis left no room for acquired characters, so improvement of the human stock could only be accomplished through selective breeding. Later the great German biologist August Weismann recognized that Galton's hypothesis was essentially equivalent to his own highly acclaimed theory separating the soma from the germ line.[17]

Meanwhile Galton was anxious to apply the properties of the normal distribution to various human characteristics such as height and to determine their heritability, but since this was difficult he chose a model system, as biologists often do, in his case sweet peas. They were self-fertilized as far as he knew; they were hardy and prolific; and seed weights and sizes varied little within individual pods. From these experiments Galton made three important observations. First, seed size was normally distributed among parents and progeny. Second, the mean size of progeny of large seeds "regressed" or returned toward the mean of original population from which they were drawn. Third, he drew the first regression line by plotting the average diameters of parental seeds against those of progeny seeds and connecting the points. He also computed the first regression coefficient. These statistical concepts have such broad application today that it is hard to imagine how unique was the process that led to their discovery. Peter Bernstein in his fascinating history of risk analysis, *Against the Gods: The Remarkable Story of Risk*, has a chapter on Galton.[18] In it he shows how regression to the mean applies to mutual funds. For example, in one period international stock funds may be up and in the next period regress toward the mean while the reverse is true for aggressive growth funds.

In 1883 Galton published *Inquiries into Human Faculty and Its Development*[19] where he pulled together the results of his twin studies, his thoughts on anthropometrics and statistics, and touched upon topics like psychometrics, psychology, race, and population. It was in this book that he coined the term "eugenics." But what he really wanted for analysis was anthropometric data. He began by obtaining heights and weights of English schoolchildren, extending his data base vastly when he set up an Anthropometric Laboratory at the International Health Exhibition in 1884. He equipped and maintained the laboratory at his own expense. After the exhibition closed, he continued to operate the laboratory in the Science Galleries of the South Kensington Museum, collecting enormous amounts of information not only on heights and weights, but on reaction time, strength of pull and squeeze, color sense, etc.

In the midst of his inquiries into heredity, anthropometrics, and statistics Galton took a detour. From 1877 until 1885 he was preoccupied with psychological studies. His real goal was to measure mental ability, but the IQ test was still far in the future. His first approach was rooted in physiognomy. If a certain group of individuals shared a particular mental trait, and this was somehow reflected in their physical appearance, the common features might be extracted by superimposing photographs of their faces on one another. This should factor out the unique features and emphasize shared attributes, a photographic mean or average as it were. With the aid of the Director General of Prisons, Galton examined many thousands of photographs of thieves, murderers, etc., hoping that composite photography would reveal features that typified different groups of criminals. These, and other studies using composite photography in race and pedigree analysis, failed to provide Galton with the key to personality type that he sought. Ironically, Professor David Hopkinson and his colleagues at the Galton Laboratory at University College, London, are currently studying the underlying relationship of genes to human facial features.[20] A principal tool is an optical surface scanner that allows the face to be digitized into approximately 30,000 coordinates. Composite photography is also a key tool in criminal identification today, but now the composite pieces are arrayed in combinations that identify unique individual features.

Galton next examined mental imagery, the nature of the evanescent impressions that pass through one's consciousness. He wanted to apply quantification to this psychological process so he used questionnaires addressed to public school boys and entreaties to friends in various learned societies to help him gain the information he sought. In a recent volume of *Advances in Psychology*[21] devoted to mental imagery, Galton's work is cited frequently and the authors of one article have this to say: "When investigating mental imagery by subjective report, researchers still tend to rely on a small set of overvalued questionnaires, the content of which is derived mainly from Galton's original study."[22] Overvalued they may be, but it is a tribute to Galton's insight that a questionnaire he designed over a hundred years ago is still the basis for inquiries into this subject today.

In searching for methods of personal identification, Galton came across the classification system of the French criminologist Alphonse Bertillon, who combined photography (full face and profile) with precise measurements (height, limb length, head width, etc.) to characterize felons. Galton became intensely interested in "Bertillonage," as it was called, as a tool for personal identification, and actually visited Bertillon to learn more about his system. But he was also on the threshold of developing the statistical concept of correlation and realized that the set of individual measurements Bertillon made on each criminal did not necessarily represent independent variables. Meanwhile he became absorbed with another method of personal identification that seemed promising, fingerprinting, publishing several major papers and books on the subject (e.g., *Finger Prints*, 1892;[23] *Finger Print Directories*, 1895[24]). He used his Anthropometric Laboratory to gather thousands of fingerprints and developed a highly sophisticated system for classifying fingerprints. His painstaking comparisons allowed him to confirm a critical point for criminal identification, first noted by William Herschel, that fingerprints do not change with time. He also made estimates showing that the probability of any two individuals having the same print on a single finger was vanishingly small.

Galton's strong advocacy of fingerprints helped to bring them into use by Scotland Yard. However, his classification system was cumbrous and difficult to apply so it fell to Sir Edward Henry to construct a workable system of fingerprint classification. Even today in the era of DNA fingerprinting, conventional fingerprinting remains a major tool in criminal identification.

In the years 1888–89 Galton completed what were probably his two most influential scientific works. Both were based largely on the great mounds of data he had collected in his Anthropometric Laboratory. The first, a paper published in the *Proceedings of the Royal Society*,[25] described the correlations he had found in arm and leg lengths. Thus a person with long arms usually has long legs. In the same paper he computed the first set of correlation coefficients. Thousands of correlation coefficients are calculated today for all kinds of variables. Galton's second major work, *Natural Inheritance*,[26] launched the science of biometrics. This important book has never received the recognition it deserves, perhaps because it is sometimes confusing and Galton's prose is often elliptical. *Natural Inheritance* inspired Galton's two most devoted disciples, the great statistician Karl Pearson and the marine biologist Walter F. R. Weldon. Pearson took the statistical tools Galton had formed clumsily and laboriously, and almost effortlessly transformed and expanded them to define many of the elements in our modern-day armamentarium of statistics. Weldon used the methods Galton and Pearson had developed and applied them successfully to data he had gathered on shrimps and crabs. Together this triumvirate of mathematically inclined scientists would launch biometrics as a science and *Biometrika* as its flagship journal.

Galton had another disciple, William Bateson. Like Pearson and Weldon, Bateson was impressed by *Natural Inheritance*, but for a different reason. Whereas biometrics concerned itself with continuously varying traits like height and length, Bateson was mightily impressed with discontinuously varying traits such as white and red flower color. He collected hundreds of examples of such traits in his lengthy monograph *Materials for the Study of Variation* (1884).[27] For Bateson, discontinuous variation posed a serious difficulty for Darwin's theory of evolution. It was unclear to him how natural selection could act progressively on selected small alterations in a species when discontinuous variation seemed a much more likely source of diversity. The same problem concerned Galton, but for a different reason, regression to the mean. Thus the progeny of tall people tended to be closer to the mean of the population as a whole. Hence, it seemed to Galton that evolution could never progress by small increments, because it would be thwarted continually by regression to the mean. Instead, he visualized evolution as proceeding by a sudden change in the equilibrium of a species "a leap from one position of organic stability to another," a discontinuous change that would prevent regression to the mean. Galton elaborated his hypothesis in *Natural Inheritance* and it appealed greatly to Bateson. Similarly, Galton was lavish in his praise for Bateson's book. They both believed that evolution must proceed via some sort of discontinuous or saltatory process and the evidence they used to support this hypothesis was complementary.

In 1897 Galton promulgated a new law of heredity[28] dubbed by Pearson "Galton's Law of Ancestral Inheritance." This law derived directly from the way in which Galton viewed his pedigree data. He contemplated a continuous series in which parents contributed one half (0.5) of the total heritage of the offspring; the four grandparents one quarter $(0.5)^2$, the eight great-grandparents $(0.5)^8$, and so forth. By adding these contributions $(0.5) + (0.5)^2 + (0.5)^3 \ldots$ the whole series sums to 1. Galton tried to apply his law quantitatively by taking into account the mean and deviation of offspring from parents, parents from grandparents, etc., an approach greatly refined mathematically by Pearson. While Galton's law described the total, average genomic contribution of ancestors to progeny, it did not consider the fate of individual, discontinuously varying traits, the centerpiece of Mendel's theory. Upon its rediscovery Bateson immediately grasped the importance of Mendel's theory and wrote Galton about it, but Galton was probably too old by then or too enamored of his own hypothesis to attach great significance to the rediscovery.

These disparate theories of heredity led to a falling out between Bateson, one of Mendel's great defenders, and Pearson and Weldor. This was not be resolved until R. A. Fisher, in a paper published in 1918, showed that continuously varying traits could be reconciled with Mendelian inheritance.[29] Between 1900 and the First International Congress of Eugenics, held in London in 1912 a year after Galton's death, the Mendelians routed the biometricians. Consequently, at the Congress large numbers of pedigrees were presented of mostly imagined Mendelian segregations for all sorts of human ailments and difficulties ranging from feeblemindedness to alcoholism to tuberculosis. This put a strongly hereditarian spin on human frailty and infirmity. The most tangible result of the Congress was to foster the spread of negative eugenics in Europe and the United States.

Much of the last decade of Galton's life was directed toward promoting eugenics, mostly of the positive sort. One of the notions he peddled was the use of eugenic certificates to vet a person's heritage. While such certificates have not come to pass, pedigree data are becoming an ever more important part of one's medical record as geneticists unearth an ever-expanding collection of disease and susceptibility genes. Insurance companies are interested in data relating to genetic maladies with regard to risk assessment. Employers worry about susceptibility genes. Sperm banks exist, as does the possibility of elective abortion of genetically defective children and, in some countries, embryos of an unwanted sex. In an eerie throwback to Galton's unpublished novel *Kantsaywhere*, the 1997 movie *Gattaca* sketches a modern version of a eugenic paradise. Sex and procreation are totally decoupled in *Gattaca*. In vitro fertilization is used instead and the genetic profile of each embryo is vetted with only those lacking genetic defects being implanted. The protagonist is a product of the old-fashioned genetic lottery by which we bear children today. He has numerous potential genetic defects and is consigned to sweep floors. The rest of the story is not important for the purposes of this book, but the analogy is. *Kantsaywhere* and *Gattaca* are very similar places. The ethical problems posed in both cases are the same. The only difference is 100 years of technological advancement.

PART ONE

Antecedents and Beginnings

ONE

An Enviable Pedigree

Few men have had more noteworthy ancestry in many lines than Francis Galton.

—Karl Pearson [1]

Francis Galton invented pedigree analysis to measure the heritability of human "talent and character." This technique caught the imagination of eugenicists in the early twentieth century and is a fundamental tool of modern human genetics. One reason Galton set such store by this method was his own sterling pedigree (Table 1-1). His ancestors, the scientific and medically inclined Darwins, and the Galtons, a family of wealthy Quaker merchants, both hailed from the environs of Birmingham. Galton's cousin Charles Darwin inspired his investigations in human heredity while his father, Samuel Tertius Galton, endowed him with a substantial inheritance. This permitted him to roam without financial constraint through his various scientific pursuits. His father may also have sparked his lifelong interest in numbers and quantification. Francis Galton indeed had an enviable pedigree, so to understand the man one must know something of his family and especially his maternal grandfather Dr. Erasmus Darwin, the common familial link between Galton and Charles Darwin (Fig. 1-1).

Erasmus Darwin was a massive figure with a prominent stomach and a jowly face surmounted by a majestic nose.[2] His cheeks, pitted with old craters and scars, bore mute testimony to a severe childhood case of smallpox. Atop his head was a wig tied up behind in a little bob-tail. Darwin stammered when he spoke, but his physical shortcomings were soon forgotten as "no patient consulted Dr. Darwin who, so far as intelligence was concerned, was not inspired with confidence in beholding him; his observation was most keen; he

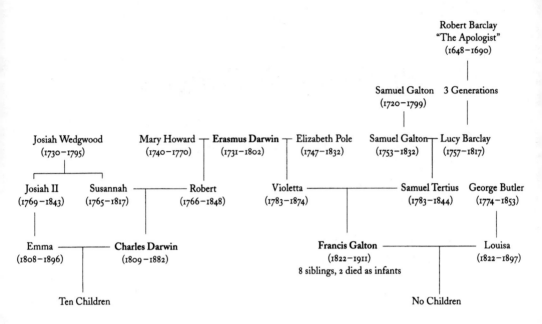

Table 1-1 Partial Pedigree of the Darwin, Galton, and Wedgwood Families

constantly detected disease, from his sagacious observation of symptoms apparently so slight as to be unobserved by other doctors."[3] Darwin's medical reputation became so great that King George III urged him to come to London to attend to his medical needs, but Darwin declined, preferring his life in Derby. Although Darwin often treated poor patients for nothing, his fees for a wealthy man like Samuel Galton were substantial. To see his patients Darwin probably bumped and jounced almost 10,000 miles a year around the countryside in Derbyshire and neighboring counties in all sorts of weather and on roads of varying quality and condition. Sooner or later he was bound to have an accident and in 1768, while riding in a two-wheeled carriage, the axletree broke, pitching him onto the road and breaking the patella of his right knee. Afterwards, Darwin always limped slightly, and it is not surprising that proper carriage design was among his many interests.

Darwin's good friend Josiah Wedgwood, Charles Darwin's grandfather (Table 1-1), was also interested in carriages, specifically their avoidance, as the china and crockery he manufactured was often smashed during overland transport on pitted and rutted roads. To minimize breakage Wedgwood was anxious to move his wares via canal to Liverpool or Hull where they could be conveniently exported. Hence Wedgwood, with Darwin's enthusiastic support, promoted extension of the Burslem-Trent canal to connect with the Mersey to make a "Grand Trunk Canal" from which other canals might later branch

Fig. 1-1 An Enviable Pedigree. *Top*: Erasmus Darwin from a print after a painting by Rawlinson of Derby. From Karl Pearson, *Life* I: plate 3. *Bottom Left*: Charles Darwin, aged 51. From Karl Pearson, *Life* I: plate 37. *Bottom Right*: Francis Galton, aged about 50, from *Life* I: plate 37.

off. Once parliamentary approval was secured in 1766, James Brindley, who designed the Burslem-Trent canal, extended it to the Mersey.

Darwin and Wedgwood belonged to the Lunar Society of Birmingham, which met monthly for discussion on the Monday afternoon nearest the full moon.[4] Its members were men of varied backgrounds united by their interest in the sciences pure and applied. The Lunar Society was born out of a more informal group, the Lunar Circle, initiated by Darwin and Matthew Boulton. Boulton, the son of a buckle maker, became the leading manufacturer in England. They were joined by Wedgwood and by Dr. William Small, a former professor of Natural Philosophy at the College of William and Mary who was a much-appreciated teacher of Thomas Jefferson. Gradually this group of like-minded enthusiasts grew to 14 and included notables like James Keir, a pioneer in the chemical industry, Joseph Priestley, the famous minister and chemist, and James Watt, one of the greatest British engineers and the inventor of the modern steam engine.

Erasmus Darwin was not only a gifted physician, but a talented inventor. One of his best known inventions was his speaking machine, which used his phonetic theory to divide the sounds of speech into four classes (vowels, sibilants, a mix of the two, and consonants). Its mouth was wooden with leather lips. An inch-long silk ribbon a quarter inch wide provided vocalization when a bellows passed an air current over it. "This head pronounced the *p, b, m*, and the vowel *a*, with so great nicety as to deceive all who heard it unseen, when it pronounced the words *mama, papa, map*, and *pam*; and had a most plaintive tone, when the lips were gradually closed."[5] Like his grandfather, Galton would also prove adept at model making. But it is for his major contributions to the natural sciences, particularly biology, that Erasmus Darwin is best remembered. In response to the great interest in plants generated by the appearance of the *Genera Plantarum* by the Swedish botanist Carl Linnaeus, who proposed the system of classification for living things we use today, Darwin published a four-part English translation between 1782 and 1785. In 1789 he published *The Loves of Plants*, part II of his encyclopedic poem *The Botanic Garden*, with part I, *The Economy of Vegetation*, appearing in 1791. *The Loves of Plants*, received with delight by the reading public, discussed the Linnaean classification of plants while at the same time humanizing them and their sex lives. *The Economy of Vegetation*, a far stronger and less frivolous poem, was divided into four cantos whose subjects were Fire, Earth, Water, and Air. Its subjects ranged widely from Watt's steam engines to Wedgwood's Portland Vase.

Darwin's next great work was *Zoonomia, or the Laws of Organic Life*, which he labored at for over 20 years.[6] By the third edition (1801) the original two volumes had fissioned into four. Volumes I and II of this edition dealt generally with medical or medically related topics (sleep, drunkenness, stomach, liver, etc.) while the other two volumes attempted Linnaean classification of the known diseases. Even though this classification system did not ultimately work very well, *Zoonomia* compiled an enormous amount of medical knowledge and personal experience. It was highly acclaimed and was translated into German, French, and Italian and there were at least five American editions. In chapter 39, "Of generation," Darwin proposed the rudiments of evolutionary theory, challenging the notion that species were unchanging. This was embedded not only in Christian teachings, according to which species were created by God and immutable, but in the Linnaean system of classification. Darwin imagined that over the millenia since the earth's creation warm-blooded animals somehow arose from "one living filament, which THE GREAT FIRST CAUSE endued with animality, with the power of acquiring new parts, attended with new propensities, directed by irritations, sensations, volitions and associations; and thus possessing the faculty of continuing to improve by its own inherent activity, and of delivering down those improvements by generation to its posterity, world without end?"[7] Later, in his poem *The Temple of Nature*, published after his death in 1802, Darwin clarified what he meant by one living filament writing that "all vegetables and animals now existing were originally derived from the smallest microscopic ones, formed by spontaneous vitality"[8] in primeval oceans.

Darwin's reputation suffered when George Canning, under-secretary for Foreign Affairs in Pitt's government, wrote *The Loves of the Triangles* (1798). This was a parody in Darwin's style of *The Loves of Plants*. He aimed to discredit the political and religious radicalism not only of Darwin, but of William Godwin, author of *Political Justice*.[9] The French Revolution was underway, inspired by the radical democratic members of the Jacobin Society. It was provoking strong anti-Jacobin sentiment in Great Britain, particularly within a government fearful of a similar uprising. Canning's poem, published in three sequential issues of *The Anti-Jacobin* under the pseudonym Higgins, implied that Godwin was the true author. It succeeded in (1) ridiculing Darwin's idea that human beings might have evolved from lower organisms, (2) that electricity might have important uses, and (3) that the Earth is much older than stated in the Bible. Galton commented years later that "Canning's parody *The Loves of the Triangles* quite killed poor Dr. Darwin's reputation."[10]

Charles Darwin, who knew his grandfather well and initially "admired greatly the *Zoonomia*," was disappointed on rereading it ten or 15 years later "the proportion of speculation being so large as to the facts given."[11] Darwin changed course again late in life when he published his 127-page biography as a "preliminary notice" to *Erasmus Darwin* by Ernst Krause whose own essay was only 86 pages long. Darwin's daughter Henrietta crossed out most references favorable to Erasmus Darwin with a thick blue pencil since, as a good Christian, she "did not wish to damage the Darwin family image by allowing her father to praise him."[12] Erasmus Darwin's prowess as a poet was assaulted by G. L. Craik in his popular *History of English Literature* in the midnineteenth century. L. V. Lucas in *A Swan and Her Friends* (1907) burlesqued Darwin by quoting Canning's poem *The Loves of the Triangles* and not one line of Darwin's. The final indignity was in the *The Stuffed Owl* (1930), an anthology of supposedly bad verse in which Darwin took a bow, but he was not lonely being joined by illustrious poets like Dryden, Byron, Wordsworth, and Tennyson.[13] Eventually others joined Charles Darwin in recognizing his grandfather as a man of great breadth and talent.

Erasmus Darwin was grandfather to Charles Darwin and Francis Galton by successive marriages (Table 1-1). He first wedded Mary (Polly) Howard following which they moved to a fine old house in Lichfield. Soon the family expanded to include four boys and a girl, two of whom expired within a year. While pursuing medical studies at Edinburgh, Charles, the eldest son and his father's favorite, died in 1778 at age 20 because he cut his finger dissecting the brain of a child who had died of "hydrocephalus internus." Mary Darwin's other two surviving children were Erasmus Jr. and Robert Waring. Following completion of his medical studies in 1786 Robert Darwin set up practice in Shrewsbury becoming an extremely successful provincial doctor. He married Susannah Wedgwood, the daughter of his father's old friend Josiah Wedgwood, and they had two sons and four daughters, their second son being Charles Darwin. Robert Darwin dominated his children as Erasmus had dominated his and, like his father, continued to increase in girth. He stopped weighing himself when he reached 24 stone (336 lbs) and had his coachman (also heavy) test the floor-boards of a new patient's house before he entered.

Mary Darwin died at 30, possibly from liver disease exacerbated by alcohol, leaving her husband, 38, with three young sons to rear. Within a year of her death he had struck up an acquaintanceship with a Miss Parker. It flowered so rapidly that before long she had born him two natural daughters who were treated as if they were his own legitimate children. However, when they grew old enough convention dictated that they would seek employment, perhaps as governesses, while legitimate daughters of the gentry prepared themselves for marriage to gentlemen of appropriate means and class. Within a few years Miss Parker and Darwin parted ways. She later married and, as far as anyone seems to know, lived happily ever after in a fine house in Birmingham.

In 1777 Darwin was smitten with a raven-haired beauty of 30, Mrs. Elizabeth Collier Sacheveral-Pole (Table 1-1), while paying a visit to the Poles' home at Radburn Hall near Derby to treat their three-year-old daughter Milly. The doctor, as he often did, prescribed a generous dose of opium and before long Milly recovered. Unfortunately Mrs. Pole was married to a man 30 years her senior, Colonel Edward Sacheveral-Pole. Given this impasse Darwin chose to court Mrs. Pole in verse. His strategy paid off since after the Colonel expired in 1780 he bid for Mrs. Pole's hand and succeeded against richer, younger, and better-looking suitors. The marriage yielded a ready-made family of three young Pole children and the Colonel's older natural son together with Erasmus Darwin's two natural daughters plus his sons Erasmus and Robert. The newly-weds, undaunted by their large flock, were prolific and added five additional sons and two daughters although one died as a baby and four in their thirties or forties. The remaining two children were Violetta, Francis Galton's mother, and Francis Sacheverall Darwin, the godfather of Francis Galton and a doctor like his father.[14]

The Galtons, Quakers who began as small businessmen, became ever more successful with each new generation.[15] The first Samuel Galton, the great grandfather of Francis Galton (Table 1-1), married Mary Farmer in 1746. The next year he became an assistant to his brother-in-law, James, and by 1753 had full partnership in the Farmer business. James possessed a large stake in the operation of his cousin, Benjamin Farmer, a merchant in Lisbon, but an earthquake struck in 1755 destroying the Farmers' Lisbon business and causing James Farmer to declare bankruptcy. Somehow, the circumstances are not entirely clear, Samuel Galton not only survived this crisis, but profited from it. The partnership between James Farmer and Samuel Galton was briefly dissolved and then reconstituted in 1757. In the process the estates of Duddeston and Saltley were assigned to Galton. By 1766 Samuel Galton's share of the business was worth £22,821. His wealth continued to grow as he accumulated more property after the deaths of his mother and brother John.

His granddaughter, Mary Anne Schimmelpenninck, loved to visit her grandfather at his fine country home, Duddeston House, with its great portico supported by four imposing Doric columns. He would call her at 6:00 A.M. to accompany him on his morning walk. They proceeded first to the little garden he had given her, then to the greenhouses, and then to a large lake with a stream running through it where seagulls swooped and wheeled while Muscovy ducks and Canada and Peruvian geese clamored and quacked near the lake's edge or swam in little convoys on its surface. Next they dropped by his nearby mill where he inquired after the health and well-being of his workers and then they breakfasted. Samuel Sr. gave each grandchild a guinea the day the child was born and on each successive birthday. He frequently added other gifts of a half-a-crown, or sometimes more. Being a good businessman, Samuel gave Mary Anne

> a little account book in which he desired I should set down accurately everything I received and expended. This was contrary to my natural taste and habits; it was also very different from my dear mother's magnificent manner of spending and acting in all that related to money: but one day my grandfather called me to him and said: "My child, thou didst not like when I advised thee, the other day to save thy sixpence, instead of spending it in barberry

drops and burnt almonds. . . . We cannot be self-denying wisely till we know the real value of what we give up; that is why I wish thee to keep exact accounts."[16]

Samuel Galton died at age 80 in 1799. His obituary in the *Gentleman's Magazine* for 1777 reported that he was a highly respected and hard-working citizen and generous with the local charities. He could well afford to be as his estate now was worth £139,000, in today's currency perhaps between £5-7,000,000.[17] Since Samuel Sr.'s six other children expired prior to his own death, the entire estate went to Samuel Jr. who had joined the firm of Galton and Farmer at the age of 17. At 21 his father transferred £10,000 to his account and made him manager of his Gun Foundry, an odd line of business in view of the pacifist teachings of the Quakers. Samuel Jr. was greatly interested in the sciences and joined the Lunar Society in 1781. He greatly admired Joseph Priestley, one of the giants of the Lunar Society, who came from a family of Calvinist dissenters and began his career as a minister.[18] Priestley's Lunar Society friends, including Samuel Jr., helped to cover the expenses of his scientific experiments with such tact that he was unaware of their support. Priestley's religious publications gained him fame in the Unitarian movement, but caused him much grief later on as his heretical religious views, prominence as an advocate for abolition of the slave trade, and his support for the American Revolution marked him as a radical. During the Church and King riots of 1791 his house near Birmingham, like those of other dissenters, was destroyed along with his apparatus and papers.

During these troubles, Samuel Jr. was one of Priestley's strongest supporters, sending him financial contributions while others provided chemicals and equipment. Priestley in turn helped Galton gain admission to the Royal Society for his one major scientific contribution, the color top. Newton had supposed that if his seven prismatic colors (violet, indigo, blue, green, yellow, orange, and red) occupied pie-shaped sections of a circle when the circle revolved swiftly around its center it would appear white. Galton deduced that blue, yellow, and red are the only true colors and mixed in the proper proportions should also produce white. He demonstrated this with spinning, circular cards. Meanwhile the Society of Friends finally took notice of Samuel Jr.'s lucrative gun trade and in 1795 he was formally disowned "for fabricating and selling instruments of war."[19] This greatly irritated Galton so he penned a rebuttal arguing that to be consistent, taxes should not be paid by any Friend to a Government that prepared for war, or riots. He observed that his grandfather, father, and uncle had been in the gun-making business for 70 years without the Friends raising any complaint. He did not offer to give up this lucrative trade and he ignored his disownment, continuing to attend Quaker meetings until his death in 1832. Since his regular donations were accepted, his "excommunication" seems at best to have been a face-saving gesture by the Friends. In 1804, three years after his father's death, Samuel Jr. wound up the gun business. The Galton-Farmer factory on Steelhouse Lane in Birmingham was converted into a bank in which his sons Samuel Tertius and Hubert Galton plus a colleague, Paul Moon James, were partners. Meanwhile the Galtons left the Society of Friends, first embracing Unitarianism, and later the Anglican or Roman Catholic faiths. Francis Galton's father Tertius was an Anglican. This had an important consequence for his son since he could apply for admission to Cambridge or Oxford when the time came, an option not open to dissenters.

Samuel Galton Jr.'s wife Lucy (Table 1-1) was the great granddaughter of the Apologist for the Quakers, Robert Barclay. They were a prolific couple whose union yielded ten children. In addition to Tertius, the father of Francis Galton, his aunt Mary Anne Schimmelpenninck deserves mention. She was an accomplished author, but considered the family black sheep by both the Darwins and the Galtons who felt she maligned them in her autobiography. On reading her account Francis Galton left this marginal note: "As though this was the only matter! Demon of mischief-making whose name was rarely mentioned by any of the family, and then only with horror!—winning confidences and then misrepresenting friends to each other! She broke off *eleven* marriages."[20] Though treated rather like Cleopatra's asp by family and historian alike, she left the only eyewitness account of a Lunar Society meeting and enduring portraits of the two eminent grandfathers of Francis Galton. Perhaps Mary Anne was vilified too much.

Pearson remarked that Samuel Tertius Galton "was not a man of the kind of note which finds its way into biographical dictionaries, but he did—what many of us everday mortals fail to do—the usual work of the everyday world and he did it well."[21] During the 1825 financial panic a bank run occurred throughout Great Britain and the Galton bank had to borrow funds to cover withdrawals from Barclay's Bank in London. The run lasted about a week, but the Galtons' friends stood by them. One even tossed a bag containing 1,000 sovereigns on the counter and asked the Galton Bank to deposit them while surrounded by panicked depositors clamoring to withdraw their funds. Tertius had actually predicted the crisis in his only known publication (1813), which attempted graphical correlations between English bank notes in circulation, the foreign exchange rate, and prices of gold, silver, and wheat. After the crisis he gradually closed the bank, completing the process in 1831, and retired to Leamington.

Tertius Galton married Violetta Darwin (Table 1-1) in 1807 and they lived at Ladywood near Birmingham where their first eight children were born. Later the family moved to the Larches, also close to Birmingham, where Francis, the youngest child, arrived in 1822. Priestley's house, burned in 1791, had previously occupied the site and was rebuilt by another Lunar Society member, William Withering, an important eighteenth-century physician. The Larches, named for two towering specimens of the species guarding its entrance, was spacious with three stories in front and two wings extending out in the rear. The left wing had a bay window and faced a garden with a terrace leading to the summer house. The right wing ran back to the stable and brewhouse that had been Priestley's laboratory. At the back of the Larches was a large yard at the end of which were poultry-, coach-, and pig-houses plus cow sheds. The cow sheds led to fields where the children could ride their ponies.

By all accounts Tertius Galton was a patient and diplomatic man who time and again had to arbitrate disputes as High Bailiff of Birmingham, and as magistrate and deputy lieutenant in Leamington following his retirement. Tertius was widely respected for his common sense and good judgment. These personal qualities extended to the home as well. His eldest daughter Elizabeth Ann Wheeler reported that when "we children quarrelled and went to my Father or Mother to complain, he used to send one into one corner of the room, and the other into the opposite corner, and at the word of command, each had to rush into the other's arms. This made us laugh and ended the dispute. My father was a true peace-maker, he always turned the matter off playfully."[22]

In his massive biography Pearson attempted to deduce the character traits that Francis Galton derived from each branch of his family and in the process makes this interesting comparison.

> We cannot fit diverse types of mind into rigid categories, but roughly we may say that Erasmus Darwin, Charles Darwin, and Francis Galton all possessed a high degree of scientific imagination. Erasmus put down his inspirations without due demonstration or effective self-criticism. Charles Darwin collected his facts before he allowed his imagination to play on them, he followed his inspirations by self-criticism and due demonstration. Francis Galton used his imagination to find his problem, then narrowed it to a small issue, and tested it by experiment and observation before publication.[23]

TWO

Metamorphosis
From Birth to Medical School

> It was strongly desired by both my parents, but especially by my mother, that my future profession should be medicine, like that of her famous father, Dr. Erasmus Darwin, F.R.S., and of her half-brother, Dr. Robert Darwin, F.R.S.
>
> —Francis Galton, *Memories of My Life*[1]

The children spent the day at grandfather Galton's mansion, Duddeston, now and then dashing into the library to clamber onto a chair near the speaking tube to shout down to the servants in the pantry, "Mama had a baby and it was a *Boy!*"[2] Little Frank, as his family called him, was born around 9 P.M. on February 16, 1822. His sister Adèle later wrote her brother that their mother "had been taken ill at eight o'clock."[3] Dr. Joseph Hodgson was called and Adèle was awoken by Dr. Hodgson around midnight to learn that she had a little brother. She fell back to sleep to find the next morning that he was "such a red little thing."[4] Francis was the last of Tertius and Violetta Galton's nine children (Table 2-1), two of whom died in infancy.[5] His brothers Erasmus and Darwin were six and eight years older while the youngest of his four sisters, Emma, was 11 years his senior. Like Emma, the other girls, Adèle, Lucy, and Elizabeth, were crazy about him. Emma recalled that he "was the pet of us all, and my mother was obliged to hang up her watch, that each sister might nurse the child for a quarter of an hour and then give him up to the next."[6] Galton's nursery was in Adèle's room. Delly, a frail creature with a spinal curvature that frequently forced her to lie on her back on a board, saw to his early education.

To instruct her brother Delly boned up on her French, Greek, and Latin and took pains to familiarize him with English verse. She gave him his lessons

Table 2-1 Francis Galton's siblings[1]

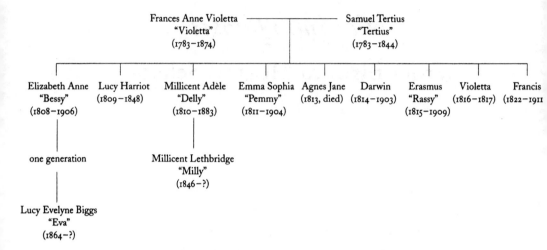

Children and grandchildren of his siblings to whom Galton was particularly close are also indicated.

bit by bit, emphasizing repetition, so that by age five Galton could recite Sir Walter Scott's lengthy epic poem "Marmion."[7] Delly became her brother's favorite sister, especially as his other sisters teased and petted him alternately.[8] She proved an accomplished instructor and her brother an apt pupil.[9] Once, when he was four, his mother asked him why he carefully saved his pennies. "Why to buy honours at the University," he answered.[10] Another time he fell off his pony into a muddy ditch and, while being extracted from the mess, spluttered out these lines from *Hudibras*, Samuel Butler's satire of the English Civil Wars.

> I am not now in Fortune's power
> He that is down can fall no lower.[11]

Galton learned the *Iliad* and the *Odyssey* well. Once, when his father's friend Leonard Horner visited, he repeatedly quizzed the six-year-old on their fine points. One day Galton said "Pray, Mr. Horner, look at the last line in the twelfth Book of the *Odyssey*," and ran off.[12] This translates as "But why rehearse all this tale? For even yesterday I told it to them and thy noble wife in thy house: and it liketh me not twice to tell a plain-told tale." Galton was familiar with the alphabet by 18 months, and had read a little book called *Cobwebs to Catch Flies* at age two and a half, but these early intellectual feats partially reflected Delly's hard pushing. Nor did she neglect the sciences, teaching him a little about birds and insects as well as geology. When he reached age five Delly, worried that her brother was becoming too attached to her, advised sending him to a nearby school with about 25 other little boys run by Mrs. French.[13] He liked the school and before long was doing so well that he was made headboy even though many of the children were older than he. Mrs. French reported that "the young Gentleman is always fond of studying the abstruse sciences," but the curriculum did not permit him to neglect the classics either and he became familiar with works like Ovid's *Metamorphoses*. On his eighth birthday Galton drew up a last will and testament. It stipulated the division of his possessions among his siblings, with Bessy getting his hygrometer and his duck, Emma his medals, etc.[14] To his "dearest sister Adele for her great kindness" in teaching him, her little brother bequeathed many items including his beetle collection. The will was drawn up on parchment with a red seal and Delly was named as executrix with his parents acting as witnesses.

Although Galton was happy at Mrs. French's school, his father packed him off at eight and a half to Mr. Bury's school in Boulogne to perfect his French.[15] The school was a converted convent near the Calais gate of the upper town with its playground being the convent's paved square. The boys were marched daily by twos around the ramparts or past the partially completed Napoleon's column. In summer they frequently bathed in the salty waters of the Straits of Dover by the boulders near the old fort, which were covered with seaweed and encrusted with barnacles and mussels. They often prised the mussels open and ate the succulent meat with bread and butter surreptitiously pilfered from the breakfast table. Except for such respites Galton was probably miserable, although his unhappiness was not reflected in his letters home, which may have been written with Mr. Bury or another master looking over his shoulder.[16] Tertius probably suspected as much for Bury appended a note to one of his son's letters saying that he understood Tertius preferred his son "writing *freely*" so he never looked over the letters and hoped they were "not very full of errors."[17] The French the boys learned was a "detestable and limited patois" spoken under penalty of a treacherous system of fines that turned one boy against another.[18] Transferable metal tags called "marks" were assigned to boys who accidentally dropped an English word or two while speaking. Since birchings were administered once enough marks were accumulated, it was advantageous to transfer marks to one's fellows. The idea was to trick another boy into saying a word of English whereupon the unlucky dupe received the mark.

Much to his delight Galton returned home in December, 1831, to find his family had moved from the Larches to 44 Lansdowne Place in Leamington. There he was sent to a private school run by the Reverend Atwood, the Vicar of Kenilworth.[19] There were only six boys in the school, which was rather free-form for the time except for a fair dose of religious teaching. For the next three years Galton was happy again trapping birds, engaging in archery, and playing cricket. Two of his classmates, Matthew P. W. Boulton and Hugh William Boulton, were grandsons of Matthew Boulton, the great manufacturer, and a founder of the Lunar Society. William died young, but Matthew remained a friend for many years. Summers were spent at Aberystwyth on the Welsh coast, where Galton learned to shoot to the detriment of the local seabirds. One day, while visiting his sister Lucy at Smethwick, Galton, aged 11, spied a bird's nest that he coveted on a tree branch next to a canal at the bottom of the garden. While stretching out to reach it, he lost his balance, fell into the water, and got stuck upside down with his feet and legs entangled in the roots.[20] He methodically worked himself free of the treacherous roots, rose to the canal surface, and swam ashore. This ability to remain composed under duress would later serve him well.

The halcyon days at Kenilworth terminated abruptly late in 1834 when Tertius decided to send his son, now almost 13, to King Edward's School in Birmingham.[21] King Edward's, a Tudor grammar school whose students were mostly the sons of tradesmen, manufacturers, and clerks, would later become a well-known public school.[22] Things started badly with a scarlet fever epidemic in January 1835. A seriously ill Galton had to return home to recover while another student, Johnny Booth, who had been with him at Boulogne, succumbed to the disease.[23] The headmaster of King Edward's, Dr. Francis Jeune, was a man with a bright future whose tenure at King Edward's was short as a more enticing opportunity soon beckoned.[24]

Galton boarded in Jeune's house in Edgbaston from whence he walked to school. Jeune was a stern disciplinarian and boys were frequently given impositions, got caned, and were sometimes expelled. Although punishment was often arbitrary, young Galton seemed to have a knack for attracting attention. One evening Earp, the usher at Jeune's house, caught Galton stealing off to school without permission. That got him an imposition of 50 lines of Virgil to recite. Later he was in a spitball fight and Gedge, his form master, gave him another 30. Another night he and his housemates became so rowdy that an infuriated Jeune flung impositions at them left and right. There must have been worse infractions, for Galton was caned several times. So he was a typical, adolescent prep school boy, and not a particularly happy one, since the curriculum at King Edward's emphasized repeated drills in grammar and classics, but gave science and mathematics short shrift. Much later Galton recalled that, while retaining Jeune's friendship until his death, he had "learnt nothing" at King Edward's. "I had craved for what was denied, namely an abundance of good English reading, well-taught mathematics and solid science."[25]

In 1838 Galton, aged 16, escaped from King Edward's. His mother and Dr. Hodgson, who had brought him into the world and was surgeon to the General Dispensary and General Hospital in Birmingham, were anxious that he study medicine.[26] His father agreed and he was apprenticed to the General Hospital as a house pupil, as apprenticeship was the commonest road to medical qualification.[27] Before he began his apprenticeship Tertius arranged for his son to team up with two young doctors, Bowman 22 and Russell 20, for a European jaunt in July 1838.[28] William Bowman had distinguished himself as Hodgson's prize apprentice at the General Hospital. In 1837 he moved to London to join the medical faculty of King's College where his reputation grew by leaps and bounds.[29] Eventually he was regarded as the leading opthalmic surgeon in the land. The purpose of the tour was to combine recreation with business by visiting several European hospitals.

The travellers landed at Antwerp and toured Belgium. In his letters to Tertius Galton remarked enthusiastically about museums, churches, and the geological and ornithological collections in Brussels. His occasional sketches included an excellent rendering of the Bishop's Gate at Liège. From Belgium, they continued to the Rhine where Galton marvelled at the Cologne Cathedral—"it is most splendid. . . . I never saw anything like it in England"[30]—and thence to Bonn, Koblenz, and Frankfurt visiting hospitals along the way. After poking about in a hospital in Frankfurt, Galton wrote Tertius that while the hospital was said to be very clean "I never fully appreciated the value of fresh air till I found myself without its wards."[31] They travelled southwards east of the Rhine to Darmstadt where he visited the natural history museum, but was not impressed by the "jawbone of the Dudotherium and all that sort of fossil nonsense (!)"[32] In Heidelberg they marvelled at the skill of "Tiedermann a top-sawyer of the medical line and a whole quantity of others."[33] From Heidelberg they headed for Stuttgart, Augsburg, and finally Munich, but jouncing around on bumpy roads in diligences took its toll. Galton wrote his sister Bessy that "I have got one boil and two blisters in such awkward positions that when sitting back, I rest upon all three; when bolt upright on two, and when like a heron, I balance myself on one side upon one!!!"[34]

From Munich they went east to the Austro-Hungarian border as they planned to take a Danube River steamer from Linz to Vienna. At the border, an Austrian customs officer with great black mustaches apprehended them. "Kein Tabac," he growled.[35] Tobacco was an imperial monopoly and could not be imported. "Kein tabac," replied each of the young men innocently. "The officer's eyes flared." He looked "awful" at Galton's "green bags with black strings, in which two or three dirty shirts were ensconced, and terrible at the other luggage; he made signs that everything must come out, when in the moment 3 Zwanzigers (a coin about 10 d) touched his hand—a galvanic shock seemed to thrill his whole system." The officer's demeanour changed suddenly and the "flare of his eye changed in an instant to a twinkle, the baggage was shut up and officer fell into a 'paroxysm of bows' and away we drove."[36]

The next morning, August 26, they reached Linz and were soon steaming down the Danube toward Vienna, probably on a vessel built by John Andrews and Joseph Pritchard. They had formed the "Imperial-Royal Privileged Steamship Company" in 1829 with exclusive rights to navigate the Danube with vessels of an improved type.[37] In Vienna they were escorted to various hospitals and museums by Dr. Seligman whom Galton fancied writing that "some of these Germans are not a bad sort of fellows."[38] In the women's wards of one Vienna hospital an attractive young lunatic wrapped her arms passionately around Galton clasping him tightly to her bosom. She insisted that he was her long-lost Fritz while a blushing Galton probably wished he could perpetuate the fiction.[39] From Vienna, they turned northwards to Prague, Dresden, and Berlin, and Galton developed an ingrown toenail that became painfully inflamed. He informed his two travelling companions and "a smile of conscious professional power illumined the face of one, a grin of delight that of the other. Both readily proferred their services...."[40] Galton accepted "the Senior hand of Bowman" who examined the infected toe and shook his head "Bad,—very. Russell, have you a pair of forceps?" "No," said Russell. The two aspiring physicians put their heads together and decided on a bent pin as a substitute instrument. With Russell pinning the flesh back and Galton "writhing," Bowman wrenched "up the nail, then cutting it snip-snap all round." But the surgery was successful and their journey ended in Hamburg in late September. This trip probably triggered Galton's lifelong interest in travel.

Upon returning to England, Galton settled in at the General Hospital where the ill entered at their peril as in most early Victorian hospitals. Overcrowding was endemic and patients with different diseases often ended up in the same wards and sometimes even in the same bed so that a man with a broken leg might find his bedmate had typhus.[41] The atrocious washing facilities were usually inaccessible to patients while the nursing staff, mostly recruited from the gutter, consisted of uneducated women who were poor, slovenly, and frequently drunks. Medicine for one patient often found its way to another and food distribution was hit or miss. Hospitals were foul-smelling and the "operating theatre" was usually a grimy, poorly lighted room of moderate size. The surgeon donned the filth-encrusted "surgical coat" he had used for operations for years. There were no surgical masks, gloves, or anaesthetics and the instruments were unsterilized. Crowding round the operating table sat colleagues and visiting doctors waiting expectantly for the patient to be dragged or carried in from the ward to be laid on the operating table and sometimes strapped down. Speed was of the essence since the unanaesthetized patient would start to shriek with pain as the knife began to cut, but as the operation proceeded cries turned to whimpers until finally the exhausted patient was returned to a ward.[42]

Galton once observed the anaesthetic properties of alcohol when a great drayman was brought in dead drunk.[43] Both legs had to be amputated as his thighs had been crushed by a heavy wagon. The drayman slept soundly through the operation awaking the next morning aghast at his loss. But he was fortunate as healing took place "by first intention,"[44] unaccompanied by inflamation or suppuration. The worry was that surgical wounds would heal "by the second intention," which was the rule. If all went well, suppuration, inflammation, and fever did not last long, but often things took a turn for the worse and one of the "hospital diseases," erysipelas, pyaemia, septicemia, or gangrene set in. By late fall in 1838 Galton was sounding a lot like a modern intern or resident.[45] He wrote Tertius about a typical evening. It began with rounds of the wards at 5:30 P.M. Afterwards he made up 15 prescriptions, entered records of his patients in the hospital books, made notes in his case book, and finally found time for dinner at 9:00 P.M. Following a hasty meal, he rounded several wards again, helped to set the broken leg of an accident case, and read medicine until 11:30 P.M. He was exhausted and about to fall into bed at midnight when the accident bell rang and he had to deal with a serious fracture. The bleary-eyed apprentice retired again around 1:30 A.M. only to be awakened by a loud knocking on his door at 3:00 A.M. This time a dreadful compound fracture kept him up until 5:00 A.M. when he nabbed a couple of hours sleep and arose at 7:00 A.M.

Early the next year Galton reported his first dentistry attempt to his father.[46] A young boy had come in with a decaying tooth so Galton confidently picked up the extractor, called a key, and inserted it backwards. Not realizing his error, he groped around for the sick tooth, got hold of it, and yanked. The boy, in excruciating pain, gurgled and kicked at the novice dentist, causing him to wrench even harder. The tooth snapped in half leaving the stump buried in the boy's jaw whereupon the boy clapped his hands over his mouth in agony. Galton offered to pull out the stump, hiding the fact that he had snapped the tooth by calling it a double tooth of which he had only got half. The boy wisely refused to let his tormentor proceed, but, unaccountably, allowed him to try pulling another rotten tooth. Galton got a good purchase on that tooth with the key and began to tug "away like a sailor with a handspike, when the boy, roaring this time like a lion with his head in a bag, broke away from me and the sawbone that was holding his head, bolted straight out, cursing all Hospital Doctors right manfully."[47]

He learned to make infusions, decoctions, tinctures, and extracts, and how to cap bottles and to roll pills.[48] He became skilled at setting bones and at one point 16 patients with fractures, dislocations, or other injuries of the arm were under his sole care.[49] One night he witnessed a remarkable operation by a young house surgeon when an unconscious man was brought into the hospital breathing in a labored, stertorous manner.[50] Some hard object had fallen on his head and depressed a small piece of the skull against his brain. It was a case of life and death. The surgeon could not await the arrival of a more highly skilled physician so he went to work with a trepan, a hollow steel cylinder with teeth cut out of its lower rim, and removed a circular section of bone adjacent to the depressed fragment. The surgeon laid a metal rod across the hole and levered up the skull fragment with a miniature steel crowbar. The snoring stopped and the man finished the sentence he must have begun before the accident. The surgeon deftly inserted a metal plate over the hole and stitched the scalp over it and his patient recovered.

In October 1839 Galton's medical career entered its next phase when, at age 17, he took up lodgings with four or five other pupils in the house of Richard Partridge, professor of Anatomy at King's College, London.[51] Partridge, who had also apprenticed at Birmingham General Hospital, was already a distinguished physician.[52] Galton was delighted with his new surroundings and his landlord, only 34, was witty and full of clever quips. Partridge's house had a fine library and a well-lighted sitting room where the students could study guarded by a skeleton hovering like a sentry in a hanging closet. In the evenings Galton often joined Partridge's bachelor dinner parties where he "listened with admiration to the brilliant talk."[53] At King's, he attended lectures and studied anatomy from 9 to 4, delighted to find that Bowman was one of his immediate chiefs in the dissection room. He did not enjoy Partridge's anatomy lectures despite his good sense of humor as his landlord was more interested in the minutiae of human anatomy than in the comparative aspects of the subject. He joked to Tertius about Robert Todd, recently appointed to the newly created chair of physiology and general and morbid anatomy. His laboratory contained a "sight which a Frenchman would give his ears to see, viz. a most splendid collection of large green frogs all alive and kicking and croaking too, kept, however, for Dr Todd's Physiological Experiments."[54] Todd and Bowman were collaborating on an *Encyclopedia of Physiology*, a highly respected work in the field. Galton reported his expenditures to his father regularly as Tertius liked to keep careful track of such things. In a letter to Bessy he waxed poetic over a plague of boils. "I have had another boil exactly by the side of the former which has partially reappeared. The new one is mountainous, but alas! *not* snow-capped like Ben Nevis, but more like Ben Lomond covered with scarlet heather. I shall have a complete Snowdonia of them soon."[55]

During the winter of 1840 Galton crammed hard preparing for the April examinations, but about two weeks beforehand he broke away and travelled by steam packet a few miles up the Thames to see the annual four-mile Oxford and Cambridge boat race from Putney to Mortlake.[56] On the return the packet careened toward the Battersea Bridge on a strong outgoing tide and a collision with the bridge's middle pier appeared imminent. Galton grabbed onto one of the steps leading up to the starboard paddle-box that housed the revolving paddle wheel expecting to have to jump overboard upon impact. Instead the paddle-box smashed into the pier, splitting it open. He was thrown headfirst through the paddles into the Thames. Momentarily knocked unconscious, he awoke to find himself submerged beneath a huge fragment of the paddle-box. He was badly in need of air and struggled vainly to dive out from under the wreckage. As he peered through the murk he finally spotted the paddle-box edge, which he grasped and used to lever himself out from underneath. He sank down again and became entangled in more wreckage, but struggled to the surface once more and crawled aboard the flotsam that had nearly drowned him. He was eventually pulled aboard a passing boat dazed and covered with blood.

Galton persevered with his exams, exulting in a note to Delly, "Hurrah! Hurrah!! I am 2nd Prizeman in Anatomy and Chemistry. I had only expected a certificate of honor. Hurrah! Go it ye cripples."[57] His chief competitor was George Johnson to whom he was second in physiology. Johnson, later a professor of medicine at King's College, became an expert on kidney disease and a leading advocate of one of two opposed methods of treating cholera.[58] His "eliminative" method prescribed castor oil for he felt that the constant diarrhea associated with the disease was nature's way of ridding the body of some toxic principle. This was precisely the wrong way to cure these severely dehydrated patients and Johnson's opponents credited him with some 11,000 unnecessary cholera deaths.

The circumstances surrounding Galton's transfer from King's College to Cambridge remain unclear, but he had apparently discussed the move with his father before he left Birmingham. At any rate shortly after settling in Partridge's house, he wrote Tertius arguing that a year at King's would be enough and once "the Dissecting season is over about June . . . my hands will not be full for three months or so before going up to Cambridge, in which time I shall hope to get up the first part of my Mathematics well, and a fair proportion of the Classics."[59] Galton had been strongly encouraged by his cousin Charles Darwin (Table 1-1), aged 30, living nearby at Macaw Cottage, on Upper Gower Street with his bride Emma Wedgwood. Darwin was a newly-minted Fellow of the Royal Society whose *Beagle Journal* had just been published.[60]

Much earlier, Darwin, like Galton, had been propelled into medicine at his father's behest.[61] Not quite 16 he arrived in Edinburgh with his elder brother Erasmus in October 1825. After watching a couple of botched operations he quickly became disenchanted with medicine. He hated anatomy, taught by Alexander Monro, and disliked Duncan on Materia Medica since he was "so very learned that his wisdom has left no room for his sense."[62] However, he enjoyed Hope's chemistry lectures, and Jameson's on natural history. To his father's disgust Darwin neglected his studies exploring natural history and shooting small animals and birds. He decided to transfer his son to Cambridge to study for an ordinary Arts degree.[63] This degree would provide Darwin senior's errant son with the initial preparation necessary for entry into a clerical career as Cambridge had a strong affiliation with the Church of England and many of its students went on to become clerics. But Cambridge proved a revelation for Darwin. He was soon befriended by John Stevens Henslow, the professor of botany, who "opened the door to Darwin's future."[64] The Beagle journey followed.

Galton and Darwin found they had much in common. Like their grandfather Erasmus Darwin they were interested in science. Darwin soon became a guiding influence as the younger man reached a critical fork in his education. Later on it would be Darwin's *Origin of Species* that would turn Galton from travel writing, geography, and meteorology toward the improvement of mankind through selective breeding. In December Galton wrote Tertius saying that Darwin had recommended that he go up to Cambridge the next fall to "read mathematics like a house afire."[65] The usual course of medical study required four years. Galton had two under his belt, but he was only 17 and he could not qualify for the Bachelor of Medicine degree until 21.[66] A Cambridge sojourn would be advantageous. This would allow him to return to his medical studies afterwards to qualify for the degree with his medical courses fresh in his mind. Should he continue his medical studies for two more years and then attend Cambridge, Darwin believed that he would "forget all the theoretical part of medicine, I mean 1/2 of Physiology, 3/4ths of Surgery and 4/5ths of Medicine, to say nothing of Anatomy Lectures, on the two last of which I shall attend next year and will be time thrown away."[67]

Darwin had also argued that "the faculty of observation rather than abstract mathematical reasoning makes a good Physician."[68] However, the "higher parts of Mathematics, which are exceedingly interwoven with Chemical and Medical Phenomena (Electricity, Light, Heat etc.) all exist and exist only on experience and observation."[69] Thus Darwin had cleverly related mathematics to observation and observation to medicine. Galton added that his ignorance of basic physics and chemistry was a positive detriment as he was unable "at present to comprehend one half of the fundamental principles which are mathematical, Light especially."[70] Besides, he continued, cadavers were rare and he was having difficulty in gaining dissection experience. Since he anticipated strenuous objections from Hodgson, he pointed out that Bowman, Hodgson's own prize apprentice, endorsed his plan to attend Cambridge as did three other physicians at King's.

Hodgson's antennae sensed danger signals immediately. He recognized that, once at Cambridge, Galton was likely to raise additional excuses and might forsake medicine altogether. Consequently, early in the winter of 1840 Galton found himself writing his father that he disagreed with Hodgson's opinion that a mathematical interlude at Cambridge would be detrimental.[71] He pledged to "work like a trooper" while finishing up at King's so he could study mathematics at Cambridge, which, he pleaded, had been his heart's desire for the last several years.

With summer approaching and Galton winding up his studies at King's, Tertius offered to stake him once again to a continental holiday. Galton decided to accompany William Miller, another aspiring doctor at King's and a Birmingham General Hospital product.[72] Miller planned to study with the great organic chemist Justus Liebig in Giessen, Germany.[73] But Galton found Giessen a depressing "scrubby, abominably paved little town—cram full of students, noisy, smoky, dirty."[74] He soon realized that Liebig's laboratory was not for him. Liebig was the general commanding a little army of chemists: he told them what to do; he analyzed their results; he told them what to do next; and he published papers on each discovery with the surrogate who had done the work. Galton, flailing about for an alternative, wrote Tertius that he had engaged a tutor to teach him German and insisted that he would "work hard at Giessen for a fortnight till I can speak it tolerably."[75] For Miller, Liebig's laboratory was perfect and he became a distinguished chemist after completing his medical degree at King's.[76]

Within three days Galton's pledge was forgotten. He wrote his father that he was "determined to make a bolt down the Danube and to see Constantinople and Athens," delighted with the thought of escaping Giessen's dirt for oriental adventure. He travelled south to Frankfurt, shipped home Liebig's organic chemistry text, and continued to Wurzburg. After making some sketches of the town, he boarded a diligence bound for Nuremberg finding that he had a little Hungarian gent and a pretty young girl for companions. He fell asleep and woke to find the Hungarian holding hands with Marie, and singing her love songs, whereupon they all burst out laughing. Not to be outdone, Galton began flirting with Marie "with much more success than my rival, at which his mustachios desponded and looked sad."[77]

He went by coach from Nuremberg to Passau and then to the twin cities of Linz and Urfahr only to find that the river steamer was under repair. Undaunted, he teamed up with an elderly British officer, Major Parry, and they engaged a boatman to take them to Vienna, over a hundred miles away. They embarked at 3 A.M. in the cold predawn hours of August 6 on the decrepit, leaking boat, so Galton alternated between helping to row and bailing. Although the current was swift, they were not lured to their destruction by the evil genii of the Danube. They are said to appear at night with sinister signal lights, will-o'-the wisps, and seductive calls enticing unwary boatmen to their deaths in the rapids, swells, rocks, and shallows of the "Struden," the "Wirbel," and the "Schwall."[78] At 2 P.M. the next afternoon they passed the ancient abbey at Melk, looking down upon them from a cliff high above the river, and entered the Wachau with its lovely vine-covered hills. At Stein, some 50 miles from Vienna, they exchanged their rower for two fresh men and finally reached the city at 2 A.M. on August 7. Galton gaily recounted these exploits to Tertius, adding that Linz was "universally famous for the beauty of its fair sex, and so is Wurzburg, and everything prosperous."[79] He spent the next three days touring St. Peter's, St. Stephen's, and Archduke Charles' collection of "etchings and sketches ... 35,000 in all," visiting the Belvedere Gallery and Schönbrunn Palace. One evening he heard Madame Lutzer at the opera who was "very pretty, but rather wicked looking" and had "a very sweet voice."[80]

On August 10 Galton embarked on a packet in whose "forefront was crammed only one pretty girl and she would hold down her eyes."[81] They sailed past Pressburg (Bratislava), where the Danube widens and fills with islets, spotting an occasional heron stalking its prey stealthily among the reeds that marked their marshy edges. They steamed past the brooding citadel at Esztergom docking eventually in Budapest where Galton sketched a Hungarian whose "hair and mustachios are no exaggeration."[82] He was off again at 4 A.M. on August 12 with five other English travellers and a long haired Wallachian who spoke "French—A Frenchman who spoke good Italian and an Italian who spoke capital French," plus two ladies and a German colonel "who flirted considerably with one of them."[83] As the packet steamed past Belgrade, Galton sketched its waterfront, noting that the city "was then in Turkish occupation and the Turks still wore turbans. The town being in quarantine, we were not allowed to land."[84] Beyond Belgrade was Romania, then a Russian protectorate. They traversed the picturesque narrows at Klisura, the islets of Moldova, Kiseljevo, and the Babakai Rock and, in the rapids beyond where the river was swift, he saw "whirlpools occasionally and splendid eagles soaring about."[85]

At Orsova the Danube funneled into the rapids of the Iron Gate, which he found disappointing. "The Iron Gate is humbug, the rapid is swift enough but the scenery nothing particular."[86] Once past there was no return without a ten-day quarantine for plague. He found "the flat shores of Wallachia most uninteresting" and looking "fever haunted."[87] East of Bucharest, the Danube bends sharply to the north before turning east once more to empty through a myriad of smaller waterways into the Black Sea. Galton debarked north of the bend at Cernavoda and journeyed east by carriage with three English travellers to the Black Sea port of Constanta. There he found a pleasant inn and a welcome "Barclay and Perkins' porter, a bottle of which I drank to the health of all at home."[88] The next day he embarked for Istanbul at noon.

Constantinople (Istanbul) was a city of vivid impressions.[89] The nightlong howling of myriads of dogs echoed through the streets and made sleep difficult, but in the morning the lucky traveller might be "greeted by the rays of the rising sun, gilding the snowy summits of Mount Olympus and the beautiful shores of the Sea of Marmora, the Point of Chalcedon, and the town of

Skutari: . . . the marble domes of St. Sophia, the gilded pinnacles of the Seraglio glittering amid groves of perpetual verdure, the long arcades of ancient aqueducts, and spirey minarets of a thousand mosques."[90] Turkish soldiers drowsy with tobacco lolled about over "the checkers of a dice-board" or listened "to the licentious fairy tales of a dervish."[91] The narrow footpaths of the city jostled with crowds of people: women wearing long caftans whose gauze provocatively hinted at their features, soldiers, government officers in gaudy uniforms, Jews, Armenians, Greeks, Albanians, Franks, and Circassians. Street vendors hawked bread, fruits, sweetmeats, or sherbet carried in a large wooden tray.

Galton visited the slave market. Black slaves were kidnapped by expeditions sent out from Egypt and the Sudan. The men often perished following castration by crude surgery to convert them into eunuchs.[92] They were sold as house-servants in Syria, Asia Minor, and Istanbul with Egypt being the main market. Lovely white Circassian and Georgian girls were in great demand. They came onto the market because of the extreme poverty of their parents, but only after their condition had been improved through proper diet, protection from the sun, and the daily Turkish bath. The Turks purchased these women as servants and they could become concubines under Turkish law. Should a woman give birth, her owner had to marry her and so infanticide was common. Galton thought the Circassian women enticing and remarked to Tertius that he wished he had an extra 50 at his disposal to purchase a particularly attractive slave.

On August 26, he embarked from Constantinople and entered the Dardanelles the next morning. He was disappointed by a short excursion to the site where Troy once stood. Homer "must have had a brilliant imagination to make a little bit of plain 2 miles long and 1 mile broad the scene of all the manoeuvres of a ten years' war. The idea too of fighting ten years for a woman!"[93] Late that evening they anchored off Smyrna (Izmir). On August 29 he set sail for the island of Syra (Syros) in the northern Cyclades on the *Dante*, a French man-of-war. Its port of Ermoúpolis was the main coal-bunkering station for packets from the Eastern Mediterranean and the chief port of Greece. He was quarantined for plague for ten days, even though the disease had nearly vanished. Afterwards, a medical officer lined up all the passengers and ordered them to stick out their tongues. He then told them to do exactly as he did and "clapped himself sharply under the left armpit with his right hand, and under the right armpit with his left hand. Similarly on the left and right groins. This was to prove that none of the glandular swellings that give the name of 'bubonic' plague were there, otherwise the pain of the performance would have been intolerable. Then, with a sudden change from a stern aspect, he put on a most friendly and courteous smile, and stepping forward he shook each of us cordially by the hand, and we were freed."[94] Galton continued to Athens and then to Trieste.

The normal ten day quarantine was shortened at Trieste by making "Spoglio," the assumption being that after a week a healthy, well washed person could be judged free of infection. After a doctor examined him, Galton struck a bargain with men selling old clothes on the next quay. His money and papers were taken from him and fumigated. He stripped, leaving his own clothes behind, and dove off the quay swimming across 20 feet of seawater to the next quay where a new set of clothes, somewhat threadbare, but serviceable, awaited him.

THREE

A Poll Degree from Cambridge

You may roam where you will through the realms of infinity And you will find nothing so great as the Master of Trinity.[1]

—Lord Kelvin on William Whewell,
Master of Trinity College 1841–1866

In October 1840 Tertius and his son went up to Trinity College, Cambridge. With blessing of Henry VIII who dedicated it to "the Holy and Undivided Trinity" the venerable college was formed by fusing King's Hall, founded in 1317 by Edward II, and Michaelhouse, established in 1324 by his Chancellor of the Exchequer, Hervey de Stanton.[2] Trinity, like King's, would fly the royal standard, but, unlike King's or any other college, the Crown appointed its master. The college endowment provided support for its fellows and a complement of 60 fellows and 60 scholars plus pensioners (students) who paid their own way were authorized. The statutes stipulated that the master and fellows must take Holy Orders and remain celibate, but exceptions were soon made for the master and marriage became the custom. The unfortunate fellows, however, were consigned to celibacy until 1882 with the marriage penalty being forfeiture of the fellowship.

Tutors, appointed by the master from among the ordinary fellows, were few in number, with Trinity having only three.[3] A tutor was the main advisor to the students and in charge of the college teaching program for which he received a significantly increased stipend. He was responsible for admissions and discipline, advised students on lodgings, and was authorized to use the college tuition fund to hire lecturers. Student caution money, put up by the student as security for good conduct, was deposited with the tutor who could invest it and keep the interest as long as the principal was returned at graduation. So tutors became surrogate fathers to students, a task not easily managed since a tutor often looked out for over a hundred students. Professors participated little in undergraduate teaching. Occasional lectures, devoted to their own original work, were generally of little value to the students.

To cope with the overload, the tutor had a staff of assistant tutors. They served as college lecturers and were attached to the side (students) for which the tutor was responsible. Because of the rather chaotic nature of teaching, students at both Oxford and Cambridge used coaches (private tutors) extensively. Coaches were particularly vital to students like Galton who intended to try for honors. They were often college fellows, lecturers, or graduates who had married. Their ability was measured by their capacity to cram students for a high place in the Tripos (honors examination). Coaching paid well, especially during the summer Long Vacation when reading parties of a few students and a coach departed for the Lake District, Wales, or Scotland to spend mornings in study and afternoons on long hikes. The Tripos, an eighteenth-century invention, was a comprehensive, written examination covering various aspects of mathematics, optics, astronomy, and the physics of Isaac Newton, one of the most distinguished earlier alumni of Trinity.[4] Although a Tripos in classics was established in 1824, the mathematics Tripos was unique for the first half of the nineteenth century as it could be taken directly by any student. Until 1857 only students of noble birth could sit the classics Tripos without having first taken mathematics honors. Most Cambridge students took the equivalent of the current ordinary degree. They were called pollmen, the hoi polloi, or rabble, from which the degree took its name.[5] Pollmen were often regarded as idle and dissolute. However the preparation required for the tripos was so grueling that many able students chose to bypass the examination meaning they graduated with the ordinary degree.

Galton thought the Cambridge curriculum narrow declaring that its "religious dogmas were of a more archaic type than I latterly learnt to hold."[6] No one seemed interested in what he had assimilated during his medical career and "what we have since learnt to call Biology."[7] Unlike Darwin, he found no mentor and friend like Professor Henslow to take him under his wing and spark his enthusiasm.[8] Perhaps this is why Galton, who always remembered Cambridge and its graduates fondly, never really found focus at the university. He set up housekeeping in rooms on the ground floor of the New Court[9] with its pleasing neo-Gothic exterior. It was the great contribution of Christopher Wordsworth, the younger brother of the poet, as Master of Trinity. It provided quarters for many Trinity men previously forced to take lodgings in town.[10] Galton's sitting room looked east into the court with a sofa in front of the fireplace over whose mantel hung a low mirror. Above the mirror, Galton mounted two pistols he had purchased in Smyrna surmounted by crossed foils. His bedroom faced west toward the banks of the river Cam and the willow and lime trees of the Avenue. It was hung with two pictures Emma had painted for his lodgings in Partridge's house in London. Adjacent to the bedroom was a small room for Galton's gyp, his bedmaker, and servant.

Galton was now an attractive young man with blond hair, high forehead, pale blue eyes, and a v-shaped mouth. He dressed like a proper Cambridge undergraduate: black frock coat, heavy gray or plaid trousers, matching vest, and a handsome cravat, often blue, fastened with a large gold-headed pin.[11] As a young man of means he needed to entertain properly. Thus he wrote Tertius that the six silver teaspoons he proffered were quite sufficient, but he was concerned that if his father could not "send wine easily from Leamington, the best plan will be to write to your London wine-merchant as there is a carrier direct from there."[12] Initially he had just two friends, his cousin Theodore Howard Galton with whom he socialized and smoked pipes in the evening, and Matthew Watt Boulton whom he knew from the Reverend Atwood's school in Kenilworth.[13] He sometimes doodled and wrote fragments of poetry on scraps of paper such as "A bugs lament over his widowed mother" featuring a mother bug dead on her back, legs erect faced by her grieving son standing on all six of his.[14]

Socializing and fine rooms were well and good, but there was work to be done and he required a good coach, so he contacted Matthew O'Brien. O'Brien, a Gonville and Caius graduate, was third Wrangler in 1838.[15] This meant that he had scored exceptionally well in the mathematics Tripos that year topped only by the Senior Wrangler, the student with the highest marks, and the first and second Wranglers. But O'Brien had fallen in love while coaching a reading party at Inverary and was dawdling over returning to Cambridge. In late October, O'Brien appeared and recommended that Galton begin to read differential calculus and its application to the physical forces governing statics and dynamics, so he now began a routine. He was up early attending divine services from 7:00 to 7:30 A.M., read and breakfasted until 9:00 A.M., heard lectures until 11:00 A.M., read by himself and with O'Brien until 2:00 P.M., and then took a long walk.[16] Such "constitutionals" were an integral part of preparing for the Tripos.[17] Regular physical exercise, particularly long distance walking, was seen as a necessary adjunct to hard study as the Cambridge undergraduate equated physical fitness with mental agility. This perceived relationship later took on particular significance for Galton as he used physical ability as a kind of surrogate indicator of mental capacity.

Galton often read through tea, later breaking for dinner, and then reading on into the evening. To stave off drowsiness, he employed the Gumption-Reviver, a contraption consisting of a large funnel supported on a stand six feet high, filled with water and fitted with a stopcock. While he sat reading it dripped at a predetermined rate onto a cloth band surrounding his head.

Rivulets of water wound down his shirt, which was all to the good since "damp shirts do not invite repose."[18] Galton's unfortunate gyp had to refill the Gumption-Reviver every quarter hour or so. In late November Galton became feverish and delirious for almost a fortnight. He ceased work for the rest of the term and wrote his father that his illness had "put a pro tempore dead stop to maths."[19] By the end of January 1841, a reinvigorated Galton was back at Trinity reading with O'Brien again. Much like his grandfather Erasmus Darwin, Galton also began designing machines of various kinds beginning with an accurate balance followed by a rotatory steam engine.[20] These were the first of many designs and the completed objects are collectively known as "Galton's toys." In March Wombell's travelling menagerie visited and Galton wrote his father that he appeared "before the eyes of wondering Cantabs, where do you think? Why right in the midst of a den containing 1 Lion, 1 Lioness, 1 huge Bengal Tiger and 4 Leopards," but the "Lion snarled awfully" and Galton "was a wee frightened for the brute crouched so."[21]

In April, preparing for the May examinations, Galton panicked when he failed to locate the mathematics notes from his tutoring sessions.[22] He thought they might be at home advising Tertius to burn "the Duddeston titledeeds if you will, but preserve these manuscripts." He cast a curse on his father should he forget to send the notes immediately "may the spirit of gout tweak your remembrance!!!"[23] Despite the precious notes, Galton only made third class in the examinations, but he was philosopical. He wrote his father that he performed as well in mathematics as he expected, but was dragged down by classics where he was competing "with men who have spent that time on them which I have employed in medicine."[24] He also discussed his future with O'Brien, who did him a favor by recommending that he tutor with William Hopkins, the greatest of the early Victorian mathematics coaches.[25] Hopkins, a born teacher, did not simply cram his students for the Tripos, but encouraged them to take a speculative and philosophical view of mathematics. His approach proved so fruitful that he became known as the "the senior wrangler-maker" and by 1849 had made nearly 200 wranglers of whom 17 had been senior and 44 in one of the first three places. He was only surpassed later on by his pupil Edward John Routh. His students included the great Victorian physicists Lord Kelvin and James Clerk Maxwell. Because he was married Hopkins could never have been a don, a position for which he was eminently suited intellectually.

During the summer of 1841, Galton repaired with a reading party to the small town of Keswick in the Lake District nestled between the Skiddaw mountains and the shimmering expanse of Derwent Water.[26] The mathematics coach, Mathison, was a Trinity Fellow, whom Galton knew. Eddis, a Chancellor's medallist, was the classics tutor. The five undergraduates and two tutors had rooms in "Browtop," a villa with a panoramic view of the surrounding countryside. In the morning the students worked with their tutors, but in the afternoon, if the weather was decent, they took long hikes up Skiddaw, with its stunning views of the Isle of Man and Ben Lomond, or perhaps ventured up the flanks of another nearby fell like Helvellyn. The local country people were good to the hikers, giving them milk, oatcakes, and homemade cheese, and Galton eyed the local girls admiringly. The boys often equipped themselves with a generous supply of spirits and one day on Skiddaw "it was very hot and we pitched into much whiskey, and on the strength of it cheered 3 times 3 for God save the Queen, Trinity etc."[27] When clouds lowered over Skiddaw and rain pelted Browtop, the students played battledore and shuttlecock or fives, a variant of handball. Dinner and conversation followed and then it was time to read again until late in the evening.

William Whewell, the new Master of Trinity College, was also nearby as he had trapped his quarry, becoming engaged to Cordelia Marshall, a wealthy Ullswater beauty, in June 1841.[28] Whewell was a prodigious scholar of humble origin who studied philosophy, read Kant, learned German, and was elected to a fellowship at Trinity in 1817 and then named tutor for one of the sides in 1823. He became a mathematics lecturer, wrote a textbook on mechanics, studied architecture on the continent, and produced a book on his theory of Gothic design. This whirlwind of activity was recognized by his induction into the Royal Society in 1820. When the professorship of mineralogy at Cambridge became vacant, Whewell announced for it and was elected in 1828, soon publishing a treatise on the classification of minerals. To Whewell we owe the word "scientist," which he coined in 1834 in his review of Mary Somerville's *The Connexion of the Physical Sciences*.[29]

Legend had it that a prize fighter once exclaimed to Whewell, a physically powerful and highly masculine man, "What a man was lost when they made you a parson!"[30] For such a man, now in his late forties, the monastic don's life became tiresome. The timing of Whewell's marriage in October 1841 was perfect, for Christopher Wordsworth had written Whewell of his planned resignation as Master of Trinity.[31] Wordsworth, an unreconstructed Tory, had refused to resign as master so long as Whig ministers were in power, for he feared they would advise the Queen to appoint a liberal in his place. So he lingered on, isolated and unhappy, until the election of a Conservative government in the autumn of 1841. Whewell was recommended to the Queen by the prime minister, Peel, and duly appointed master. Simultaneously, Whewell had advanced his own career and obtained a position for which celibacy was not a prerequisite.

During that deliciously pleasant summer of 1841 at Browtop, Galton and his fellow students observed Whewell's courtship at close range. He briefed his father on the impending marriage, remarking that he and his comrades were endlessly speculating "how Whewell would set to work to make love, he is nearly 50, she a little more than 20."[32] Later Galton described Whewell's courtship behavior as reminding him "of a turkey-cock similarly engaged. I fancied that I could almost hear the rustling of his stiffened feathers, and did overhear these sonorous lines of Milton rolled out to the lady à propos of I know not what, 'cycle and epicycle, orb and orb,' with hollow o's and prolonged trills on the r's."[33]

In late July Tertius took his family to vacation at Scarborough on the Yorkshire coast. Galton invited them to visit Browtop,[34] but afterwards Tertius became seriously ill and his son hastened to Leamington to be with him. With Tertius sufficiently recovered, Galton returned to Cambridge in late October where he began tutoring with Hopkins. He loved the experience. "Hopkins to use a Cantab expression is a regular brick; tells funny stories connected with different problems and is no way Donnish; he rattles us on at a splendid pace and makes mathematics anything but a dry subject by entering thoroughly into its metaphysics. I never enjoyed anything so much before."[35] Not being donnish was a compliment, for Cambridge undergraduates stereotyped dons in one of two ways, both emphasizing the don's separation from his students.[36] The genial don, frequently an aging, often eccentric bachelor who participated in undergraduate functions, might be known for an interest in antiquarian history or the history of the college plate. The best of these were witty conversationalists while the worst had lost all sense of responsibility, often gambling and drinking excessively. One Master of the 1850s was referred to as "an ancient megatherium, who liked his bottle in the evening and asked only to be left in peace."[37]

The other kind of don could be pompous, arrogant, authoritarian, or sometimes morose. Wordsworth and Whewell fitted this description, but in different ways. While Wordsworth was authoritarian, Whewell could be pompous and arrogant, but then who could blame a man of such great intellectual achievement? He did much for Trinity as its master, with his gifts and bequests in connection with Whewell's Court being valued at £100,000.[38] This gloomy building, constructed during the worst period of Victorian architecture, provided a hundred sets of rooms to the College. In a manner he felt befitted the new Master and his bride, he returned the facade of the Master's dwelling to its original Tudor-Gothic character. Although renovation of the Lodge was Whewell's conception, he carried it through with the College's money. To restore one of the two bow windows, or oriels, he put up £250, while an old Trinity man, A. J. Beresford Hope, contributed £1,000. Whewell, proud of his handiwork, caused an inscription to be made claiming that he had restored the antique beauty of the Lodge with the aid of Beresford Hope. However, the merciless eye of the undergraduate is quick to detect pretentiousness and the Seniors were up in arms. They knew that while Hope's contribution was substantial, the College had provided two-thirds of the renovation money. Their ire was described by Tom Taylor, a young fellow of Trinity College, in the following lines:

These are the Seniors who cut up so rough,
When they saw the inscription or rather the puff
Placed by the Master so rude and so gruff
Who lived in the house that Hope built

Early in his second year, Galton became good friends with two Etonians, Henry Hallam and Frederick Campbell.[39] Hallam was a man of gentle disposition and the brother of Arthur Hallam, an earlier Trinity graduate in whose rooms Tennyson spent many happy hours. He and Campbell, later Lord Stratheden, had set a goal of public service. Through introductions via Hallam and Campbell, Galton's circle of friends expanded rapidly. They included Henry Maine, winner of an exhibition (scholarship) to Pembroke College, who became a distinguished jurist.[40] Maine, later knighted, rose to the highest legal post in India, but unadvisedly embarked on a study of the customs of the so-called Aryan races, which mired him in much controversy. William Johnson Cory was another.[41] Cory, a brilliant student, elected King's scholar at Eton at eight and Newcastle scholar at 18, received a scholarship to King's where his academic triumphs multiplied. Johnson of King's, as he came to be known, was named First Classic of his year. He later returned to Eton where he became Master. Then there was Tom Taylor, whose lines are quoted above.[42] Taylor eventually moved to London where he simultaneously studied law at the Inner Temple and was a professor of English literature and the English language at London University. Afterwards, he became a famous playwright and held down the editorship of *Punch*. So Galton's friends were uncommonly talented and it is no wonder that he regarded his university, not to mention his fellow students, highly. He was establishing a Cambridge connection that he valued for the rest of his life.

The first real test of Galton's mathematical abilities came in the spring of 1842 when he took the Little Go, the popular name for the first major examination for the Bachelor of Arts degree with the Great Go being the final examination.[43] To amuse himself while he studied, he tried his hand at the occasional poem and, with the birth of the Prince of Wales on March 31, penned an honorific that began

Sleep thou royal child, take thy calm rest
Pillowed in the quiet of thy mother's breast[44]

Despite intense preparation, Galton only made the second class along with his friend Joseph Kay, another of Hopkins's pupils. Kay would later be called to the bar at the Inner Temple and subsequently made a Queen's Counsel and then a judge.[45] Seven of Hopkins's other students were in the first class list in the Little Go, including another friend Charley Buxton, an embryonic politician. He would eventually sit in the House of Commons successively for Newport, Maidstone, and East Surrey.[46] But Galton considered himself lucky to have got a second class as he wrote Tertius. "I have I consider had 3 grand escapes in my lifetime: 1st walking into a Lion's den and coming out undigested, 2ndly bathing in a frosty stream in moonlight and not remaining at its bottom in apoplexy, 3rdly going into the Little Go when I had not read over half of my subjects and coming out unplucked, not, however, that the pluck would be of any consequence."[47] Although Hopkins, the ever-caring tutor, complimented Galton on his abilities in mechanics, he was also a realist. He counseled his pupil not to subject himself to the rigors of preparing for the Scholarship Examination as was he very unlikely to receive an award.[48]

In June Galton voyaged to Dundee with his reading party. Their tutors were Cayley in mathematics and Venables in classics and the students included Galton's friends Charley Buxton, Joseph Kay, and Joseph's half-brother Eben. Cayley, the Senior Wrangler and first Smith's Prizeman in 1842, was a former Trinity graduate and now a fellow of the College. He was destined to become one of the leading English mathematicians of the period.[49] He supported himself by practicing at the bar. Venables, a recent Pembroke graduate, Third Wrangler and fifth in the second class of the classical Tripos, would forge a career as a distinguished clergyman.[50] The voyage was rough and everyone was seasick some of the time, but, despite their discomfort, they all grew to like Cayley who was "unanimously voted a *brick* and a most gentlemanly-minded man."[51]

From Dundee they travelled up the Firth of Tay to Perth where Galton made note of its attractive young ladies. From Perth they journeyed overland along the River Tay to Aberfeldy set in the high moorlands of Scotland where they found the local residents most friendly.[52] Galton had an introduction to Sir Neil and Lady Menzies who lived in Menzies Castle, a sixteenth-century edifice in the tower house style. This proved propitious as Galton and his friends were invited on July 24 to a Highland wedding. They danced from three in the afternoon until four the next morning with an intermission of little more than an hour, often doing Scotch reels, which Galton loved. In gratitude for all the hospitality they received the boys gave a ball at the end of August "on Wednesday the 31st instant."[53] Twenty-nine "Dancing Ladies" and 22 "Dancing Men" attended. They danced the Wallsette, the Eisenbahn, the Elizabethan, the Mosaique, and so forth. At 12:30 the guests broke for a midnight supper where tables groaned with chicken roast boil, grouse roast pie, wild roast of duck, rabbit pie, tongue, pickled salmon, cold beef, and assorted sandwiches followed by a plethora of sweets and desserts, jellies, creams, custards, apple pie, plum cake, cheese cake, and tartalleti. That month the Queen and Prince Albert visited Lord Breadalbane at nearby Castle Taymouth and Galton's friends, the Menzies, took him to view the pageantry. The Highlanders, dressed in the Menzies and Cameron tartans, were drawn up in four files round a large quadrangle. During the proceedings, which were accompanied by fireworks, the crowd frequently broke out in cheers. For Galton the only blemish on a blissful summer occurred when he returned from a walk one day to find a cavalry officer in his rooms.[54] His books, papers, clothes, etc. had been removed and replaced with the officer's gear. Galton was furious and said so to the offending officer who, though amused, stood firm. Orders were orders and he was billeted in Galton's rooms and Galton was quartered elsewhere. "This little incident made me realise the odiousness and too probable insolence of military rule and the lesson sank deep."[55]

Although Galton greatly enjoyed the social life at Aberfeldy, he complained that his "head scarcely improves. I have been able to do but little reading since I have been here and altogether am very low about myself."[56] Things got worse when he returned to Trinity in the fall and by early November, he was writing Tertius that his "head is very uncertain so that I can scarcely read at all."[57] He remarked that he was not alone. In the class above him, the top three men in their college examinations were settling for the poll degree with two suffering from bad health and the third unable to continue to handle the reading. His friend Charley Buxton was going out in the poll and Joseph Kay withdrew for a semester unable to withstand the academic pressure any longer. Consequently, Hodgson sent the sputtering young Cantab "a certificate for degrading." Degrading in Cambridgespeak meant that Galton could put off taking the honors examination for the B.A. for a year. Even this was not enough. Galton wrote his father in late November saying that he was getting steadily worse; that reading in mathematics made him dizzy; and that heart palpitations would come upon him without forewarning. Galton, like his friend Joseph Kay, was in trouble and he dropped out for the rest of the semester setting the stage for him, like Darwin, to go out in the poll.

Galton was now free, like the rest of the hoi polloi, to enjoy his last year and a half at Cambridge with his now numerous friends. He filled his diary for the winter of 1843 with invitations to dinner parties, balls, and wine parties.[58] Together with Charley Buxton he founded a debating group called the "Historical Society"[59] that started out with nine members and eventually grew to 60. The member elected president for the week had to put out a spread plus wine and suggest the debate for the following week. Galton began to play hockey. He continued to write short poems and founded an "English Epigram Society" that met three times a term.[60] He was having a good time at Cambridge, but he was not exaggerating in calling attention to his own problems and those of his friends trying to go for honors. Preparation for the Tripos was very strenuous work and bets were laid as to probable winners since "true to their sporting instincts the English had contrived to turn even the university examinations into an athletic contest."[61] But the constant cramming could and did injure a student's health or at least his state of mind to the point where he thought he was ill.

There were no reading parties the next summer, which was spent with sister Emma in Germany where Galton sketched churches in Hamburg, a spire in Magdeburg, and cathedrals in Dresden and Ratisbon.[62] He planned to return to King's for his final year to complete his medical studies after which he would move to St. George's Hospital in London in the summer of 1844 to finish his clinical work. Meanwhile Tertius's health was slowly failing, his severe asthma was worse, and his gout was acting up. A worried Emma wrote her brother in early March that "My father says over and over again 'Give my affectionate love to my dear Francis.'"[63] She implored her brother to write his father. "It would please him *very much*, if in a day or two . . . you would write him an affectionate . . . letter a letter from you is as good as a dozen draughts."[64] In September Tertius travelled with his son to St. Leonard's on the coast. He enjoyed himself joking to Emma that his son was an excellent travelling physician. He felt better saying he would love to plunge into the sea if it were not for the dread of facing Dr. Hodgson's wrath, but it was a false sense of well-being. After the trip Tertius's health declined rapidly and, on October 23, he died.

Tertius had adored his youngest son. He had indulged him in education and travel while providing a firm guiding hand and prodding him to keep accurate account of his expenditures. With Hodgson's help he kept his son steadfastly aimed toward a career in medicine. With his death the family began to break up, and Galton lost direction so completely that it would be six years before he once more regained a sense of purpose at the age of 27.

FOUR

Drifting

> Charles Darwin was a student and naturalist from his College days; Francis Galton's six fallow years threw back his work in life, so that much of it was achieved at an age when most minds grow quiescent.
>
> —Karl Pearson[1]

In the fall of 1844 Galton took rooms in London near Hyde Park with two Cambridge friends, W. F. Gibbs and H. Vaughan Johnson.[2] Gibbs was later tutor to the Prince of Wales[3] while Johnson became involved in legal work with Galton's friend Eben Kay. During 1845 the affairs surrounding Tertius's estate were wound up and the family began to scatter. Emma, a spinster, lived with their mother either at Leamington or at Claverdon, a Warwickshire estate Tertius had purchased in 1824 as a summer residence.[4] Delly married Robert Shirley Bunbury on May 18, 1845, only to be widowed a year later and left with their baby, Millicent.[5] Milly would become close to Galton late in his life. In December Bessy married Edward Wheler. Lucy, wed for some time to James Moilliet, was suffering from the effects of rheumatic fever contracted as a child and had only three years to live.[6] Darwin, Galton's eldest brother, was married to his second wife. They were living with her mother at her country house near Stratford-on-Avon. Erasmus, Galton's other brother, was farming his estate at Loxton in Somersetshire.

Galton, free of the constraints imposed by education and family, "had many 'wild oats' yet to sow."[7] He wrote Henry Hallam proposing a Nile expedition in search of big game, but Hallam demurred so Galton set off by himself for Egypt in the fall of 1845. He stopped off in Malta and encountered his friend, Robert Frere, a King's graduate. He was looking after his uncle John Hookham

Frere, a former diplomat, who was ailing badly and could not receive company.[8] This was a disappointment, for Hookham Frere, with George Ellis, and George Canning, had written *The Loves of the Triangles* that so cruelly parodied Erasmus Darwin's poem *The Botanic Garden*.

Upon leaving Malta on a steamer bound for Alexandria, Galton luckily encountered two old Cambridge acquaintances, Hedworth Barclay, a distant relative, and Montagu Boulton, the younger brother of his friend Matthew Boulton.[9] Boulton and Barclay had just toured Greece and now intended a Nile adventure so the three banded together.[10] Barclay had a very smart Greek courier (a travelling servant) named Christo who would serve as cook. Boulton's courier, Evard, once groom of the chamber in the service of Lady Jersey, acted as butler and kept track of all accounts. Galton hired a dragoman (guide) named Ali who spoke Arabic. The three travellers lingered in Cairo,[11] a city of high, narrow houses with projecting upper stories from which jutted delicate lattice-work wooden windows.[12] Its streets were roofed over with long rafters and pieces of matting to shield the throngs below from the merciless desert sun. The unpaved thoroughfares were narrow, rutted, and lined with little wooden shop-fronts where cross-legged merchants sat amongst their goods silently smoking and watching. Men on horseback and carriages forced their way through the thick crowds and swirling dust. Syrians in baggy trousers and braided jackets sauntered by; barefooted Egyptian fellaheen in ragged blue shirts and felt skull-caps; Greeks in stiff white tunics; Persians with woven caps shaped like mitres; Bedouins in headshawls and flowing white robes with chocolate stripes a foot wide; Englishmen in palm-leaf hats and knickers mounted on tiny donkeys; women covered in blue or black striped cotton with their black veils slit only for their eyes; dervishes with matted hair streaming from under extravagant head-dresses; fine-featured Ethiopians with slender, bowed legs; Armenian priests in long black gowns and high square caps; Arabs passing by like white wraiths; Janissaries with gold-embroidered jackets and jingling sabres towering above the crowd on their horses and; merchants, beggars, soldiers, boatmen, laborers, workmen, in every variety of costume, some white, some black and every shade of color in between.

Sated with the sights and smells of this jostling oriental city the three adventurers were ready to move on, but first they needed a firman (an edict) from the sovereign. This would permit them to travel freely and allow them to impress local inhabitants to manhandle their boat up the Nile cataracts. Barclay obtained an audience with the long-reigning Muhammad Ali Pasha, the Ottomans' viceregent in Egypt.[13] He had extinguished the last of the Mameluke nobles after their defeat by the French in the Battle of the Pyramids bringing their dynasty to a close. He did so through a treacherous deceit by luring the unsuspecting Mamelukes to a banquet supposedly being held in their honor on March 1, 1811. Once they entered the Citadel through the Bâb el-Ázab gate, flanked by its lofty and menacing towers,[14] they were locked in and "shot down like mad dogs in a trap."[15]

Barclay, firman in hand, rejoined his friends who went shopping for a *dahabeyah*, a lateen-rigged Nile sailing vessel, plus crew. These flat-bottomed craft had a big mast near the prow, and a smaller one at the stern and could be sailed or rowed. Cabins occupied the vessel's aft and its roof formed a raised deck reached from the lower deck by two short flights of steps.[16] This was the exclusive territory of the passengers with the lower deck being occupied by the crew. The kitchen, a sort of a shed located between the big mast and the prow as far as possible from the passengers' cabins, contained a large charcoal stove and a row of stewpans. In this position the cook was protected from a favorable wind by the shed, but if the wind came around he was screened by an awning. Despite their similar ground plans, a confusing number of decisions was required in choosing a vessel. Some had six cabins and others eight; some could get up the cataracts and others could not; and some were twice as expensive as they should be and others five or six times.[17] Choosing a captain, or *Rëis*, was equally problematical. Certificates from former travellers attesting to the qualifications of one *Rëis* had a habit of turning up on board different boats and in the hands of different captains. Despite these hurdles the Cantabs hired their *dahabeyah* plus a crew, and an Arab boy of ten named Bob, as an all around helper. "Barclay put on board a keg of his own porter, and so we started, intending to live luxuriously and in grand style."[18] Evard made sure his young charges ate well, purchasing mutton, bread, onions, cream, fish, and eggs when and where the opportunity arose.[19] He also supplied them with local beer, coffee, tobacco, and even a few bottles of dubious wine now and then.

The trip up the Nile was relaxing and especially "the pleasure of living all day barefoot and only half dressed, and of waking oneself by a header in the river, clambering back by the rudder."[20] Next would come the first cup of coffee and a pipe. The three travellers would chatter with Bob, the captain, and the sailors in rudimentary Arabic paying occasional attention to the majestic ruins of the great Pharaonic temples they were passing. Galton recorded his impressions in watercolors and sketches of the Elephantine Island, the Temple of Osiris at Philae, an Arab Plow, a solitary heron, Barclay reclining hands behind his head with knees pulled up, Ali in profile, Bob from behind, and so on.[21] At the First Cataract the Nile, diverted from its original course, spread over a rocky basin bounded by sand-slopes on one side, and by granite cliffs on the other. Islets abounded and numberless channels foamed over sunken rocks and eddied among water-worn boulders. Only the Shellalae, or Cataract Arab, possessed the key to this labyrinth whose passage was arranged by the Sheikh of the Cataract.[22] When he cried "*Roôh*" (forward), his men suddenly appeared laden with coils of rope shouting and gesticulating as they fought

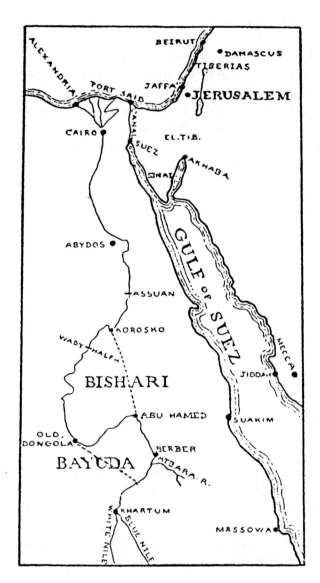

Fig. 4-1 Map showing Francis Galton's trip up the Nile to Khartoum and a few of the points he visited in the Near East. From Francis Galton, *Memories*, 88.

through the rapids and manhandled two ropes from the *dahabeyah* to the nearest island, making them fast to the rocks. Two ropes from the island were then pulled aboard the *dahabeyah*. A double file of men on deck, and another on shore scattered themselves along the ropes and on a signal from the Sheikh began chanting as they double-hauled the *dahabeyah* slowly and steadily up the cataract.

Near Korosko, where the Nile makes a great bend westward (Fig. 4-1), the travellers again encountered rapids. They needed some men to pull them forward, but they found that the available labor had already been "impressed by the owner of a small and dirty looking Egyptian boat, who they told us was a Bey."[23] The impudent Englishmen loudly threatened to pitch the Bey into the Nile. The Bey, a handsome man with a full gray beard, answered them courteously in perfect French. They realized that he was far more cultured than they had supposed so they bantered politely with him as they strode together toward Korosko, having finally moored the *dahabayeh*. Joseph Pons d'Arnaud Bey, trained as a civil engineer possibly at the Ecole Polytechnique, was a follower of the Comte de Saint-Simon whose socialist credo advocated state control of all property and distribution of produce according to individual vocation and capacity. D'Arnaud Bey had entered Muhammad Ali Pasha's service where he was appointed to a commission to study the best method of extracting gold dust from the bed of the Khawr al-Adi, a river flowing northward into the Blue Nile near the border between Sudan and Ethiopia.[24] He had only recently returned from a trip to the gold deposits.

He invited the travellers into his simple, well-ordered mud house. Galton was charmed, spotting a barometer, a thermometer, and other scientific gear as well as many maps and good books. As they became better acquainted, d'Arnaud Bey gently reproved his guests. Although his English was not too serviceable, he had understood their rude remarks at the Nile rapids, but forgave them, something a remorseful Galton never forgot. From long experience d'Arnaud Bey knew well the barren country that lay beyond so he said "Why do you follow the English routine of just going to the second cataract and returning? Cross the desert and go to Khartoum."[25] In an instant the expedition's goal was transformed. The caravan route east of the Nile across the Bishari (Nubian) desert to Abu Hamad at the southern end of the great Nile bend avoided the second, third, and fourth cataracts (Fig. 4-1). Since d'Arnaud Bey knew the Sheikh of the Bishari Desert, he volunteered to arrange an escort for the travellers across the desert and thence along the Nile to Berber beyond the fifth cataract where they could hire a vessel to continue to Khartoum. Galton and his friends were delighted and invited d'Arnaud Bey to join them for dinner. Evard did his best to put on a sumptuous meal accompanied by frequent draughts of wine, for the always resourceful courier had somehow obtained a copious supply. The young Cantabs listened excitedly as d'Arnaud Bey described the country ahead, interrupting him frequently with questions and, as the evening wore on and more wine was consumed, the proceedings became ever more boisterous until, with every one feeling very gay and the cabin reeking of alcohol and cooking smells, the door suddenly opened. There silhouetted before them, with the cool, dry night air pouring in, stood the Sheikh of the Bishari Desert, a tall, dignified stranger on whose forehead was impressed a circle of sand indicating that he had recently prostrated himself before Allah. Despite their obvious mortification, they managed to come to terms with the Sheikh. With Bob in command, the *dahabeyah* was to proceed to Wadi Halfa below the second cataract and wait for the expedition's expected return in early March.

The next day their convivial guide appeared, a son or nephew of the Sheikh, and, mounted on camels, the travellers embarked. The first day they rode only an hour, camping on the sand in the cold clear air only three miles from Korosko, a customary procedure so they could take stock and make sure that nothing of importance had been forgotten. Having accounted satisfactorily for their supplies and equipment their real journey began the following day. "It *was* a desert, like the skeleton of the earth, with sand blown clean away from the bare stones or lying here and there in drifts, table topped drifts."[26] Over the centuries, the caravan track had been well used and the way was marked by the whitened bones of camels and human beings, presumably slaves, who had perished during the desert crossing. As they progressed others "joined our caravan; a man, his wife, baby and donkey just like Joseph's flight. Also another man on foot, with no possessions but an old French cuirassier sword, wherewith he was going to join slave raids in Abyssinia."[27] After eight days they reached the Nile at Abu Hamad where men and beasts, grateful for the unlimited supplies of fresh water, drank deeply, bothered only by the clouds of midges that spun round their heads.

After another three days the caravan arrived in the dusty river town of Berber with its four squalid villages of ramshackle huts.[28] They were greeted hospitably by the Governor who treated them to sherbet made from his own limes, found them a mud house to lodge in, and gave Galton a monkey, after which Galton purchased a second as a companion for the first.[29] The Governor granted them permission to hire a boat to proceed to Khartoum so they obtained a small, one-masted vessel with a cabin about four feet high. As they clambered aboard, cockroaches scuttled everywhere, but truculent Berberines gathered menacingly about the mooring, seizing a rope secured to the boat to prevent its embarkation. With most of the ship's company still on shore and afraid to board, the situation threatened to turn ugly when Barclay cast off the rope and ordered the two or three crew on deck to set sail. The rest of the party ran along the shore, leaping one after another into the Nile and swimming out to their vessel. On the way south they passed the ruined mud pyramids at Meroë, site of the capital from which the Ethiopian Pharoahs and their Queens had governed Egypt almost to the delta in Roman times. They passed the town of Shendi with its large open marketplace where grain, cotton, liquid butter, cord, locally woven cotton piece goods, and slaves were on sale twice a week.[30] Four miles beyond Shendi on the opposite bank was al-Matamma where caravans from Dar Fur, Kordofan, and Ethiopia frequently arrived. The three Cantabs likely heard tales from the crew of the elegant courtesan named Amna who lived there. "No traveller to the Sudan will have failed to know of Amna bint 'A'isha. Her piquant beauty attracted many admirers. Her hospitality towards all merited the gifts with which several of her lovers recompensed her."[31]

Khartoum was built on a promontory of land at the confluence of the two Niles, an area known to the local Arabs as El Khartoum for its presumptive resemblance to the trunk of an elephant.[32] The town was a collection of squalid huts athwart narrow, filthy streets with a single great hall where the Pasha held audience. There was as yet no British consul in the city. Because of the primitive surroundings the Europeans in residence rarely brought their wives with them, buying girls from Ethiopia or central Africa on the slave market instead to serve as concubines.[33] They lived, dressed, ate, and drank *al-laturca*, with their Syrian, Armenian, Egyptian, and Turkish business associates. Khartoum was a rich source of white gum, ivory, rhino horns, ostrich feathers, beeswax, and hippo hides. Slaving expeditions regularly made off for the mountainous region of the Ethiopian border, but their prey usually fled with their families into deep caverns in the rocks hoping to avoid discovery.[34] However, the children's cries often gave them away so the slavers would flush the poor souls out of their hiding places by loading their muskets with red pepper and firing the guns into the cavern entrances. The "thick and pungent pepper dust" forced "the victims to plead for their lives."[35]

Galton's party settled into a mud house overlooking the Blue Nile across which they could see dust devils dancing on the wide plain.[36] From the locals they learned there was an Englishman in Khartoum. He was Mansfield Parkyns, a student at Trinity while Galton, Boulton, and Barclay were there, who had got into some sort of a scrape and left college prematurely. Afterwards he had wandered in Africa alighting first in Ethiopia and later moving to Khartoum. Parkyns, reeking of rancid butter, had gone native. His head was shorn except for a Muslim tuft and he was nearly naked with a leopard skin casually thrown over his shoulder. He knew all of the seamy and disreputable characters in Khartoum including the slavers and the outlaws. "The saying was that when a man was such a reprobate that he could not live in Europe, he went to Constantinople; if too bad to be tolerated in Constantinople, he went to Cairo, and thenceforward under similar compulsion to Khartoum. Half a dozen of these trebly refined villains resided there as slave-dealers; they were pallid, haggard, fever-stricken, profane and obscene."[37]

Parkyns accompanied his friends on a short voyage up the White Nile that, in stark contrast to the swift-flowing Blue Nile, was stagnant and blanketed with fetid air having filtered through that endless swamp, the Sudd.[38] At night the cook pitched the offal from dinner overboard where it could be seen the next morning bobbing about the anchored boat. On the banks and sandbars of the river were great flocks of flamingos and pelicans. Galton blazed away at herds of distant hippopotami without success. One night Parkyns and

Boulton stole out to lie in wait for a hippopotamus and, as their eyes became accustomed to the darkness, they spied their quarry on the edge of the river, raised their guns and fired only to find they had killed a cow. They sprinted back to the boat and urged the captain to weigh anchor before the cow's owner discovered his loss and stormed down upon them.

On the return they dropped Parkyns off at Khartoum and disembarked at al-Matamma where they hired camels to cross the Bayuda Desert west of the Nile (Fig. 4-1), a much pleasanter overland route than the Bishari. Wells were plentiful, there were periodic rainfalls, and scraggly mimosas dotted the desert landscape. After six days they reached the Nile at Meroë and rode along the riverside to Dongola where a hospitable Pasha greeted them and held a review in their honor. They continued to Wadi Halfa to find their *dahabeyah* awaiting them with Bob thoroughly in control. Bob, only a boy, had bested the captain in some dispute, presumably about who was in charge, and had ordered the crew to flog him. The *khamseen*, the hot south wind from the desert often thick with sand and dust, was blowing on their way to Cairo where Evard, having kept meticulous records of expenditures, billed each of his masters for one-third.[39]

The friends parted, with Barclay returning to Scotland and Boulton heading for Syria over the desert. Galton sailed by steamer for Lebanon accompanied by Ali and the two monkeys plus an *ichneumon* or "Pharoah's Rat," a relative of the mongoose with a strong predilection for crocodile eggs.[40] The first stop seems to have been the seaport town of Acre, now Akko in northern Israel, where they went into quarantine. Again, there was a helpful Pasha. He marvelled over Galton's journey to Khartoum. This was considered a great feat and Ali helpfully embellished, implying that his master was very important. The quarantine was magically relaxed and they left Acre for Beirut, a thriving commercial center, from which raw silk, cotton, olive oil, fruit, sesame, sponges, cattle, and other goods were exported. Its verdant hills and green slopes stood in marked contrast to the starkness of the Egyptian desert. Galton bought horses for Ali and himself, a fancy tent, a canteen, excellent coffee, and apparently a hookah.

They set off for Damascus, spending a night with a Druse chieftain to whom Galton had an introduction from the Pasha of Acre. The next day they caught sight of the minarets of Damascus sparkling above the orchards and plantations surrounding the city. Galton arranged to board in an English doctor's house where the faithful Ali unexpectedly caught dysentery, became delirious, and shortly thereafter died. His body was washed and covered in shrouds and Galton followed the burial procession at a distance since, as a Christian, his presence would have polluted the Muslim funeral. Afterwards he made arrangements for Ali's tombstone and sent his wife his belongings plus some wages due him.

The weather got stiflingly hot as summer approached so Galton left Dr. Thompson's Damascus house and moved to the suburb of Salahieh stretched out at the foot of the Jebel Kasyun, which rises sheerly from the plain to an altitude of 3,960 feet. The coffee houses were delightful and the lovely public gardens were laced with little streams of clear water diverted from the river. He often rode horseback through the outskirts of town passing lush apricot orchards where he watched the fruit being boiled in great cauldrons and the resulting mush being flattened out into sheets to dry. These were then rolled up like pieces of oilcloth. As fall approached and the weather became more tolerable, Galton returned to Lebanon where he was a guest of the Sheikh of Aden. He stayed for a whole week in spite of counting 97 flea bites on his lower right arm upon awaking the first morning.

Late that summer, Galton must have enjoyed brief ecstasy with a woman of easy repute. Her secret later became apparent when an anguished Galton wrote about his experience to Montagu Boulton. Boulton commiserated from Damascus on September 30. "What an unfortunate fellow you are to get laid up in such a serious manner for, as you say, a few moments' amusement."[41] Boulton continued that he planned to purchase a slave-girl and had several Ethiopian women brought in for show, but none were pretty enough to suit his tastes. He remarked that the "Han Houris [a virgin of the Muslim paradise, a great beauty] are looking lovelier than ever, the divorced one has been critically examined and pronounced a virgin."[42]

Galton travelled from Aden to the ancient Phoenician city of Tripoli. The city was built on either side of the river Kadîsha, with its fine harbor and its 18 churches, and dominated by that great relic of the First Crusade, St. Giles citadel.[43] While there he camped near some ditches filled with stagnant water, and contracted a sharp intermittent fever that plagued him on and off for several years. Then he returned to Beirut, but along the way, in mountainous terrain, one of his horses stumbled and fell to its death off the edge a cliff. He sold the other horse in Beirut and met Boulton for the last time. Boulton ended up in the Punjab as an onlooker with the British Indian Army forces besieging Multan during the Second Sikh War (1848–49). While peering out through a loophole in a turret, he was fatally shot through the eye with a matchlock ball.

Galton next voyaged by collier to Jaffa, now modern Tel Aviv, a place of singular beauty. He rode inland by camel, passing through groves of orange, lemon, and pomegranate trees into the plain of Sharon on his way to Jerusalem. Before long he could see the olive groves surrounding Lydda in the distance on his left. Approaching Jerusalem the next day, he could see that it was a walled town standing on four hills separated by valleys partially filled with the debris of successive destructions and surrounded by glaringly white limestone hills.[44] The city walls were surmounted by numerous towers and Galton entered through the Jaffa Gate. Perhaps he was disappointed, as were many travellers then, to find that Jerusalem, with its grand history of sacred events, was nothing more than a little town around whose walls he could walk in an hour. Having toured Jerusalem, Galton decided to sail down the Jordan from the Lake of Tiberias (Sea of Galilee) to the Dead Sea on a small raft made of water skins.[45] He headed north on horseback accompanied by an escorter, Sheikh Nair Abu Nasheer, spearmen, his native cook, and one or two others. One night, after camping, they were attacked by raiders who cut the tails off some of the horses, but they were uninjured. Once at his destination Galton built his raft and launched it onto the Jordan where the river issued from the lake. It was a harebrained scheme for the river was narrow, he capsized twice, and then was caught in the swirl of a constricted channel and knocked into the river by overhanging boughs. Galton abandoned the raft, mounted his horse, and outfitted with Arab headdress rode south followed by the spearmen, their long lances topped with ostrich feathers, and visited a great bedouin encampment ruled by the Emir Ruabah whose sister was married to Galton's escorter. Everything went well until Galton made an unintentional mistake when he shot but failed to kill a desert partridge. He finished the bird off English-style by knocking its head against the stock of his gun, insulting the bedouin as Muslim custom was to cut the bird's throat while repeating a certain incantation.

Upon returning to Jerusalem, he was greeted by an official letter in Arabic demanding that he make restitution not only to Ali's wife, but to numerous other kin under threat of legal action. Galton consulted his banker who advised him to leave the Levant as soon as possible. At Marseilles he was detained in the Lazarette (quarantine station) for about ten days before sailing home with his little menagerie in November 1846. Upon his arrival, an old friend with a flat agreed to board the monkeys. He turned them over to his landlady with detailed instructions for their care, but she disliked the animals and locked them in a cold scullery where they were found dead in each other's arms the next morning.

To a paleontologist seeking to understand the evolution of a group of brachiopods or the demise of the dinosaurs the worst thing that can happen is the interposition of a hiatus in a continuous series of strata. The evolutionary record is summarily cut off and the paleontologist is reduced to hypothesis and speculation about what happened next. Just so the biographer depends on letters, journals, notes, and contemporary accounts to trace the life of his quarry. There is a nearly complete hiatus in Galton's life from 1847 to 1849 save for what he recorded in his *Memories*, published when he was 86.

Galton spent most of the next three years hunting and shooting, the preparation for which he characteristically went about systematically. His eldest brother Darwin advised him on the purchase of a hunter, as well as a hack for ordinary riding, and he joined a small "Hunt Club" in Leamington. Foremost among its members was the junior Jack Mytton who, with his cronies, liked to party, carouse, and gamble. He, at least, came by it honestly. His father had commenced his morning by consuming port while shaving, having downed between four and six bottles by the time he wove his way unsteadily toward bed late in the evening.[46] After wasting a fortune, and hounded by creditors, he wound up in the King's Bench prison where he died of convulsions from delirium tremens at the age of 37. His son followed in his footsteps dissipating yet another fortune. With friends like these, it is no wonder that Galton tiptoed so lightly over these fallow years.

One day another Leamington friend invited Galton to shoot with him in the Highlands. At the end of the season, the weather was still fine so Galton took the postboat from John o'Groat's across the Pentland Firth to the Orkneys, enjoying the fine views of the treeless hills of Hoy, South Ronaldsay, and the numerous islets in between. He settled in the town of Kirkwall, nestled on the isthmus that divides Orkney into two unequal land masses and dominated by St. Magnus, its towering twelfth-century Norman cathedral. He must have enjoyed these windswept northern islands for the next summer he travelled beyond the Orkneys to Shetland where he explored the coastline, one day coming upon a shoal of pilot whales that had beached themselves. He stalked guillemots, razorbills, puffins, and perhaps even skuas and gulls, not to mention the seals that basked on rocky ledges. He learned how to steal storm petrels by digging rapidly into the shingle, where the birds tunneled to make their nests. "Its oily smell is very strong and rank. The popular belief is that if you cram a wick between the beak and down the gullet of a dried-up petrel and light it, the bird will burn like a lamp."[47] Galton left Shetland in December with a great crate of live seabirds for his brother Darwin, but the crate was set in an exposed position on his southbound train and three-quarters of the birds died of the cold.

Lacking a mental compass to direct his future, Galton consulted a phrenologist hoping that the resulting personality profile would steer him in the right direction. He had long been interested in phrenology as it seemed to represent a "scientific" approach to forecasting ability and may even have served as a precursor to Galton's later system of head measurements.[48] As a boy at King Edward's school, he first encountered phrenology when a Cambridge examiner asked to study the boys' heads the day before a test to see whether his phrenological predictions would be verified by their scholastic performance the next day. After pressing his fingers over the contours of Galton's head, he reported to Dr. Jeune that this "boy has the largest organ of causality I ever saw in any head but one, and that is the bust of Dr. Erasmus Darwin."[49] This meant that Galton should have an acute ability to discern the connection between cause and effect. Dr. Jeune revealed to the examiner that Galton was the grandson of Erasmus Darwin. During his Cambridge summer reading party at Keswick, Galton met a famous German phrenologist named Schmidt. He got Schmidt "to paw my head, he gave me I think a very true character (self-esteem was remarkably full). I have not now the bump of constructiveness very large though he says it is large."[50] In his 1843 trip to Germany with Emma, Galton visited the house of a man named Noel who made highly accurate casts of living heads, specializing in the extremes, individuals of noteworthy ability, and ones who were notorious criminals.[51] The famous Scottish phrenologist George Coombe was staying there as well.

So it is not surprising that in 1849 the listless Galton visited the chief phrenologist at the London Phrenological Institute, a man named Donovan. Donovan's interview was conducted in a business-like manner, much like a modern physical examination, except that Galton's head was the only part of his anatomy over which Donovan's skillful fingers marched. Based on his examination Donovan filled out an extensive, printed checklist and wrote a report of several pages to accompany it.[52] The checklist contained 18 personality characteristics divided into four categories (propensities, moral sentiments, perceptive faculties, reflective faculties). Galton was likely pleased to learn that his "amativeness" was "large," his "concentrativeness" was "full," but that his "alimentativeness" was moderate. Donovan deduced that Galton had lots of self-esteem, was cautious, benevolent, conscientious, but possessed a good sense of humor. He confirmed the large size of his organ of causality, concluding:

> Men of this class are likely to spend the earlier years of manhood in the enjoyment of what are called the lower pleasures, and particularly of those which the followers of Mahomet believe to form the chief rewards of virtue in the realms above. . . . Self-will, self-regard and no small share of obstinacy form leading features in this character. . . . There is much enduring power in such a mind as this—such that qualifies a man for roughing it in colonising. . . . He is not calculated to gain good will on a brief acquaintance. . . . As regards the learned professions I do not think this gentleman is fond enough of the midnight lamp to like them, or to work hard if engaged in one of them.[53]

These rudderless years embarrassed Galton, who justified his seeming indolence in his autobiography writing that one would think "I was leading a very idle life, but it was not so. I read a good deal all the time, and digested what I read by much thinking about it. It has always been my unwholesome way of work to brood much at irregular times."[54] But Karl Pearson thought otherwise, remarking that "Galton was never a great student of other men's writings; he was never an accumulator like his cousin Charles Darwin; and the most well-read and annotated books in his library certainly belong to a later date and to periods of definite lines of research."[55] In these aimless years, the wealthy young Galton caroused, he travelled, and he undoubtedly sought out the fair sex where possible but, like his grandfather Erasmus Darwin, Galton was also an inventor. In 1849 he published a pamphlet describing his design for a printing telegraph machine that Pearson thinks represents the reawakening of his scientific interests.[56] Perhaps the phrenologist helped too.

PART TWO

Geography and Exploration

FIVE

South Africa

> My own inclinations were to travel in South Africa, which had a potent attraction for those who wished to combine the joy of exploration with that of encountering big game.——But I wanted to have some worthy object as a goal and to do more than amuse myself.
>
> —F. Galton, *Memories*[1]

Francis Galton was 27 in 1849 when the idea of a South African expedition came to him. Except for North Africa, parts of West Africa, and the southern rim of the continent, extending eastward from the Cape of Good Hope, little was known about this vast land mass. Galton learned that David Livingstone, a young Scottish medical missionary, had travelled far north in South Africa past the Kalahari Desert to a lake called Ngami. Since "the well-watered districts beyond this desert could now be reached by wagon from the Cape," he "felt keenly desirous of taking advantage of this new opening, and inquired much of those who had recently returned from South Africa concerning the conditions and requirements of travel there."[2]

Exactly how Galton got wind of Livingstone's discovery of Lake Ngami is unclear. It must have been before Arthur Tidman, the Foreign Secretary of the London Missionary Society, published an account of the lake's discovery in the March 1850 edition of the *Missionary Magazine*, but after Livingstone returned from the Lake in October 1849.[3] Livingstone's discovery piqued Galton's curiosity and he discussed his plans initially with his cousin Captain Douglas Galton of the royal engineers.[4] Douglas Galton had achieved early recognition for his role in demolishing the wreck of the *Royal George* at Spithead in 1842 using an electric current to detonate the explosive charges for the first time. Later he would gain fame as a sanitary engineer. Douglas Galton suggested that his cousin contact the Royal Geographical Society. This organization could provide the kind of influential backing he would need for his expedition even if its support was moral rather than financial. Galton's old Cambridge friend Robert Dalyell knew Sir Roderick Impey Murchison, the current vice-president of the Royal Geographical Society, and could provide an introduction.[5] Meanwhile, Darwin and Douglas Galton put Galton up for membership and he was elected a member in the spring of 1850, beginning a long and active association with the Society.

Galton could not have proposed his expedition to Lake Ngami at a better time. The hard-charging Murchison, at the height of his career, was about to become president of the Society for the second time. He had a personal interest in South Africa as his friend, Sir John Herschel, the famous astronomer, had sent him trilobite fossils from there.[6] Earlier Murchison had made a major geological discovery while working through a succession of sedimentary rocks in Wales underneath the Mesozoic series called the "Grauwacke."[7] Murchison's Silurian System, named for the Silures, a British tribe indigenous to the region in Roman times, was underlain by his friend Adam Sedgwick's Cambrian System and would constitute part of what is now called the Paleozoic series. Murchison's magnum opus, *The Silurian System*, published in 1839, brought him worldwide fame.

Society members were helpful in many ways to Galton whose "vague plans were now carefully discussed, made more definite and approved."[8] He was introduced to many persons useful to him including the "Colonial Secretary, Lord Gray, who gave instructions in my favour to the Governor of the Cape." A particularly important addition to Galton's expedition was Charles J. Andersson, a Swede brought up in England, who was a keen observer intensely interested in natural history, not unlike Galton. He had sailed from Gothenberg to Hull in 1849 with a large collection of living birds, mammals, and preserved specimens intending to dispose of his collection before travelling and collecting elsewhere around the globe.[9] By happy circumstance Andersson and Galton were introduced by Sir Hyde Parker, a scion of the distinguished naval family and a rear admiral in the Royal Navy.[10]

Galton next outfitted his expedition and, considering he was a novice, he was meticulous in his planning. Wagons and beasts of burden would be purchased in South Africa. Supplies and equipment were collected "on the principle of having them as light as possible, and in duplicate, the half of which" Galton could leave in a cache when "I had to quit my waggons, as a store to fall back upon should I happen to meet with robbery or accident."[11] He was uncertain what presents to carry with him for the local chiefs he encountered so he bought guns, beads, knives, gaudy calico, mirrors, accordions, hunting-coats, old uniforms, burning glasses for concentrating the sun's rays, bracelets, anklets, Jews' harps, rings, and a faux crown that was to prove handy. He also possessed some charts to aid him in his exploration.[12] One, a detailed map of the Cape of Good Hope and surrounding areas from John Arrowsmith of the well-known family of cartographers, rapidly became devoid of any geographical features north of the Orange River. Another, which he apparently planned to use in proceeding to Lake Ngami, displayed the East Cape region as far north as Delagoà Bay in what is now Mozambique with the mission at Kurumen, Livingstone's jumping off point, clearly marked.

On April 5, 1850, Galton and Andersson embarked on the *Dalhousie*, a slow three-masted East Indiaman commanded by Captain Butterworth, which was "quite incapable of beating against a head wind."[13] The ship had rough accommodations for the British emigrants it normally carried at inexpensive rates plus a few cabins for more affluent passengers. Galton hastily wrote his mother on May 9, as a homeward-bound vessel was to pick up the mail, remarking that they had taken heavy seas on the ten-day passage of the Bay of Biscay and that one of the emigrants, a young clergyman's daughter, had perished from a lung infection.[14] On the long and tedious trip, Galton became quite attached to his seemingly accident-prone second-in-command who on one occasion succeeded in ramming a harpoon through his hand and on another had an old musket burst on him while he was firing it. Once Andersson clambered to the maintop chased by a sailor who planned to bind his feet with a piece of twine when he had gone as high as he dared so he could make Andersson "pay his footing."[15] But Andersson, with simian agility, descended from the heights via the mainstay, a feat even the sailor would not attempt, confirming Galton's opinion that he was of the mettle necessary for their expedition.[16]

Gentle breezes carried the *Dalhousie* so close to the island of Madeira that vineyards and handsome cottages were visible scaling the mountainside all the way to its summit.[17] Later uncooperative gales blew them so far to the west that they sighted the South American mainland. Meanwhile Galton passed away the monotonous hours reading and learning to use a sextant. After 86 days at sea, a third longer than average for the voyage, the travellers spied the hulking massif of Table Mountain hovering like a landlocked aircraft carrier over Cape Town and, after rounding Robben Island, the *Dalhousie* entered Table Bay where she anchored. The travellers disembarked "among the white stone and green shuttered houses of Cape Town."[18] They viewed with interest the pentagonally shaped castle, built between 1666 and 1679, with its thick walls and 500-foot-long sides. It stood behind the Grand Parade where Jan van Riebeeck, who established the first Dutch East India Company trading post, had built his first primitive earthwork in 1652.

They lodged in Welch's Hotel.[19] Galton was delighted to find his old friend Hedworth Barclay from his Nile adventure in residence and that Sir Hyde Parker's ship was in port trailing some prizes behind it.[20] Andersson observed that Cape Town was laid out in a regular pattern with broad, unpaved, rubbish-littered streets set out at right angles to each other. The diversity of Cape Town's populace was striking— "Indians, Chinese, Malays, Caffres, Bechuanas, Hottentots, Creoles, 'Afrikanders,' half-castes of many kinds and negroes of every variety from the east and west coasts of Africa, and Europeans of all countries."[21] Except for the Europeans, the Malays seemed the most capable residents being "distinguished for their industry and sobriety," and while the women wore no head covering, the men tied red handkerchiefs around their crowns over which they wore enormous umbrella-shaped straw hats.[22] Galton had originally planned to stay in Cape Town for several weeks and then sail eastwards to Algoa Bay proceeding northwards from Port Elizabeth on a route similar to Livingstone's. However, the Governor of the Cape Colony, Sir Harry Smith, dissuaded him, for the Boers had occupied the habitable land north of the Orange River through which he would have to travel. "The Boers," said Harry Smith, "are determined men; and although I have no fear for the safety of your lives, they will assuredly rob you of all your goods and cattle, and thus prevent your proceeding further."[23] Nor could Galton outflank the Boers to the west for that would bring him face to face with the vast Kalahari Desert.

Smith was the quintessential Victorian soldier and the cause of the problem.[24] He had fought under Wellington in the Peninsular War (1808–14) meeting his future wife Juana during the seige of Badajoz the Proud. In 1814, he was posted to America and, following the defeat of Windham's Yankee militia in front of Washington, rode up to the White House with his victorious comrades. The dining room table was laid with 40 settings, the food was warm and the wine chilled as President James Madison and his entourage had moments earlier beat an unexpected and hasty retreat. Smith found "a supper already which was sufficiently cooked without more fire, and which many of us speedily consumed . . . and drank some very good wine too."[25] The soldiers picked up souvenirs ranging from the president's love letters to a pair of rhinestone buckles while Smith contented himself with a presidential shirt, but he was shocked at the expedition's goal with its "barbarous purpose of destroying the city."[26] He "had no objection to burn arsenals, dockyards, frigates, buildings, stores, barracks, etc., but well do I recollect that, fresh from the Duke's humane warfare in the South of France, we were horrified at the order to burn the elegant Houses of Parliament and the President's house."[27]

Smith was with Wellington again at Waterloo, and after serving in a succession of posts, was sent to the Cape Colony in 1828 as deputy quarter-master-general, arriving during the lull between the fifth and sixth frontier wars against the Xhosas to the east. In early 1840 he was posted to India where his greatest moment came in the battle of Aliwal, when his forces defeated a Sikh army of 10,000 under Ranjur Singh. Smith, the "Hero of Aliwal," was honored by Wellington in the Lords and Peel in the Commons. Late in 1847, Smith became governor of the Cape Colony as the Seventh Frontier War, or War of the Axe, wound down. He was so highly regarded that the South African towns of Harrismith and Smithfield were named for him, Ladysmith and Ladismith for Juana, and his victory over Ranjur Singh was celebrated by naming two towns, Aliwal North and Aliwal South.

Smith, having dealt with the Xhosa, turned his attention to the Boers north of the Orange River. But he had not reckoned on Andries Pretorius trekking out of Natal toward Trans-Orangia with a large party of Boers to escape British rule. Pretorius wanted to establish a Boer homeland in Natal and became a national hero in the process by leading a band of 470 voortrekkers to victory over a Zulu impi of 10,000 warriors in the battle of Blood River in 1838. In 1843 the British annexed Natal, so Pretorius with his followers trekked out of Natal to join their comrades north of the Orange River. After much discussion Smith got Pretorius to agree to canvass the Boer communities to see whether they would accept the Union Jack as their ensign. A misunderstanding arose and Smith proclaimed British sovereignty over the Orange River territory before Pretorius reported his results. They revealed that the majority of Boer communities rejected British rule. The Boers revolted, so Smith marched on Boomplaats defeating Pretorius after a furious battle in August 1848 and restored order in the Orange River territory. But the calm was only on the surface, as many of the Boers had retreated northwards and there they would harass the British in the future, rendering Livingstone's route to Lake Ngami unsafe.

Galton got on famously with Smith, who, after a glorious dinner at Government House, stood up and made a speech proclaiming to all what a good fellow Galton was.[28] He asked Galton to aid him by establishing friendly relationships with any local chiefs he might meet and to persuade them that the Boers were up to no good. He provided Galton with an enormous parchment passport inscribed in large letters in English, Dutch, and Portuguese. From the parchment dangled a huge seal, eight inches in diameter, set in a tin box. After years of experience the governor knew how to impress the natives, so to add an extra flourish he had cut the seal from the royal mandate creating Galton a lieutenant governor of the colony and attached it to the parchment.

There were two routes that would gain Galton access to Lake Ngami while avoiding the Boers. He initially considered sailing up the east coast of Africa and landing near the southern of tip of Mozambique at the port of Lourenço Marques, now Maputo, the capital of Mozambique.[29] He soon abandoned this notion because of the pestilential conditions existing there, as this was country where malaria and sleeping sickness abounded. Then he met a distinguished Portuguese gentleman, Signore Isidore Pereira, who advised him that his father had crossed Africa from east to west travelling from Mozambique to Benguela in Angola and that he himself had travelled extensively in Mozambique and knew the native chiefs well. Pereira suggested landing further north in Mozambique at Quillimane near the Zambezi River delta. However, Galton abandoned this plan too when he found that the only means of carrying baggage into the interior was on men's backs with the travellers themselves being transported on palanquins. This would not do since Andersson was to assemble a large natural history collection and beasts of burden would be needed to carry it. The other way round the Boers was to travel up the African west coast. Some merchants suggested that Galton's expedition sail to Walfisch, now Walvis, Bay in what is presently Namibia and proceed inland across the desert. Wagons and oxen could be used for transport and there were missionary stations nearby. Furthermore, fertile, inhabited land lay east of the desert. This was the plan Galton settled on and he began gathering the necessary beasts and manpower for his expedition, buying two wagons, nine mules, and two horses. He knew the horses would eventually succumb as horse distemper was prevalent in Namibia, but the mules were more disease resistant. He took only a few sheep for meat, thinking wrongly that there would be plenty of game.

He next set about hiring the personnel he would require. A Portuguese named John Morta, a Madeira native, signed on as headman and chef. This was a coup as Morta was the cook at the club in Cape Town where he had won high praise. He was also honest, frugal, hardworking, and a great story teller who could dissolve his audience in laughter. His only fault was irritability, which Andersson passed off by saying "this in a cook, always excusable."[30] Next Galton hired Timboo, a fine looking black man and a lady-killer, to do various odd jobs. Timboo's childhood had been cruelly interrupted by the spectacle of his tribe being attacked by rival warriors who carried off his parents after killing many of his kinsmen. Later Timboo was sold as a slave to the Portuguese, escaped, was recaptured, and put on board a Portuguese slaver. The slaver fell into the hands of a British cruiser and Timboo, together with many other slaves, was brought to Cape Town and released. While Galton was pleased with both Timboo and Morta, Gabriel, a fine-looking Cape native with a ready smile, was another story. He attached himself to Galton offering to be his agent in gathering horses, dogs, and anything else. Galton assented, much to his later regret, for Gabriel was a troublemaker. He also hired two wagon drivers and two leaders for the oxen. John St. Helena, a man of mercurial disposition, was hired as head wagoner and a relative of his, John Waggoner, as one of the leaders. Waggoner was a slacker whose imagination was infinite when it came to finding excuses for not doing things. The other wagon driver, Abraham Wenzel, a Cape Town wheelwright, was worse. He was discovered pilfering various articles from the expedition's stores for which he was punished. John Williams, the other leader, rounded out the group. Williams, a short, stout, merry lad, was a jack of all trades who cooked, washed clothes, and generally made himself useful in addition to leading the oxen. Galton chartered the *Foam*, a small schooner, to take them to Walfisch Bay as ships only called there every one or two years. The kicking mules, whinnying horses, boxes, axle trees, wagon wheels, etc. were manhandled aboard and the schooner embarked in mid-August 1850. The 60 or so oxen required to draw the wagons in spans of 14 to 16 would be purchased when they landed.

SIX

Making Peace with Jonker Afrikaner

The great man of all the country, who could do what he liked, and of whom everybody stood in awe, was Jonker Afrikaner.

—Francis Galton, *Tropical South Africa*[1]

On August 20, 1850, the *Foam* rounded Pelican Point, a fingerlike sandspit forming the western rim of Walfisch Bay, so named by the Dutch for the humpback whales that were abundant there in breeding season. The schooner edged in gingerly for no proper nautical charts of the bay existed and the explorers glimpsed a desolate, sandy beach dancing giddily in the mirage. They anchored about a mile offshore as nightfall approached. On the east side of Pelican Point a shallow lagoon teemed with fish that were often stranded at low tide, becoming prey for the local natives who speared them on the tips of gemsbok horns affixed to slender sticks. Walfisch Bay was home to immense numbers of geese, ducks, countless flocks of sandpipers, myriads of flamingos, white pelicans, and several species of cormorants. American and British whalers frequented the bay and British ships, collecting guano on the small island of Ichabo to the south, provisioned there.[2] Although fresh water was absent near the beach, it was available three miles inland and supported abundant pasturage. Some enterprising individuals from Cape Town had established a facility for salting and curing beef, and they furnished cattle to guano traders and to Cape Town and contracted with the British Government to supply the island of St. Helena with livestock.

The next morning the schooner sailed closer in and the captain and the explorers disembarked to be greeted by seven natives drawn up in a line, three of whom brandished guns. Galton wrote that they had "a most ill-looking appearance; some had bad trousers, some coats of skins, and they clicked and howled, and chattered, and behaved like baboons ... but the time came when, by force of comparison, I looked on these fellows as a sort of link to civilization."[3] They were used to sailors and exchanged goat's milk and some scruffy oxen for tobacco, clothes, and other luxuries. A Rhenish missionary named Schöneberg had accompanied the expedition from Cape Town. He sent a letter via native courier announcing their arrival to the Reverend Mr. Bam whose mission at Scheppmansdorf was 25 miles distant on the Kuisip (Kuiseb) River, a euphemism as its bed contained water only once every four to five years.

Galton with several others scouted inland a short distance to the Kuisip River, flanked on either side by great shifting sand dunes and located a hole six inches in diameter full of green, stagnant, undrinkable water called Sand Fountain. Even following purification in Galton's copper still, the water tasted so foul that it could only be used for cooking or making strong tea or coffee. Sea breezes cooled Sand Fountain, whose seemingly pleasant environment proved deceptive, as Andersson soon discovered when he was assailed by myriads of fleas and bitten badly a couple of times by bush ticks. Dabby bushes eight to 12 feet high grew everywhere and a vine called the 'nara, possessing a prickly gourdlike fruit about the size of a turnip, covered the sand hills nearby. Its deep orange fruit was the staple food of native and wild animal alike.

Galton spent the night on the schooner while a fire sprang to life on shore signalling Mr. Bam's arrival. He was accompanied by a cattle trader named Stewartson who lived at the mission. The next day the explorers disembarked and Galton camped on shore with several of his men to guard the animals and stores, but during that windy, chilly night the two horses broke loose and bolted off into the darkness. He sent two men after them, but being unfamiliar with the terrain, they returned empty-handed. Hence, he motioned to "Frederick," the leader of the natives on the beach, to see whether he could retrieve the errant animals. Frederick, a seasoned bargainer, offered their capture in exchange for a good coat and a nice pair of trousers. By the next day Frederick had caught and tethered the horses at Scheppmansdorf, relieving a grudging Galton of a fine dress coat brought along for special occasions. Galton soon busied himself with observations of latitudes, longitudes, and altitudes, and drew a sketch map of Walfisch Bay, covering it with topographical illustrations and notes.[4] Although he recorded many observations in ordinary notebooks, he possessed a supply of small pocket-sized notepads available at a moment's notice, each equipped with a metal-tipped pencil attached by a thong. Surveying the bleak desert with its great sand dunes he contentedly wrote his mother on August 22 that "I am sure I have selected a far better route than my first one, for now I am quite near the undiscovered country as I should have been after 3 months land journey from Algoa Bay."[5]

With the baggage landed and accounted for, the expedition was ready move on to Scheppmansdorf (Fig. 6-1) as soon as the oxen arrived. Mean-

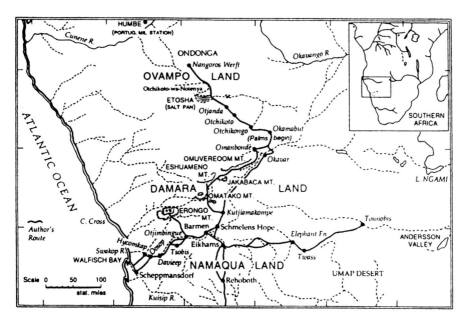

Fig. 6-1 Map of northern Namibia showing Galton's routes of exploration in 1850-52. From Francis Galton, *Narrative of an Explorer in Tropical South Africa*. London: Ward, Lock, 1889. First published under the title of *Tropical South Africa*. London: J. Murray, 1853.

while Mr. Bam offered Galton the use of his wagon and team of oxen to haul some of the provisions to the mission, so he set off leaving Andersson behind to transport the remaining baggage later. From the ends of the wagon protruded muskets and other items intended for barter.[6] The beasts, unwilling to budge, kicked violently to the right and left as the sweating wagon driver cracked an immense Cape whip along their flanks. The man sitting next to him swore violently at the recalcitrant animals in a mixture of Dutch and Khoikhoi, the language of the people referred to as Hottentots by the Dutch, a word meaning "stammerer" or "stutterer." When the wagon finally lurched forward, two goats tied to its backboards reared and bleated before trotting along while Galton, smoking a clay pipe, trudged serenely through the sand beside the wagon followed by a bevy of ever-hopeful mongrel dogs.

Scheppmannsdorf, named for the Reverend Mr. Scheppmann, was prettily sited on the left bank of the Kuisip River amongst a mixed grove of acacias, ana, and camelthorn, framed behind by enormous sand dunes. A little stream unexpectedly burst forth from the ground and meandered to a reedy pond full of wildfowl. The whitewashed mission chapel was flanked on one side by Mr. Bam's house and on the other by Stewartson's, and beyond were native huts. Mr. Bam loaned Galton a building adjacent to the mission for his use. A well-known lion had migrated from south from the vicinity of the Swakop River and was prowling around the mission attacking livestock, so Galton gathered a hunting party and shot it, discovering one of Stewartson's dogs inside. Meanwhile, the oxen having arrived at Walfisch Bay, Andersson conveyed the rest of the baggage to Schepmannsdorf.

Stewartson agreed to guide the explorers to the next mission, Richterfeldt, at Otjimbinguè on the Swakop. Galton bought several oxen including Ceylon who would become his faithful ride ox. The explorers started off with Stewartson guiding from the back of his ox, most of the entourage walking, and Andersson and Galton riding the two horses except when allowing the men to ride now and then. Peas, sugar, rice, biscuits, coffee, water, ammunition, spears, tents, instruments, clothes, tools, trinkets for barter, Andersson's natural history implements, etc. were carried on pack oxen or in a cart drawn by the mules. A few goats were driven ahead for milk and some sheep for meat. Galton was confident that after traversing the sterile Naarip plain they would find game on the Swakop, which flowed every year during the rainy season. The river had cut a deep gorge and its moist bed, along which a rivulet still trickled, was smooth and covered with grass, creepers, and ice plants, and fringed by giant reeds. There were clumps of camelthorn trees here and there and the explorers discovered a pool of excellent water under a projecting rock where, hot, dirty, and thirsty, they could drink and bathe along with their animals. Andersson collected a redbilled francolin, a quail-like bird, and a couple of species of flycatchers.

The next day, September 21, 1850, the expedition turned east on the Naarip plain paralleling the Swakop. Following a magnificent sunrise that tinted the distant mountains vermillion and caused the dewdrops on the pebbles beneath their feet to glitter like diamonds, the blinding, pulsating disk rose ever higher searing the landscape below so by noon the air was deathly still and the sand so hot that it cruelly burned men's feet. The animals were suffering too, heads drooping and tongues hanging out, with the mules being the most distressed, having gorged on grass from the riverbed instead of their usual ration of dried fodder. Three of the miserable creatures lay down refusing to budge, so their cart had to struggle on without them. After camping that evening, Andersson and two others returned to capture the recalcitrant animals, but the shadows were lengthening and they came back after a couple of hours empty-handed. On Stewartson's advice the remaining mules and the two horses were allowed to forage and rest overnight in the river bed. The next morning Galton sent a man after the animals, but he returned without them. He excitedly reported finding their galloping hoofprints flanked by the pawprints of several lions after which he found a hyena gorging itself on a half-eaten mule and nearby the carcass of the largest horse guarded by a ferocious-looking lion.

Timboo and a companion were sent to fetch the other horse and remaining mules that, miraculously, had been joined by two of the three mules abandoned the previous day. The explorers hacked as much flesh off the two dead animals as they could, for they had seen little game and lacked fresh meat. After dinner Galton returned to the scene of the carnage determined to watch for lions, but with Stewartson and the men still carrying back horsemeat, he decided to have another go at the mule, clambering up the side of the gorge where the dead animal lay. As he tugged out the mule's last shoulder, the men below spied a lion crouched on a ledge above him and shouted him a warning. "I did feel queer, but I did not drop the joint. I walked steadily down the rock, looking frequently over my shoulder; but it was not till I came to where the men stood that I could see the round head and pricked ears of my enemy peering over the ledge under which I had been at work."[7]

The caravan pressed on with Andersson spotting a flock of grey louries high up in the trees displaying their prominent crests and calling out distinctively "go-way-y-y." Delicate and pretty butterflies danced everywhere and Andersson was badly stung trying to capture a brilliant blue wasp. The weather became intolerably hot with Galton recording 143°F in the sun and 95°F in the shade. Andersson fell behind while pursuing some interesting birds and suddenly realized that his comrades were nowhere to be seen. He hurried to catch up, but just as he spied the party he began feeling giddy and barely managed to rejoin the expedition, where Galton propped him up on the horse while he gradually recovered his senses. The expedition proceeded along a tributary of the Swakop, the Tjobis. Guinea fowl were abundant and Galton shot a giraffe whose meat was cut up and jerked before they continued to Tjobis Fountain where there was water and they could "outspan" for the night. They remained there for almost two days and were visited by several Hill Damaras who provided them with some ostrich eggs after learning via appropriate gestures that they could take any meat left on the giraffe. John Morta whipped up a superb ostrich-egg omelette by cutting a small hole in an egg, adding salt and pepper, and shaking the egg violently to mix the yolk and the white.

The country was less bleak than before with a thin grass ground cover dotted with small shrubs and occasional aloes and thorn bushes. Galton learned to ride an ox, soon to become his sole means of transport. On September 30 the entourage arrived at Richterfeldt, situated amongst abundant water and grass, where the Reverend Rath and his wife welcomed them. They camped in a stand of tall shade trees adjacent to a spring. Behind Rath's house were three small villages where some 200 members of the Damara tribe lived. Andersson thought them a fine-looking race. The men were often over six feet tall and well proportioned with good and regular features, but while their outward appearance denoted great strength, they could "by no means compare, in this respect, with even moderately strong Europeans."[8] The women seemed delicate with small hands and feet and full forms, but their beauty was fleeting in the harsh conditions under which they lived so "in a more advanced age many became the most hideous of human beings."⁹ He deplored the Damaras' dirtyness that made "the color of their skin totally indistinguishable" and "to complete the disguise" they smeared their bodies with red ochre and grease so that "the exhalation hovering about them is disgusting in the extreme."¹⁰ Married women wore a picturesque helmet-like headdress and women who could afford to wore "a profusion of iron and copper rings—those of gold or brass are held in little estimation—round their wrists and ankles."¹¹ Damara warriors sported the ubiquitous *assagai*, the slender iron-tipped spear with a hardwood shaft favored by southern African tribesmen, plus bows and arrows and a few guns. But their favorite weapon was the *kierie*, a knobstick used dextrously for purposes as diverse as laying an enemy low or knocking down a francolin on the wing. Galton jotted down a few Damara words like bone (*etuba*) and bruise (*omasuro*) preparatory to learning to carry out a rudimentary conversation in the language.

At Richterfeldt, Galton met Hans Larsen, a fair-haired blue-eyed Dane and ex-sailor, who had jumped ship seven years earlier. Immensely strong with a reputation for courage, Larsen came highly recommended by Mr. Bam, so Galton was anxious to hire him. Larsen did odd jobs around the missions to make ends meet, but he was accumulating a substantial herd of cattle that he eventually hoped to drive to Cape Town. Fortunately, Galton succeeded in hiring Larsen for not only did he know the country well, but he was steeped in bush lore. He also provided ride oxen for the trip, Galton's being exhausted.

Galton now learned of a potential threat to his expedition. About a hundred miles further up the Swakop was Schmelen's Hope, a mission with a large Damara encampment that had been brutally attacked by Namaquan tribesmen led by Jonker Afrikaner. Jonker's warriors had fallen upon the mission murdering and mutilating the Damaras and pillaging their cattle. The Rev. Kolbe and his wife had been forced to flee to Barmen, a mission between Richterfeldt and Schmelen's Hope. The attack was a surprise and Galton decided to ride ahead to Barmen with Stewartson and Larsen to see whether his future plans might be affected. After a couple of days, they covered the 40-odd miles to Barmen, which was situated about three-quarters of a mile from the Swakop. Toward the west behind the mission irregular formations of low, broken rocks ended abruptly at a bluff a thousand feet high, and over this tortured landscape grew a profusion of shrubs and thorn trees. To the east, the Swakop's course was marked by handsome black-stemmed mimosas and beyond the river a range of mountains rose majestically to a height of six to seven thousand feet.

At Barmen, Galton met the Reverend Hugo Hahn, a Rhenish missionary of Russian extraction married to an Englishwoman. Hahn and another missionary, F. H. Heinrich Kleinschmidt, had come to Africa in 1842 at the urging of the Reverend Hugo Schmelen.¹² Schmelen seems to have hoped that they could make peace between Jonker and the Damaras. Their first impression of Jonker was favorable since, while he was passing through Okahandja where they planned to establish a new mission, he advised them to move a little further to the north to Otjikango where there were hot springs. The missionaries did so, naming their mission Barmen after the German town where they had studied theology. Later Kleinschmidt trekked south to create Rehoboth, a new mission among the Namaqua tribesmen, leaving the Reverend Hahn in charge at Barmen.

The Reverend Kolbe and his wife had taken refuge at Barmen following Jonker's massacre and Galton learned the details of what had happened. Schmelen's Hope had seemed a very promising mission since a Damara tribe led by Kahikenè had set up camp there. He was the richest and most powerful of their chiefs or *kapteins*, as measured by his sheep, cattle, and oxen. But Jonker's warriors had destroyed this vision, slaughtering and mutilating with horrifying results. Later Galton himself saw two Damara women who had crawled the 20 miles to Barmen. Their legs ended in bloody stumps because the Namas had severed their feet so they could steal their iron anklets. After a day of brutal carnage Jonker and his blood-sated Namas celebrated through the night. The next morning a thoroughly intoxicated Jonker came weaving up to the mission door, banged on it loudly, and ordered the cowering Kolbes to unbar it. To their relief he neither ran them through with an *assagai* nor raped Mrs. Kolbe, but simply demanded breakfast and, after gorging himself, departed unsteadily with his marauding followers and the Damara cattle.

Who were these warring tribes? The Damaras or Hereros, were cattle herders living north of the Swakop who apparently migrated southwards from central Africa at the turn of the eighteenth century.[13] In Namibia they encountered a people called by Galton the Ghou Damup, a corruption of the Khoikhoi name Xou-Daman meaning "filthy black people." The Afrikaaners called them Bergdamas, Mountain, or Hill Damaras, to distinguish them from the cattle Damaras. South of the Swakop was Namaqualand. The Nama, or Namaquan, tribesmen living there were Khoikhois of yellow or reddish complexion whose women often possessed very prominent buttocks (steatopygia). The largest of these tribes was called the Red Nation because of the complexion of its members. The most powerful Nama chieftain, Jonker Afrikaner, was not a Nama at all, but an Orlam, most of whom derived from unions between Afrikaaner masters and Khoikhoi slaves. They began migrating north from the Cape Colony about 1800. The Basters, another colored people of mixed parentage, even more Europeanized than the Orlams, later trekked north of the Orange River in search of freedom, with the main group settling in 1870 in the vicinity of Rehoboth.

Jonker was the son of Jager Afrikaner, who had murdered his tyrannical Afrikaaner master. Jager Afrikaner had stolen his cattle, and fled northwards out of the Cape Colony terrorizing the countryside until Johann Leonhard Ebner, a fearless missionary, came to live with him and his family.[14] Jager was baptized, presumably forgiven his many sins, and died a Christian. The first major clash between the Damaras and the Namaquas occurred about 1820, when severe drought caused the two peoples to intrude on each other's territory in search of grazing land. The Red Nation appealed to Jonker for help in defeating the Damaras. In return he would receive his choice of land for a residence and for cattle grazing. Jonker and his well-armed men defeated the Damaras in three successive battles, stole their cattle, and settled in the midst of the Red Nation's tribal domain at a place where hot springs provided a plentiful supply of water. The Khoikhoi word for the springs was "Ai-gams" or fire water, corrupted to Eikhams by Galton's time. It is now Windhoek, the capital of Namibia. In 1840 missionaries arranged a three-year peace between Jonker and the Damaras, setting up a blacksmith's shop at his headquarters where quantities of *assagais*, hatchets, and beads were made and sold for cattle.[15] The cattle were exchanged with Cape traders for clothes, guns, and the like, but in the process Jonker got deeply into debt. He resorted once more to cattle rustling from the Damaras and it was in one of these raids that the Damaras encamped at Schmelen's Hope were massacred.

To explore northwards to Lake Ngami Galton had to secure his rear against Jonker, meaning that he must arrange another peace. He recognized that Jonker, born a British subject in the Cape Colony, still feared and respected the colonial government. As Governor Sir Harry Smith's deputy, he wrote Jonker early in October demanding that he cease raiding the Damaras. His letter left no doubt that the colonial government would be greatly displeased if he continued to attack the Damaras. He concluded by saying Jonker's "past crimes may profitably be atoned for by a course of upright wise and pacific policy, but if the claims of neither humanity, civilisation or honour have any weight with you perhaps a little reflection will point out some danger to your personal security."[16] Galton had his letter translated into simple Dutch, rewritten on a magnificent sheet of paper, and sent off to Jonker by messenger.

There was nothing to do but wait and much to do in preparation for the upcoming expedition north, so Galton returned to Richterfeldt where he would remain with Timboo and John Morta. Meanwhile Andersson, Stewartson, and Larsen continued on to Scheppmansdorf to break in more oxen. On the way they killed a black rhinoceros and cut off much of the beast's hide to make *shamboks*, wicked whips capable of inflicting severe wounds. At Scheppmansdorf, Andersson observed the fiscal shrike or butcher bird impale a small, frantically kicking mouse on a thorn. He found that its name did not signal monetary prudence, but derived from the Afrikaaner term *fiscaal* used for a magistrate. The Cape people believed that the bird administered justice to smaller creatures much as a judge does to mortals.

After three weeks at Scheppmansdorf, Andersson's party started back to Richterfeldt with the remaining stores and wagons, with the oxen partially broken in. Crossing the Naarip plain they startled an enormous black rhinoceros with her calf. Andersson hit the mother with a musket ball, but she seemed unfazed, galloping away at high speed. Andersson and Larsen gave chase, but suddenly, the rhinoceros wheeled around and came to a dead stop facing them. With Larsen hanging back, Andersson walked within 15 or 20 paces of the massive creature, cocked his musket, and pulled the trigger. His first barrel misfired whereupon the rhinoceros made an about face as he fired his second barrel hitting her in the hindquarters, but she charged off again. When Andersson chided Larsen for not coming forward, Larsen was indignant:

> Sir, when you have had my experience you will never call that man a coward who does not attack a wounded black rhinoceros on an open and naked plain. I would rather face fifty lions than one of those animals in such an exposed situation; for not one in a hundred would take it as quietly as this one has done. A wounded black rhinoceros seldom waits to be attacked, but charges instantly; and there would not have been the least chance of saving one's life in an open place like this.[17]

They pursued the rhinoceros and her calf hitting her once again with a musket ball, but she finally escaped. Andersson, showing some remorse, wrote that he felt "sorry for the poor rhinoceros; for, though she was lost to us, I felt certain it was only to die a lingering death at a distance. From experience, indeed, I should say that a similar fate awaits a large proportion of birds and animals, that escape us after being badly wounded."[18]

After suffering from intolerable heat, the exhausted men and animals camped near Onanis, where a small, periodical stream flowed from which they could drink. Onanis was home to a community of Hill Damaras. They raised a little tobacco, for which they had a perfect mania, and also some *dacka* or hemp whose young leaves and seeds they sometimes substituted. Their pipe holder was the long, gently spiralling horn of the kudu, to which they added a little water. Near the horn's tip was a small hole into which a clay pipe containing burning tobacco or *dacka* was inserted. The tribesmen would sit in a circle with the chief enjoying the first pull. "As little or no smoke escapes from his mouth, the effect is soon sufficiently apparent. His features become contorted, his eyes glassy and vacant, his mouth covered with froth, his whole body convulsed, and, in a few seconds, he is prostrate on the ground."[19]

While awaiting Jonker's reply, Galton tried to keep busy copying Mr. Rath's dictionary of Damara words, but the days dragged by and no answer came so he decided to return to Barmen. As luck would have it Jonker's reply reached him there. It was rambling and evasive so Galton wrote him an even more strongly worded letter. While waiting for a response, Galton got to know the Reverend Hahn better. His large household was served chiefly by Namas who had migrated with him from Eikhams amongst whom were an interpreter and subinterpreter who had learned Damara. The subinterpreter "was married to a charming person, not only a Hottentot in figure, but in that respect a Venus among Hottentots. I was perfectly aghast at her development, and made inquiries upon that delicate point as far as I dared among my missionary friends."[20] Being a "scientific man" and also, one suspects, a randy one after several months in the bush, Galton wanted to measure her heroic buttocks. He dared not ask the lady's permission, especially as his Damara was fragmentary, nor could he ask the Reverend Hahn to do it for him. Galton, momentarily at a loss, "gazed at her form, that gift of bounteous nature to this favoured race, which no mantua-maker, with all her crinoline and stuffing, can do otherwise than humbly imitate."[21] Then he spied his trusty sextant as the "object of my admiration stood under a tree, and was turning herself about to all points of the compass, as ladies who wish to be admired usually do."[22]

He grabbed the instrument and recorded "a series of observations upon her figure in every direction, up and down, crossways, diagonally, and so forth, and I registered them carefully upon an outline drawing for fear of any mistake; this being done I boldly pulled out my measuring-tape, and measured the distance from where I was to the place where she stood, and having thus obtained both base and angles, I worked out the results by trigonometry and logarithms."[23]

One day the Reverend Hahn remarked that he had learned Damara from a man who had lost half his nose to a hyena while sleeping on his back. Galton would have been skeptical had he not seen something similar happen to an old Bushwoman. One night a hyena caught hold of her heel while she slept, but her anguished cries drove it off. The next day she came into the mission to have the wound bandaged, but that evening, as she slept coiled up close to the fire, the terrifying hyena again attacked. The third evening Galton and one of Mr. Hahn's men lay in wait and shot the hyena when it made its appearance on cue.

While awaiting Jonker's reply, Galton wrote each Damara chief. He explained that he represented the monarch of a great nation who wished to send traders to the Damaras to purchase cattle in exchange for iron implements, but who did not rob and plunder as the Namas did. Hahn translated the note into Damara, but to convince the chiefs the messenger needed some token from Galton to demonstrate that he really represented the great white chief. He rummaged around in his kit and found a "great French cuirassier's sword in a steel scabbard" bought years ago in Egypt. "This was just the thing. The Damaras adore iron as we adore gold; and the brightness of the weapon was charming to their eyes."[24] With nothing to do but wait, the impatient Galton returned to Richterfeldt bent on doing a little exploration. On December 11, he rode off with Larsen, John St. Helena, and Gabriel to the broad table mountain called Erongo north of Otjimbinguè. Its surface was composed of huge, smooth white rock slabs hundreds of feet in length with great fissures in between them. The climbers moved gingerly up them, removing their shoes for fear of slipping. Leopards were numerous and baboons and steinboks provided them with wild prey enriched with the occasional goat or sheep filched from the Hill Damaras. The summit of Erongo was dissected with ravines and clothed with thickets of camelthorn and abundant cactus-like Euphorbias. Having satisified their desire for some action they returned to Richterfeld the day after Christmas with 25 oxen and 30 or 40 sheep that Galton had managed to purchase.

But this was merely a diversion. Galton knew he must force the issue with Jonker or the expedition would remain mired indefinitely in the necklace of Rhenish missions along the Swakop. On December 30 the expedition set off again for Barmen taking a week to make the journey because of frequent timeouts for freeing stuck wagons. On the last day they camped a few hours from Barmen and Galton rode ahead to see if anything had been heard from Jonker. Although Jonker's raiders had not swooped down on Barmen, there was an ominous feeling in the air as if something could happen at any moment. A guide Galton had hired at Richterfeldt refused to accompany him. His other Damaras were also becoming more restive, fearful that the Nama leader and his horde would fall upon them. Galton knew he must act soon or his expedition might disintegrate.

Jonker's letter had invited Galton to visit his headquarters in Eikhams so he decided to accept, recognizing that he could be walking into a deliberate trap, but gambling that Jonker would defer to him as the governor's representative and a white Englishman. On January 13, 1851, the expedition moved to the devastated mission sited at Schmelen's Hope on the right bank of the Little Swakop, its banks lined with majestic acacias now in full bloom. Leaving Andersson in charge, he set out for Jonker's lair on January 16 with Larsen and John Morta and several others. He brought along his red hunting-coat, cap, corduroy breeches, and jackboots that he had shipped to Africa perhaps expecting to hunt with the governor in the Cape Colony. It proved to be a stroke of genius. After riding for three days until within a few hours of Eikhams, they stopped to rest their oxen. Galton donned his "official" costume and rode into Eikhams where Larsen pointed out Jonker's large hut. Galton dug his spurs into Ceylon's ribs causing the great beast to do his best imitation of a canter. A deep ravine about four feet wide crossed in front of Jonker's hut, but the intrepid Ceylon leaped over it. The trusty ox thrust his head, mounted with its two formidable horns, through the doorway of Jonker's hut where the astonished chief was smoking his evening pipe.

Galton's ruse had worked. He had made detailed notes on what he planned to say. He read his riot act loudly in English glaring menacingly at Jonker who dared not look up. Since Jonker did not understand English, Galton had an interpreter make his meaning clear in Dutch. Then he turned Ceylon's head and left in a great huff to make camp in Jonker's village, after which the humbled chief sent Galton several notes begging rapprochement. Capitalizing on the ferocious impression he had made, Galton forced Jonker to sign letters of apology both to the Kolbes and to the British Government with the latter acknowledging the wrong he had done and pledging his word to refrain henceforth "from all injustice to the Damaras." He promised to do his best to "keep the peace with them" and to use his influence to persuade other Nama chiefs to do the same.[25] Jonker signed the letter and Galton and another member of his party witnessed the signatures. The amused explorer later remarked condescendingly that this "may seem laughable, but Oerlams [Orlams] are like children, and the manner which wins respect from them is not that which has most influence with us."[26] Galton also drew up a 15-point code of conduct for the Nama chiefs. After the chiefs were assembled, Jonker read them the code and Galton lectured them sternly. The code held for a year, enough to see Galton safely through his explorations. Later, Jonker resumed raiding the Damaras again and did so until his death in 1861 following which his son Jan Jonker continued the tradition.[27]

SEVEN

Expedition to Ovampoland

In 1850 the famous English explorer, Sir Francis Galton, landed at Walvis Bay, and set off for Ovamboland. He reached a point some seven miles short of his goal— Lake Ngami—which had recently been discovered by Livingstone.

—O. Levinson, *Story of Namibia*[1]

With Jonker under control, Galton was ready to proceed in March 1851.[2] But first there were personnel problems to deal with. John Waggoner, whom Galton had fired at Barmen, was pretending he was Galton's representative and inveigling cattle, horses, wagons, etc. from their unsuspecting owners.[3] Galton pursued Waggoner in a strenuous 24-hour chase, failing to overtake him. Waggoner returned to Cape Town with his booty and conned a trader into advancing him a large sum of money with which he vanished. Gabriel also decamped for the Cape leaving a trail of insolence and violence while Abraham Wenzel, already caught for stealing once, got into another scrape so Galton fired him. Fortunately, the Nama chief Swartboy, with whom Galton was friendly, provided two Damaras. One, Onesimus, who spoke fluent Damara and Namaqua, had been captured by the Namaquas as a child and brought up by them. The other, Phillipus, had forgotten his native tongue, but could speak Namaqua and Dutch fluently.

Galton's largest wagon, containing spare guns, canvas bags full of books, and other items, plus artifacts for barter, was divided in two by a curtain so when it was wet Andersson could sleep in front and Galton in back. The smaller wagon was filled with other freight and no one slept in it except in driving rain. Because of their weight not many provisions could be carried, and biscuits and vegetables were long gone. There was coffee, tea, and a little sugar, but the expedition's food supply consisted of the sheep and oxen they drove before them plus any game they shot. Galton, who loved making calculations, estimated that one sheep fed ten people for a day, that an ox equaled seven sheep, that a hartebeest provided as much meat as two sheep, and a giraffe the equivalent of two oxen. A white rhinoceros made a feast equal to four oxen.

The large wagon, driven by John St. Helena, was led by Onesimus while Phillipus drove the small wagon led by "any odd Damara." Hans, John Morta, and Timboo were the other regulars, with Damara and Ghou Damup servants, trailed by wives and children, changing often. Since the Europeans had difficulty pronouncing their names, they often gave them nicknames, some odd like "Grub," "Scrub," "Moonshine," "Rhinoster," and others ordinary like "Bill." On March 4 the expedition traversed the Swakop through a narrow, boulder-strewn gorge dissected with ravines and armed with abundant thorntrees. The oxen, not yet fully broken in, were wild. Galton remarked that if "I had to undergo two or three more such days of journeyings, the waggons would have to be left behind."[4] The safari was now in the hands of its Damara guides in uncharted territory. Galton became distinctly uneasy, remarking that "they have no comparative in their language, so you cannot say to them, 'Which is the *longer* of the two, the next stage or the last one?' but you must say, 'The last stage is little, the next stage is great?' The reply is not, it is a 'little longer,' 'much longer,' or 'very much longer;' but simply 'it is so,' or 'it is not so.'"[5]

Nor did the Damaras distinguish days, weeks, or months, reckoning instead by the dry season, the rainy season, or the pig-nut season. They had no system of counting, driving the numerically oriented Galton to distraction as they used "no numeral greater than three. When they wish to express four, they take to their fingers, which to them are as formidable instruments of calculation as the sliding-rule is to an English schoolboy. They puzzle very much after five, because no spare hand remains to grasp and secure the fingers that are required for 'units.'"[6] However, the Damaras seldom lost oxen because they knew them all by their faces. Galton was scornful:

> Once, while I watched a Damara floundering hopelessly in a calculation on one side of me, I observed Dinah, my spaniel, equally embarrassed on the other. She was overlooking half a dozen of her new-born puppies, which had been removed two or three times from her, and her anxiety was excessive, as she tried to find out if they were all present, or if any were still missing. She kept puzzling and running her eyes over them backwards and forwards, but could not satisfy herself. She evidently had a vague notion of counting, but the figure was too large for her brain. Taking the two as they stood, dog and Damara, the comparison reflected no great honour on the man.[7]

So Galton dismissed the average Damara as of little worth. This, plus his success at cowing Jonker and later hoodwinking the Ovampo chief Nangoro with a fake crown, caused him to dismiss black Africans as not having much ability. This opinion was reflected later in his book *Hereditary Genius* where he assigned blacks to the bottom rung of the ladder. The "mistakes the negroes made in their own matters, were so childish, stupid, and simpleton-like, as frequently to make me ashamed of my own species."[8] But as he acknowledged telling oxen apart was more important than their total number not only to the Damara, but to his own expedition since "it is perfectly essential to a traveller here that some trustworthy persons of his party should be able to pick out his own oxen from any drove in which they have become mixed; for, depend upon it, the strange Damaras will give no help on these occasions."[9]

Galton respected the native chiefs up to a point, remarking in the same chapter of *Hereditary Genius* that he "has as good an education in the art of ruling men, as can be desired; he is continually exercised in personal government, and usually maintains his place by the ascendency of his character, shown every day over his subjects and rivals."[10] But then came the put-down. "A traveller in wild countries also fills, to a certain degree, the position of a commander, and has to confront native chiefs at every inhabited place. The result is familiar enough—the white traveller almost invariably holds his own in their presence."[11] Despite this he respected Kahikenè who was "the only friend among the Damaras the Missionaries ever had, and his friendliness and frankness to me, and my men interested all of us without exception most thoroughly in his favor."[12]

As they travelled north, a messenger arrived from Kahikenè inviting Galton to visit. He took advantage of the opportunity to question the chief about the country beyond. Kahikenè reported that he had sent trading expeditions to the Ovampo people across whose land Galton must proceed to Lake Ngami, but west of the direct route Galton proposed through the village of Omobondè (Fig. 6-1). This avoided the territory of an unfriendly Damara chieftain, Omagundè. After Jonker had decimated Kahikenè's people at Schmelen's Hope, Omagundè's son, like a jackal, had preyed on the leavings, making off with some cattle, killing several of his children, and stealing one or two more. Galton offered to mediate with Omagundè's son to recover the children and some cattle as it was common custom among the Damara for the conquering tribe to return part of the spoils to their victims. But Kahikenè was too proud to accept Galton's aid even though, as he explained in front of his remaining warriors, his best men had been killed and those he would take with him were likely to scatter at first blood. Sadly, the chief's prediction proved correct. Galton later learned that Kahikenè attacked Omagundè's son shortly after they parted company. In the thick of the fight, his men abandoned the chief who was overwhelmed by a shower of arrows and speared to death as he fell. A son, who rushed to his defense, was summarily cut down too.

At first sight, Kahikenè seems a tragic figure, but he could administer fierce justice on occasion. One morning Galton found that three of his best front oxen and a slaughter ox had been stolen by a band of Damara marauders. A posse recovered three of the oxen and captured six of the rustlers. Kahikenè proposed lynching them on the spot, but Galton, disliking violence, temporized so the chief made a case that Galton felt he could not refute. The thieves were not only guilty of stealing Galton's cattle, but had perpetrated a crime against the chief. The expedition was in Kahikenè's protection so, although Galton might choose not punish the rustlers, Kahikenè must. While two of the culprits escaped, four were beaten with *kieries*, speared with *assegais*, and left for dead, with one surviving in a horribly mangled state. Soon one escapee was captured and brought to Galton for punishment. He flogged the prisoner before releasing him to prevent him from suffering a far worse fate at the hands of Kahikenè's warriors.

To circumvent Omagundè's territory the explorers marched west past the high cones of Mt. Omatako rising two thousand feet above the plain (Fig. 6-1). After rounding its escarpment they found that while the periodic river to the north was dry, a pool remained where they could water, but then they would have to cross a largely waterless plain to Mt. Omuvereoom. They had no estimate of distance as their Damara guides first said the journey would take ten days, but later decided that three would be sufficient. Given this ambiguity, Galton rode out about 20 miles with a couple of men to Mt. Eshuameno which had an excellent view of the surrounding countryside. They ascended the mountain and he ascertained, by means of rough triangulation, that Mt. Omuvereoom could probably be reached in 12 to 14 hours. After heavy rains during the night, they set out across the plain on March 22, discovering a fine temporary pool after several hours. The next day they arrived at some large wells with plentiful water, but beyond there was no guarantee they would find more. On March 24 the expedition camped in the narrow valley between Mt. Ja Kabaka and Mt. Omuvereoom near a wretched pool of abominable water stirred up by animal herds. Since finding an acceptable source of water was becoming a priority again, Galton and Larsen made an exhausting climb up a steep hill near Mt. Omuvereoom the next day. After scanning the desolate landscape with their telescopes they finally spotted water in the distance.

Getting to the water proved singularly unpleasant because thorn trees were everywhere. Andersson counted seven species, noting that each "was a perfect 'Wacht-een-bigte,' or 'Wait a little,' as the Dutch colonists very properly called these tormentors."[13] The oxen bucked and thrashed violently as the thorns tore at their flanks and "got their heads out of the yokes; and often the waggon-men could not get up to the fighting creatures on account of the thorns."[14] The water Galton had spied was a magnificent fountain called Otjironjuba, the calabash, on the flank of Mt. Omuvereoom. Its source was two hundred feet above the base of the mountain where several rivulets united into a stream that cascaded merrily down the mountainside. At the fountain's edge stood an enormous fig tree whose gnarled roots entwined scattered boulders and whose broad and leafy branches afforded welcome protection from the noonday sun. Here the explorers gratefully bathed their grimy bodies and washed their filthy clothes using soap made by the cook John Morta by ladling a mixture of wood ash and water from one pot into a second pot of simmering fat sitting atop a fire. "This ash-water is sucked up by the grease; and in ten days the stuff is transformed into good white soap."[15] The trick was to make ash from the right kind of wood since ash from some bushes made the soap too hard while that from others was too soft. As usual Galton scrupulously filled several notebook pages with masses of measurements accompanied by sketches of the mountain peaks they had seen and records of their altitudes.

By the end of March the expedition had covered about 150 miles. Their next destination was Omanbondè at the north end of the Omuvereoom escarpment, but to reach it they had to navigate an undulating plain covered with more thorn bushes. On the second day out they came upon some Bushmen digging for wild roots, capturing a man and a woman. After much gesturing they learned that at Omanbondè the " 'water was as large as the sky' and that hippopotami existed there."[16] The Bushman and his wife escaped that night, but Galton and Andersson now excitedly anticipated another excellent water hole so the expedition picked its way along a dry river bed hemmed in by a thorn tree jungle, halting at occasional small watering places. On April 2 they came across ox tracks indicating the presence of a native village and spotted some Damara men and women who tried to escape. The women, heavily laden with iron anklets, were caught and soon the men came after them. The explorers made friendly gestures to the Damaras, plying them with tobacco, and eventually, one enormously tall Damara volunteered to lead them to the great lake of the hippopotamuses at Omanbondè.

On April 5, a year to the day after Galton departed from England, the expedition reached the brow of a hummock overlooking the broad, grassy Omoramba river bed. On the far bank, beyond a projecting rock, was a hill topped with a grove of camelthorn trees from which the name Omanbondè derived. The explorers' spirits soon sank "as the water as large as the sky" proved to be a nine-mile dry reach of the river devoid of hippopotamuses. Briefly disappointed, Galton considered turning back, but he decided to reconnoiter northwards with several others to ascertain whether the country was passable. They returned three days later, reporting that the terrain ahead looked promising and that they had located another Damara village. On April

12 they started north parallel to the Omoramba and then headed east past huge herds of giraffes. As the day progressed, tall and graceful fan palms became more abundant and that evening they arrived at the Damara village. Galton found that these Damaras intended to deceive them by sending the expedition east rather than north, their wives revealing the plot to the wives of his Damaras. The tall guide proclaimed his innocence saying that he would happily take the explorers to the Ovampo and do anything else they wanted in exchange for a calf. Timboo was taken in. The Damara got his calf and Timboo lent him his horse rug to sleep on, but that night he decamped with both, further reducing Galton's low opinion of the Damara.

Finally, Galton obtained a reliable guide who led the expedition to Okamabuti on the northern edge of Damaraland, the village of the great chief Chapupa where they arrived on April 17. An impatient Galton wanted to press on toward Ovampoland, but Chapupa, after many excuses, flatly refused to provide a guide. Making the best of Damara estimates of questionable reliability, Galton guessed that the journey to Ovampoland would take about 20 days. He decided to delay briefly in favor of a shooting expedition to some wooded knolls a few hours distant where a fountain springing from a limestone bed supposedly served as a drinking place for elephants. They rode through countryside that contrasted favorably with the barren, thornbush-studded terrain they had grown tired of. It was marked by savannas of grass so tall that the blades reached above their heads, alternating with magnificent forests of straight-trunked stinkwood trees with great spreading limbs and dark foliage. At first all went well, but then a calamity struck. The oxen pulling the largest wagon unexpectedly bolted down an incline, careening it into a stump so hard that a front wheel spun off, and the axletree broke. This was the kind of disaster Galton constantly feared, as fashioning an axletree for the long return trip required seasoning the wood for several weeks. He decided to proceed north with a reduced party including Andersson, leaving Hans Larsen, the handiest of his companions, in charge of repairs.

While waiting at Okamabuti to bribe a guide, Galton learned from Chapupa that his people carried on a lively trade with the Ovampo. Every year or so their caravans arrived to barter beads, shells, *assagais*, axes, etc. for cattle. Chapupa was also greatly indebted to Nangoro, the Ovampo ruler. He had allied himself with Chapupa, then a minor chief, to seek revenge against the principal Damara chief of the region, who had betrayed the Ovampo by stealing back all the cattle he had bartered with them. After eliminating the principal chief, Nangoro and Chapupa split up his cattle and Chapupa became the dominant ruler. The reason for his reluctance to supply Galton a guide now became clear. Chapupa, knowing nothing of Galton and his men, feared they might be spies and that he would incur Nangoro's wrath if he showed them the way to Ovampoland. He requested that Galton await the expected arrival of the Ovampo caravan. This would free Chapupa of any possible recriminations from Nangoro. While the days marched by in slow procession, a frustrated Galton diverted himself briefly by convincing Chapupa and his wife to pose for sketches.[17] She was bare-breasted and crowned with the typical helmet worn by Damara women, her neck surrounded by a wealth of long necklaces, with a simple skirt around her waist. Chapupa was clad in a loincloth and from his neck hung a single necklace with a pendant. Both the chief and his wife had valued iron bands on their upper arms and many iron bracelets circling their wrists.

Chapupa's reasoning was impeccable, but Galton, in a hurry as the season was advancing, took advantage of a Damara's offer to guide him. They rumbled off on April 25, but three days later the guide confessed he was hopelessly lost. Fortunately, the explorers ran into some Bushmen who guided them to a series of wells named Otchikongo, which they christened "Baboon Fountain" for the troops of baboons that frequented it. Their Damara guide recognized this as the place he had aimed for originally and promised no further mistakes, but the next day they were lost again. Toward evening Andersson came across another party of Bushmen and coaxed them into camp. Galton and Andersson lavished favors on them and showed them their faces in a mirror that Galton kept for this purpose. The Bushmen were won over and agreed to lead the expedition to Otchikoto.

Early on May 2, they were overtaken by several men the Damara recognized as Ovampo, the vanguard of the expected caravan. They were tall and scantily clad with shaven heads and one front tooth chipped out. Each carried a dagger at his waist. In their hands they held light bows and a short, well-made *assagai* while on their backs were quivers holding ten to 20 barbed and poisoned arrows. Around their necks were strung quantities of necklaces for trading. Each man carried a narrow pole across his shoulders from the ends of which dangled small square palm leaf baskets containing items for barter such as spear-heads, knives, and copper and iron beads. Galton won their hearts by providing them with meat that they greatly appreciated, having eaten only kaffir corn, a variety of sorghum, since leaving home. Their leader was a tall young man named Chikorongo-onkompè whom Galton nicknamed Chik. He tried to convince Chik to loan him a guide, but Chik firmly refused, saying the expedition must return to Okamabuti while the Ovampo bartered, after which they could accompany Chik and his men home. Galton, now quite skilled at Damara, noted that Chik "spoke the Damara language perfectly, but with an accent, and so did Kaondoka and Netjo, the next in command, but the others could barely make themselves intelligible."[18] He was impressed that the Ovampo, unlike the Damara, could count adding up his oxen as quickly as he could and numbering Nangoro's wives at 105.

On May 23 the caravan was ready to return, having added Galton's party plus numerous Damara men, women, and children, and 206 head of cattle.

They passed through the Baboon Fountain and continued to Otchikoto, a deep, bucket-shaped hole scooped out of the limestone terrain some 400 feet across. Thirty feet below its rim Otchikoto was filled with water to a depth of 186 feet. Dirty and badly needing baths, Galton and Andersson plunged in to the horror of the tribesmen who believed that any man or beast falling into Otchikoto would perish, a myth that had arisen because neither the Ovampo nor Damara could swim. The water was cold and sea green in color and Galton and Andersson, joking and laughing, paddled over to a cavern in the rim startling a couple of owls, but the myriad of bats clinging to the rocks never moved, for they had died years earlier and been mummified in the dry climate. Galton, like many a modern tourist, scratched his name on a great boulder that jutted out into the lake.[19]

After Otchikoto, there were more thorn-tree forests to navigate, but on the afternoon of May 29 they reached Omutchamatunda, an Ovampo cattle post, swarming with several thousand people, where there was a fountain luxuriously overgrown with tall reeds. Vast herds of cattle grazed on the surrounding plain together with troops of zebra and springbok. The explorers were soon surrounded by mobs of curious Ovampo who marvelled at their white skins. They were most hospitable and seated Galton's party on the ground, following which an immense dish of butter was brought out. The head man proceeded to smear the face and chest of each individual with butter. Galton, his turn nigh, held out both hands and exclaimed, "Oh! for goodness' sake, if the thing is necessary, be it at least moderate"[20] so the head man gingerly daubed Galton's cheeks once or twice to everyone's amusement. The explorers relaxed for a couple of days at Omutchamatunda shooting ducks, geese, and francolin. On May 31 they moved on, marvelling at the great Etosha salt pan, the "big white place"[21] shimmering with the mirage. They departed Etosha for Ondonga, where Nangoro's village was located, crossing the edge of a boundless savanna called the Otchikoto-wa-Notenya and passing a majestic tree that, according to Damara belief, was the parent of all Damaras, Bushmen, oxen, and zebras. Then they were ensnarled in endless thorn-tree forests once more until suddenly, on June 2, "the charming corn-country of the Ovampo lay yellow and broad as a sea before us. Fine dense timber-trees, and innumerable palms of all sizes, were scattered over it; part was bare for pasturage, part was thickly covered with high corn stubble; palisadings, each of which enclosed a homestead, were scattered everywhere over the country."[22] On these fertile plains the Ovampo also cultivated millet, calabashes, watermelons, pumpkins, beans, and peas. To his friend Dr. William F. Campbell, Galton wrote "they have poultry and pigs and live right well."[23]

On the way to Ondonga the explorers were put up by old Netjo, then by Chik, and then by a friend of Chik's, but despite this hospitality, Galton was ill at ease. "Everybody was perfectly civil, but I could not go as I liked, nor where I liked; in fact I felt as a savage would feel in England."[24] Finally they came to a big clump of trees, a quarter of a mile from Nangoro's dwelling, where Chik ordered them to halt, but Galton was deeply concerned as there was no place for the oxen to graze. He pleaded with Chik for better pasturage, but Chik refused saying he must wait for Nangoro who would arrange everything.

Who was Nangoro and from whence came the Ovampo? Theories of Bantu migration are fraught with controversy, but linguistic evidence suggests that Bantu speakers, including the Ovampo, probably migrated south from the Niger-Congo region.[25] Nangoro succeeded his uncle Nembungu as king, founding his royal capital at Ondonga around 1820. At his accession, Ovampo power was dispersed among princelings who ruled small and scattered wards and did not recognize the king's power. Nangoro first made peace with the oldest inhabitants of Ovampoland, the Aakwankala, or Bushmen who became members of his bodyguard. Having achieved internal stabilization in his own realm, he entered an expansionary phase, attacking and defeating the Ovampo kingdoms of the Aakwanyama, Askwambi, and Aangandjera. His expertise in rainmaking endeared him to his subjects, as crops were plentiful and his subjects experienced little hunger. He bartered with the Damara for cattle and other commodities, using salt, iron ore, and finished products in exchange. He also sent ivory and slaves (his own subjects) north to a Portuguese trading post south of the Kunene River in exchange for glass beads and pearls. Hence, through negotiation and war, Nangoro's kingdom increased in population, natural resources, and wealth and he was at the height of his power when Galton's expedition arrived.

Galton's pasturage predicament was becoming increasingly desperate. Nangoro failed to come on June 6 as promised, but sent some corn as a present and asked Galton's party to fire their guns so he could hear the explosions. They obliged, shooting musket balls into the sky with loud reports. The next day Nangoro, an enormously fat old man short of breath, appeared in the midst of a large bodyguard accompanied by his miniature court of well-appointed Ovampo men attending to his every need. He waddled up to Galton who bowed elaborately, but Nangoro took no notice and simply stared at him. Galton, not knowing what to do, sat down and began making notes in his journal. After a few minutes Nangoro gave Galton a friendly poke in the ribs with his staff and Galton gave the king his presents apologizing that he did not have more. Unfortunately, he had gilt finery and not beads. "The sway of fashion is quite as strong among the negroes as among the whites; and my position was that of a traveller in Europe, who had nothing to pay his hotel bill but a box full of cowries and Damara sandals."[26]

Galton compounded his mistake by displaying bad manners. The Ovampo were a superstitious people "as are all blacks, and most whites."[27] An Ovampo man believed that if he supped with a stranger, his guest could exert a powerful magic against him and charm his life away. A countercharm was needed and Nangoro devised one. "The stranger sits down, closes his eyes, and raises his face to heaven; then the Ovampo initiator takes some water into his mouth, gargles it well, and, standing over his victim, delivers it full in his face."[28] The dripping stranger was now in the king's good graces and all proceeded decorously, but Galton refused to be splattered by Nangoro as he had previously when the guest of Chik and of Netjo. This was bad form just as it would be today if the proferred cheek of one's hostess remained unkissed, but the Ovampo, of course, believed the consequences could be far more serious.

Despite Galton's miscues the king was a good sport and said he would forgive him provided he donated a cow to accompany the ox he had earlier presented. Galton acquiesced and Nangoro requested the travellers to shoot their muskets again, the loud explosions delighting the king. They continued chatting via an interpreter, although Galton suspected that the king knew Damara, and Nangoro eventually decreed that Galton's party was free to trade. This was the signal the Ovampo awaited and they crowded around the travellers ready to do business, a jolly people full of good spirits. The women "were decidedly nice-looking; their faces were open and merry, but they had rather coarse features and shone all over with butter and red pigment. They seemed to be of amazingly affectionate dispositions, for they always stood in groups with their arms round each other's necks like Canova's graces."[29]

There was nightly dancing to tom-toms and a guitarlike instrument and Nangoro invited Galton and Andersson to attend. Andersson, greatly bored, amused himself by ogling the young Ovampo women, many of whom had exceedingly good figures. But these social occasions did not mark a warming of relations with the Ovampo monarch who visited Galton rarely. The oxen remained a sore point since Galton's animals, lacking access to Nangoro's stubble fields, were beginning to starve although they could drink at Nangoro's watering places once his cattle were finished. One day, when Nangoro seemed in a good mood, Galton presented him with the faux crown he had bought in Drury Lane explaining that the great chiefs in England wore such headdress. He begged Nangoro to honor him by donning it and the flattered chief assented so Galton adjusted it to its maximum size and crowned the Ovampo monarch. His courtiers were overjoyed as was the king who viewed himself with great satisfaction in Galton's mirror. While Galton feigned delight at the king's pleasure, he wrote his mother that "I . . . crowned him straightaway with that great theatrical crown I had" although "he was a brute fat as a tub."[30]

Nangoro, a man of proper manners, wanted to reciprocate and to present Galton a valuable gift. Shortly after Galton crowned Nangoro, he entered his tent dressed in his one well-preserved suit of white linen and there in one corner was Chipanga, heiress to the Ovampo kingdom, clothed in her scanty finery and painted with red ochre and butter. She was "as capable of leaving a mark on anything she touched as a well-inked printer's roller."[31] Without further adieu Galton ejected his temporary wife, grievously insulting her and the king. Nangoro was now thoroughly fed up with Galton. He told him on June 13 that he could trade that day, take his leave the next day, and depart the day after. Galton longed to journey northwards four days to the Kunene River, but this would involve temporizing further with Nangoro with no guarantee of success. Since the river was already frequented by Portuguese traders, Galton "could not help feeling that Nangoro's refusal to let me proceed was all for the best."[32] His oxen were in such bad shape that he would have to cross Omagundè's pasture lands on the way back. Galton had not reached Lake Ngami nor had he cast his gaze across the swift Kunene River, but he had carried out one of the first great African explorations of the nineteenth century. He left with a much higher opinion of the Ovampo than he had of the Damaras.

> I should feel but little compassion if I saw all the Damaras in the hand of a slave-owner, for they could hardly become more wretched than they are now, and might be much less mischievous; but it would be a crying shame to enslave the Ovampo.... They are a kind-hearted, cheerful people, and very domestic. I saw no pauperism in the country; everybody seemed well to do; and the few very old people that I saw were treated with particular respect and care.[33]—The Ovampo have infinitely more claims on a white man's sympathy than savages like the Damaras, for they have a high notion of morality in many points, and seem to be a very inquiring race.[34]

The journey to Schmelen's Hope took nearly seven weeks, but nothing was seen of Omagondè's warriors. One day Andersson observed that the oxen began to careen about "cutting the most ridiculous capers"[35] their antics catalyzed by the arrival of a large flock of yellowbilled oxpeckers that alighted on the beasts to dine royally on the ticks infesting their hides. By the time they arrived at Schmelen's Hope on August 3, the wagons were unfit for overland journey to the Cape. Since the next ship was not expected in Walfisch Bay until December, Galton decided to explore toward the fringes of the Kalahari Desert splitting his party in two. One group led by Hans Larsen headed west to Walfisch Bay while Galton marched eastwards with Andersson toward Elephant Fountain and Tounobis through drought-stricken country with very little grass (Fig. 6-1). His route took him through Eikhams where he was courteously received by Jonker whom he thanked for keeping the peace. Andersson, who had not accompanied Galton on his previous visit, noted that it was very prettily situated on the slope of hill whose summit was bare, but whose base was adorned with fine stands of mimosas. The land was fertile and well-supplied with water from several copious springs.

On August 30 they set out for Elephant Fountain, named for the vast numbers of elephant tusks and bones discovered there, shooting hartebeest, impala, and zebra along the way. After a fortnight of difficult travel, they arrived at Elephant Fountain, a copious spring on a thorn-tree-covered hillside where animal herds came to drink, but from which the elephants were long gone. Amiral and about 40 of his Nama tribesmen were encamped there, returning from a shooting expedition further east where they had bagged 40 rhinoceros, but they decided to retrace their steps with Galton to engage in further sport. On September 19 they left Elephant Fountain with the land soon becoming sandy, bushy, and devoid of prominent landmarks. At 'Twass they came upon a large encampment and Galton hired an Afrikaaner named Saul, an expert shot who spoke perfect Namaquan, to accompany him. On September 24 they left 'Twass for their "shooting excursion" and two days later camped where Amiral's men had slaughtered the black rhinoceros, seeing skulls all around. On October 1 they started out for 'Tounobis, which proved to be overrun with game. "The river-bed was trodden like the ground in a cattle fair by animals of all descriptions."[36] There were large herds of gnu and troops of zebra, and the hunters slaughtered rhinoceros, both white and black, with abandon, avoiding the elephants for fear of being trampled on. After a week of shooting Amiral's men were agitating to return to their wives and Galton had tired of "massacreing the animals."[37] By November 5 he was back in Eikhams where he parted with Jonker for the last time. The hunting party arrived in Walfisch Bay in early December, but Galton waited until the next month for a schooner to appear. He sailed first to St. Helena arriving in England on April 5, 1852, two years after his departure on the same day of the same month.

What happened to the *dramatis personae* after Galton departed? Andersson travelled back through Eikhams and Tounobis and thence to Lake Ngami.[38] In 1856 he published *Lake Ngami*, chronicling his travels with Galton and his subsequent expedition to the lake. He assiduously collected flora and fauna, subsequently enriching the British Museum among other institutions. While recovering from a serious leg injury suffered in a battle with Jonker and his men, Andersson wrote a book on the birds of Namibia. In 1866 he ventured once again to Ovampoland and succeeded in reaching the Kunene River, but, suffering from poor health, he died on the return trip.

Jonker soon began marauding again and by 1857 his repeated raids had left Damaraland desolate.[39,40] One day he seized Andersson's entire herd of cattle on the way to the Cape, murdering all but one of his men, so Andersson allied himself with Maherero, a great chief who had begun to rebuild the Damara nation. When the Namas attacked Otjimbinguè in 1860, Maherero with Andersson's help defeated Jonker, killing his son Christian. The next year they marched with 3,000 Damaras and stormed Jonker and his Namas in a mountain lair south of Rehoboth, defeating him once again, but they allowed his son Jan to escape. Jonker died that year after contracting a fatal disease following a raid on Ovampoland, but Jan continued the family tradition until an uneasy peace was signed between Jan Jonker and Maherero in 1870. That peace, which lasted for ten years, was brokered by the Reverend Hugo Hahn.

And what happened to Nangoro? In 1857 the missionaries Hahn and Rath endeavoured to extend their good works north into Ovampoland following the trail blazed by Andersson and Galton.[41] Along the way they were joined by a hunter named Green. Just before reaching Nangoro's palace their guide told Green that the king wished their assistance in a war against a small neighboring tribe, but they turned him down so Nangoro refused to see them for five days, after which they got a chilly reception. The missionaries sent beads to Nangoro, which were returned with the demand that all presents be given him at the same time. This annoyed Green and the missionaries, and according to Green, they told Nangoro that this was their custom in sending gifts and "he must conform to it" or else he would appear to be on unfriendly terms. The problem of the presents was settled, but Nangoro failed to show up so three days later the travellers sent him an ultimatum and he appeared.

As the expedition started north, Ovampo tribesmen rushed out and surrounded it. The Reverend Hahn, recognizing one of Nangoro's sons in the crowd, complained about their detention. For a moment they were quiet, but then the son plunged his *assagai* into the back of a Damara. His gun discharged as he fell, killing another of Nangoro's sons and wounding his assailant. Green then shot an Ovampo approaching him with a javelin and took command, holding 800 Ovampo at bay.[42] There were more deaths and Nangoro, on hearing of the demise of his son and several leading followers, reportedly succumbed after a stroke. Galton was disgusted on hearing Green's account. He believed that Green should have placed himself in the position of the Ovampo who felt their land was "almost invaded" by foreigners, who from their color, language, and intermarriages, must be related to the marauding Namas. Furthermore, "these foreigners are fully armed and dictatorial in their ways; they refuse to give those presents which are well described as taking the place of customs duties in African nations. They show scant courtesy to the king, and they very probably trespass in not a few of the many requirements of the witchcraft ceremonial."[43]

The Galton who sailed home to England had undergone the transition to maturity. He had planned, paid for, and executed a major expedition and shown bravery and clear thinking when faced with adversity. He also launched his scientific career as he began to write a sober report of his journey for publication in the *Journal of the Royal Geographical Society*, to be followed in quick succession by his popular account of the expedition, *Tropical South Africa*. In fact, throughout his career he would write articles for general audiences as well as more complex manuscripts aimed at professionals. His African experience left him with the prejudice, when he thought about it at all, which was not often, that blacks were in general savages although some like the Ovampo were quite civilized. Even so their chieftains were no match for a European explorer like himself.

EIGHT

Fame and Marriage

The lion-killer certainly seems smitten.
—Emily Butler writing to her brother Arthur Butler[1]

On February 23, 1852, while Galton was sailing home, part one of the paper describing his expedition was read before the Royal Geographical Society.[2] It began by tacitly acknowledging that it was the Society and African exploration that catalyzed his transformation from fun-loving idler to serious scientist. The reading was completed on April 26, shortly after he set foot in England. The paper was workmanlike, describing his journey, the places he visited, and the native peoples he met. Altitudes of mountains were given based on boiling point thermometer readings and there were two dense tables of data extracted from the masses of numbers accumulated in his notebooks. One gave latitudes for many of the towns and landmarks he had visited and the other longitudes for carefully selected locations across the entire East to West transect he had covered. They were calculated by the lunar distance method of Neville Maskelyne, the Fifth Astronomer Royal, and by triangulation with respect to Walfisch Bay whose longitude was known. Galton's presentation of precise data essential for accurate mapping stood in marked contrast to the other two papers on African geography in the same volume of the *Journal of the Royal Geographical Society*.[3] Henry Gassiott's merely summarized a hunting trip to South Africa while that by David Livingstone and W. C. Oswell described their Central African explorations beyond Lake Ngami, but lacked any quantitative data.

Galton's Namibian adventure turned him into an instant celebrity, but he was exhausted from his long African sojourn and desperately desired to escape "being lionised which is exceedingly wearisome to the lion after the first excitement and novelty of the process has worn off."[4] Hence, he was delighted to be invited by his friend Sir Hyde Parker, whom he saw last in Cape Town, to sail and fish in Norway. He relaxed on Parker's splendid yacht enjoying the panoramic scenery of the fjord country, the excellent fishing, and long, sunlit evenings full of storytelling as Parker's boat was joined by vessels piloted by a couple of friends. Despite this pleasant diversion Galton, who had not been feeling well for some time, became seriously ill with fever later in the year. He was nursed back to health by his spinster sister Emma at his mother's estate, Claverdon.

To complete his recuperation the family felt a change of scenery was needed. Hence, they settled in Dover for the winter, where he met Louisa Butler at a Twelfth Night party on January 5. Loui, as Galton called her, hailed from a distinguished academic family. While a student at Sidney Sussex College, Cambridge, her father, George Butler, had been all that Galton was not.[5] At age 19 Butler emerged victorious from the Tripos as Senior Wrangler and was also named First Smith's Prizeman. Nevertheless, the competitive stress affected even this prodigious scholar, who had a serious breakdown making him unable to compete for the Chancellor's Classical Medals. He was named headmaster of Harrow in 1805 and held that position for 24 years. He subsequently served as parish priest at Gayton, Northamptonshire, for six years before being named dean of Peterborough Cathedral in 1842. In typical Victorian fashion the Dean and his wife Sarah were prolific, producing four boys and six girls. Three of the boys had distinguished academic careers, with Henry Montagu Butler following in his father's footsteps as headmaster at Harrow. Later he was appointed Master of Trinity College, Cambridge.

The courtship progressed rapidly, facilitated by several small dinner parties at the Butlers' house in Dover.[6] These probably stood out in Galton's mind amongst the panoply of balls and evening parties he attended that gay season.[7] Meanwhile, in early March, he completed the first part of a popular account of his Namibian adventures, *Tropical South Africa*, and submitted it to the publisher John Murray. Murray apparently liked what he saw so Galton persevered. A few weeks later he met Louisa in London and they visited the Crystal Palace, one of the architectural and engineering miracles of the Victorian era.[8] This glittering ediface, built to house the Great Exhibition of 1851, was a marvel of glass and iron that had covered 19 acres in Hyde Park and enclosed 33 million cubic feet of space. There were thousands of exhibitors from around the world and the complete catalog listed over 100,000 objects. When the exhibition closed in late 1851, the Crystal Palace was dismantled, moved, and reassembled at the summit of Sydenham Hill. This lovely site covered with evergreen plantations and a park sprinkled with trees overlooked a natural panorama extending as far as the hills of Surrey and Kent.[9] The lovers probably spent their time strolling the grounds with its woods, lawns, and picnic places contemplating hundreds of workers resurrecting the Crystal Palace and digging beds for its plantings. This was a romantic enough setting for his proposal of marriage, which she accepted. Accordingly, in mid-April the prospective groom wrote the dean asking for his daughter's hand, and the dean replied with his blessing.

The next week Galton dropped off the second half of his book with Murray and left for Peterborough in high spirits to visit Louisa and her family. Galton arrived in Peterborough "to look only on the dead face of the man, who should have welcomed his daughter's future husband."[10] His prospective father-in-law, who was 79 at his daughter's engagement and suffering from heart disease, had died suddenly at lunch that day. Out of respect for the grieving family, the couple postponed their wedding by two months.

Meanwhile Galton's reputation was being further burnished. The Royal Geographical Society annually conferred two medals of equal merit at its Anniversary Meeting in late May.[11] The Founder's Medal was emblazoned with the profile of King William IV, the royal patron of its precursor, the London Geographical Society, while the patron's medal featured Queen Victoria. Galton won the Founder's Medal. The official award praised him for financing his own expedition and for enabling the Society to publish a description and map of "a country hitherto unknown; the astronomical observations determining the latitude and longitude of places having been most accurately made by himself."[12] Murchison, the president, presented the medal emphasizing the importance of Galton's measurements in the eyes of the awards committee and the Society's Council that "saw in this fact, a special reason why the journey of Mr. Galton should be preferred to all other enterprises now on foot in the interior of Africa; none which had . . . determined such positions in other tracts of that continent."[13] Hence, "standing alone in this respect Mr. Galton had a distinct claim on us above his African fellow travellers."[14] Murchison's message was clear. The Society was no talking shop for tourists visiting interesting places, but a serious scientific organization dedicated to the presentation of accurate geographical data.

Murchison ended mixing flattery and bombast saying "so long as England possesses travellers with the resolution you have displayed, and so long as private gentlemen will devote themselves to accomplish what you have achieved, we shall always be able to boast that this country produces the best geographers of the day."[15] Galton accepted his medal thanking Murchison for the honor. However, Emily Butler wrote one of her brothers that Murchison had also remarked that he "regretted that so spirited an adventurer was going to be spoilt and married. Mr G. says it was very well put or he would have thrown the decanter at the worthy President."[16] Murchison was right, as Galton never again engaged in serious exploration although he and Louisa became inveterate European travellers.

Tropical South Africa, published in the summer of 1853, was well received. The *Westminster Review*, a distinguished periodical, claimed enthusiastically that

Galton's book "describes tribes, customs, animals, and scenery very effectively, and is altogether worth reading."[17] *Tropical South Africa* rejuvenated Galton's old friendship with Charles Darwin who wrote on July 24, 1853, that Galton would

> probably be surprised, after the long intermission of our acquaintance, at receiving a note from me; but I last night finished your volume with such lively interest, that I cannot resist the temptations of expressing my admiration at your expedition, and at the capital account you have published of it.... If you are inclined at any time to send me a line, I should very much like to hear what your future plans are, and where you intend to settle.... I live at a village called Down near Farnborough in Kent, and employ myself in Zoology; but the objects of my study are very small fry, and to a man accustomed to rheinoceroses (sic!) and lions, would appear insignificant.[18]

Francis Galton and Louisa Butler were married on August 1, 1853 (Fig. 8-1). They honeymooned in Switzerland and Italy, spending the winter in Florence and Rome where Louisa indulged her interest in the fine arts.[19] The young couple was home in March 1854, camping in various temporary lodgings until they settled in a Victoria Street flat where Louisa began a diary. With his help she reconstructed short, separate entries for "Frank's Life" and "Louisa's Life" going back to 1830 when Galton, aged eight, went off to school in Boulogne.[20] Of their marriage Galton's entry read "Left Dover in March. engaged in April. Spirit rapping mania. Married in August." Louisa was slightly more voluble. "Twelfth night party. First saw Frank there.... Met Frank in London in April, went with him to Crystal Palace and engaged returned to Peterborough April 27. Saturday April 30th, Papa suddenly seized and died.... Married on August 1st. Tour in Switzerland and in Italy. Winter in Florence and Rome." Thereafter Louisa combined their yearly entries. She recorded family events like births, deaths, and marriages along with short accounts of their annual continental excursions and notes on significant events in her husband's life including books and papers published, important meetings, etc. But as the years marched by no little Galtons appeared, while the Darwins produced new progeny regularly until they totalled seven. Why the marriage was barren is unknown, but it probably troubled Galton, especially as he began thinking about improving mankind through selective breeding. After all, his own marriage represented the union of two distinguished pedigrees. For Louisa the absence of children was perhaps a greater disappointment. She led the life of a typical Victorian wife, devoid of opportunity for individual advancement. Her loneliness was probably exacerbated by her husband's near total absorption with his various scientific pursuits. Friends, her close-knit family, her devotion to her husband, and their annual expeditions to the continent were vehicles that likely made it possible for her to cope.

Fame and Marriage 97

Fig. 8-1 Francis Galton and his wife, Louisa Jane Butler. From Karl Pearson, *Life* I: plate 61.

After three years in Victoria Street, the Galtons moved in 1857 to 42 Rutland Gate, a well-appointed town house, that would be their home thenceforward. The drawing room was white enamelled, light, and airy with a hodgepodge of furniture from different periods.[21] The long dining room had a bookcase at the back and Galton's working table was next to the front window. On the walls were prints depicting his friends including Darwin, the botanist Joseph Hooker, the philosopher Herbert Spencer, and the mathematician William Spottiswoode. The back room, which was rather dark, contained shelves stacked with cases for holding pamphlets, letters, and manuscripts. There were boxes loaded with Galton's many mechanical models, "Galton's toys." Although the house had a fine general reference library, Galton's personal scientific library was small, consisting chiefly of books presented him by their authors and papers sent him by admiring colleagues or lesser lights trying to attract his interest.

At Rutland Gate the Galtons "followed the usual routine of social life for persons of our class, making tours every year, usually abroad."[22] Their doctors sometimes sent one or both to take the cure at watering-places like Spa, Vichy, Contréxéville, Wildbad, Baden, Royat, and Mont Dore les Bains. Over the years they accumulated a glittering and diverse circle of friends. Some were scientists, explorers or ex-Cantabs that Galton knew, while others were acquaintances of his wife and her family. Besides those mentioned above whose portraits adorned the dining room, they included famous botanists like

George Bentham and Joseph Hooker;[23] intrepid explorers like Sir Richard Burton; distinguished scholars in mathematics, physics, and astronomy including Sir John Lubbock and William Spottiswoode; the philosopher Herbert Spencer, architects and architectural writers like Thomas Atkinson and James Fergusson; diplomats, colonial servants, and travellers like Laurence Oliphant, Sir Lewis Pelly, and Sir Rutherford Alcock; and Louisa's good friends the Russell Gurneys. Gurney, trained for the bar, served as "recorder" for the City of London and as a Member of Parliament from Southampton, where he shepherded through numerous legislative measures.

At the Galtons' dinner parties the conversation likely sparkled. Afterwards, some guests would have needed to make their way to the loo at the top of the stairs. To signal occupancy Galton had replaced one of the wood door panels with frosted glass across which the user slid a rod to lock the door, this barrier being clearly visible from downstairs.

The Galtons often travelled with friends. On one trip to the South of France with Sir Lewis and Lady Pelly, Galton enjoyed a long day by himself in a flat bottomed boat cruising the canyon of the clear Tarn River. But some friends were more important than others and Charles Darwin especially. Galton often visited Down, "usually at luncheon-time, always with a sense of the utmost veneration as well as of the warmest affection, which his invariably hearty greeting greatly encouraged. I think his intellectual characteristic that struck me most forcibly was the aptness of his questionings; he got thereby very quickly to the bottom of what was in the mind of the person he conversed with, and to the value of it."[24]

Meanwhile Galton was solidifying his reputation as a geographer and travel writer. He contributed a paper for a special edition of the *Journal of the Royal Geographical Society* called *Hints to Travellers*, listing the various scientific and cartographic equipment that would be most useful in unexplored country. *Hints to Travellers* became the Society's most successful publication and was regularly revised and reissued for over a century. Galton coedited the second and third editions, was sole editor of the fourth edition and dropped his editorship with the fifth edition in 1883.[25] Simultaneously he was completing his most popular book, *Art of Travel*, which went through eight editions between 1855 and 1893 and was reprinted again in 2001.[26] It was no gripping narrative like *Tropical South Africa*, but a "how to" book for amateur traveller and experienced explorer alike, filled with practical tips for surviving in the bush. He carefully studied *Pinkerton's Travels* (1808), a 15-volume collection of narratives from every continent that included accounts as disparate as those of "missionaries on the borders of distant Tibet, and of holidaymakers on the Isle of Man."[27] He intended no successor compendium, but only a small practical guide that could be slipped in the pocket of a bush shirt.

The Galtons spent the summer of 1854 in the Chateaux Country visiting enormous Chambord, with its multiple sixteenth century towers and spires; Blois, where they attended a fête; Amboise, sitting majestically above the Loire looking down on the town it dominates; and elegant Chenonceaux soaring outward over the river Cher on graceful supporting arches, the gift of Henri II to his mistress "the ever beautiful" Diane de Poitiers. These splendrous sights they viewed sometimes on foot, sometimes on horseback, or straddled across the wide flanks of a donkey. Galton made sure that he did not lack for material pleasures, buying fly-fishing tackle and provisioning himself amply with cigars, spirits, and wine. But all the while the book was wandering in and out of his mind. What should he call it?[28] He tried "Bushcraft or Science of Travel" and discarded that. What about "Bushcraft or the Shifts and Science of Travel in Other Countries"? That seemed too cumbersome so he tried "The Craft Shift, and the Philosophy of Travel in Other countries." Confusing and pretentious to say the least. Slowly he worked his way toward his final title through "Handbook of Hints for Rude Travel," "Hints on Rough Travel," and finally "Art of Travel."

As the leisurely tour progressed, Galton realized he needed Andersson's help, for he could amplify with new details and methods. Andersson had travelled for another year, studying the natural history of the regions they had explored together, and actually reached Lake Ngami. On his return to London he was approaching destitution. He several times requested loans from Galton who steadfastly refused because "I have nothing like fortune sufficient to do so. If you had struggled hard with a scrupulous economy, and if as Sir James Brooke did, you had even worked your passage home like a common sailor, if you had lived thriftily and frugally determining to keep as much as possible of what you had so well earned in order to win more, the world would have respected you the more highly."[29] Galton, indulgent of himself, was being preachy and mean-spirited with his former right hand. But now he needed help so he invited Andersson to join him in Avranches and sent £35 to cover the journey.

The first edition of *Art of Travel*, 196 pages long and published in 1855, drew largely on Galton's South African experiences supplemented by Andersson. Later editions included ever more material, much obtained from fellow travellers and explorers. For one edition he approached Samuel Baker, the great Nile explorer who discovered Lake Albert. Baker offered some useful tips.[30] He described a saddle constructed by the Nubians by attaching two tree forks at each end of a single long pole. He also instructed Galton on the fine points of using an inflated antelope skin as a float. The mathematician Arthur Cayley supplied a method for finding the distance to an inaccessible point. Splicing in suggestions like these with scissors and paste Galton gradually enlarged

Art of Travel so by the fifth edition, published in 1872, the volume had grown to 366 pages, still small enough to fit in a large pocket, after which no further changes were made.

The book was revealing in several respects. Thus Galton's section on cattle (which also included horses, mules, elephants, dogs, etc.), drew directly on personal experience vividly describing how to break a pack ox.[31] His penchant for numbers and calculations bubbled up time and again. In discussing the weights that pack animals could carry "in trying, long continued journeys" he estimated an ass could manage 65 lbs, an ox 120 lbs, and an elephant 500 lbs. He indulged his fondness for mathematics by explicating a theory of loads and distances and burnishing it up with some simple explanatory equations. Elsewhere he showed the traveller how to compute the length of a journey by time, rate of movement, measurement of angles, etc. He told the explorer "that a capital substitute for a very rude sextant is afforded by the outstretched hand and arm. The span between the middle finger and the thumb subtends an angle of about 15° and that between the forefinger and the thumb and angle of 11 1/4°."[32] There was a section on finding one's way that ended by advising the traveller who lost direction to "set systematically to work to find it."[33] Don't panic, but "calculate coolly how long you have been riding or walking, and at what pace, since you left your party; subtract for stoppages and well-recollected zigzags; allow a mile and a half per hour for the pace when you have been loitering on foot, and three and a half when you have been walking fast."[34] Once the lost, but unflappable, traveller answered three questions Galton posed, he could use some geometry and trigonometry outlined by Galton to find out where he was.

Galton's section on the "Management of Savages" reflected Namibian impressions that would later be influential when his interest shifted to human heredity. Sometimes he was approving: "a sea-captain generally succeeds in making an excellent impression on savages: they thoroughly appreciate common sense, truth, and uprightness; and are not half such fools as strangers usually account them."[35] Other times he was not. "If a savage does mischief, look on him as you would on a kicking mule, or a wild animal, whose nature is to be unruly and vicious, and keep your temper quite unruffled."[36] He was practical about theft. "If all theft be punished, your administration will be a reign of terror; for every savage, even your best friends, will pilfer little things from you, whenever they have a good opportunity."[37] Native women were kind-hearted and important additions to an expedition, but in discussing hostilities Galton advised that "a skulking negro may sometimes be smelt out like a fox."[38]

Publication of *Art of Travel* won Galton accolades. A reviewer, writing in *Blackwoods Magazine* under the pseudonym Tlepolemus, remarked that the book was "a good manual for travellers of the more serious and desperate case, if they must by a necessity of their constitution seek difficulties and dangers. . . .

Another title, which should seem to suit it equally well, would be, 'Hardships made Easy.' "[39] The anonymous critic for the *Westminster Review* wrote that Galton did

> not profess to give hints for a tour up the Rhine, or a visit to Paris. Travel is with him a much more serious affair. It consists in marches through the interior of Africa, where the wanderer may have to catch a wild beast for his dinner, in a pitfall and then boil it in its own skin: where he must know how to secure himself at night from storms and natives; and how, if he gets lost in a trackless wilderness, he may best succeed in regaining his companions.[40]

But the reviewer added there were "not a few paragraphs containing instructions useful to ordinary tourists; as for instance, the directions for fording a river, and for protecting a boat in rough water." *Chamber's Magazine* was also favorable saying it "is a small book, but is stuffed full of facts; and many of these facts are not only of great value to a traveller, but are worth knowing by those whose travels extend only a little way beyond their own firesides."[41] Galton gave Darwin a copy, prompting his cousin, who liked it, to write perspicaciously that "I hope that your volume will have a large sale, but what I fully expect is that it will have a long sale, and if you save from some disasters half a dozen explorers, I feel sure that you will think yourself well rewarded for all the trouble your volume must have cost you."[42]

The Crimean War was in full swing when *Art of Travel* came out, and Galton felt that his practical knowledge would be invaluable to the soldiers in the field. The conflict began in March 1854, when the French and British allies declared war on Russia.[43] A major aim was to crush Russian naval power in the Black Sea by neutralizing the naval base at Sebastopol. The allies landed at Eupatoria in early September, defeating the Russians in a major battle on the River Alma some 15 miles north of Sebastopol on September 20. The allied commanders disagreed on whether they had sufficient troop strength to invest Sebastopol and compromised by deciding to capture Balaclava on the south coast of the Crimea. Its harbor would provide anchorage for their fleet and a springboard from which to attack Sebastopol. Balaclava was taken, but the Russians attacked in late October, with the battle being accompanied by the famous charge of the Light Brigade. The outcome was mixed, with the allies holding the town and the Russians capturing several strong points above Balaclava. The allies won a final battle in early November at Inkerman overlooking Sebastopol, but they failed to take the port and the notoriously variable Crimean winter set in.

The winter of 1854–55 featured periods of extreme cold broken by occasional mild and sunny days and there was lots of sleet, rain, and snow. Due to poor planning life became misery for the infantry, the sea transport system broke down, and food and forage spoiled. The British press reported harrowing tales of squalor, disease, and lack of food on top of which the terrible storm of November 14 (see chapter 11) caused numerous ships to founder. The *Times* organized a *Comfort Fund; Gift Funds* and *Hospital Funds* also proliferated that were handsomely supported by public subscription. In Parliament the Sebastopol Committee began investigating the winter miseries of the British troops.

Galton wrote the War Office in May offering to give free lectures on the art of survival at the new army camp at Aldershot.[44] If he was successful, Galton continued, the Army might wish to adopt such survival courses more widely. His letter was greeted by dead silence, so he wrote directly to Lord Palmerston, the prime minister, and that got results. General Knowles, in command at Aldershot, made two huts available to Galton and he let a small house nearby. He lectured three times a week for three months starting in July before breaking off to visit Paris with Louisa and his sister Emma until Christmas. The course was repeated in the spring of 1856. It was highly practical in content. Thus Galton discussed methods for finding, purifying, and filtering water and for storing water in greased canvas bags or skins. He considered various kinds of tents, where to pitch them, how they were affected by rain and dew, etc. He used illustrations, models, and did experiments to emphasize his points. However, despite the potential importance of the information Galton dispensed, lecture attendance was sparse with a maximum of 15 officers present and the number often dropping to three.

His lectures caught the attention of the press. The *Times* sent a reporter to Aldershot to investigate. He wrote an approving article on Galton's course and reproved the army brass since with all his "experiences, savage and civilized, it is clear that Mr. Galton is entitled to some attention."[45] If those in command had "any misgiving as to the dignity or usefulness of such studies, let them read that elaborate despatch of *the* Duke (of Wellington who had died two weeks earlier) on the importance of 'the little teakettle,' . . . which shows how details apparently trifling are of the greatest significance in the eyes of a great commander. We thus present them with 'a teakettle precedent,' which they would do well to study at their leisure."[46] But there was little time for the British generals to consider "a teakettle precedent," for on February 28, 1856, the allies and the Russians reached an armistice and Galton's lectures terminated. Finally realizing the value of Galton's methods of survival indoctrination, the War Office, two years later, ordered ten sets of his illustrative models to be distributed to various training centers throughout Great Britain.

Galton's reputation as a travel writer caused Alexander Macmillan to invite him to edit and contribute to a new series, *Vacation Tourists and Notes of Travel*. The first volume, published in late 1860, included articles by Galton on a total solar eclipse in northern Spain that July; by George Grove on the town of Nablus and the little community of Samaritans inhabiting a corner of the town; and by W. G. Clark who, tiring of the incessant Scottish rain, journeyed to Naples where Garibaldi was about to wrest power from the Bourbons. Galton's account was lively. Sir George Airy, the Astronomer Royal, organized the expedition and Galton was accepted as one of the party. Spain was new to him and his eye lighted on many things ranging from the "chiaroscuro tint of everything I saw" to the public promenades that occurred in each town in the evenings.[47] Despite precautions, Galton broke the actinometer given him by Sir John Herschel to measure delicate temperature changes. This left him free to observe the eclipse unimpeded so he made detailed sketches of the corona and solar prominences. Afterwards he joined Louisa in Bordeaux and they travelled to the jagged crests, snowy peaks, great amphitheatres, and high lakes of the Pyrenees.

The *North American Review* liked *Vacation Tourists*, which afforded "a striking illustration of the prevalent taste for travel and adventure,"[48] but the reviewer warned that the essays were "marked by great inequalities of style, and by the defects incident to the hasty preparation of such a volume." The review of volume two was far less laudatory, predicting the demise of *Vacation Tourists*. Unless Galton was more fortunate "in his selection of authors and subjects, he will scarcely be gratified in his hope of seeing a long series of annual volumes."[49] The quality did not improve, the reviewer's prediction proved correct, and *Vacation Tourists* expired after volume three, its continuation failing to make financial sense. Then John Murray asked Galton to compile a walker's guide to Switzerland. He tramped around the lakes, high passes, and valleys of that country several times with Louisa on holiday, and made one trip of his own in 1863. He published the *Knapsack Guide for Travellers in Switzerland*, which was promptly panned in the *Alpine Journal*.[50] In spite of these minor setbacks, Galton's reputation was established.

Galton invented useful devices for the adventurous, including a hand held heliostat enabling its operator to catch a flash from the sun and direct it accurately to a distant point.[51] A version of his heliostat, manufactured commercially as "Galton's Sun Signal," was used in late-nineteenth-century British nautical surveys to "enable shore parties to make their exact whereabouts visible to those on the ship."[52] He also wrote occasionally on exploration. One article, on the exploration of arid countries,[53] something Galton knew a lot about, illustrated his penchant for collecting numerical data, making calculations, and developing simple equations. Much of it was devoted to rations. How much water and food can a horse, an ox, or a man carry, and for how long? What if wheels are used as opposed to the back of an animal or a person? He originated methods for determining the number of daily rations that an individual should carry under different conditions and tabulated the results. It was all very practical. He sometimes wrote popular articles on geography like one in *The Cornhill Magazine* in 1862 on recent geographical discoveries in Australia.[54]

Galton's inexhaustible energy was soon directed toward another literary venture, *The Reader*.[55] It was launched by the pioneering Christian Socialist J. M. Ludlow and a group of like-minded men including prominent writers like Charles Kingsley and Tom Hughes, the author of *Tom Brown's Schooldays*.[56] The periodical's ambition was to command the services of distinguished contributors in every branch of literature and science to review contemporary progress in the evolution of human culture. Ludlow and Hughes had earlier lived together in Wimbledon where one of their neighbors was Norman Lockyer, a young astronomer supporting himself as a clerk in the War Office. The three often travelled to London together, frequently with the humorist Tom Taylor. Lockyer was good at popularizing scientific discoveries and became a contributor to the *Spectator*. Ludlow and Hughes recognized he would make an excellent scientific editor for the periodical. Meanwhile Galton wrote Hughes offering his editorial services. Hughes, rushing to get out the first issue in early 1863, thanked him. While the periodical's financial condition prevented him paying a salary, he was pleased to offer Galton an unpaid position as "special editor for science and travels" and an allowance of "three guineas a week to edit and superintend" the "department."[57] He would pay Galton for any articles he wished to contribute and offered him the services of "a young scientific man, Lockyer by name, who has done the work for No. 1 and who would work under you."[57]

The next year *The Reader* was purchased by a group of shareholders. The two principals put up £500 apiece while 13, including Galton, Huxley, Spencer, and Tyndall, each put up £100. Others donated lesser amounts. In laying out the magazine's goal the editors made explicit their desire that science be better served than it currently was in the popular press.[58] *The Reader* would rectify the problem, bringing science into proper balance without neglecting literature, art, music, or drama. To support its science writing a star-studded cast was assembled including Darwin, Hooker, Huxley, Bentham, Tyndall, Murchison, and Lyell. The first meeting of the reconstituted *Reader* was held at Tom Hughes's rooms in Lincoln's Inn Fields on November 15, 1864. The shareholders were to contribute £2300 to cover the cost of paper plus the printing plant and its lease. Cairns would take charge of Political Economy, Galton of Travel and Ethnology, Huxley of Biology, Spencer and Bowen of Philosophy, Psychology, and Theology, Lewes of Fiction and Poetry, and Seeley of Classics and Philology. Hughes recommended to Huxley that Lockyer do the general editing and the matter was decided. Each issue would contain ten pages of Literature, three of Miscellanea, eight of Science, two of Art, and two of Music and Drama. Four thousand copies would be published weekly and the trustees did their sums, estimated costs, calculated advertising revenue, and so forth.

Despite its noble aims and distinguished editorial board, *The Reader* was a failure. Spencer wasted the committee's time by squabbling endlessly about "first principles," prominent scientists, then as now, failed to produce promised reviews and articles, the professional subeditor employed by the group got into a flap with the governing committee about his methods for procuring ads while learned, but illegible contributions poured in. When reviewers were overly critical, authors got hurt feelings and sent in angry letters. In 1866 the periodical foundered, but, phoenix-like, flew out of its ashes three years later as *Nature*, one of the greatest scientific journals of all time, under Lockyer's editorship. It continued under his leadership until shortly before his death in 1920.

Meanwhile Galton rapidly climbed the rungs of the British scientific hierarchy and into its major institutions. He regularly attended the meetings of the British Association for the Advancement of Science being named a general secretary (1863-67), president of the Geographical Section (1862, 1872), and president of the Anthropological Section (1877, 1885). He twice excused himself from being considered for its presidency on account of health. The British Association, formed in 1831,[59] was a "Parliament of Science," where specialists discussed common interests and popularizers promoted the importance of science in everyday life.[60] It was the principal vehicle for general dissemination of science in Great Britain, and met annually for a week in one or another of a circuit of provincial cities. The Association was governed by a General Committee consisting of past and present members of its Council, officers of Sections, and specially elected individuals. This structure determined that the Council, the General Committee and the Sections dominated society affairs. The president was often a figurehead, the first ten including a duke, an earl, two marquises, and a viscount. From 1836 real power was vested on a day-to-day basis with the secretary. After 1862, two general secretaries were appointed one representing the natural sciences and the other the physical sciences. As William Pope wrote in 1920 the Association was "really run by the General Secretaries and Treasurer."[61] These men were "chosen from the most eminent scientific talent in the country" constituting "a body even more exclusive than the Council of the Royal Society."[62] This may explain why Galton, while glad to serve as a general secretary of the Association, demurred for "health reasons" when twice asked to stand for the presidency.

The honors for Galton's geographical exploits continued to flow in. In 1854 he received the Silver Medal of the French Geographical Society and in 1856 he was elected to the Royal Society, probably the most famous academy of distinguished scientists in the world, whose genesis occurred during the English

Civil War in 1645.[63,64] But he apparently derived his greatest satisfaction from his election to the Athenaeum Club the same year on the fast track "under Rule II, which provides that the Council may elect not more than nine persons in each year on the ground of distinction in Science, Literature, Art, or Public Service, being at the average rate of a little more than two elections annually, under each of these broad heads. The recipient is thereby saved many, sometimes sixteen or more, years of waiting, before his turn would arrive to be balloted for in the ordinary course of election."[65] Its elegant club house was at No. 12 Waterloo Place next door to the Traveller's and other prominent London clubs that parade majestically along. Pall Mall.[66] Theodore Hook, a member of the club, caught the essence of the place when he referred to the United Service Club as "the regimental" club and the Athenaeum as "the mental" club.

Membership in the Atheneaum was set at 1,000. The ordinary route of entry was by vote of the membership, where one black ball in ten was enough to defeat a candidate. Benjamin Disraeli's father Isaac put his son up for the club in 1831, withdrawing the nomination in 1837 after the younger Disraeli was repeatedly blackballed because his malicious wit, extravagances, and strong party sentiments had made him many enemies. However, as Conservative Party leader in the House of Commons in 1866, he was easily elected under Rule II as were Galton, Henry James, Thomas Carlyle, Lytton Strachey, and W. B. Yeats. Galton's new club was indeed distinguished with a superb library and glittering membership, but perhaps not the place to dine. In G. W. E. Russell's opinion the club was an institution "where all the arts and sciences are understood except gastronomy."[67]

NINE

Riding High with the Royal Geographical Society

I. The Great Lakes of Africa

> The travels of the successive explorers of Eastern Africa who started from the Zanzibar Coast were watched by geographers with the keenest interest. I was in one way or another somewhat closely connected with the principal actors, and may therefore speak about them with propriety.
>
> —F. Galton, *Memories*[1]

In May 1854, married and enclosed in a carapace of Victorian respectability, Galton was elected to the Council of the Royal Geographical Society, the Society's command and control center. The hard-charging Murchison was ever-present either as president, vice-president, or in some other capacity. Two unpaid honorary secretaries, appointed from the Council, supervised Society meetings. The Council's Expeditions Committee provided instructions to explorers, which were then published. Once the Council decided favorably on a project, members sought funding through public subscription and from the Foreign Office. The Council also nominated candidates for the two gold medals the Society awarded annually. From his Council perch Galton watched a great drama unfold as Baker, Burton, Grant, Livingstone, Speke, and Stanley in search of the Nile's source discovered Africa's Great Lakes in the process. But Galton was no bystander. He wrote instructions for explorers, helped to raise subscriptions for their expeditions, and questioned measurements they reported.

The Council Galton joined included his old master at Trinity, Whewell, and his Cambridge drop-out friend Mansfield Parkyns, whom he had last seen in Khartoum. It also boasted luminaries like Lt. Col. George Everest, the surveyor general of India, for whom the world's tallest mountain is named, and Rear Admiral Francis Beaufort, hydrographer for the Navy, who devised the scale by which windspeed is measured. According to Sir Clements Markham, a future president of the Society, the ambitious young Galton, with his fondness for mathematics and statistics, impressed him positively as being clever, lacking in vanity, straight in his dealings with others, and having a strong sense of duty.[2] But equally striking were Galton's tendency to hold his opinions too tenaciously, his lack of imagination, his inability to allow for the failings of others, and his tactlessness.

Galton's active involvement in Society affairs led to a collision with its assistant secretary, Dr. H. Norton Shaw, a trained surgeon and medical doctor. Shaw was the focal point for correspondence emanating from the Society's far-flung explorers and he oversaw publication of its journal where their reports appeared in print. He was the one constant in the hierarchy and "very popular among the Fellows."[3] The contretemps began as an outgrowth of the Society's publication policy. The *Journal of the Royal Geographical Society*, which appeared annually, was regarded as rather dull failing to deliver the excitement of new explorations with any immediacy. Galton felt a second journal was needed with an accelerated publication schedule to highlight new geographical findings. He agreed on its structure with Shaw and in early 1856 the *Proceedings of the Royal Geographical Society* was born. Shortly thereafter Shaw got sick, so on March 3 Galton wrote to commiserate adding that "until you are well, I will make a point of taking the whole trouble of" editing the *Proceedings* "so you need not be anxious on that score."[4] Over the summer the Society was quiescent, but with fall it was time to gear up the *Proceedings* again. On October 22 Galton wrote Shaw offering "to undertake the chief share in the management of them according to the provisional plan of last summer." He did not "covet the task" nor did he "suppose it would prosper better under my charge than anyone else's, but I am very anxious for their success and my engagements leave me time."[5] Shaw replied unequivocally that he was "always glad of any assistance, but the Secretary must superintend the editing of the Society's papers."[6]

Although much of their correspondence was routine, Galton couldn't keep his fingers out of the pie. He informed Shaw on November 28 that he would be at the Society's offices the following Tuesday with a colleague to sort through some papers, but they also wanted "to come to a more distinct understanding than at present about the future management of the Proceedings so far as it concerns ourselves."[7] The meeting resulted in a compromise. In early December Galton notified the president and Council that Shaw was "willing and desirous to undertake the sole management"[8] of the reports of meetings except for the "Additional Notices." Hence, Galton and friends would merely suggest inclusion of short reports of travel and exploration they thought merited expedited publication. Shaw was now in charge of the *Proceedings* editorial work, making him highly vulnerable to criticism.

The situation deteriorated following Galton's appointment as an honorary secretary in 1857. On February 22, 1858, Galton requested that Shaw furnish him "with a written reference to the various laws and bye-laws by which the duties and title of your office are defined, especially in reference to myself."[9] On July 10, Galton complained about the "great dearth of books at the present moment which would turn proper subjects in the *Proceedings*."[10] Only Logan's "Canada" occurred to Galton, "but it has been thoroughly reviewed in all the newspapers, etc." Shaw retorted to himself on Galton's letter, "Logan's work is fully mentioned in the President's Address." By 1861 Shaw was being so ceaselessly bombarded by Galton on minutiae related to the *Proceedings* he was having trouble keeping up with his many other duties so he threw up his hands and turned over its editing to Galton. The other honorary secretary, Dr. Thomas Hodgkin, tried making peace, but to no avail, so he exasperatedly resigned in favor of William Spottiswoode, a friend of Galton's. Now both honorary secretaries were chivvying Shaw and the fetid swamp of mistrust and ill feeling became such that Galton, on April 4, 1862, accused Shaw of failing to read a specific letter at the previous Council meeting.[11] Shaw complained to himself in his spidery handwriting on Galton's letter that the letter "was down in the agenda as No. 17—Mr Galton was present—was also read by Mr Galton at the Council Meeting on the 10th of March 62 by Mr Galton—himself."

The quarrel escalated. Galton and Spottiswoode urged Murchison to force certain reforms they were pushing on the assistant secretary's office. Murchison dithered, so they tendered their resignations as honorary secretaries. Galton's was accepted, but Spottiswoode was persuaded to stay on. Shaw was coerced into resigning with a year's salary and a new assistant secretary was appointed whose duties were limited to editorship of the Society's publications.[12] Many Geographers were outraged as they regarded Shaw highly. They felt that Galton should resign from the Council and play no further role in Society affairs. Murchison called a general meeting for damage control. Galton must have been grateful since he defused the issue, but several years passed before Galton again became an important voice in Society affairs. The Shaw incident embarrassed Galton, who referred obliquely to his Council resignation his autobiography.[13]

Meanwhile an exciting period in African exploration was unfolding in which Galton was intensely interested. South of the Sahara and North of the Zambesi River most of the continent was a geographical void. The White Nile's source was cloaked in mystery, the Blue Nile having been traced to the Ethiopian highlands by James Bruce in the 1770s.[14] In the first century A.D. the Greek merchant Diogenes reportedly claimed to have "travelled inland for a 25-days' journey and arrived in the vicinity of two great lakes, and the snowy range of mountains whence the Nile draws its twin sources."[15] This story was later passed on to the ancient geographer Ptolemy. His map, showing two lakes watered by a high range of mountains, the Lunae Montes, remained the subject of speculation and dispute for 1,700 years. By the 1850s there was reason to believe that Diogenes might have been onto something, as Arab slave and ivory traders returning to Zanzibar from the interior kept reporting the existence of two vast lakes far to the west. One was called the Ujiji and the other the Nyanza. There were also rumors of a third lake further south, the Nyasa. Missionaries returned with intriguing findings. In 1848 Johann Rebmann reported that he had seen a towering snow-capped mountain called Kilima-Njaro (Kilimanjaro). The next year a second great peak, Mount Kenya, was observed from a distance by another German missionary, Johann Ludwig Krapf.

In 1855 a third missionary, James Erhardt, submitted a short communique with a startling map showing an enormous freshwater lake, the Sea of Uniamesi, which Galton arranged to have published in the first issue of the *Proceedings* (Fig. 9-1).[16] Kilimanjaro featured prominently and the country north of the lake's approaches appeared very mountainous. Had the legendary Lunae Montes been found? Were the Sea of Uniamesi, the Ujiji, the Nyanza the same or different? Galton was urged to mount an expedition to confirm Rebmann's report on Mount Kilimanjaro, but he declined on the grounds that his health was not sufficiently restored for such an undertaking. Perhaps and perhaps not. Galton was married, living comfortably in London, surrounded by powerful friends and acquaintances, and deeply involved in Society affairs.

Meanwhile Captain Richard Burton requested funds for an expedition to seek out the Sea of Uniamesi.[17] To consider his proposal, the Society formed a committee that included Galton, Murchison, and Colonel William Henry Sykes, chairman of the Court of Directors of the East India Company. They approved the project and persuaded the Earl of Clarendon, the foreign secretary, to appropriate £1,000 to support it. Burton chose John Hanning Speke as his second in command. Speke, a lieutenant in the 46th regiment of Bengal Native Infantry, had earlier accompanied Burton on an expedition to Somalia. They sailed early in December 1856 for Zanzibar, arriving there on December 20. Burton was an exotic choice. A great linguist and prolific author, he was permanently frozen at the rank of Captain in the Bombay Native Infantry for his unpublished manuscript on pederasty in Karachi. His interest in Islam, particularly Sufism, led him to pass himself off as a character of half-Arab, half-Persian descent, giving him an excuse for any imperfections he exhibited in pronouncing either language. In 1853 Burton made the Haj to Medina and Mecca in disguise.

Meanwhile Galton was immersed in Society affairs while the *Proceedings* highlighted new explorations and discoveries worldwide.[18] The articles and letters combined to give a dynamic description of intrepid adventurers and expeditions fraught with peril that were sometimes capped by sublime moments of discovery all in the name of geography. On April 18, 1856, Thomas Hopkins

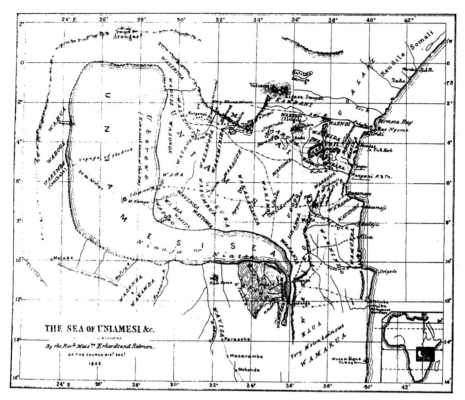

Fig. 9-1 The Sea of Uniamesi. From the *Proc. of the Royal Geographical Society* 1 (1855–1857): after p. 26.

read a paper on the causes of aridity, speculating that the mountainous and rainy promontories of South America and South Africa influenced the dryness of the air blowing over them, extracting much of the water vapor.[19] Desertification would occur north of these "areas of concentration" because they would be traversed by dry montane winds. Afterwards, Galton objected to Hopkins's statement that the African continent, up to the 6th degree of south latitude, was low and arid, and gave his own detailed account of the region's geography and aridity. Hopkins replied he had stated that South Africa was not so dry as Patagonia or Peru, and so its character was not so strongly marked as in South America. At the Society's final meeting before summer on June 23, a letter of Livingstone's was read, accompanied by a sketch map of the African countryside he had been exploring. To illustrate the confusion currently plaguing African geography, Galton spoke up asking the audience to glance at three maps hanging in different parts of the room. Two represented "the respective opinions of Mr. Cooley and Mr. McQueen, two of our best informed African geographers, the third was the compilation of Mr. Erhardt, from most abundant native testimony; and yet these three maps were as utterly dissimilar in all their physical features as it was possible to imagine."[20]

On November 25, 1857, Galton presented a paper on the exploration of arid countries.[21] He intended to provide the traveller with methods for calculating the number of rations he should carry, the number to be included for the supporting party assuming a certain fraction returned to base camp at each stage, and the length of individual stages given the loads carried, etc. As always his approach to the problem was quantitative and mathematical. Afterwards, Dr. Barth remarked that Galton's plan "was suitable to a country like Australia, where the danger of the câches being destroyed by barbarous tribes"[22] was remote, but inapplicable in North Africa where they would likely be discovered and pillaged "by tribes who constantly infest the roads." But Count Strezelecki could "bear testimony to the value of Mr. Galton's suggestions."[23]

Burton's expedition left the coast in June 1857, along a route well known to Arab traders (Fig. 9-2). In spite of countless difficulties and repeated illness Burton and Speke made two major discoveries. On February 14, 1858, they entered the ivory and slaving town of Ujiji some 600 miles from the coast to view the wide expanse of Lake Tanganyika. On voyaging almost to the northern end of the lake they were told by the natives that a river, the Rusizi, flowed into the lake, but did not personally confirm this. On the return to the coast illness became rife, forcing them to halt at Tabora (Fig. 9-2) where they learned from some Arabs of another lake to the north. Burton sent Speke in search of the lake. On August 3, 1858, Speke found himself on the shores of an enormous glittering sheet of water. He concluded it must be the Nile's source and promptly named the lake Victoria in the Queen's honor. Upon returning to Tabora he told Burton excitedly about his find, but Burton was unimpressed, believing the Nile's source lay further east. He argued that their report should focus strictly on their survey of the northern part of Lake Tanganyika. On February 2, 1859, they reached the Indian Ocean and arrived in Aden on April 16 where Burton met an old friend. Burton and Speke were offered transportation back to England on HMS *Furious* a few days later, but Burton dallied in Aden while Speke journeyed directly home.

Upon returning to England, Speke wrote Shaw that he firmly believed "that the Nyanza (Lake Victoria) is one source of the Nile if not the principal one."[24] Meanwhile Burton had written the Society that he would be delayed in returning to England, but that Captain Speke would present their observations and maps. Significantly, he had changed his mind about Victoria Nyanza, adding that "there are now reasons for believing it to be the source or the principal feeder of the White Nile."[25] The Society met on May 9, 1859, and James McQueen, who frequently commented on African explorations, spoke first.[26] Next, presumably to the surprise of many, Murchison introduced Speke, who had just returned to England. Speke briefly recounted his adven-

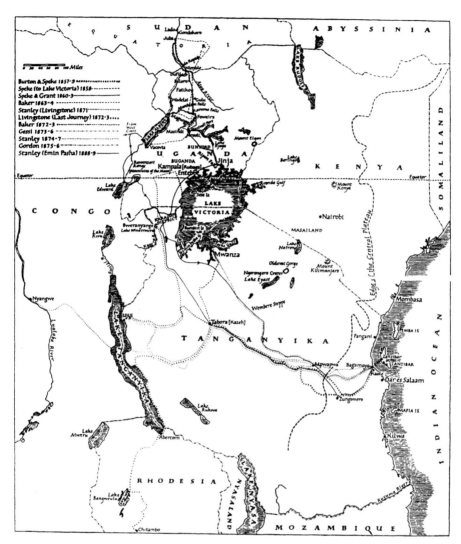

Fig. 9-2 Central and East Africa showing the routes of the principal expeditions. From Alan Moorehead, *The White Nile*, Harper & Brothers, New York, 1960, after p. 324.

tures with Burton to the "Sea of Ujiji" (Lake Tanganyika), but the thrust of his remarks related to the vast Nyanza. He gave the longitudinal and latitudinal coordinates for the position he had reached at the southern extremity of the lake. From Arab information in which he had "implicit confidence," he estimated that it extended five to six degrees northwards, concluding that the Nyanza was "the great reservoir of the Nile."[27] McQueen disputed Speke's claim based on other reports, but Colonel Sykes argued that the difficulties raised by McQueen were reconcilable with the facts set forth by Burton and

Speke. The press caught wind of the Speke's discovery, and on Burton's return 12 days later, Speke was the hero of the moment, causing Burton to write sourly that "my companion now stood forth in his true colours, an angry rival."[28]

Speke was boyish, attractive, a proper Englishman in contrast to the exotic Burton. Galton knew them both and seems subtly to have favored Speke. "Burton was a man of eccentric genius and tastes, orientalised in character and thoroughly Bohemian. He was a born linguist, and ever busy in collecting minute information as to manners and habits. Speke, on the other hand, was a thorough Briton, conventional, solid, and resolute."[29] But Galton understood Burton's great disappointment at Speke's scoop. "Speke got the credit, for without him the lake would not have been reached; but the disappointment to Burton at being superseded in solving the problem of ages by discovering the source of the Nile was very bitter and very natural."[30]

Despite the excitement generated by Speke's discovery, Murchison presented the Founder's Medal to Burton at the Anniversary Meeting on May 23, 1859, as the leader of the expedition that discovered Lake Tanganyika.[31] But he praised Speke, saying a "marked feature of the expedition is the journey of Captain Speke . . . to the vast interior Lake of Nyanza."[32] Speke's disappointment was partially alleviated by Burton's gracious acceptance speech in which he credited Speke for "those geographical results" to which Murchison had "alluded in such flattering terms."[33] Later Murchison looked admiringly in Speke's direction again saying "let us hope that when re-invigorated by a year's rest, the undaunted Speke may receive every encouragement" to demonstrate "the view he now maintains, that the Lake Nyanza is the source of the Nile."[34]

Galton was less convinced that Speke had found the Nile's source, but, in view of Murchison's marathon speech, he waited until the next meeting to voice his concerns. At that meeting on June 13, spears, clubs, pottery, and other examples of native craft brought from East Africa by Burton and Speke were on display. The Earl of Ripon, the new president, hurried through his brief remarks so as not to delay his audience from listening to the presentations by Burton and Speke. After Speke finished, Ripon opened the meeting to questions. MacQueen did not find Speke's claim to have discovered the Nile's source compelling and was joined in his skepticism by Galton. Galton "was particularly struck with the difficulty of accounting for the escape of the large quantity of water which is said to be poured from the lake into that river which is commonly accepted as the true White Nile."[35] This river had been crossed by M. Brun Rollet on a fallen tree trunk, suggesting it was too narrow to accept the volume of water that supposedly issued from the lake. However, Galton acknowledged that the river forming the White Nile might leave the Nyanza west of M. Brun Rollet's stream.

Speke and Burton hastily submitted proposals for further exploration of Victoria Nyanza, which were presented to the Expedition Committee on June

27. Galton was one of the two members present.[36] Possibly because he was partial to Speke, Speke's proposal was accepted. Galton's friend Laurence Oliphant, a correspondent for *Blackwood's Magazine*, had read Speke's diaries and urged John Blackwood to publish them. The edited diaries appeared in the fall of 1859, meaning Speke had stolen a march on Burton and the Royal Geographical Society. Burton's issue-long account of the expedition in the *Journal of the Royal Geographical Society* would be read mainly by specialists, but Speke's claims were in a widely read popular magazine. Burton was furious and in *The Lake Regions of Central Africa* (1860) he attacked what he asserted were Speke's inaccurate data, hastily gathered evidence, and outrageous speculations. Claim and counterclaim only added to the confusion over the Nile riddle.

Speke chose another Indian Army officer, Captain James Augustus Grant, to accompany him. Their instructions were drawn up by Galton and Findlay.[37] They were to journey to Victoria Nyanza, around the lake to the Nile's presumed source, and down the river to Gondokoro near the Sudanese border to rendezvous with John Petherick, an ivory trader and British Vice-Counsel in Khartoum. Galton, ever vigilant for opportunities to gather numerical observations, wrote additional instructions for Consul Petherick on various measurements he should make while proceeding up the Nile to meet Speke.[38]

The explorers sailed from England in late April 1860. They left Zanzibar for the interior on October 2, suffering from all the usual problems of disease, dying animals, torrential monsoon rains, etc. Meanwhile Galton proposed Speke for a gold medal, which was awarded in absentia by Murchison at the Anniversary Meeting on May 27, 1861.[39] After overcoming herculean difficulties, the explorers arrived in Gondokoro on February 13, 1863, nearly three years after their departure. Grant being ill had not accompanied Speke as he travelled along Lake Victoria's shore. Speke had discovered the Victoria Nile and proceeded upstream to a magnificent cascade that he named for Ripon, the Society's president, and then beyond to the lake itself. Afterwards Grant joined Speke and they paddled down the Victoria Nile by canoe to Karuma Falls. From there they continued on foot, missing a crucial link in the Nile story, Lake Albert.

At Gondokoro Speke was greeted by his friend Samuel Baker and his wife, who had financed their own expedition up the Nile. Baker filled Speke in with news from home including the passing of H.R.H. Prince Albert and reports on the progress of the American Civil War. Speke regaled Baker with tales of his adventures and his discovery of the Nile's source at the northern end of Lake Victoria. Baker asked somewhat plaintively, "Does not one leaf of the laurel remain for me?"[40] Whereupon Speke handed Baker a map he had sketched from native accounts of another large lake west of the Victoria Nyanza called the Luta Nzigé. Meanwhile, a false rumor reached Speke that

Petherick was engaged in slaving and collecting ivory west of the Nile rather than attending to his relief. When Petherick and his wife arrived on February 18, Speke was convinced that Petherick, financed by public money including £100 from Speke's father, had neglected his sworn duty to pursue his own commercial interests. In truth the Pethericks had spent nearly a year struggling to reach Gondokoro, nearly dying in the attempt.

Among the letters and papers Petherick brought Speke was a report to the Royal Geographical Society he inferred Galton had written. It implied that Victoria Nyanza was not the Nile's source, but perhaps of the Congo or another river system. This prompted Speke to write Galton from Gondokoro on February 26.[41] He conjectured that Galton had missed his earlier letter, written after reading Burton's report in the *Journal of the Royal Geographical Society*, or Galton would not have arrived at this erroneous conclusion. He implied that Burton had chosen to "hide" the importance of Victoria Nyanza "to excuse himself for not visiting the Nile." Speke groused "it is a pity that my geographical papers *read* before the Society were not put into the Societies (sic!) Journal in preference to Burton's papers which were *not read* and therefore not commented on for that alone has put everybody wrong. Burton's geography was merely a copy of my unfinished *original* maps left open until I reached England for further information."[42] Burton had "wanted me to instruct him acknowledging that he knew nothing whatever of the topographical features of the country. He could not have written one word unless I had instructed him but he gave up his lessons too soon, imagined largely upon the nucleus I gave him, and fell into error accordingly."[43]

The explorers reached Khartoum to find Speke had been awarded the Founders Medal in 1861 for his discovery of Lake Victoria. Murchison, president once more, reported on May 11, 1863, that Speke had telegraphed Mr. Layard at the Foreign Office of his arrival at Khartoum.[44] Murchison believed Speke had located the White Nile's source since Speke asked Layard to inform him "that the Nile is settled." But in the question period following, Galton worried that Speke's conclusions were too hasty, since the "reported size of the river above Gondokoro appeared to him too small to be commensurate with so great a source."[45]

The Anniversary Meeting was held on Monday, May 25, at Old Burlington House in St. James's. The *Times* reported the portion of Murchison's "address which was listened to with most attention was the narrative of the recent discovery of the sources of the Nile by Captain Speke and Captain Grant, compiled from their journals."[46] The evening festivities took place in Willis's Rooms, on King St., St. James's. This fashionable watering place, with its magnificent ballroom decorated with gilt columns and pilasters, classic medallions and mirrors, and lit by gas jets in cut-glass chandeliers, was the home of the Assembly Balls held every Wednesday night during the season.

Meanwhile Speke and Grant returned to England to be fêted the next month at a special meeting of the Society on the evening of June 23 at Burlington House. Numerous distinguished guests were present including a gaggle of Members of Parliament. Murchison accorded the two explorers an admiring introduction. He recounted the profound impression their discovery had made not only upon the Queen and the Prince of Wales, but upon Victor Emmanuele, King of Italy, who had "taken the lead amongst foreign sovereigns in the expression of his desire to commemorate this great discovery, and has directed two gold medals to be struck in honour of our heros of the Nile."[47] Galton, not to be denied the opportunity of making use of the explorers' quantitative data, appended a paper to the *Proceedings* on the climate of Lake Victoria as deduced from their observations.

The following evening the festivities continued and Speke recounted his adventures before H. R. H. The Prince of Wales and a select audience at the Royal Institution. As a reward for his achievement, Speke was permitted to incorporate the phrase "Honour est a Nilo" into his coat of arms,[48] but his relationship with Burton continued to be venomous. Viscount Strangford, tired of the animosity, penned an anonymous article in the *Saturday Review* on July 2, parodying Speke with the title "Dishonour est a Nilo," where he wrote "Burton and Speke are so blind with rage and bitterness that they fight like untrained street boys."[49] But he came down on Burton's side. "So far as it is possible to see the points at issue through the haze of sneer and wrath with which they are encompassed, we believe him to be mainly in the right."[50] James McQueen attacked Speke's book, *Discovery of the Source of the Nile* (1863), with a devastating series of reviews in the *Morning Advertiser*. The Geographers themselves were skeptical since Speke's latitudes for Lake Victoria's northern shore seemed wrong and because he had failed to trace the river flowing from the lake with precision. He had not even shown that Lake Victoria was a homogeneous body of water. Speke's report to the Society in the spring of 1864 was not a carefully reasoned paper, but a hurried summary of the hydrography of the Upper Nile that failed to address the criticisms already raised. Murchison was shocked by the casual nature of Speke's paper, deciding to publish it in the *Journal* with a preface written by the Council regretting that such an important discovery was represented by so incomplete a memoir. Here was Speke's chance to answer his detractors with a serious geographical article, but instead he played into their hands. Livingstone returned to England in July 1864, after six years' exploration in the Lake Nyasa and Zambezi River regions.[51] He favored Speke initially and had exchanged friendly letters with him, but now he lent his enormous prestige to Burton because Speke, intent on demolishing any claim of Burton's that Lake Tanganyika might be the Nile's source, asserted, based on native reports, that a river at the lake's southern end flowed toward Lake Nyasa.

In August 1864, Burton returned to England on home leave, having been consul on the tiny island of Fernando Po off the African west coast from 1861 to 1863 and subsequently on a special mission to the West African kingdom of Dahomey.[52] His return set the stage for a Burton-Speke debate on the Nile's sources. As Galton remembered it, Burton was to present a paper in Section E "Geography and Ethnology" on September 16 at the British Association meeting in Bath, severely critical of Speke's claims. On hearing this, Speke remarked indiscreetly that "if Burton appears on the platform at Bath, I shall kick him."[53] Speke's haughty boast was relayed to Burton by Laurence Oliphant in front of his wife Isabel who remembered her husband saying "Well, *that* settles it! By God, he *shall* kick me."[54] Speke's remark was ill-timed, for he knew he was an indifferent orator while Burton was a master of debate who could marshall history and language at his fingertips. Galton recalled that Speke, staying nearby with a shooting party, was invited to participate in the discussion of Burton's paper.[55]

The meeting convened Wednesday evening, September 14, in the New Theatre Royal completed in the spring of 1863 on the site of the old theatre, which had been destroyed by fire a year earlier. Its auditorium commodiously held 1,750 people above which were three tiers of elaborately ornamented boxes and galleries.[56] Four great chandeliers hanging from the high, vaulted ceiling lighted the stage on which were seated various notables including the Cornish Bishop of Natal, the learned John William Colenso, the Earl of Cork and Orrery, the American ambassador, the Mayor of Bath, and several Geographers including Murchison, Burton, and Galton, who had hurried home with Louisa from their summer holiday in Switzerland. The reknowned geologist Sir Charles Lyell, incoming Association president, was introduced and strode up to a table near the front of the stage covered with a heavy drapery to read his inaugural address. Before him were throngs of well-dressed ladies, some with elegant shawls covering the bustles of their long dark dresses, and gentlemen fitted out in smartly cut suits, many clutching top hats. Some were seated, but most were still standing or milling around. Lyell's speech was a masterful combination of geology and diplomacy. He chose not to provide a "synoptical view of the progress of all branches of science," but focused instead "on the Bath Waters: their history, the geological theories of their origin."[57] The delighted mayor purred contentedly over Sir Charles's ability to impart "the charm of lucid and elegant language to the communication of ideas."[58]

On the morning of September 15, Section E convened at the Royal Mineral Water Hospital, whose foundations, excavated in 1738, revealed parts of the old Roman city including the Praetorium, mosaic pavements, and an altar.[59] Burton walked in with his wife on his arm and sat down near Speke. Isabel recorded what happened. "He looked at Richard, and at me, and we at him. I shall never forget his face. It was full of sorrow, of yearning, and perplexity.

Then he seemed to turn to stone."[60] Meanwhile Murchison was delivering the introductory address, a wide-ranging affair, coming to the Nile with particular enthusiasm.[61] Later he alluded to Speke and Grant and their report of the mysterious Luta Nzigé whose relationship to the Nile drainage system Baker was investigating. He concluded his African survey, noting that "the return of Captain Burton insures for us some fresh and pregnant communications respecting Western Africa; and when we know that Francis Galton, though much occupied with the duties of general secretary, will always take part in our discussions, and that Barth is likely to be present, it is probable that we shall have a concourse of travellers capable of illustrating the geography and ethnology of Africa, such as was never assembled at any former meeting."[62] As he listened to Murchison unfurl the Union Jack in one far corner of the globe after another, Speke began to fidget, exclaiming half aloud, " 'Oh, I cannot stand this any longer.' He got up to go out. The man nearest him said, 'Shall you want your chair again, Sir? May I have it? Shall you come back?' and he answered, 'I hope not' and left the Hall."[63] And with that Speke went off shooting with his cousin George Fuller.

Galton recorded what happened the following day. The president and committee of each section customarily met in the morning to discuss pressing matters and select papers for presentation the next day. They had completed their business and

> Sir James Alexander was urging that the Council of the association should be requested by the Committee to bring Captain Speke's services to the notice of Government and to ask for their appropriate recognition, when a messenger brought a letter for the President, Sir Roderick Murchison. He motioned to the Secretary, who was seated at his left hand, to read it, while he, the President continued to attend to Sir James. The countenance of the Secretary clearly showed that the letter contained serious news. Sir James Alexander went on speaking, the letter was in the meantime circulated and read by each in turn, including Captain Burton, who sat opposite me, and I got it the last, or almost the last of all before the President. It was to say that Speke had accidentally shot himself dead, by drawing his gun after getting over a hedge.[64]

Despite the shocking news Murchison convened Section E, apologizing for being a little late because "in the committee they had been profoundly afflicted by a dreadful calamity, an accident that had befallen his friend Captain Speke."[65] He described the circumstances of Speke's death and moved that the "geographers and ethnologists of the British Association" resolve to offer "their most heartfelt condolences . . . to his relatives on his being thus cut off in so awful a manner in the fulness of his strength and vigour, and while animated by an unabated spirit of enterprise."[66] Murchison's motion passed unanimously and he introduced Burton who read an uncontroversial paper on the ethnology of Dahomey. After Burton sat down, Petherick reported the latest news of Baker's African discoveries. This was followed by a roundtable on the Nile's sources from which Livingstone was absent. Perhaps Speke's departed spirit anxiously hovered nearby. On Saturday only a few sections met so most participants took advantage of excursions arranged for the occasion.[67] The geologically inclined visited the Mendip Rocks, near Frome, to examine the Carboniferous limestones, and then toured nearby Nunney Castle, the bastion erected by the De La Meres and subsequently garrisoned by Charles I. They next tramped through the quarries at Holwell. Another excursion visited Stanton Drew inspecting its Druidical circles. On Monday evening Livingstone's anticipated presentation in the theatre on his explorations of the Zambezi River and Lake Nyasa proved so popular that it was read separately in a nearby room to those unable to gain access to the main event. And Speke's memory gradually began to fade away.

At the Society's first fall meeting on November 14, 1864, Murchison mourned Speke's passing, "the loss of that gallant spirit."[68] Burton followed with a paper that recognized "the many noble qualities of Captain Speke," but could not accept his 'settlement' of the Nile."[69] He listed five reasons why he believed Victoria Nyanza was not the Nile's source, arguing that his discovery, Lake Tanganyika, was the "top head," though not the source of the great river, concluding that the "Arcanum Magnum of old world geography has not yet been solved."[70] Livingstone concurred, but Galton disapproved of the disparaging remarks concerning Speke's explorations, saying he "was sure that all who might take part in the discussion would feel themselves embarrassed by the reflection that, although Captain Speke had been amongst them for more than a year and a half, this was the first time his conclusions had been criticised in this room."[71] Galton acknowledged that Speke's conclusions could not have been questioned sooner as his data had not been in the Society's possession. However, he hoisted Burton by his own petard, saying that while Burton's current theory "of the Tanganyika having a northern outlet, had very much to commend it," after his first journey "he would not hear of the possibility of a river running out of the lake. But *nous avons changé tout cela*," he meant no disparagement of Burton, but only that they should be tolerant they "of any mistake into which poor Captain Speke might have fallen."[72] Burton admitted Galton was correct, but reminded Galton that he had "drained the Tanganyika into the Nyassa, which is also not the case."[73]

Galton gathered with Murchison and several others to raise subscriptions for a public memorial. Murchison announced this goal at the Anniversary Meeting on May 22, 1865, at which the Patron's Medal was awarded to Samuel Baker in absentia. Sufficient funds were collected to erect a simple obelisk of red polished granite in Kensington Gardens, but its inscription proved problematical as there was still major controversy over whether Speke had discovered the White Nile's origin. Lord Houghton solved the problem by proposing the inscription "In memory of Speke—Victoria Nyanza and the Nile 1864."[74] Many years later following Burton's death in 1890, Galton, being fair-minded, attempted to have a second memorial to Burton with appropriate inscriptions erected near Speke's and again after Stanley's death, but the scheme came to naught because of opposition from the families of Speke and Grant. On Grant's death in 1892 Galton's friend General J. Shaw Stewart proposed to the Society's president, Sir Mountstuart E. Grant Duff, that Grant's name be added to the Speke memorial. Galton worried that Speke's family might object and suggested consulting them. He also wrote to Grant's widow who with great discretion replied that, despite Speke's attachment to Grant, his family might prefer not to divide the honor as they were jealous of Speke's fame.

Lord Houghton's proposal for the Speke Memorial inscription proved prophetic as the riddle of the Nile's source was still not completely solved. On March 14, 1864, Baker and his wife sighted a great glittering body of water cradled by high granitic precipices, the Luta Nzigé, which Baker rechristened Lake Albert for Queen Victoria's beloved, departed consort. They obtained dugouts and paddled north for two weeks until they could see the lake's end in the distance from which a river appeared to flow. But another river entered the lake from the east and they followed it until they came to a spectacular waterfall, which they named after Murchison. This was the river that Speke and Grant had earlier crossed near Karuma Falls which flows out of Lake Kyoga, with that lake being fed by the Victoria Nile rushing out of the great Nyanza over Ripon Falls. After much travail the Bakers reached Gondokoro in February 1865.

Baker had not solved the Nile problem for, like Speke before him, he failed to circumnavigate the body of water he discovered, so whether there were more lakes than the two they had named was uncertain. But he could conclude that the river Speke had observed leaving Lake Victoria and pouring over Karuma Falls must flow into Lake Albert and another river flowed out of Lake Albert to the north, which he believed was the White Nile. When the Geographers met on November 13, 1865, at Burlington House, Baker reported on his expedition alluding to the enormous complexity of the Nile question.[75] While he had solved the problem of the Victoria Nile he had observed the exit point of the White Nile from Lake Albert with a telescope from 20 miles away, never actually tracing its route from the cataracts south of Gondokoro to the lake itself. Following enthusiastic applause and Murchison's complimentary remarks, Galton got to his feet and asked about "the relative sizes of the river that runs into the Albert Nyanza and the river that runs out of it: by knowing this, a good idea might be obtained of the proportion of water afforded the Nile by either source."[76] Baker's careful answer began that "Mr. Galton as an old African traveller, must know how very difficult it is to form any opinion, without actually measuring a river and the force of its stream, as to the quantity of water which it may carry down in a given time."[77]

Now it was Livingstone's turn.[78] At the end of his eulogy to Speke on May 22, 1865, Murchison had proposed sending Livingstone to Africa one last time to clear up the mystery of the Central African lakes, their watersheds, and the Nile's source. Livingstone believed the Nile's true source lay southwest of Lake Tanganyika. He suspected that a river issuing from the northern end of Lake Tanganyika flowed into Lake Albert while the one emptying from the northern end of Lake Albert became the Nile in line with Baker's conjecture that Lake Albert was a Nile reservoir. Livingstone arrived in Zanzibar in January 1866, but he travelled south to the mouth of the Rovuma River and proceeded inland toward Lake Nyasa. From there he planned to head northwest to a lake he had heard about earlier called Bangweolo or Bemba where he expected to find the Nile's source. He would trace the connecting river to Lake Tanganyika and then follow the river flowing north from Lake Tanganyika to Lake Albert, and finally the Nile itself issuing north out of Lake Albert. It was not to be. Lake Bangweolo does not connect to Lake Tanganyika, but rather via Lake Moero to the mighty Lualaba River that drains into the Congo.

In a letter from John Kirk, his old naturalist friend and now vice consul of the British Agency in Zanzibar, Livingstone was erroneously reported murdered by a band of Mavite tribesmen in early December 1866.[79] But a January letter from G. Edwin Seward, acting political resident in Zanzibar, was more hopeful, observing that the natives who reported his murder had simply used this excuse to explain why they deserted him.[80] The Royal Geographical Society mounted a search expedition and on April 27, 1868, Murchison announced it had located the explorer, still very much alive.[81] Fearing that he was running out of money, Livingstone wrote Murchison chastising the Government for their meanness. He attacked the Society's Council, saving his worst for John Arrowsmith, the cartographer, and Galton. Arrowsmith and Galton, sitting comfortably in London, badgered Livingstone to make regular observations that, in view of the dense cloud conditions of the last few months, was not an option. "Put Arrowsmith and Galton in a hogshead," Livingstone wrote Murchison, "and ask them to take bearings out of the bunghole. I came for discovery and not for survey, and if I don't give a clear account of the countries traversed, I shall return the money."[82] On November 8, 1869, letters from Livingstone to Kirk and others written the previous July, were read indicating that he believed Ptolemy's springs, from whence the Nile gushed, were much further south than previously supposed, some four hundred miles from Lake Victoria.[83]

TEN

Riding High with the Royal Geographical Society

II. Stanley Faces Off with the Geographers

> One wishes that [Stanley's expedition to relieve Livingstone] could have been effected with less secrecy in the beginning, and less ostentation and comparison of Americans and English to the prejudice of the latter.
>
> —Francis Galton, *Memories*[1]

The friction between Galton and other members of the Royal Geographical Society and Henry Morton Stanley was, surprisingly enough, precipitated by Stanley's relief of Livingstone. Galton and others came to regard Stanley as a headline-grabbing reporter with no interest in collecting quantitative geographical facts. In a sense, the whole problem could be traced to the Society's decision to mount its own relief expedition, only dimly aware of Stanley's existence not to mention his intentions. The story begins shortly after Murchison's death when the accomplished Assyriologist and Indian Army veteran, Major General Sir Henry Rawlinson, opened the Society's 1871–1872 session on November 13, 1871.[2] He reminded the membership "of the irreparable loss" that Murchison's recent demise was for science and the Society.[3] Livingstone had vanished again and was keeping the Geographers "in a state of painful suspense."[4] Kirk had reported sending a batch of supplies for Livingstone to Ujiji on Lake Tanganyika and that an "American traveller, Mr. Stanley" had proceeded inland from the coast. But later Kirk wrote that a conflict in the interior had "cut off Ujiji from the coast" and it might be a long time before they knew much of Livingstone's whereabouts.[5] Stanley had also got caught in the meleé and while he had escaped was apparently no closer to rescuing Livingstone. Hence, the Council had decided to mount a relief expedition. Funds would be raised by subscription, as the Treasury declined to foot the bill even though Livingstone was Her Majesty's Consul for the interior of Africa. From the scores of applications that poured in the Council selected Lieutenant Llewellyn Dawson, R.N., to lead the expedition. It sailed on February 9, 1872, with Lieutenant William Henn as second in command, and Oswell Livingstone, the explorer's son, on board. Public subscriptions now amounted to £5,000, with the populace of Glasgow having contributed £1,000 alone to succour their fellow Glaswegian.[6]

In late January Kirk wrote Rawlinson that Stanley was an American reporter.[7] He suspected Stanley's objective was to obtain Livingstone's story for his paper, the *New York Herald*. So there was "little chance of getting much from Mr. Stanley as original news" unless through his paper.[8] Rumors soon began filtering in to the London papers that Stanley had relieved Livingstone. On May 3 the *London Daily Telegraph* observed that it was sad Murchison was gone for he "would have hung delighted over the brief telegrams which came to us yesterday from Bombay and Aden" reporting "that the Doctor has been found by Mr. Stanley at Ujiji."[9] The same day the *London Standard* chimed in with "the good news about Livingstone." However, it was embarrassed "to record—that the assistance and the encouragement which the government of England denied for the expedition bound in search of the man whom England so profoundly admires was given ungrudgingly, lavishly, by America."[10]

Rawlinson encapsulated the Geographers' frustration at the May 13 meeting when he remarked that while most people believed Stanley had relieved Livingstone, without disparaging Stanley "it was Dr. Livingstone who had discovered and relieved Mr. Stanley. Dr. Livingstone, indeed, was in clover while Mr. Stanley was nearly destitute."[11] The difference in response of the British press and the Geographers was not lost on the *New York Herald*, which sniffed on May 20 that "British geographers would rather that Livingstone should not be rescued than that he should be rescued by an American. Journalists themselves have spoken honestly and fairly of Stanley and the Herald expedition."[12] The Anniversary Meeting was held on May 27 and, after the medals and prizes were presented, Rawlinson read the Annual Address on the Progress of Geography.[13] When he came to Stanley he remarked that he "put the most natural 'English' construction" on Stanley's African objective, believing he was motivated "by the mere spirit of discovery."[14] But now he realized Stanley had been dispatched "by our Transatlantic cousins, among whom the science of advertising has reached a far higher stage of development than in this benighted country," to interview Livingstone, and communicate his findings to the *New York Herald*, one of the most popular American papers. He thought this "highly complimentary" to the British "geographical reputation." His remarks were reported in the *Times* on May 28.

Early in June telegrams began leaking out of Aden and Bombay confirming that Stanley had not only found Livingstone at Ujiji, but they had travelled to the north end of Lake Tanganyika and found that the Rusizi River flowed into the lake. Rawlinson inserted a postscript to this effect in the proofs of the *Proceedings*.[15] A confusing situation was now sorting itself out. The problem arose largely because of the contradictory information contained in detailed letters, travelling slowly from origin to destination over periods of months, and telegrams that moved like lightning, but were limited in content. The upshot was that Stanley apparently had found Livingstone and would soon be on his way to England with the doctor's letters and diaries, so the Livingstone Search and Relief Expedition had been preempted.

This was confirmed by Dawson's May 19 letter from Zanzibar, which constituted his official report.[16,17] His expedition arrived at Bagamayo on the mainland on April 28 only to find that Stanley, having relieved Livingstone, was about to appear.[18] Dawson left for Zanzibar the next day, reporting to Henn and Oswell Livingstone that "nothing remained to be done but to forward"[19] any stores Livingstone might still need. Oswell Livingstone got Dawson's permission to take a caravan inland with the supplies. Meanwhile Stanley arrived in Zanzibar with written instructions from Livingstone making him his agent. Dawson believed he possessed Livingstone's correspondence and journals, but Stanley declined "to answer any question relating either to his own or Dr. Livingstone's travels."[20]

News of Stanley's exploit spread. James Gordon Bennett, Jr., the *Herald*'s owner, cabled him in Aden that he was now as famous as Livingstone. However, Stanley's new celebrity raised curiosity about exactly who he was. While most took him to be an American from Missouri, the Welsh papers began alleging that he was actually one of their own.[21] *Yr Herald Cymraeg* reported that Stanley's birthplace was Denbigh, North Wales, that Stanley was his adopted name, and that his mother ran a pub, the "Cross Foxes" in St. Asaph. It incorrectly gave his real name John Thomas, but the error was rectified in a letter from a reader who knew Stanley's true name was John Rowlands and said his mother kept a public house, the "Castle Arms," in Denbigh.

In the sweltering months of July and August, while most American newspapers were busily chronicling the presidential contest between Horace Greeley and Ulysses S. Grant, the *New York Herald* on July 2 began spinning out the story of Livingstone's rescue complete with his dispatches and large speculative maps.[22-25] It was soon apparent that Livingstone was considering a radical new theory for the White Nile's origin. Out of a series of interconnected rivers and lakes, some real (e.g., Lake Bangweolo), some not (e.g., Lake Lincoln), flowed a mighty river, the Lualaba, which eventually connected with the westernmost branch of the White Nile, the Bahr-el-Ghazal. This put the Nile's main course west of the great African lakes. But Livingstone never traced his river all the way to the Bahr-el-Ghazal. He and Stanley had also concluded that the Rusizi flowed into Lake Tanganyika, eliminating the possibility that it flowed into Lake Albert (Fig. 9-2). Both points caught the attention of the Geographers. With Livingstone in Africa, the only way they could settle the matter was through close examination of Livingstone's dispatches and by questioning Stanley upon his return to England.

Meanwhile the British press was avidly following each development, with reporters swarming around the *Herald*'s London bureau in Fleet Street anxious to garner any new tidbits of information. Livingstone's first and second dispatches were released to the London newspapers a day or so following their appearance in the *Herald*. In contrast to their raucous cousin across the Atlantic with its great speculative maps and teasers in different type faces designed to tempt the reader into the article, the buttoned-up British papers simply reprinted the letters with a minimum of separate commentary. The London papers heaped praise on Stanley. The *Telegraph*'s reporter wrote that Stanley had accomplished "a work as daring in its execution as that of Vasco da Gama, as solitary in its accompaniment as that of Robinson Crusoe, and quite as romantic in its progress as that of Marco Polo."[26]

Stanley arrived in Paris on July 28 and was fêted the next evening at an elegant banquet with the cream of Paris society. On July 31 the American colony in Paris gave him a magnificent farewell feast in the new dining room of the Hotel Chatham. A bleary-eyed reporter for the *Daily News* was "beginning to write his column as the clock strikes two" on Thursday morning, the party having ended an hour earlier.[27] Covers had been laid for 80 and the capacious windows opened wide to help ameliorate the heat. Ambassador Washburn was master of ceremonies. The chef had outdone himself with a new dish "Poularde à la Stanley aux truffes." But across the English Channel the Geographers were enduring rough sailing. Stanley had made mockery of their Livingstone relief committee. On July 27 the *Times* printed an article by the Reverend Charles New, who had witnessed the arrival of the relief expedition. A sentence at the outset said it all. "But just as the Expedition had assumed these full-blown proportions—not unlike an immense balloon trembling through all its gigantic bulk, almost breaking its bonds, and ready to leap on its unknown course the moment its bonds should be severed—it received a staggering blow and instantly collapsed."[28]

On August 1, Stanley arrived in England greeted only by his uncle Moses Parry and his stepbrother Robert.[29] Stanley called at the Royal Geographical Society on August 5 and made an appointment for the next day when he was greeted by Galton, Admiral Sir George Back, and other Council members.[30] Galton, currently president of the Geographical Section of the British Association, invited Stanley to read a paper at its meeting in Brighton the following week on Lake Tanganyika and his journey with Livingstone to its northern end. Stanley accepted and travelled to Brighton on August 14 as the guest of its genial mayor, Cordy Burrows. He hadn't much time to prepare his speech, but a big audience was anticipated and, Geographers excepted, they wanted to hear about how Stanley found Livingstone. This was the season's height at this seaside city of 86,000.[31] Skiffs and bathhouses dotted the shingle beach and throngs of gaily dressed vacationers on foot, in carriages, and on horseback milled about on the King's Road and in the Esplanade. The Congress opened that evening in the enormous mock-Moorish Pavilion with its onion domes and decorative minarets. A large audience including the exiled Emperor Napoleon III and Empress Eugenie was present for the opening speech by the president, the distinguished neurophysiologist Dr. W. B. Carpenter.

The following day Galton delivered the opening address for the Geographical Section at the Concert-room in Middle Street. He announced that Stanley's address would be delayed until the next day because interest in Stanley's speech was so great it threatened to interfere with the attention paid to other papers. These duly followed. The soireé that evening was overly crowded and very hot.[32] Stanley put in an appearance, but the Emperor and Empress did not. This disappointed a milling assemblage of bystanders gathered on the Parade. Stanley was entirely "hemmed in by portly gentlemen in white cravats who shook hands with him solemnly, and while one put him through a sort of viva voce examination upon the whole of Africa, the rest made mental notes of his replies, and stared steadfastly at his nose and eyes and mouth as if to see if he were real."[33]

The Concert-room was over 200 feet long with a gallery running around the sides. At nine in the morning on August 16, seats in front of the stage began to fill. By 10:30 A.M. the room was packed with over 3,000 people. At 10:45 A.M. whispers were heard and heads turned as the Emperor, Empress, and Prince Imperial entered the room and made their way with difficulty to their seats at the front. Several Members of Parliament, the Bishop of Chichester, and Stanley's patroness, the Baroness Burdett-Coutts, joined the ex-royals. A great map of Africa faced the audience. At 10:50 A.M. Rawlinson, Galton, the mayor, and Stanley strode across the stage. Galton stopped briefly before the Emperor and Empress to introduce Stanley.

Galton's introductory remarks were brief as he knew little about Stanley.[34] Stanley would give an account "of the parts of Africa visited by him—that is, the northern part of Tanganyika and the River Rusizi" and then read extracts from Livingstone's dispatches "bearing solely on the geographical aspect of the question."[35] The extracts would be used to add to and correct a nearby map. Galton briefly reviewed the history of the Society's several supply and relief expeditions to Livingstone. He explained that only when it appeared that Stanley's rescue attempt had failed had they sought subscriptions for the relief expedition. But by the time the expedition was mounted "Mr. Stanley had actually shaken hands with Dr. Livingstone at Ujiji."[36] The audience broke in with loud hurrahs for Stanley and Galton introduced him.

Stanley, tousle-haired with moustache and goatee and looking very young, peered out at the sea of faces and began speaking. He paused and began again. He did this once more. Did he have stage fright or was he deciding whether to give the assembled Geographers and scientists a proper, dry account of his explorations or the general audience and the press a stem-winding speech about finding the good doctor? Stanley, the journalist, chose the latter course, departing from the prepared text of his paper. "I consider myself in the light of a troubadour, to relate to you the tale of an old man who is tramping onward to discover the source of the Nile."[37] In a twinkling, Stanley had the rapt attention of his audience as he regaled them about his adventures, about meeting Livingstone, and about his high regard for the "old man." The audience loved it and Stanley was repeatedly interrupted by laughter and cheers as he told his remarkable story. Then Stanley launched into his prepared address to the "Gentlemen of the Royal Geographical Society" and began including geographical details. He described his explorations with Livingstone with verve. The explorers key objective was to reach the Rusizi River at the lake's northern end and determine its direction of flow. They entered the Rusizi by following "some canoes which were disappearing mysteriously through some gaps in the dense brake (of papyrus)."[38] They soon found themselves in what "proved to be the central mouth of the river. All doubt as to what the Rusizi was vanished at once and for ever before that strong brown flood, which tasked our exertions to the utmost as we pulled up."[39] So the Rusizi flowed into Lake Tanganyika. A local chief told them it rose to the north in Lake Kivu. Stanley concluded confidently that the Rusizi was solved, meaning that Lake Tanganyika was not a Nile reservoir. Stanley's thanks to the audience were drowned out by prolonged cheering.

Galton was probably offended because Stanley failed to present precise geographical data, but, bowing to the crowd's enthusiasm, he thanked Stanley graciously. He took the chairman's prerogative of asking the first questions. He wondered how much further north Lake Tanganyika extended beyond the point reached by Burton and Speke. Stanley answered that the two explorers were within 13 miles of its northern end and would have seen it "had they gone half-way up the mountain"[40] referred to in his address. Then Galton asked how brackish the lake water was since whether the Rusizi is "an affluent or an effluent depends upon the character of the water."[41] This really nettled Stanley because it reflected skepticism about the Rusizi's direction of flow, so he answered cheekily he "could not wish a nicer or sweeter water to make a cup of tea or coffee than the water of Lake Tanganyika."[42]

Next Markham read Grant's paper, a detailed geographical critique, prompted by Livingstone's dispatches. As Speke's former partner, Grant had a vital interest in the Nile's source. He did not believe that Livingstone's proposed drainage system fed into the Bahr-el-Ghazal. For one thing there was a

1,000-mile separation between Livingstone's "most advanced position" and the Bahr-el-Ghazal. Furthermore, Schweinfurth's observation "fully satisfied geographers" that the Bahr-el-Ghazal's source was north of the equator "not, as Dr. Livingstone supposes, 11 degrees south of it."[43] Consul Petherick, who had explored the Bahr-el-Ghazal, sometimes called Petherick's Nile, followed. He was sure Livingstone was in error. Later the Ethiopian explorer Charles Beke spoke. Beke thought, correctly as it turned out, that Livingstone's Lualaba did not connect to the Bahr-el-Ghazal, but was a tributary of the Congo. Rawlinson then congratulated Stanley on his feat, but was also dubious about Livingstone's hypothesis.

The nit-picking rankled Stanley who, having thanked Rawlinson, launched into a rebuttal. Grant argued that Livingstone was wrong about the Lualaba, but Stanley wondered "how a geographer resident in England can say there is no such river when Dr. Livingstone has seen it?" Then he got after Beke. He briefly outlined Livingstone's reasons for thinking the Lualaba flowed into the Bahr-el-Ghazal and asked how Beke who was never "within 2,000 miles" of the river could say that Livingstone had not discovered the Nile's source. The audience cheered the elderly "armchair" geographer's putdown. Then he took a swipe at Rawlinson who doubted that Livingstone's "great system of river drainage" related to the Nile. "If the Nile has not been discovered what ... has been discovered?" Did the mighty Lualaba "flow into a lake as Sir Henry Rawlinson supposes? What! the Lualabu (sic!) flow into a lake!—into a marsh!—into a swamp!"[44] "Why you might just as well say that the Mississippi flows into a swamp!"[45] Stanley belittled Beke and Grant for using Schweinfurth's findings as evidence Livingstone had not located the Nile's source. He had "never yet heard of an Englishman who had discovered anything but a Herr of some sort came forward and said he had been there before."[46] The audience howled. Stanley defended Livingstone's actual observations against the Geographers theorizing. "I think if a man goes there and says, 'I have seen the source of the river,' the man sitting in his easy chair or lying in bed cannot dispute the fact on any ground of theory."[47]

Soon thereafter Galton halted the proceedings. He thanked Stanley for his passionate appeal on Livingstone's behalf, but admonished him not to think "that because a man had not been in a country he therefore knew nothing about it."[48] He disagreed with Stanley about theory, saying the Geographers who had spoken were all competent to give opinions on the topic that "was one as much a matter of theory as anything could be."[49] But Stanley was particularly incensed that Galton "with remarkable suavity charged me with being a sensationalist."[50] The next day Stanley slipped away with the *Daily News* correspondent.[51] They evaded the admiring crowds by dodging up the back streets of Brighton and then cantered over the nearby Downs. The day was bright and breezes blew in off the water. That evening a dinner party was held in the Banqueting Room of the Pavillion. It had soaring, 23-foot high ceilings, cornices inlaid with pearl and gold in the form of lotus leaves and pendent trefoils alternating with silver bells, great spacious windows facing east with draperies of crimson silks, adorned with gold and sustained by flying dragons. Stanley was at the mayor's right at the head of the table. Carpenter was also there together with various other dignitaries. The mayor proved an adroit host, there were short speeches and toasts, and vocalists provided musical interludes. Carpenter, asked to give an after-dinner address, "commenced by saying he could not make an after-dinner speech" and "demonstrated the truth of his assertion for three-quarters of an hour."

Toward the banquet's end Stanley was genially relating some of his experiences in searching for Livingstone when "three or four incredulous laughs were heard."[52] Stanley paused and "after a few words of indignant protest, left the room."[53] His reaction surprised the audience, but his sensitivity to criticism from the British scientific elite had become acute. The next day, Sunday, August 18, he left for London with Kalulu, the seven-year-old Lunda slave boy given him by an Arab during his expedition. There were no further geographical fireworks on Monday and Tuesday, but all were concerned, especially the mayor, that Stanley might not reappear for the final banquet on Wednesday.[54] But he returned, was welcomed enthusiastically, and seated once again to the mayor's right at the head of the table together with Carpenter. The proceedings began with the usual toasts including one by the mayor to the British Association. Carpenter responded, thanking the local authorities for their hospitality, praised Stanley's rescue of Livingstone, and absented himself to deliver his closing plenary lecture. Now it was Galton's turn. He was to toast the foreign visitors, mentioning Stanley and the respected spectroscopist Professor Jansen specifically. The *Daily News* reporter was shocked because "Galton has been credited here with having given the American a scanty welcome."[55] Galton appreciated "the great geographical facts" that Stanley made known to them, but he was curious about Stanley's origin since one of the papers had claimed he was a Welshman. He "hoped Mr. Stanley would tell them whether this rumpus was or was not well founded."[56]

Stanley paused as the banquet hall filled with one round of cheers after another. "Before I went to Central Africa," he began, "it was supposed that Dr. Livingstone was the most interesting topic that could be discussed in this country before an assembly of Englishmen. Mr. Galton seems to think that the most interesting topic that could be discussed before this assembly is my own life."[57] Stanley was only 30 and had done nothing "to justify giving you my biography." If Galton wished to learn more, he could read Stanley's book, *How I Found Livingstone*, which would soon be published. There were "interesting facts concerning the geography of Englishmen and the geography of Americans, and geographical facts, and nothing shall be sensational."[58] Galton's use of the word had really irritated Stanley. Stanley thanked the mayor, whom he "found to be the most amiable of men" and made some other off-the-cuff comments to frequent cheers and laughter.

Galton disliked Stanley, regarding him as an opportunistic journalist rather than a serious scientist. He had no place among the pantheon of great nineteenth-century British explorers. In his *Memories* Galton remarked that Stanley "was essentially a journalist aiming at producing sensational articles."[59] Stanley probably knew little of Galton before the fateful British Association meeting, but he certainly did afterwards. On August 27 he wrote the *Daily Telegraph* saying that he greatly resented "all statements that I am not what I claim to be—an American; all gratuitous remarks, such as 'sensationalism', as directed at me by that suave gentleman, Mr Francis Galton."[60] He followed this up in a speech to the Savage Club, reported in the *Manchester Examiner* on September 2, where he remarked "it was at the British Association where Mr Francis Galton, F.R.S., F.R.G.S. and God knows how many letters to his name, said 'We don't want sensational speeches.' That does stick in my throat."[61] "I suppose that when I spoke of all the long names of which he had never heard before, I touched his technical heart...." "He wanted facts. I gave him facts."[62] He was interrupted by cheers. "There was no gilding."[63]

In the adjacent column was an article on Stanley's nationality reprinted from the *Carnarvon and Denbigh Herald*. The reporter had interviewed Stanley's mother, now Mrs. Jones, at the Cross Foxes Inn that she managed near St. Asaph in Wales. She confirmed she was Stanley's mother and that his real name was John Rowlands, providing photographs he had given her to prove it.[64] The *Manchester Examiner* also reprinted a letter published in the *Rhyl Journal* from Mr. E. P. Hipperley, who had discussed the matter with Stanley's half-brother, Robert. He provided independent confirmation that Rowlands and Stanley were identical. Carpenter had forwarded the article to Galton as it contained "full particulars of Stanley's early years," most of which Carpenter had confirmed independently.[65] He had urged Stanley to explain his background in his forthcoming book, writing that if he did he "would take no further steps." Stanley had ignored Carpenter's letter and renewed his "attack upon you (Galton) and the R.G.S. in a thoroughly ungentlemanly spirit."[66] Stanley was a man "on whose word no reliance whatever can be placed, except so far as his statements are supported by collateral evidence."[67] Carpenter had related all of this to the foreign secretary, Lord Granville, because of "the way in which Stanley has been complimented by the Queen."[68] He offered "to take any action" Granville thought he "ought *officially* to take" as president of the British Association "to make the Public aware"[69] that Stanley was lying about his American origin and concealing his illegitimacy.

Meanwhile Dawson arrived in London on August 24 and was asked to appear before the Livingstone Search and Relief Committee on September 2 to explain his abandonment of the expedition.[70] On the long journey home he had come to believe the Society had tried to use him as a dupe to retrieve Livingstone's journals for its members' examination. This view was likely reinforced by Galton's letter to the *Daily News* on September 7 stating flatly that the Geographical Society was "not a humane society, established to succour persons in distress, but for the promotion of geographical science."[71] The public had "sent their subscriptions mainly, but not entirely, for the relief of a man personally dear to the nation, and the two objects of relief and geography admitted of being simultaneously fulfilled."[72]

But Dawson had a mysterious defender. On August 30, the *Daily News* published a letter written on the Reform Club's stationery signed "P."[73] P. summarized the story of Dawson's ill-fated expedition, concluding that while Livingstone's relief was supposedly the expedition's main purpose, Dawson would have fared badly "if he had not come back well stored with 'geographical facts.'"[74] P. also described Livingstone's antipathy to the Geographers. He thought they occasionally doctored his letters before publication. Livingstone sometimes believed his explorations were carried out "in accordance with the plans of the all-embracing Providence," but at other times he felt *"as if serving a few insane geographers."*[75] P. begged the public not to draw negative conclusions about Dawson's behavior before knowing Livingstone's views. Dawson had not realized that the Geographers expected "far more" from him than Livingstone's rescue "and that 'more' was in direct opposition to Livingstone's wishes."[76]

The September 2 *Daily News* contained an article, probably written by Stanley's reporter friend with Stanley's input, that anticipated the upcoming meeting of the Royal Geographical Society "to consider its final judgement" on Dawson's expedition.[77] The columnist was highly critical of the Geographers, claiming that if Murchison had been alive their "anxiety to obtain Livingstone's notes, coupled with the apparent indifference to Livingstone's welfare, would have been far less manifest."[78] He trotted out Stanley's perceived slights including Galton's comment that Stanley produced "sensational stories" instead of "geographical facts" and ridiculed the greedy Geographers' attempt "to extort from Livingstone that which he had determined not to part with before the proper time."[79] He accused them of concealing from Dawson the fact that Livingstone did not wish his travel notes to fall into the hands of the Royal Geographical Society "as on former occasions his notes had been altered to meet the private views of individual members." Dawson believed like many that Livingstone was employed by the Society, "and had only been prevented communicating his journals and notes by the difficulty of transmitting from the interior."[80] Dawson's published orders made clear that, while his main purpose was to relieve Livingstone, he was to try to obtain copies of Livingstone's geographical data for the Society should the explorer choose not to return with the Relief Expedition. Hence, "Dawson was expected to come home well stored with 'geographical facts' whether Livingstone liked it or no." Stanley's purpose, in contrast, was to find and succour Livingstone and request "the old traveller" to write some letters to his newspaper. Dawson had not wasted the subscribers' money by attempting to reach Livingstone to bully "him into submission to the behests of the Royal Geographical Society."[81] Instead, on learning Livingstone had been relieved, Dawson "chose a nobler course" and resigned his command.

The *Daily News* article was potent poison and the Geographers needed to administer an antidote swiftly. It came two days later in a letter from Clements Markham, one of the honorary secretaries.[82] Markham observed that the reporter's argument was based on two assertions: that Livingstone's dispatches were doctored before public dissemination and that he was on poor terms with the Geographical Society. Markham rejoined that Livingstone's dispatches were published in the *Proceedings* exactly as written except for a paragraph in one dispatch eliminated by Murchison, "Livingstone's oldest and truest friend." This paragraph was of no public concern as it related to a matter between Livingstone and his publisher. Livingstone was on good terms with the Society, which had "rewarded his labours with its highest honours; has warmly advocated his cause with the Government, and procured a consular position and a grant of 1,500£; has itself granted him sums amounting to 1,000£,"[83] provided him with instruments, and kept an eye on his activities for 25 years. As for the accusation concerning Dawson, the Society was devoted to advancing geographical science, and would "always zealously strive to further the objects for the pursuit of which it exists by every means in its power." This did not mean it would neglect Livingstone, to whom it had been "too true a friend ... ever to injure his interests in any way."[84] Furthermore, the appearance in the Society's publications "of strictly geographical data, has always had the effect of stimulating rather than injuring" the sale of Livingstone's works. As for Dawson he would have "to show that his proceedings are worthy of approval."[85]

At some point during the controversy, Stanley apparently remarked, probably to Markham, that, as Livingstone's rescuer, he was a worthy candidate for a gold medal. Markham, who likely suspected Stanley's behind-the-scenes role in the *Daily News* article, wrote Stanley on September 4 rejecting the claim because a "medallar of the RGS is expected to be able to fix latitudes, longitudes, heights above the sea etc. with scientific precision, by observing meridian altitudes, lunar distances, etc. etc."[86] This put-off very likely whipped the volcanic embers of Stanley's resentment into an eruption. On September 5, he wrote Markham a long letter complaining about his treatment by the Society.[87] He regarded himself "as a most sociable man, rather inclined to make friends, than make enemies" who was unlikely to be "hostile to any one member of the R.G.S." By rescuing Livingstone he had "performed what the R.G.S desired to do." Yet, his critics including "your Sir H. Rawlinson's, your Francis Galton's omit no opportunity to cast doubts, slurs, or ridicule." He accused the Society of raising doubts about his integrity and honesty. His job was to do his duty by the *New York Herald*. He would have done the same for the Society were he its employee. Since Stanley's "strict adherence to duty" yielded the information the Geographers wanted why had they "cavilled" at him.[88]

Stanley complained of the "very limited recognition" Galton had given him "with such bad grace" at the British Association meeting, referring to his work as "sensational"[89] and insulting him "at the Farewell banquet by questioning" his nationality. He demanded that Galton "write to the newspapers and disclaim that word 'sensational.'"[89] He would immediately withdraw any intemperate remarks he may have made and "shall be the first to let byegones be byegones." If Galton did not, all Stanley had to do was "retort upon him and the Society in the same vein, by pen, by speech in every way I can, and never to give it up until I shall have perfect satisfaction." Markham, trying to defuse a nasty situation, replied somewhat patronizingly the same day. "I am very glad you have done such good work by dead reckoning (as we call it at sea) ... next to actual observation dead reckoning is most valuable. I will also give you full credit respecting the map."[90] In private Markham's reaction was very different. He sent Stanley's letter to Galton with a note containing the remark, "The blackguard took me in. What a gull I am."[91]

Markham and P. hurled accusations back and forth in the *Daily News* for several more days.[92–94] Markham's suspicion that P. was Stanley was confirmed in one of P.'s letters in which he referred directly to Markham's reply to Stanley's letter while concealing the letter's existence.[95] Stanley, knowing that Markham must realize that he and P. were one, combined fabrication with chutzpah, and wrote Markham denying he was the source of the leak and urged him not to publish their correspondence.[96] Markham wrote back loftily: "I must remind you that when you write to honourable men and mark your letter private, it is quite unnecessary to urge them not to publish it as retaliation."[97] In the *Daily News* on September 9,[98] Markham acknowledged that he had written "a private letter" to Stanley. It was marked "'private' or 'personal'" and was not intended for publication, although its "publication is quite indifferent to me." He called off the acrimonius exchange saying that "having been driven from all his positions, your Correspondent 'P.' resorts to incivility and abuse, and there I must leave him."[99] The paper agreed, saying "this correspondence must now cease."

However, Markham, the honorable gentleman, had no compunction about forwarding to Galton Stanley's letter that was marked "Personal" and twice underlined. In fact Markham, Galton, Carpenter, and other influential Council members were sneakily exchanging letters whose goal was to ensure the public, and particularly the Queen, learned Stanley was no American, but a lower-class Welsh bastard. But it was already too late. On August 27 the Queen sent Stanley a gold snuffbox inlaid with blue enamel with the legend "VR" in diamonds, emeralds, and rubies. Stanley was invited for an audience with Her Majesty at Dunrobin Castle, the home of the Duke of Sutherland, on September 8-9. Rawlinson accompanied Stanley, informing him of proper etiquette in the Queen's presence. He introduced Stanley to the Queen. He knelt, kissed her hands, and answered her questions about his journey to rescue Livingstone. Stanley was enchanted by the Queen, "this lady to whom in my heart of hearts next to God I worshipped."[100] The Queen, less enamoured of Stanley, wrote the Princess Royal that "I have this evening seen Mr. Stanley who discovered Livingstone, a determined ugly little man—with a strong American twang."[101]

Meanwhile Carpenter sent Galton a copy of Lord Granville's response to his query. The Queen had already received and recognized Stanley's great service and did not wish to enter the controversy concerning Stanley's birth.[102] Granville understood "Stanley's real character," but advised Carpenter not to take any public action. Carpenter considered Granville's opinion decisive unless a move was made "to bring in Stanley as a Foreign Member into the Association, in which case I should be called on to prevent such a false step from being made." Nevertheless the diehard Geographers, including Galton, Markham, and Grant, continued to insist that Stanley was unworthy of a gold medal. But Rawlinson was deeply concerned. In the eyes of people like Lord Granville, the Duke of Sutherland, and others of the elite, the hard liners' intransigence was damaging Britain's reputation abroad. Markham was warned by "Charlie Forbes, a great friend, who pointed down with his stick and said 'the Society is going down, down, down in public estimation.' "[103] Markham had retorted, "Damn public opinion. The fellow has done no geography."[104] Rawlinson approached Markham about the medal. Markham replied "give the fellow a jolly good dinner for finding Livingstone . . . but it would be a desecration to give him a Royal Award."[105] In the end the opposition to Stanley involved not only the mean-spirited controversy over his origin, but genuine concern that an award would compromise the Society's own standards over what constituted proper geography.

Rawlinson overruled the diehards and called a special meeting of the Council, which voted to award Stanley the Patron's Medal. On October 21 the Geographers held a banquet honoring Stanley at Willis's Rooms.[106] Two of the vice-presidents were present, but Galton was not one of them. Other Council members were in attendance, as were a number of Fellows including Burton. The Lord Mayor of London was seated at the table as was Mr. Moran, representing the American Embassy in London, and the former secretary of the Treasury Mr. Hugh McCullough. After the ritual round of toasts by Rawlinson and the Americans he recounted the events leading up to Stanley's rescue of Livingstone, observing that the Geographers had no clear picture of what had been happening. He apologized for hurting Stanley's feelings earlier that year and awarded him the Patron's Medal for 1873 well in advance of the next Anniversary Meeting. Then Stanley spoke. He could not resist twitting the Geographers. Once when he and Livingstone "had killed and salted a bullock, the Doctor exclaimed 'this is our Geographical Society.'"[107] The audience laughed heartily. But then he expressed his delight at being awarded the Patron's Medal even though Rawlinson was unable to deliver it as such was his haste to honor Stanley that the medal had not yet been struck. Mark Twain, also present at the banquet, wrote "Rawlinson stood up and made the most manly and magnificent apology to Stanley for himself and for the Society that ever I listened too; I thought the man rose to the very pinnacle of human nobility."[108]

Stanley never forgave Galton for accusing him of sensationalism and questioning his origin at the British Association meeting in Brighton. Most of all Galton's failure to apologize for what Stanley regarded as his intemperate, not to mention unjust, remarks rankled Stanley. His book, *How I Found Livingstone* (1872) recounted his perilous adventure and its success, and included a chapter on geographical and ethnographic facts that probably pleased the Geographers. But they must have disliked his final chapter "Valedictory." There Stanley scornfully reviewed all the slights he had received, but for Galton he reserved a special insult. Galton had "with a remarkable suavity" charged him "with being a sensationalist."[109] He belittled Galton's failure to reach Lake Ngami, saying Livingstone had found the lake because he had "held on his way dauntlessly, and his efforts were crowned with its discovery."[110] He mocked Galton by quoting from Andersson's account of the expedition. Andersson was surprised when he later found that Galton really cared little whether he reached the lake or not. He "appeared delighted with the prospect of soon returning to civilized life. Though he had proved himself to be capable of enduring hardships and fatigue as well as any of us, it was evident that he had had enough of it."[111]

Dawson appeared before the Livingstone Search and Relief Committee on September 2 and was asked whether he wished to add anything to his earlier report.[112] He replied negatively, preferring to answer written questions on any points the Committee felt required further explanation. It forwarded him five questions relating to his reasons for terminating the expedition. He replied on September 6 with detailed answers to each. The Committee, including Galton and Rawlinson, weighed Dawson's answers. On September 14 it issued a statement saying it did not believe "Dawson was justified in breaking up" the expedition "on the grounds stated by him" and expressed "their disapprobation of his conduct."[113] Dawson's censure and its basis were published by the *Daily News* on September 17.[114] He fought back in a letter that appeared in the paper the following day.[115] Having tried to rebut each accusation, Dawson charged "the Committee with having played false to me; with having accepted my services in a responsible post, while at the time they concealed from me important truths relative to Dr. Livingstone and his correspondence."[116] Had he known beforehand that the Society's main object was really "to obtain copies of Dr. Livingstone's notes against his will, I should have declined the office; and so would, I think, every officer and gentlemen in her Majesty's service."[117] He also took a swipe at Galton. Livingstone had received the instruments he had requested, but "Mr Galton's field-gun telescope, shipped at his request for a speculative observation of Jupiter's satellites, was shelved as useless lumber, cumbrous in transport and unavailable in survey."[118] The dirty laundry of the Dawson controversy continued to be aired in the press for many months thereafter.

Galton never lost his antipathy to Stanley. From 1874 to 1877 Stanley traversed the entire African continent from Zanzibar to the mouth of the Congo River financed by the *Daily Telegraph* and the *New York Herald*. In 1878 Galton discussed Stanley's *Daily Telegraph* dispatches in a long, unsigned article in the *Edinburgh Review*. He contrasted modes of African exploration. The earliest explorers had travelled with small retinues and used the art of persuasion with native rulers to progress through their domains. As new expeditions ventured into more populous parts of Africa in search of the Nile's source, explorers like Burton, Speke, and Baker brought with them larger and better armed groups of followers, but they always adopted a conciliatory attitude to avoid bloodshed. However, Stanley was "travelling with an armed retinue on a much larger scale than any of" the others and had "carried, by these means, a great expedition successfully through Africa."[119] He had used his large force to conduct "a geographical raid across the middle of Africa, which has led him into scenes of bloodshed and slaughter, beginning in the Victoria Nyanza, and not ending until he arrived in the neighborhood of the Western Coast."[120] At this point, the editor, fearing trouble, inserted a sentence. "This achievement undoubtedly places Mr Stanley in the foremost rank of African discoverers and ensures him a hardly-earned lasting fame."[121] This was probably wise, for Galton persisted asking why "a newspaper correspondent, has a right to assume such a warlike attitude, and to force his way through native tribes regardless of their rights."[122] Stanley had assumed "sovereign privileges, and punishes with death the natives who oppose his way."[123] And Galton's criticism was on the mark.

In the summer of 1878, on a steamer to Boulogne, Galton learned the answer to a question that had long perplexed him.[124] A Welshman named Yates, a Fellow of the Royal Geographical Society, approached him and related the story of Stanley's origin and the events leading to his emigration to America. Yates explained that Stanley's "refusal to acknowledge his real origin" stemmed from his desire "not to make his mother's shame public, and he feared that those who had been kind to him would be scandalized if they knew he was a workhouse boy."[125] Yates told Galton that Stanley had been very good to his mother and she doted on him. So at last Galton understood why Stanley had been so loath to explain his origin.

Galton served on the Society's Council until 1893. He remained engaged even though his main interest had shifted to human heredity. Galton and his friend Colonel Richard Strachey, a retired Indian engineer officer, pushed hard to focus the Society on science. They got the president, Rutherford Alcock, to agree to setting aside £500 a year for scientific goals. The Scientific Purposes Committee awarded the funds and invited lecturers who were paid £50 a piece.[126] Strachey gave the first address on February 12, 1877, but the series proved unpopular because the talks were overspecialized for a mixed audience, and they folded two years later.[127] Galton's competitive school examinations in geography continued until 1884.[128] Markham regarded them as "mischievous" because unscrupulous masters would "run boys for the medals, while conscientious masters" declined to compete.[129] Among 45 schools invited to participate only ten accepted and "Dulwich and Liverpool were the chief runners of crammed boys. Out of 62 medals they got 30."[130] The examinations' demise was precipitated by a letter from the headmaster of Clifton School which exposed the problem.[131] In 1886 Galton and his allies hatched a new scheme for geographical instruction at the university level observing that the subject was widely taught on the continent. Douglas Freshfield, an honorary secretary, arranged conferences between Oxford authorities and several Society representatives including Galton.[132] A Reader in Geography was elected at Oxford for which the Society paid half of the £300 stipend. In 1892 a similar appointment was negotiated with Cambridge, so Galton and his colleagues were succeeding in their admirable goal of promoting geography as a worthy academic subject.

Strachey became president in 1887, putting Galton and his allies in firm control. This was the year of the Jubilee celebrating the fiftieth year of Queen Victoria's reign and the question of admitting women arose as the Queen was the Society's Patron.[133] Strachey moved for admission, but the Council was divided and the topic was dropped. In contrast the Scottish Geographical Society had from its inception in 1884 admitted women on the same terms as men. One of their fellows was Mrs. Isabella Bishop, née Bird, a famed traveller and author.[134] In early 1892 the Royal Geographical Society's Council invited her to read a paper at one of its evening meetings. She declined, not wishing to speak before a society of which she could not be a member. Instead, she lectured before the local branch of the Scottish Society. This branch folded shortly thereafter when the Royal Geographical Society admitted members of all British geographical societies to its meetings. So by a fait accompli Mrs. Bishop became a Fellow. Many members were perturbed, as it seemed absurd that a woman should be admitted to their society by virtue of her membership in another organization. Consequently, she was formally proposed for admission on July 4, 1892, and the president, Sir Mountstuart Grant Duff, moved in Council that women should be eligible for membership on the same terms as men. The motion, which carried almost unanimously, was strongly supported by Freshfield, Strachey, and Galton.

A protracted dispute resulted that split the Council and eventually the Society itself into two factions. In August 1892 the Council's decision to admit women was published in the *Proceedings*. As no objections were voiced, Grant Duff, with Galton, Strachey, and Freshfield as allies, proposed to the Council the election of 15 highly qualified women as Lady Fellows. The motion, loudly opposed by Markham, George Curzon, and others, carried. The dissidents refused to accept the result and requested a polling of all Fellows at the general meeting on April 24, 1893. As they had foreseen, the membership was not yet ready to admit the opposite sex and the motion failed. By now 22 women had been admitted, so the Council, in a decision that pleased no one, voted to retain these Lady Fellows, but to admit no more. The bickering between the factions became public in the *Times*. Curzon, later viceroy of India and president of the Society, wrote that his faction contested the "capability of women to contribute to scientific, geographical knowledge. Their sex and training render them equally unfit for exploration, and the genus of professional female globe-trotters with which America has lately familiarised us is one of the horrors of the latter end of the 19th century."[135] Curzon apparently did not consider the likes of Mrs. Bishop as worthy explorers. At the Anniversary Meeting in May the dissidents demanded the exclusion of the 22 women already admitted, but the challenge was beaten down. A general referendum of Fellows in the British Isles was taken, producing a substantial majority favoring the admission of women, but the dissidents forced yet another meeting in July at which the membership of women was once more rejected. At this point Grant Duff resigned the presidency and the following year Galton and Freshfield resigned from the Council. Strachey retired from the Society and Markham was made president. Markham offered Galton the trusteeship of the Society, but Galton declined. He elected to play no further role in the organization's affairs. Later Curzon changed his mind and his advocacy resulted in the admission of women to the Society in 1913.

ELEVEN

Weather Maps and the Anticyclone

> Francis Galton, member of the Meteorological Committee of the Royal Society, coined in 1863 the word "anticyclone" for areas of high pressure.
>
> —Kutzbach, *The Thermal Theory of Cyclones: A History of Meteorological Thought in the Nineteenth Century*[1]

Francis Galton's interest in meterological patterns derived directly from his own experience of striking climatic variations in Namibia and indirectly from his service with the Royal Geographical Society. There he would often have been reminded of the effects of the vagaries of the weather on the various far-flung expeditions sponsored by the Society. Closer to home he would frequently have been made aware of the disastrous consequences of foul weather on the merchant shipping of the world's greatest maritime nation. Among the storm-induced ship founderings reported in the press one in particular must have caught his attention.

Early on October 12, 1853, the *Dalhousie*, on which Galton had sailed to South Africa three years earlier, cleared the docks at Blackwall in East London and headed down the Thames for Gravesend. The ship was bound for Sydney, Australia, with a cargo of merchandise valued at £100,000.[2] Captain Butterworth was still in command and his wife and two of their sons were on board. The ship was manned by a crew of 61 and there were about 20 passengers in chief cabin class, including two families with three children each. On October 17 the *Dalhousie* made port at Deal near Dover, where the Butterworths' eldest son disembarked to return to school nearby. His mother and the other son stayed on board as they planned to debark at Plymouth. The next day began uneventfully as Captain Butterworth guided his ship westward behind a fresh northwest breeze that weakened off Dungeness about 10 A.M. Around 8 P.M. ten miles west of the Dungeness lighthouse the wind shifted to the south southeast and began to freshen. By 10 P.M. the captain had ordered the topgallants taken in and at midnight all hands were called to reef the topsails. The sea was heaving alarmingly and from time to time the light at Beachy Head, a few miles off the starboard beam, could be seen twinkling invitingly. At 4 A.M. the maintop-sails were double-reefed, and mizentop-sail stowed. About 5 A.M. a roiling wave broke over the decks, carrying away the starboard-quarter boat, and the crew began clearing the decks, throwing sheep pens, water-casks, and lumber overboard to prevent their careening back and forth. Before long another great sea crashed across the stricken ship, carrying away the long boat. By now many had taken refuge in the maintop and Captain Butterworth and several crew members dragged a man, his wife, and children in through the gallery window. Then another huge sea broke over the ship, carrying the whole family away into the shrieking darkness of that awful night. The ship was settling fast when a schooner was sighted nearby, but she was so intent on her own survival in the maelstrom that she passed the foundering *Dalhousie*. By now many passengers and crew members had been swept to their deaths in the churning waters, with the rest desperately clutching at broken spars and masts as the *Dalhousie* disappeared below the waves. As dawn broke, Seaman Joseph Reed helplessly watched several vessels pass by while the remaining survivors gradually let go of the flotsam they clung to, perishing in the cold gray swells of the English Channel. Finally, at 1 P.M. Reed signalled the *Mitchel Grove*, which threw him a line, and pulled him aboard. He was the sole survivor.

Galton must have been dismayed upon reading of the East Indiaman's sinking. He had spent almost three months aboard her on his way to South Africa and was well acquainted with Captain Butterworth. Perhaps this event helped to stimulate further his interest in comprehending the events governing weather patterns. Maps providing numerical data (temperature, barometric pressure, etc.) concerning the weather might provide the key to understanding how specific meteorological events came about. This would combine Galton's lifelong interest in quantitative analysis of scientific data with his current interest in revolutionizing geographical map making.[3] Recent geographical maps were "little more than an abstraction, or a ghost of the vivid recollections with which the memory of the traveller is stored ... when it attempts to image to itself the features of a once-visited country."[4] How could maps be made to convey these feelings? With help from his cousin R. Cameron Galton, he took a stab at the problem, attempting to create stereoscopic images giving an impression of the contours of the terrain, particularly in mountainous regions.[5] So Galton's foray into meteorology, beginning as an outgrowth of his interest in climatic data and map making, led him in 1861 to attempt to create retrospective maps of the English weather. The first, privately printed on June 12, 1861, by his friend the mathematician William Spottiswoode, who made his living as the head of an important family printing firm, charted the English weather at 9:00 A.M. on February 9. He continued to refine similar maps (Fig. 11-1).

Galton soon realized that the key to making proper weather maps was to obtain meteorological observations simultaneously from a wide variety of sources.[6] Then data like barometric pressure, wind speed and direction, clouds, and precipitation could be plotted precisely for a given time on a specific day for many geographically separated locations. On July 16, hoping to cover a large swath of northern Europe, he published a circular in English, French, and German, accompanied by an earlier map, and sent it to meteorologists throughout Europe requesting data for the entire month of December 1861. The data were to be collected synchronously at 9 A.M., 3 P.M., and 9 P.M. The responses were disappointing since most of the data came from only a few localities: Belgium, Holland, Austria, and Berlin. Nevertheless he made some sense out of these results, publishing *Meteorographica* in 1863 with a whole series of weather maps for Europe and the British Isles for December 1861. Each day was represented by a series of nine tiny charts, three for each of the three times of day when data were gathered. One chart presented barometric pressure, the second wind direction and precipitation, while the third diagrammed temperature trends.

Galton was by no means the first to construct synoptic weather charts.[7] The use of maps to study storm characteristics was introduced as early as 1820 by H. W. Brandes. From the data available he was able to construct weather charts depicting deviations from normal barometric pressure, temperature, and wind field for a major storm that had howled over Europe in 1783. The first international effort to collect synchronous and uniform meteorological observations worldwide was sponsored by the British Association in 1835. The astronomer Sir John Herschel requested that hourly observations be made at as many stations as possible during the equinoxes and solstices. The response was gratifying and Elias Loomis, an American, took advantage of the results to analyze a major storm occurring during the winter solstice of 1836. Later, Loomis, in 1843, studying two great winter storms of 1842 that roared up the East Coast of the United States, depicted the distribution of clouds, rain, and snow, used arrows for wind direction, and contour lines for regions of equal barometric pressure and temperature. Loomis's weather chart was a very important new tool in meteorology and it helped to resolve the conflicting theories of two other Americans, James Pollard Espy and William C. Redfield, on the nature of cyclonic storms.

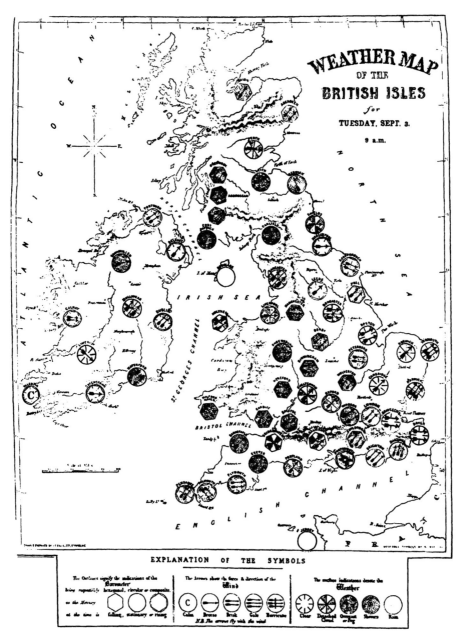

Fig. 11-1 One of Francis Galton's earliest synchronous weather maps, probably for Sept. 3, 1861, showing the use of circular and hexagonal symbols to indicate wind direction, barometric pressure change, and the weather at different locations in the United Kingdom. From *Life* II: plate VI.

Additional study of such charts proved to Loomis that surface winds in cyclonic storms had a tendency to blow both toward the center and in a counterclockwise direction.

During the 1840s and 1850s efforts to gather weather information over broad geographic regions became more systematic. By 1849 Joseph Henry, director of the Smithsonian Institution, had put together a network of stations in the United States that transmitted weather information daily by telegraph. He began publishing weather reports and charts in numerous newspapers from 1854 onwards, being interrupted only by the Civil War. In seafaring northern Europe, exposed to snarling winter storms and ship-breaking gales, systems for forecasting storms were badly needed for the protection of mariners and fishermen. In the Netherlands, A. C. Buys Ballot, in 1852, began painstaking analysis of weather patterns using charts, and in 1854 he was made director of the Nederlandsch Meteorologisch Instituut, subsequently publishing weather reports and storm warnings.

Systematic attempts at understanding weather patterns in Britain and France received a boost as a direct result of the Crimean War.[8] After the battle of Inkerman on November 5, 1854, the allies settled into winter quarters at Balaklava, preparing for a long siege of Sebastopol. The weather was deceptively calm and balmy. Many ships were at anchor both inside and outside Balaklava Bay and at Katcha Bay and Eupatoria to the north of Sebastopol.[9] But beginning on November 13 and until the afternoon of November 16 "the most terrific gale ever known in that part of the world continued to rage throughout the length and breadth of the Black Sea."[10] At Katcha Bay several supply transports ran aground and Her Majesty's steam frigate *Samson* was dismasted. At Eupatoria the French battleship *Henri IV*, pride of the French Navy, ran aground "and near the beach the tricolour" floated "mournfully over the *Henri Quatre*, strong and erect as ever, but never again to carry the flag of France to victory."[11] But the British Navy suffered its greatest losses at Balaklava Bay, where numerous ships were driven onto the rocks or dismasted. In all cases most of the crews perished. Worst of all was the loss of the "*Prince*, a splendid new screw steamer of 2700 tons, on her first voyage."[12] She hit the rocks on the morning of November 14, and sank within 15 minutes. Only six of the crew of 150 survived, and "the whole of the winter clothing for the men went down—40,000 suits of clothes, with under-garments, socks, gloves, and a multitude of other articles of the kind; vast quantities of shot and shell; and, not least in consequence, the medical stores sent out in consequence of the deficiencies which formerly existed."[13]

On shore the "hurricane" caused less loss of life, but spectacular damage and some embarrassment. Because no Russian attack was expected the night of November 13, many officers treated themselves to a night's rest in the only night-gown known in the Camp—a red flannel shirt of very scanty dimensions. The hurricane turned them out, and left them even less time to dress than the Russians would have done. With their tents twirling high in the air, or running a mad race across country, these men were seen standing, while all their household goods lay shivered around them, gazing at the scene and at themselves with intense horror and astonishment. Some were actively employed in keeping their shirts in a position somewhat compatible with decency, while others stood on one leg, the tails of their shirts flapping about their ears, while the other leg was inserted into the corresponding half of a pair of refractory trousers that would swell out with the wind, and betrayed a strong inclination for an early trip to the Russian lines.[14]

The French were shocked the weather could have unexpectedly caused such a great loss. Napoleon III's minister of war, Marshall Jean Baptiste Philibert Vaillant wrote to the director of the Observatoire de Paris, Urbain Jean Joseph Le Verrier, asking that he investigate the circumstances leading to the disastrous storm. Le Verrier had a sharp mind and proceeded to reconstruct the events leading up to the storm, writing to all the European observatories requesting their weather observations from November 11-16, 1854. By piecing the data together, Le Verrier showed that the storm had been observed in the Mediterranean on November 10, so storm warnings could have been sent to the troops in the Crimea the day before. He presented his findings before the French Academy on January 31, 1855. Le Verrier's analysis carried the day. On February 16, 1855 he submitted a proposal to the Emperor outlining methods for warning of impending storms. On February 17 a cabinet order was issued authorizing Le Verrier to organize a storm warning service he would operate jointly with the director general of telegraphic lines. By 1857 he was publishing the *Bulletin International de l'Observatoire de Paris*, which by 1863 included daily weather charts.

The British reaction was equally precipitate. A month after the Crimean hurricane the Lord Commissioners of the Admiralty ordered Captain Robert FitzRoy, R.N., and a Fellow of the Royal Society, to take the helm of a Meteorological Department recently formed under the Board of Trade. Under a Parliamentary Act of 1850, the Board of Trade was authorized to undertake general supervision of all matters relating to the British merchant fleet.[15] FitzRoy, a former governor of New Zealand, was well known for his interest in meteorology.[16] As commander of the *Beagle* on which Charles Darwin had sailed, he had made careful reports of the effects of weather. At the Board of Trade FitzRoy set up shop with two officer-assistants, a draftsman, and three clerks. The impact of telegraphy on the collection of weather data was enormous, and in 1859 FitzRoy's office began to transmit meteorological data by telegraph. The urgency of receiving foul weather warnings promptly was underlined by the *Royal Charter* disaster on October 26, 1859.[17] She had departed from Melbourne, Australia, on August 26 with 388 passengers and a crew of 112 and at least £500,000 of gold and specie. While sailing in the Irish Sea off Wales, she was driven by a violent gale onto a shelving ledge of limestone rock on the northeast coast of Anglesey where she soon broke up. From Monday night into Wednesday of the next week there were more furious gales accompanied by torrents of rain.[18] The storms caused widespread coastal damage and at Yarmouth "a perfect hurricane" blew in from the south-southeast.[19]

The snaking of telegraph lines to reporting stations across the British Isles and under the English Channel and Irish Sea meant an ever more complete picture of weather the previous day could be pieced together. FitzRoy began making maps on which he plotted wind, barometric pressure, and temperature using hatched boxes to indicate clouds, rain, snow, etc., for which he coined the phrase "synoptic chart." He instituted the use of conical storm symbols (FitzRoy's cone still being the standard gale warning) and he invented the sturdy FitzRoy barometer. In 1863 he published the *Weather Book*, a popular account of his meteorological views. FitzRoy's job at the Board of Trade was to compile statistical information on the weather, but he felt it was not enough to characterize previous events. What was desperately needed was a method for predicting oncoming storms and other weather systems so that calamities like the *Royal Charter* could be avoided. So FitzRoy's Meteorological Department began issuing the first weather forecasts. He did well enough initially and one day in 1860, shortly after he began to forecast, his daughter Laura answered a knock at the front door to find emissaries from the Queen standing before her. They were there to query the Admiral concerning the forecast, for Her Majesty was to cross the next day from England to Osborne on the Isle of Wight.

Until now weather had been presented retrospectively in the daily papers. In the *Times*, for instance, weather parameters such as barometric pressure, temperature, wind, and wind direction were reported for many different locations in Great Britain for the period between 8 and 9 A.M. the previous day. The qualitative character of the weather was also noted using terms like cloudy, foggy, fine, and dull. But on August 1, 1861, a *Times* reader would have noticed a new feature attached to the weather report, a general forecast for the next two days based on the predictions of the Meteorogical Department. The north was expected to have fine weather with a moderate westerly wind; the weather in the west would also be excellent with a moderate south-westerly wind flow; while the south, also would experience lovely weather, and a fresh westerly wind.

Meanwhile Galton's analysis of meteorological data for the month of December 1861, led to an important discovery. The month "began under cyclonic conditions" and "was followed by a condition . . . the exact opposite to the cyclone, and supplementary to it. The cyclone . . . is an uprush of air, associated with a low barometer and clouds, due to the hot and moist air becoming chilled as it rose . . . with an anti-clock-ways twist in the northern hemisphere. That which I now found . . . was a downrush of air associated with a high barometer and a clear sky, and with an outflow having a clock-ways twist" so "I called the newly discovered system an 'Anti-cyclone.'"[20]

Unfortunately FitzRoy's forecasts were often flawed and members of the Royal Society and the public at large became increasingly critical of them. In 1862 the *Times* disowned any responsibility, writing that while "disclaiming all credit for the occasional success, we must however demand to be held free of any responsibility for the too common failures which attend these prognostications. During the last week Nature seems to have taken special pleasure in confounding the conjectures of science."[21] On June 18, 1864, the *Times* published a critique of a report FitzRoy had written recounting accomplishments of the Meteorological Department.[22] FitzRoy admitted his warnings were occasionally too late. The reporter remarked that the Admiral sometimes got the wind coming out of the wrong quarter, but there was no doubt that when "Admiral FitzRoy telegraphs something or other is pretty sure to happen."[23] The theoretical underpinnings of the two-day forecasts "on which the Admiral prides and piques himself" are disputed "by many scientific men," partly, the *Times* suggested, because of the "singularly uncouth and obscure dialect" he employed to explain his methods.[24] Apparently the strain of criticism, probably exacerbated by other factors, became too much for FitzRoy. On Sunday morning, April 30, 1865, he "went to his dressing-room, for the purpose of getting ready for church," locked the door, and slashed his throat with his razor.[25] Nevertheless, forecasts from FitzRoy's office continued to appear in the *Times* until May 28, 1866, their demise presumably resulting because of the publication of a condemnatory report by a special committee appointed by the Board of Trade and chaired by Francis Galton.

Following FitzRoy's death, this blue-ribbon committee was appointed to review the Meteorological Office and to make recommendations for its management. Galton was nominated by the president of the Royal Society.[26] The other two members were T. H. Farrar, Esq., one of the secretaries of the Board of Trade, nominated by the Board of Trade, and Staff Commander Frederick Evans, chief naval assistant to the Hydrographer of the Admiralty, recommended by the Admiralty. Their fivefold charge was to examine the "meteorological observations at sea" already collected by the Meteorological Department; to determine what should be done about analyzing and publishing the data; to recommend whether meteorological observations should continue to be made at sea and in what manner; how data transmission could be improved assuming telegraphy would be used; and what staff would be necessary to achieve these ends.

A major section of the resulting report, "Foretelling weather," was damning of FitzRoy's attempts in this regard. "As early as the year 1857 the late Admiral FitzRoy's attention had been directed to the daily observation of the changes of weather over the British Isles, with a view to the prediction of such changes."[27] Even a distinguished meteorologist like Le Verrier had not attempted this, though he "had established a system of telegraphing the state of the weather daily, not only in the various ports in France, but also from other ports in Europe."[28] However, Le Verrier "confined himself to the communication of the actual state of the weather, and apparently deprecated any premature attempt to foretell anything except the approach by telegraph of storms known to exist elsewhere."[29] FitzRoy, believing meteorologists spent too much time collecting and publishing observations, "persevered in his attention of foretelling, or, to use his expression, *forecasting*, not only storms announced by telegraph as already existing, but weather generally."[30]

The committee questioned the basis of FitzRoy's analytical methods, writing that "the maxims on which the Department acts in foretelling weather" were not shown to have been deduced "by means of accurate induction from observed facts."[31] After comparing FitzRoy's daily forecasts with the actual weather occurring on the same dates, the committee concluded that his forecasts were "wanting in all elements necessary to inspire confidence."[32] They recommended their discontinuance because they were inaccurate and had "as yet no scientific basis."[33] On the other hand, FitzRoy's methods for announcing impending storms was viewed favorably since their predictions of the force "of coming gales, have been sufficiently correct to be of some use, and ... their utility is widely admired."[34]

The report resulted in the appointment of the Meteorological Committee in 1868 to oversee the issuing of storm warnings to sea ports and procuring of data for marine weather charts. Galton was a member and would remain on the Committee, and its successor the Meteorological Council, for nearly 40 years. He busied himself with designing more of "Galton's toys," this time directed toward meteorology, with the aim of organizing and equipping self-recording meteorological observatories. He, together with the Meteorological Committee, began developing weather maps once again, this time greatly simplified and directed at the public. The first was published in the *Times* on Thursday, April 1, 1875, with a general description of the weather the previous day. These descriptions gradually became more elaborate and were accompanied by graphical depictions of diurnal variations in temperatures and barometric pressures based on measurements made using instruments at seven observatories. By 1879 the Meteorological Committee felt confident enough to begin to publish forecasts again, but they were issued at 8 P.M. and only for the next day.

Being directly familiar with the effects of winds on the progression of sailing ships, Galton also wrote several papers relating to this topic, beginning in 1866 with a critique of the statistical methods used by the old Meteorological Office.[35] In one paper, "Barometric Predictions of Weather,"[36] published in 1870, Galton sought to develop a formula for predicting wind velocity based on barometric reading, temperature, and humidity. As Pearson later pointed out, "the interest of his paper lies in the evidence that he was feeling his way towards 'correlation.' "[37] As he strove to perfect his methods, the day of the sailing ship was reaching its twilight, but had these papers been published in 1850 they "would probably have been followed by the universal construction and use of isochronic charts, and Galton's name would have been honoured in the history of navigation."[38]

In 1858 Galton was asked to join the Management Committee of the Kew Observatory by his friend General Sir Edward Sabine.[39] Sabine, an artillery officer, was interested in astronomy, ornithology, and especially terrestrial magnetism. As astronomer to the Arctic expedition of Commander John Ross in 1818 seeking the northwest passage, he managed to indulge his passion for ornithology writing a paper on the birds of Greenland that included a new species, now known as Sabine's Gull. Sabine was intensely interested in terrestrial magnetism and, like Murchison, was deft at the politics of science. He promoted a world-wide effort to gather terrestrial magnetic observations dubbed the "magnetic fever" or "magnetic crusade" by scientific wags. In Germany, Gauss and Weber had overseen the establishment of a 16-station net of magnetic observatories stretching from Dublin to St. Petersburg on an east-west axis and from Uppsala to Catania in the north-south direction. This engendered a certain amount of magnetic chauvinism, and Sabine and his colleagues politicked the government and the Royal Society for a similar set of observatories in the British colonies for both magnetic and meteorological observations. They coordinated their scheme with the Germans and a final letter from Alexander von Humboldt to British officials launched the magnetic crusade.

As the observations began to flow in, Sabine organized a staff at Woolwich for data analysis, but he soon had bigger plans. The King's Observatory at Kew, adjoining Kew Gardens and erected through the largess of George III, had been unused since 1839 and Sabine wanted the facility. He envisioned making Kew into the basic geophysical observatory for the empire. It would provide standardized data and equipment for the colonial observatories. He placed the issue before the Royal Society, but Herschel objected saying the observatory would be tied too closely to a particular objective, and the Society turned down the offer of Kew in 1842. Undaunted, Sabine went to the British Association, which at his behest acquired the facility the same year, managing it until 1871 when a more receptive, or perhaps chastened, Royal Society accepted it. Kew quickly became a center for testing scientific instruments and this was, of course, right up Galton's alley. He was chiefly responsible for selecting the methods and instruments to be scrutinized and sextants, compasses, thermometers, watches, telescopes, field glasses, and photographic lenses were all put through their paces. He went about this work with his usual enthusiasm, devising an apparatus that allowed rapid and accurate verification of thermometers and suggesting a method of rating the performance of watches so successful that "a Kew certified watch has a special and recognised value, and the makers of valuable watches are far more on their mettle than they used to be."[40] In 1895 he was appointed chairman of the Management Committee, but he shrank from financial matters preferring to leave them to his colleague, General Sir Richard Strachey.[41]

In looking back over his life Galton regarded the period from 1853, the year of his marriage, to 1866 as being of unusual significance.

> This interval of thirteen years occupies a fairly well defined part of my life owing to two reasons, namely, that my scientific interests during the latter half became concentrated on heredity, and because it was in 1866 that my health suffered a more serious breakdown than had happened to it before. During the whole of this interval I find from old diaries that I frequently suffered from giddiness and other maladies prejudicial to mental effort, but that I invariably became well again on completely changing my habits as by touring abroad and taking plenty of out-of-doors exercise.[42]

His breakdown, just before his 44th birthday, bore strong resemblance to the one he had at Cambridge. His worried mother wrote admonishing him that "I am quite unhappy about your health. Do take your mother's advice and give up all writing and all head work for a year. Remember you have not got a strong constitution and overworking your mind falls upon your bodily health."[43] She folded him verbally under her wings. "Come to me whenever you like and you shall have perfect quiet." Sister Bessy chimed in advising her brother to take a good long holiday so he could "fully reap the benefit in the autumn and be brisk and well for the British Association. Your heredity will also be better by returning to it with a fresh eye and refreshed mind."[44] And Bessy put in a plug for his wife. "It will do Louisa good also."[45]

The vacation went well at first. The Galtons made an extended trip to the magnificent lakes of northern Italy, but returning to England in June to visit the Lake District proved an error, as sheets of windswept rain from a lowering sky pummeled their windows and kept them indoors. At the end of August

1866, Galton, a general secretary, left for the British Association meeting at Nottingham still in a precarious condition. He was scheduled to read a paper on the conversion of wind charts to passage charts for sailing ships in the section on Mathematical and Physical Sciences. However, he felt under such strain that he left the meeting abruptly after having arranged to have his paper read for him. This time he went off to find shelter in sister Delly's house at Edgemead where he was treated by Dr. Jephson, whose prescriptions only seemed to make him weaker.[46] Eventually he was well enough to return home in September for few weeks. Afterwards until the end of January 1867, the Galtons lodged in Hastings, where Galton rode constantly. From there they went to Rome, Naples, Venice, and, via the South Tyrol, to Switzerland and Germany, avoiding England until October. Louisa noted in her *Annual Record* that Galton was still not well early in 1868. They spent the winter at home dining out rarely and probably got on each other's nerves, but that summer they spent a couple of months in Switzerland. By early 1869 Galton was feeling much better, but now it was Louisa's turn, as her "digestion was very troublesome till June and a great hindrance to my doing much."[47] Galton, having spent almost two years recovering from his state of mental imbalance, realized that for his own good he would have to be more careful in the future:

> The warning I received in 1866 was more emphatic and alarming than previously, and made a revision of my mode of life a matter of primary importance. Those who have not suffered from mental breakdown can hardly realise the incapacity it causes, or, when the worst is past, the closeness of analogy between a sprained brain and a sprained joint. In both cases, after recovery seems to others to be complete, there remains for a long time an impossibility of performing certain minor actions without pain and serious mischief, mental in the one and bodily in the other.[48]

PART THREE

The Triumph of the Pedigree

TWELVE

Hereditary Talent and Character

> I am inclined to agree with Francis Galton in believing that education and environment produce only a small effect on the mind of anyone, and that most of our qualities are innate.
>
> —Charles Darwin[1]

For Francis Galton, approaching middle age, the publication of Darwin's *On the Origin of Species* in 1859 "made a marked epoch in my own mental development, as it did in human thought generally."[2] He "devoured its contents and assimilated them as fast as they were devoured, a fact which may be ascribed to an hereditary bent of mind that both its illustrious author and myself have inherited from our common grandfather, Dr. Erasmus Darwin."[3] Galton was encouraged to investigate topics that had long interested him, which "clustered round the central topics of Heredity and the possible improvement of the Human Race."[4] From the outset Galton seemed to have been convinced that nature, and not nurture, determined human ability, but how was he to show it? He hit upon a fairly simple device, the pedigree, one that would remain an analytical mainstay for the rest of his life. He reasoned that if ability was determined by nature, a great man's closest male relatives were the most likely to exhibit exceptional qualities, with ability diluting out with hereditary distance. Women were omitted in his analysis because his Victorian mindset viewed notable achievement as principally a male prerogative. This had certainly been true in Great Britain and elsewhere until that time, largely because opportunities for female advancement beyond the home were virtually absent.

Galton's first statement on the subject, "Hereditary Talent and Character,"[5] was a two-part article published in *Macmillan's Magazine* in 1865, which opened a debate that continues to this day on the heritability of intelligence. In choosing *Macmillan's* Galton showed he intended to reach a wide, intellectually challenging audience.[6] By the time his article was published in volume 12, the magazine had many distinguished contributors including Matthew Arnold and Herbert Spencer, and prominent scientists like Huxley and Lyell. Tennyson, Henry Wadsworth Longfellow, and Christina Rossetti, the "High Priestess of Pre-Raphaelitism," published poetry in *Macmillan's*. Richard Blackmore, who would later write *Lorna Doone*, pleaded successfully with Macmillan to serialize *Craddock Nowell: A Tale of the New Forest*, while Henry Kingsley's second Australian novel, *The Hillyars and the Burtons: A Story of Two Families*, was appearing in monthly installments. The nonfiction articles published in *Macmillan's* ranged widely dealing with topics as diverse as Buddha and Buddhism, American humorous poetry, the Suez and de Lesseps, American protectionism, and the natural history of oysters. Thus *Macmillan's*, with its great breadth of coverage, would be an excellent vehicle for Galton's message.

He recognized he was proposing a heretical idea which would probably shock most of his readership. While most would agree that physical and some mental traits were inherited in animals, they were unprepared to acknowledge this to be true of human beings. The thesis Galton promoted was that human talent and character differed little from the more mundane traits discussed by Darwin to illustrate the selection and breeding of domestic animals and cultivated plants. They should therefore be subject to selection themselves. One imagines he would have noted this statement from Darwin's book. "We cannot suppose that all the breeds were suddenly produced as perfect and as useful as we now see them; indeed in many cases, we know that this has not been their history. The key is man's power of accumulative selection: nature gives successive variations; man adds them up in certain directions useful to him."[7]

Galton's belief in the heritability of talent and character was reinforced not only by his own distinguished pedigree, but by Louisa's, and "by many obvious cases of heredity among the Cambridge men who were at the University about my own time."[8] To establish pedigrees for men of accomplishment, Galton examined works like *The Million of Facts* by Sir Thomas Phillips. From this he culled a select biography of 605 notable persons who lived between the years 1453 and 1853. He exulted because there were 102 notable relationships for a frequency of 1 in 6. He extended this analytical method to other lists and biographies, concluding that no less than eight out of every hundred sons of distinguished men were of equal eminence. Despite a strong prejudice in nature's favor, Galton acknowledged that nurture might also play a role, since the son of a great man "will be placed in a more favourable position for advancement, than if he had been the son of an ordinary person."[9] For comparison he tried estimating the frequency of men of ability in the population as a whole by rough determination of the number of students educated in Europe during the four preceding centuries. He calculated that only 1 in 3,000 of these "randomly" selected individuals achieved eminence, concluding that "everywhere is the enormous power of hereditary influence forced on our attention."[10]

The second part of Galton's article was a discursive and rambling attempt to build upon his "demonstration" that talent is heritable. Here he began developing the notion that selective breeding could be used to enhance a "caste" having advantageous qualities, but to discourage propagation of a second caste with less desirable qualities. These notions were later to be embodied in the concepts of positive and negative eugenics. One of the most remarkable ideas elaborated in this paper, for which no scientific justification was presented, was that the embryos of the next generation sprang forth from the embryos of the preceding generation.[11,12] This anticipated by almost 20 years August Weismann's experimentally supported theory of the continuity of the germ line.[13] This theory, central to modern biology, assumes that little passes between parent and child except that which is contained in the sperm and egg, leaving scant room for hereditary transmission of acquired characteristics unless, by some mysterious process, these congregate in the germ cells (see also chapter 13). Galton extended his view of the paramount role of heredity to racial differences having "collected numerous instances where children of low race have been separated at an early age from their parents, and reared as part of the settler's family, quite apart from their own people. Yet, after years of civilized ways ... they have abandoned their home, flung away their dress, and sought their countrymen in the bush, among whom they have subsequently been found living in contented barbarism without a vestige of their gentle nurture."[14]

Galton thought highly of his own handiwork. Over 40 years later in his autobiography he wrote that "on re-reading these articles ... considering the novel conditions under which they were composed ... I am surprised at their justness and comprehensiveness."[15] Karl Pearson agreed. It "is really an epitome of the great bulk of Galton's work for the rest of his life; in fact all his labours on heredity, anthropometry, psychology and statistical method seem to take their roots in the ideas of this paper. It might almost have been written as a résumé of his labours after they were completed, rather than as a prologue to the yet to be accomplished."[16] But the article evoked hardly a blip on the contemporary radar screen. He sent his friend Frank Buckland, a popular writer on natural history, an advance copy and Buckland thanked him profusely, saying his theory was "most excellent."[17] Galton was particularly pleased by one citation to "Hereditary Talent and Character," as it came in Darwin's book *The Variation of Animals and Plants under Domestication* (1868).[18]

The *Macmillan's* papers were the precursors for Galton's book *Hereditary Genius* (1869). There he used the same general method of gathering data on a much grander scale and applied the "bell curve" as an evaluative technique for the first time. He had been introduced to "the Gaussian Law of Probable Error" by his old friend William Spottiswoode.[19] In 1861 Spottiswoode published a paper in which he attempted to fit a normal curve to the distribution of direction of orientation of 11 mountain ranges to see whether they corresponded to a common "type."[20] Not unexpectedly, the fit was questionable, but the enthusiastic Spottiswoode concluded the agreement "although not perfect" was sufficient to conclude that the directions of the mountain ranges were "not accidental, and that the geologist and the physical philosopher will at least have good grounds for seeking some common agency which has caused their upheaval."[21] Galton was undoubtedly aware of Spottiswoode's paper and, when Spottiswoode explained the normal curve to his friend, he was delighted by the "the far-reaching application of that extraordinarily beautiful law which I fully apprehended."[22]

Galton now familiarized himself with the work of the Belgian scientist Adolph Quetelet who first applied the normal distribution to human measurements. Although Quetelet was the Astronomer Royal of Belgium, he gained international reputation not so much for astronomy, but as a statistician and population biologist.[23] In his first major attempt to fit the normal distribution to human data, Quetelet used published data on chest measurements taken from 5,738 Scottish soldiers to calculate the proportion of soldiers in each size class.[24] He estimated the expected probability for each size class using a symmetric binomial distribution. The agreement between the two distributions impressed Galton who, like Quetelet, was an incorrigible bean-counter always searching for the proper analytical tool with which to interpret his results.

In *Hereditary Genius* Galton used two systems of classification. The first categorized men by reputation, the method he had begun to develop in "Hereditary Talent and Character." He gathered data from the 1865 edition of *Dictionary of Men of the Time*, a biographical handbook. Since many of these individuals were past middle age, he decided to take as his baseline eminent men over 50 and compared them with men of similar age from the British population as a whole. He also employed the *Times* obituary list for 1868 to determine the number of eminent men who died during the previous year, choosing to exclude "old men who had earned distinction in years gone by, but had not shown themselves capable at later times to come again to the front."[25] Lastly, he consulted obituaries from many years back. Miraculously, each estimate gave approximately the same proportion of eminent men in the British population, 1 in 4,000. Thus, he had established a baseline against which he could compare his eminent men and their families.

Galton's second system classified men according to their performance on examinations. From a Cambridge mathematics examiner, he obtained sets of marks given over a period of two years for mathematical honors exams and compared their distributions. He found that mathematical ability was distributed over a wide range although the data showed a very distinct skew towards the low end. Next he introduced the normal curve (Fig. 12-1), citing Quetelet's

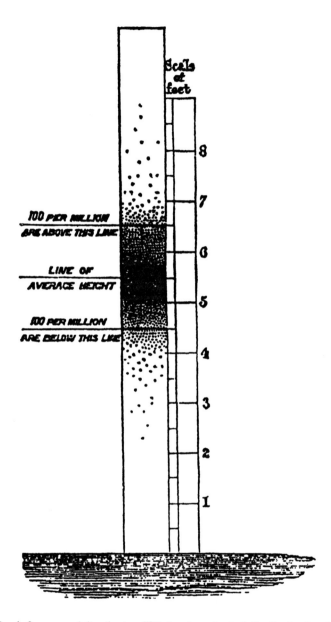

Fig. 12-1 Galton's first normal distribution. This is a hypothetical distribution in which Galton imagines that a million men have stood in turn with their backs to a board and that their heights have been dotted off on it. From Francis Galton, *Hereditary Genius*, second American edition. New York: Appleton, 1879, 28.

data on the Scottish soldiers and on the heights of French conscripts to show that observation fit expectation. But how could he apply the normal distribution to data for mental ability? There were no measurements of this faculty unless one believed in craniometry. As a proxy he used examination marks for admission to the Royal Military College at Sandhurst in 1868. Inspection of the data revealed a clear fit to the normal distribution at the upper tail of the curve and in its center, but for the dunces getting low scores there were no numbers as they had either eschewed competition or been "plucked." Assuming the male population of the United Kingdom to be 15 million, he next employed his earlier figure for the frequency of men of reputation (ca. 1 in 4,000) to establish both his highest and lowest (idiots and imbeciles) grades of natural ability and used a Gaussian distribution to calculate the expected number of individuals in each of twelve classes intermediate between the high and the low ends (Table 12-1).

CLASSIFICATION OF MEN ACCORDING TO THEIR NATURAL GIFTS.

Grades of natural ability, separated by equal intervals.		Proportionate, viz. one in	In each million of the same age.	Numbers of men comprised in the several grades of natural ability, whether in respect to their general powers, or to special aptitudes. In total male population of the United Kingdom, viz. 15 millions, of the undermentioned ages:—					
Below average.	Above average.			20—30	30—40	40—50	50—60	60—70	70—80
a	A	4	256,791	651,000	495,000	391,000	268,000	171,000	77,000
b	B	6	162,279	409,000	312,000	246,000	168,000	107,000	48,000
c	C	16	63,563	161,000	123,000	97,000	66,000	42,000	19,000
d	D	64	15,696	39,800	30,300	23,900	16,400	10,400	4,700
e	E	413	2,423	6,100	4,700	3,700	2,520	1,600	729
f	F	4,300	233	590	450	355	243	155	70
g	G	79,000	14	35	27	21	15	9	4
x all grades below g	X all grades above G	1,000,000	1	3	2	2	2	—	—
On either side of average . .			500,000	1,268,000	964,000	761,000	521,000	332,000	149,000
Total, both sides			1,000,000	2,536,000	1,928,000	1,522,000	1,042,000	664,000	298,000

The proportions of men living at different ages are calculated from the proportions that are true for England and Wales. (Census 1861, Appendix, p. 107.)

Example.—The class F contains 1 in every 4,300 men. In other words, there are 233 of that class in each million of men. The same is true of class f. In the whole United Kingdom there are 590 men of class F (and the same number of f) between the ages of 20 and 30; 450 between the ages of 30 and 40; and so on.

Table 12-1 Classification of Men According to Their Natural Gifts

But other than the Sandhurst examination marks, Galton had no way to determine whether ability in the population actually fitted a normal distribution, so he returned to his first classification method using reputation to measure ability. This allowed him to investigate whether ability had a heritable component. He asked rhetorically whether reputation was "a fair test of natural ability? It is the only one I can employ—am I justified in using it? How much of a man's success is due to his opportunities, how much to his natural power of intellect?"[26] To ward off the objection that "opportunity" (i.e., nurture) was a significant component, Galton made three points. First, a man of natural ability would succeed even if brought up under humble circumstances. Conversely, a man of moderate ability would be unlikely to achieve eminence even if raised with great social advantages. This argument was undergirded with a strong hereditarian assumption. Second, while culture was more widespread in America than in England and education of the lower and middle classes more advanced, "America most certainly does not beat us in first-class works of literature, philosophy, or art. The higher kind of books, even of the most modern date, read in America are principally the work of Englishmen. The Americans have an immense amount of the newspaper-article-writer, or of the member-of-congress stamp of ability; but the number of their really eminent authors is more limited even than with us."[27] Third, Galton compared sons of eminent men with adopted sons of Popes and other dignitaries of the Roman Catholic Church and thus anticipated the future use of adoption studies to study the heritability of intelligence. He asked "are, then, the nephews, etc., of the Popes, on the whole as highly distinguished as are the sons of other equally eminent men? I answer decidedly not."[28]

Having argued that reputation is a measure of natural ability, Galton was ready to analyze pedigrees of well-known statesmen, peers, military commanders, etc. English judges led off. This was appropriate as Galton had a useful reference, the *Lives of the Judges* by Foss, which covered the Judges of England from the Restoration in 1660 to 1865. The section on Judges was also the place where Galton honed his analytical tools for picking the eminent man and determining which male relatives were also eminent. His key assumption was that those male relatives most closely related to the eminent man (i.e., fathers, sons, brothers) were the most likely to be eminent with the probability of eminence decreasing with hereditary distance (e.g., uncles, grandfathers, grandsons). Although the pedigrees also contained information on female relatives, women were largely excluded from the analysis because Galton felt he could not compare "relations in the first degree of kinship—namely, fathers with mothers, sons with daughters, or brothers with sisters, because there exists no criterion for a just comparison of the natural ability of the different sexes."[29]

Once Galton had collected his raw data he conformed the observations from different generations and groups so they could be compared. He did this by taking the number of eminent men (column A, Table 12-2), in this case for 85 families, and multiplied each number by approximately 1.18 (there were some small arithmetical errors) to adjust his results to 100 families (column B, Table 12-2). But to calculate the percentage of eminent men among fathers, sons, brothers, etc., he needed a denominator that could be divided into the figures in column B. This was easy for fathers, grandfathers, and great-grandfathers since they form a geometric progression. A hundred eminent men have a hundred fathers, two-hundred grandfathers, and four hundred great-grandfathers (column C, Table 12-2). However, sons, brothers, uncles, and nephews vary in number. How was Galton going to solve this problem? Based on the data available to him he made a series of assumptions. For instance, he calculated that his families consisted "on average of no less than 2 1/2 sons and 2 1/2 daughters each consequently each judge has 1 1/2 brothers and 2 1/2 sisters."[30] That is, "100 judges are supposed to have 150 brothers and 250 sisters,

Degrees of Kinship.					A.	B.	C.	D.	E.
Name of the degree.	Corresponding letter.								
1 degree { Father	22 F.	—	—	—	22	26	100	26·0	9·1
Brother	30 B.	—	—	—	30	35	150	23·3	8·2
Son	31 S.	—	—	—	31	36	100	36·0	12·6
2 degrees { Grandfather	7 G.	6 g.	—	—	13	15	200	7·5	2·6
Uncle	9 U.	6 u.	—	—	15	18	400	4·5	1·6
Nephew	14 N.	2 n.	—	—	16	19	400	4·75	1·7
Grandson	11 P.	5 p.	—	—	16	19	200	9·5	3·7
3 degrees { Great-grandfather	1 GF.	1 gF.	0 GF.	0 gF.	2	2	400	0·5	0·2
Great-uncle	1 GB.	2 gB.	0 GB.	0 gB.	3	4	800	0·5	0·2
First-cousin	5 US.	2 uS.	1 US.	1 uS.	9	11	800	1·4	0·5
Great-nephew	7 NS.	1 nS.	7 NS.	0 nS.	15	17	800	2·1	0·7
Great-grandson	2 PS.	2 pS.	1 PS.	0 pS.	5	6	400	1·5	0·5
All more remote	…	…	…	…	12	14	?	0·0	0·0

A. Number of eminent men in each degree of kinship to the most eminent man of the family (85 families).
B. The preceding column raised in proportion to 100 families.
C. Number of individuals in each degree of kinship to 100 men.
D. Percentage of eminent men in each degree of kinship to the most eminent member of distinguished families; it was obtained by dividing B by C and multiplying by 100.
E. Percentages of the previous column reduced in the proportion of (286 − 24,[1] or) 242 to 85, in order to apply to families generally.

---

[1] That is to say, 286 Judges, less 24, who are included as subordinate members of the 85 families.

Table 12-2

PERCENTAGE OF EMINENT MEN IN EACH DEGREE OF KINSHIP TO THE MOST GIFTED MEMBER OF DISTINGUISHED FAMILIES.

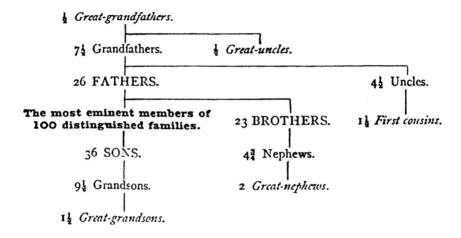

Fig. 12-2 Combined pedigrees for judges showing the percentage of eminent men in each degree of kinship to the most gifted member of these distinguished families. From Francis Galton, *Hereditary Genius*, second American edition. New York: Appleton, 1879, 83.

and each brother and each sister to have, on the average, only one son; consequently the 100 judges will have (150+250, or) 400 nephews."[31] He divided the numbers he got (column C, Table 12-2) into the corrected number of eminent persons in each category (column B, Table 12-2) and multiplied by 100 to get the percentages shown in column D.

Galton's results seemed consistent with his hypothesis (Fig 12-2). He concluded that a close relative of the eminent man had a much higher probability of being eminent than one who was remote. "Speaking roughly, the percentages are quartered at each successive remove, whether by descent or collaterally."[32] This meant that there was an "average increase of ability in the generations that precede its culmination, and as regular a decrease in those that succeed it."[33] So after "three successive dilutions of the blood, the descendants of judges appear incapable of rising to eminence."[34] His explanation was that an able man had to "inherit three qualities that are separate and independent of one another: he must inherit capacity, zeal, and vigour; for unless these three, or, at the very least, two of them, are combined, he cannot hope to make a figure in the world. The probability against inheriting a combination of three qualities not correlated together is necessarily in a triplicate proportion greater than it is against inheriting any one of them."[35] So Galton had made a novel prediction arguing that "capacity, zeal and vigour" segregate like genetically independent traits. This may help to explain why he became so infatuated with Darwin's "Provisional Hypothesis of Pangenesis" (chapter 13). Pangenesis, in contrast to the popular blending or "paint pot" hypotheses of the day, supposed that particulate elements determined the inheritance of different traits. Ability, Galton believed, was a complex trait dependent on several hereditary elements whose behavior was not correlated.

Galton knew he must dispose of the conundrum of parental influence before he could proceed further. That is, a father aids his son in garnering a plum position and Galton scores the boy as eminent although his opportunities are enhanced by his environment and not necessarily because of a sterling hereditary endowment. To counter this objection Galton told a complex story about the Norths and the Montagues where the tendrils of influence intertwined around the pillar of ability so all encompassingly that one was difficult to separate from the other. However, Galton, pedigree and notes in hand, attempted the delicate job of disentangling the two. His second argument was that heredity must play the dominant role since ability was more frequent in near than distant kin of an eminent man.

Most other categories of eminent men were analyzed similarly. The pedigrees were not confined to Englishmen. Among commanders, for instance, one finds Caesar, Charlemagne, Bonaparte, and Hannibal. In one chapter Galton warned of the hazards of being made a peer to one's fecundity.[36] Lord Campbell in his *Lives of the Chancellors*, the Lord Chancellors being the highest judicial officers in Great Britain, observed that when he was first acquainted with the English Bar, half of the judges had married their mistresses, since when a barrister was elevated to the Bench he was expected either to marry his mistress or give her up. Hence, half the judges had no legitimate offspring and either married their girlfriends when both were getting on in age or discarded them. What puzzled Galton was that Lord Campbell's observation implied that judges should have small families while his own research indicated just the opposite. As he dug through his data he stumbled on something surprising. Being elevated to a peerage was a mixed blessing for a judge, as 12 of 31 peerages he examined had become extinct. Why had they? Galton examined his notes and "found a very simple, adequate, and novel explanation . . . stare me in the face."[37] Many new peers married heiresses. Although they were titled and perhaps had "a sufficient fortune to transmit to their eldest son" they needed additional funds "for the endowment of their younger sons and their daughters. On the other hand, an heiress has a fortune, but wants a title. Thus the peer and heiress are urged by the same issue of marriage by different impulses."[38] The reason why such marriages were peculiarly unproductive of children was "that an heiress, who is the sole issue of a marriage, would not be so fertile as a woman who has many brothers and sisters. Comparative infertility must be hereditary in the same way as other physical attributes, and I am assured it is so in the case of domestic animals."[39]

So marriage to an heiress, while financially advantageous, brought with it the potential incubus of a barren union. In the Additional Notes to his epic poem, "The Temple of Nature," Galton's grandfather Erasmus Darwin made a similar point using somewhat different reasoning. "As many families become gradually extinct by hereditary diseases, as by scrofula, consumption, epilepsy, mania, it is often hazardous to marry an heiress, as she is not unfrequently the last of a diseased family."[40]

The religious press was critical of *Hereditary Genius* and it is easy to understand why. He concluded that most Divines

> are not founders of families who have exercised a notable influence on our history, whether that influence be derived from abilities, wealth, or social position of any of their members. That they are a moderately prolific race, rather under, than above the average. That their average age at death is a trifle less than that of the eminent men comprised in my other groups. That they commonly suffer from over-work. That they have usually wretched constitutions. That those whose constitutions are vigorous, were mostly wild in their youth; and conversely, that most of those who had been wild in their youth and did not become pious till later in life, were men of vigorous constitutions. That a pious disposition is decidedly hereditary. That there are also frequent cases of sons of pious parents who turned out very badly.[41]

The reason why the children of Divines often turned out poorly was that, while the parents were "naturally gifted with high moral characters combined with instability of disposition," "these peculiarities" were not correlated. Therefore, a child would often "inherit the one and not the other. If his heritage consist of the moral gifts without great instability, he will not feel the need of extreme piety; if he inherits great instability without morality, he will be very likely to disgrace the name."[42] Galton dismissed Divines, concluding that they were not "an exceptionally favoured race in any respect; but rather, that they are less fortunate than other men."[43]

While completing *Hereditary Genius*, Galton used his data on Divines for an article entitled "Statistical Inquiries into the Efficacy of Prayer,"[44] which he shipped off to the *Fortnightly Review*. This periodical, established in 1865, was, according Anthony Trollope, one of its founders, a forum that would allow any man "who had a thing to say and knew how to say it, speak freely. But he should always speak with the responsibility of his name attached."[45] In the arena of science the *Fortnightly* contained popular articles on topics as diverse as the nature of rainbows, atoms, and force. It also presented advances in medicine and meteorology. Darwin's theory of natural selection also held a prominent place in scientific discussions in the *Fortnightly*. Despite its reputation for openness, George Lewes, the well-respected critic, writer, and first editor of the *Fortnightly*, found Galton's article too hot to handle. He wrote Galton that if he owned the *Fortnightly* he would not hesitate to publish the paper, but it would so offend his Christian proprietors that he had to turn it down as the manuscript was "too terribly conclusive and offensive not to raise a hornet's nest."[46] After a couple of more rejections Galton set his paper aside to gather dust until 1872 when he resubmitted his manuscript to the *Fortnightly*, whose new editor, John Morley, accepted it.

Galton began by trying to demolish the argument that prayer must be efficacious because it is so generally used. "The argument of universality either proves too much, or else it is suicidal. It either compels us to admit that the prayers of Pagans, of Fetish worshippers, and of Buddhists who turn praying-wheels, are recompensed in the same way as those of orthodox believers; or else the general consensus proves that it has no better foundation than the universal tendency of man to gross credulity."[47] Having washed his hands of universality Galton addressed the efficacy of prayer directly: "Are prayers answered or are they not?"[48] Stripped of its Victorian gentility, Galton's article took an in-your-face approach beginning with the ill. Did they recover more rapidly on average if they prayed or were prayed for? "There is not a single instance, to my knowledge, in which papers read before statistical societies have recognised the agency of prayer either on disease or anything else."[49] He hammered his point home, observing that medical men failed to use prayer in trying to cure people. "Had prayers for the sick any notable effect ... doctors, who are always on the watch for such things, should have observed it, and added their influence to that of the priests towards obtaining them for every sick man."[50] What about life insurance? Insurers make lots of pointed inquiries, but do they ever ask how much the prospective client prays? You bet they don't. What about kings and queens? Did public prayer for the royals really help increase their longevity? This time Galton produced real data from an article in the *Journal of the Statistical Society* demonstrating that "sovereigns are literally the shortest-lived of all who have the advantage of affluence."[51] So it went for case after case.

Galton wrote that many articles "of ancient faith have been successively abandoned by the Christian world to the domain of recognised superstition."[52] Witches were one example. But he raised a cautionary finger. He had not argued that praying would fail to make a person feel better, nor had he said anything about the degree to which a man can communicate with his God. For scientists like himself he sketched a silver lining to the clouds of doubt hanging over God's existence. They were not excluded from the "confident sense of communion" and well-being a believer could muster, since the beauty of understanding the physical laws, among which Galton included hereditary influence, "may not equally rejoice the heart, but it is quite as powerful in ennobling the resolves, and it is found to give serenity during the trials of life and in the shadow of approaching death."[53]

The article might have engendered little commentary had not an anonymous writer for the *Spectator* quickly penned a withering critique published on August 3. Evidently confusing Francis Galton with his cousin Captain Douglas Galton, the author pointed to the hidden agenda he suspected scientists like Galton harbored, who argued "that if prayer is not answered, and cannot be answered, then there is in the Christian, or rather the religious, sense of the word no God."[54] But "we are not bound to submit patiently to arguments such as those by which Captain Francis Galton . . . thinks he has disposed of the efficacy of Prayer."[55] He quickly summarized Galton's evidence "which we will not dispute" referring to Galton's argument as "a direct attempt to weight mental consequences in a pair of brass scales."[56] Then he pitched in with his own counterarguments, ending his attack by recognizing the enduring power of the cross. "If the absence of protection for churches from lightning and of kings from early death are proofs that prayer is useless, then the victory of Christianity and durability of the Popedom are greater, because more certain and visible proofs that prayer is useful."[57]

The *Spectator* article provoked such a torrent of correspondence that only a selection of letters could be printed. On August 17 the magazine felt compelled to publish another piece that acknowledged "a heap of letters, all about prayer, sent us for publication in two days, which would fill, as nearly as we can calculate, sixteen pages of this journal."[58] But there was a curious thread in many letters. While most were "written, as was natural, from the supernatural side" a great many bore "a trace of feeling we had scarcely expected to find, a strong desire on the part of many persons who believe in a sentient God, and some of who are apparently Christians, to get rid of the difficulties of the subject by reducing without denying the efficacy of prayer."[59] Darwin, hugely entertained by the row his cousin's article caused, congratulated Galton on the "tremendous stir-up your excellent article on 'Prayer' has made in England and America."[60] Louisa was probably not amused, as she failed to mention her husband's paper in her *Record*.

Overall, Galton's results in *Hereditary Genius* seemed to support his thesis that talent and character were largely determined by nature as the approximately 300 families he had studied contained nearly 1,000 eminent men compared to the frequency of 1 in 4,000 he estimated for the population as a whole. Furthermore, the closer the kinship to the eminent man the higher the probability of distinction. To check whether he had weighted his results toward cases favorable to his hypothesis, he sought a set of eminent names gathered by an independent method embracing the list of the French philosopher Auguste Comte, the founder of the school of positivism. This was a clever idea for Comte, desirous of forming a "Religion of Humanity," selected a series of names he thought represented those to whom human progress was most indebted and incorporated them into the Comtist Calendar. The elite were assigned months, the next lower class weeks, and the third class days. Comte's calendar contained 13 months with each having four weeks. Galton was highly pleased with the degree of overlap between his list and Comte's, as Comte's list depended on perceived merit independent of heredity.

He also tried ascertaining the relative contributions of male and female lineages to the transmission of ability, reporting that the male line contributed 70 percent and the female line 30 percent except, of course, in the case of Divines where the reverse was true. He suggested that the explanation for this strong male bias was that "the aunts, sisters and daughters of eminent men do not marry, on the average, so frequently as other women."[61] He theorized that the underlying reasons were that these privileged ladies were "accustomed to a higher form of culture and intellectual and moral tone in their family circle, than they could easily find elsewhere" especially since "one portion of them would certainly be of a dogmatic and self-asserting type, and therefore unattractive to men" while "others would fail to attract, owing to their having shy, odd manners, often met with in young persons of genius."[62] This logic is, perhaps, more revealing of Galton and prevailing Victorian views about women than of his peculiar findings.

Galton tried generalizing from individuals to races but, lacking data, attempted logic. He compared "the negro race with the Anglo-Saxon, with respect to those qualities alone which are capable of producing judges, statesmen, commanders, men of literature and science, poets, artists, and divines."[63] He had earlier calculated a theoretical normal distribution that classified Englishmen according to their natural gifts and now stated that the curves for blacks and Anglo Saxons do not superimpose, but that the curve for blacks is shifted downward by "not less than two (of Galton's) grades . . . and it may be more."[64] One can't help but be reminded of two similar normal distributions of IQ for blacks and whites in *The Bell Curve*[65] (1994). They fail to superimpose because of a perceived 15 point mean IQ differential favoring whites. Furthermore, wrote Galton, an explorer "has to confront native chiefs in every inhabited place. The result is familiar enough—-the white traveller almost invariably holds his own in their presence. It is seldom that we hear of a white traveller meeting with a black chief whom he feels to be the better man."[66] And he again restated his observation that the proportion of half-witted blacks is very large. Thus did Galton extrapolate his results from individuals to races. This temptation to leap from trying to understand and explain actual data to the grand and sweeping generalization whose basis derives only from personal observation and prejudice has often been a hallmark of studies on genes, intelligence, and behavior.

In what would probably have been the book's last chapter, were it not for the publication of Charles Darwin's "Provisional Hypothesis of Pangenesis" (chapter 13), Galton marched grandly onwards to the natural abilities of nations. His theme was straightforward. The average age of marriage has a threefold effect on a population. Since those marrying young have larger families, produce more generations in a given period of time, and more generations are alive at the same time, the wisest policy is one that retards "the average age of marriage among the weak, and ... hastens ... it among the vigorous classes; whereas most unhappily for us, the influence of numerous social agencies has been strongly and banefully exerted in the precisely opposite direction."[67] In this statement Galton encapsulated an argument he would return to later in his own writings about eugenics and one which would be repeatedly voiced by eugenicists in the early twentieth century. He also excoriated the church once more, blaming it for the twin evils of the dark ages and for blighting the hereditary endowment of future generations because of celibacy requirements. These meant that men and women inclined to charity, meditation, literature, or the arts would often be childless, ensuring that "the rudest portion of the community" would be "the parents of future generations."[68] Equally serious for Europe's intellectual stunting were the religious persecutions that had brought thousands of the most able to the scaffold, to lengthy imprisonment depriving them of the opportunity to have children, or to attempt escape via emigration to more tolerant lands.

What was the contemporary reaction to Galton's book? "Frank's book not well received, but liked by Darwin and men of note"[69] was Louisa Galton's laconic comment to her diary. Indeed Darwin did like it, for on December 3, 1869 he wrote that he had

> only read about 50 pages of your book (to Judges), but I must exhale myself, else something will go wrong with my inside. I do not think I ever in all my life read anything more interesting and original—and how well and clearly you put every point! George [Darwin's son George Charles Darwin], who has finished the book, and who expressed himself in just the same terms, tells me that the earlier chapters are nothing in interest to the later ones! It will take me some time to get to these latter chapters, as it is read aloud to me by my wife, who is also much interested. You have made a convert of an opponent in one sense, for I have always maintained that, excepting fools, men did not differ much in intellect, only in zeal and hard work; and I still think this is an *eminently* important difference. I congratulate you on producing what I am convinced will prove a memorable work. I look forward with intense interest to each reading, but it sets me thinking so much that I find it very hard work; but that is wholly the fault of my brain and not of your beautifully clear style—Yours most sincerely, (signed) Ch. Darwin.[70]

Among the letters Galton received were two from Miss Emily Shirreff who, with her sister Maria Grey, was a pioneer in the cause of women's education in Great Britain.[71] She wrote fervently of the miserable social system existing in Victorian England that drove "women to marry for subsistence or position."[72] Fathers supposed that most of their daughters were willing to live in idleness "till a husband takes them off their hands ... while the abler, the more energetic, the most fit to be the mothers of a better generation will revolt against the injustice of our social arrangements, and struggle singly for an independent position; thereby sacrificing at once the interests of society and some of the highest cravings of their own nature." Emily Shirreff had made a key point that recurs repeatedly in the eugenics literature. Because they were ambitious, the fittest women eschewed marriage in favor of a career, thereby leaving production of the next generation to women less well endowed intellectually.

Hereditary Genius was widely reviewed in British newspapers and periodicals. The *Daily News* commented that "Mr. Galton undertakes to show, and to a large extent undoubtedly succeeds in showing, that genius is equally transmissible, and that ability goes by descent."[73] The *Times* was more critical, observing that "Darwinian theories are capable of infinite expansion" and Galton asserted that "mental and moral, as well as physical, phenomena may be controlled by their application."[74] The paper differed strongly with his view that heredity predominated in determining genius. "Mr. Galton is a little too anxious to array all things in the wedding garment of his theory, and will scarcely allow them a stitch of other clothing."[75] The long review in *Chambers's Journal* began flatteringly that "whoever likes a 'book with a purpose' will welcome Mr. Galton's work on *Hereditary Genius*."[76] But, as Galton later recognized, the writer correctly pointed out that genius was the wrong word, as he really meant talent. The reviewer also observed that ability appeared more frequently among descendants rather than progenitors of the eminent man, suggesting that he might "have stretched out to them a helping hand."[77] The *Morning Post* began skeptically that "no proposition is so extravagant as to be without some portion of truth" and concluded that "the author's statistics only recapitulate the numerous individuals who have distinguished themselves in every walk of life ... but they fail altogether in attempting to confirm the continuous descent of genius."[78] The *Saturday Review* took Galton to task for having "bestowed immense pains upon the empirical proof of a thesis which from its intrinsic nature can never be proved empirically."[79] He had spread "his net so largely" that he succeeded in securing "evidence which we can but characterize as largely mediocre," which pointed "with infinitely greater truth" to the influence of culture "than to anything of the nature of inherent genius following upon a strain of blood."[80] One of the most perceptive reviews was by the political economist Herman Merivale.[81] While acknowledging the role of heredity in determining ability, Merivale, writing in the *Edinburgh Review*, identified the central weakness in Galton's thesis. Using judges as an example, Merivale observed that some 100 out of the 250 eminent relatives tabulated by

Galton were lawyers themselves. This had little to do with the inheritance of "a special talent of the lawyer, but much to do with the ability of a judge to influence his son to enter a legal career."[82] Overall most reviewers felt that Galton had overstated the case for heredity while insufficiently emphasizing the role of environment.

Victorian scientists were the most receptive to Galton's book.[83] The codiscoverer of the theory of natural selection, Alfred Russel Wallace, wrote in *Nature* that many "who read it without the care and attention it requires and deserves, will admit that it is ingenious, but declare that the question is incapable of proof. Such a verdict will, however, by no means do justice to Mr. Galton's argument."[84] The religious press was negative, as was to be expected since Galton was quite comfortable treading on the soul. One can imagine Galton, but perhaps not Louisa, chuckling at scathing, but anonymous reviews in the *Catholic World* and the *British Quarterly Review*, a Congregationalist/Baptist journal of criticism.[85] Another group of reviews fell in the middle, finding Galton's work interesting and valuable, but criticizing the exclusiveness of Galtonian hereditarian views over social and educational factors.[86] How did Galton react to these criticisms? In the prefatory chapter to the 1892 reissue of the book, he commented that the "fault in the volume that I chiefly regret is the choice of its title of *Hereditary Genius*, but it cannot be remedied now. There was not the slightest intention on my part to use the word genius in any technical sense, but merely as expressing an ability that was exceptionally high, and at the same time inborn."[87]

Right or wrong, Galton had launched a revolutionary new theory into the public arena that propounded a strict hereditarian view of intellectual capacity, and with it a methodology that would become a mainstay in human genetics, pedigree analysis. When *Hereditary Genius* was reissued, unchanged except for a new preface, almost a quarter of century later in 1892, it was warmly praised in the popular press. As the *Nation* put it, when Galton first published *Hereditary Genius* "it was commonly believed that the human mind had something supernatural in it" and that "children were born similar in mental ability, subsequent differences being due to surroundings and training."[88] But Galton had set out to show "that individuals inherit different intellectual capacities" and irrespective of environmental influences, "nature limits the powers of the mind as definitely as those of the body. On these points, among thinkers everywhere, the author's opinions have prevailed."[89] The *Blackburn Standard* echoed this view, sternly warning fathers that heredity was a science they "should know something of, to aid them in determining what pursuits and careers their sons are most likely fitted for."[90] And the *National Reformer* approvingly chorused "what was a good book on its publication, is a good book still."[91] The *Daily Chronicle* wrote the epitaph for "the old notion of the 'freedom of the will,' which is still assumed in belated treatises," but which now "in conformity with the explanation of mental phenomena given by evolution, had been displaced by 'determinism,' or the doctrine that our actions are 'determined'; that fate, chance, and accident are as fully excluded from the operations of the mind as they are from those parts of the body and universe of which man is a part."[92] So Galton, solid as a rock, had stuck unwaveringly to his hereditarian conviction for a quarter of a century and popular opinion had bent round so far that *Hereditary Genius* was recognized as a prophetic classic.

THIRTEEN

Gemmules, Rabbits, Germs, and Stirps

> Speaking generally, most authors agreed that all bodily and some mental qualities were inherited by brutes, but they refused to believe the same of man. Moreover, theologians made a sharp distinction between the body and mind of man, on purely dogmatic grounds.[1]
>
> —Galton writing in his autobiography on the state of hereditary research in about 1868

Galton was putting the finishing touches on *Hereditary Genius* when he began to thumb through Darwin's new work *Variation of Animals and Plants under Domestication*. The first volume was replete with examples of the results of artificial selection in producing new animal breeds and cultivated plants, providing Galton with no great new insights, but the second volume was a different story. The first three chapters dealt with inheritance, the fourth with the laws of variation, and the last presented Darwin's "Provisional Hypothesis of Pangenesis." Darwin's inheritance chapters analyzed many puzzling phenomena associated with heredity, notably reversion, the occasional appearance of an ancestral character in a pure breeding strain, or the unexpected debut of a parental character in the progeny of hybrids. Galton began scribbling furiously, leaving marginal notes beside compelling passages.[2] But it was the pangenesis chapter that particularly excited him. The always methodical Darwin summarized the facts he felt he must account for, carefully laid out his hypothesis, examined its assumptions, and showed how it could account for the observations. Pangenesis was the next logical step in

Darwin's theory of evolution, for he needed to explain how the variations arose upon which natural selection acted. This question seriously concerned him, especially since most contemporary notions of inheritance involved blending of hereditary determinants.[3] The difficulty this "paint pot" view of heredity presented was that the variations on which natural selection was supposed to act would be lost. If a variant is likened to a few drops of black paint and the predominant form to a bucket of white paint, the variant will vanish when mixed (crossed) into the bucket. So how could the small changes upon which natural selection acts accumulate? Darwin assumed the hereditary determinants were particulate.

Darwin recognized that cells "or units of the body" increase by division and "while retaining the same nature become converted into the various tissues and substances of the body."[4] He assumed "that the units throw off minute granules which are dispersed throughout the whole system; that these, when supplied with proper nutriment, multiply by self-division, and are ultimately developed into units like those from which they were originally derived."[5] He named these hypothetical particles "gemmules," proposing that they were gathered from all parts of the organism "to constitute the sexual elements, and their development in the next generation forms a new being."[6] To account for reversion Darwin assumed the existence of dormant elements that might be expressed in future generations. This notion intrigued Galton and would become embedded in his own theory of heredity. However, the mechanism by which each "unit of the body casts off its gemmules" or how they were subsequently collected in the sex organs was unclear to Darwin. He assumed that in complex structures like feathers "each separate part is liable to inherited variations" so "each feather generates a number of gemmules; but it is possible that they may be aggregated into a compound gemmule."[7] But how did variations arise? He imagined two mechanisms. First, when the reproductive organs were "injuriously affected by changed conditions," gemmules from different parts of the body might fail to aggregate properly, so some were in excess while others were in deficit, resulting in modification and variation. Darwin's second mechanism assumed that gemmules could be modified "by the direct action of changed conditions." This caused the affected part of the body to "throw off modified gemmules, which are transmitted to the offspring."[8] Although Darwin postulated that exposure to modified environmental conditions had to occur over several generations for a change to become heritable, this was acquired characteristics pure and simple and this bothered Galton.

Given his penchant for quantification, Galton was immediately attracted to the notion that particles were the hereditary factors. Hence, he scrambled to add a chapter on the subject in *Hereditary Genius*, remarking approvingly that the hypothesis "gives excellent materials for the mathematical formulae, the constants of which might be supplied through averages of facts, like those contained in my tables."[9] Since his own data "were too lax to go upon,"[10] he constructed a hypothetical example in which he assumed a child acquires "one-tenth of his nature from individual variation and the remaining nine-tenths from his parents,"[11] after which he derived formulae to yield the predicted results. Fascinated by the quantitative possibilities of pangenesis, Galton wanted to put the hypothesis on a proper mathematical footing. But Huxley was concerned and, having donned his "sharpest spectacles and best thinking cap"[12] to analyze Darwin's hypothesis, flashed the amber caution light concerning its publication. George Lewes, the writer and critic living happily in sin with George Eliot, was just finishing his sober four-part critique of "Mr. Darwin's Hypothesis"[13] for the *Fortnightly*, when Darwin's *Variation* began appearing at the booksellers. He scurried to include pangenesis, begging the editor's indulgence, and apologized for having "already trespassed on the comparatively scanty space which the Review can afford."[14] However, for the sake of completeness he set down "a few words,"[15] actually several pages, on Darwin's hypothesis. Having acknowledged Darwin's formidable reputation and that pangenesis surpassed "all previous attempts in the same direction,"[16] Lewes couched his skepticism carefully. Prudence suggested "that an hypothesis carefully worked out by such a thinker should be criticised with something of a corresponding hesitation, and not dismissed if it fails to carry conviction with it at once."[17] But pangenesis did not have the advantages of natural selection and could not "hope for so ready an acceptance. It has the disadvantage of not being readily grasped, nor easily brought into confrontation with facts. It has the still greater disadvantage of being hypothetical throughout: not being one supposition put forward to harmonise a series of facts, but a series of suppositions, every one of which needs proof."[18]

Galton was less concerned and anxious to test his cousin's hypothesis. He wrote Darwin on December 11, 1869, asking where he might obtain pairs of rabbits "of marked and assured breeds" for experimental purposes, failing to mention he intended to test the hypothesis by transfusing blood from one strain into another.[19] Darwin probably guessed Galton's intent because in *Hereditary Genius*, Galton had interpreted his cousin's hypothesis as meaning that "each cell, having of course its individual peculiarities, breeds nearly true to its kind, by propagating innumerable germs, or to use his expression, geommules, which circulate in the blood and multiply there."[20] That is, Galton interpreted Darwin's hypothesis as meaning that the gemmules were transmitted in the bloodstream. This, of course, meant that gemmules in organisms lacking bloodstreams—such as plants—must transmit the particles by some other mechanism, or that a general mechanism for gemmule transmission existed that did not involve the bloodstream, a point that would later become a bone of contention between Darwin and Galton.

Galton chose a strain of pure-breeding rabbits called silver-greys to serve as recipients for blood transfused from donors having different characters, such as color. One or both parents were transfused and then crossed to see whether their progeny had inherited any of the characteristics of the blood donor. Since Galton was uncertain of his own manual dexterity, he secured the cooperation of Dr. Murie, prosector at the Zoological Gardens, who dissected animals in preparation for autopsy, anatomical research, and demonstration. Murie was a dab hand at his job, as Galton observed one day. "A dead cobra was lying on his table, and on my remarking that I had never properly seen a poison fang, he cooly opened the creature's mouth, pressed firmly at exactly the right spot, and out started that most delicate and wicked looking thing, with a drop of venom exuding from it just in front of his nail."[21]

Sometimes the experiments seemed to go well and other times not. On March 15, 1870, Galton ruefully related to Darwin that "my most hopeful doe was confined prematurely by three days having made no nest and all we knew of the matter was finding blood about the cage and the *head* of one of the litter."[22] Two days later he wrote that he hoped by using younger rabbits and technical improvements, "to get a great deal more of alien qualities into their veins." But the experiments were not yielding positive results. Darwin's wife, Emma, wrote her daughter, Henrietta that "F. Galton's experiments about rabbits . . . are failing, which is a dreadful disappointment to them both."[23] Galton reported a flicker of hope on May 12, having observed a white forefoot in one of the litters. Unfortunately, white foot later turned out to be a normal variant of rabbits of solid color. Then Galton had an idea. He had been defibrinizing the blood to prevent clotting and he began to suspect the fibrin fraction might be the gemmules' source. On June 25 he wrote Darwin that three males transfused with defibrinated blood were sterile. Maybe "the reproductive elements are in the portion of the blood I did *not* transfuse;—to wit the *fibrine*."[24]

But then vacation intervened and on July 15 the Galtons departed for Paris, getting as far as Grindelwald, Switzerland, when the Franco-Prussian War broke out. Wishing to avoid the hostilities they returned to England and bumped around the countryside until mid-October, when he resumed his experiments, having devised a way of getting around the clotting problem. He would exchange whole blood directly between rabbits. To do this, he put the rabbits breast to breast and connected their carotid arteries by a cannula. These "operations were exceedingly successful; the pulse bounded through the cannulae with full force"[25] although clot formation often blocked the blood flow so the procedure had to be repeated. Even so he was gaining confidence since the "experiments were thorough, and misfortunes were very rare. It was astonishing to see how quickly the rabbits recovered after the effect of the anaesthetic had passed away. It often happened that their spirits and sexual aptitudes were in no way dashed by an operation which only a few minutes before had changed nearly one half of the blood that was in their bodies."[26]

By mid-winter Galton had bred 124 offspring in 21 litters without a single "mongrel" appearing. On March 30, 1871, he reported his negative results before the Royal Society with his old friend from the Kew Observatory Committee, General Sir Edward Sabine, in the chair. The paper was published in *Proceedings of the Royal Society*.[27] Galton wrote that his aim was to test his cousin's hypothesis experimentally. He recapitulated its main points, making the reasonable, but crucial interpretation that the hereditary units, the gemmules, "swarm in the blood, in large numbers of each variety, and circulate freely with it."[28] He deduced from Darwin's "reasoning and illustrations that two animals, to outward appearance of the same pure variety, one of which has mongrel ancestry and the other has not, differ solely in the constitution of their blood, so far as concerns those points on which outward appearance depends."[29] Hence, "the gemmules in each individual must therefore be looked upon as entozoa of his blood, and, so far as the problems of heredity are concerned, the body need be looked upon as little more than a case which encloses them, built up through the development of some of their number."[30]

Based on this interpretation the clear experimental test of the hypothesis was to transfuse blood from a "mongrel" strain into a purebred strain, in this case the silver-grey, and to look for the appearance of "mongrel" traits among the progeny. Judging from the correspondence, Darwin not only seemed to go along with this but was anxious the experiments succeed. They did not so Galton concluded the pangenesis hypothesis was wrong and he said so in unmistakable terms in the introduction to his paper. "I have now made experiments of transfusion and cross circulation on a large scale in rabbits, and have arrived at definite results, negativing, in my opinion, beyond all doubt, the truth of the doctrine of Pangenesis."[31] But in his discussion he temporized, writing that the "conclusion from this large series of experiments is not to be avoided, that the doctrine of Pangenesis, pure and simple, as I have interpreted it, is incorrect."[32] Too bad that he didn't say "as I have interpreted it" at the beginning of his paper, as it was the introductory sentence that Darwin latched onto and he was uncharacteristically angry. Here was his cousin, who had reported his experimental results faithfully to him for months, dashing off a paper in the *Proceedings* saying pangenesis was hogwash without even the courtesy of asking Darwin's opinion. Oddly enough, Galton apparently thought Darwin would not mind having his hypothesis publicly mutilated without warning. He wrote George Darwin a week after his paper was read concerning additional rabbit experiments. He said the manuscript would be published in the next number of the *Proceedings* and added breezily that he would send a copy to Darwin with the pedigree marked of the rabbits he wanted Darwin to care for.[33]

Darwin had already taken flack about his hypothesis so he was sensitive to criticism.[34] Galton's paper was the last straw and he obviously made this clear to his cousin, who apologized on April 25. "I am grieved beyond measure that I have misrepresented your doctrine, and the only consolation I can feel is that your letter to 'Nature' may place that doctrine in clearer light and attract more attention to it."[35] Although Galton was anxious to get his letter in the morning post, he took the time to pinpoint the sentences that led him astray in hopes that Darwin would mention them in *Nature*. "In 'Domestication of Animals etc.' p. 374 . . . 'throw off minute granules or atoms, which circulate freely throughout the system. . . .' And p. 379 ' . . . the granules must be thoroughly diffused; nor does this seem improbable considering . . . the steady circulation of fluids throughout the body.' "[36]

Darwin's letter in the April 27 issue of *Nature* dissected Galton's interpretation of his hypothesis. He quoted passages from Galton's recapitulation concerning circulation of gemmules in the blood, stating categorically he had "not said one word about the blood, or about any fluid proper to any circulating system. It is indeed obvious that the presence of gemmules in the blood can form no necessary part of my hypothesis; for I refer in illustration of it to the lowest animals, such as the Protozoa, which do not possess blood or any vessels; and I refer to plants in which the fluid, when present in the vessels, cannot be considered as true blood."[37] Later he was more gallant, admitting that when he heard about Galton's experiments he had not "sufficiently reflected on the subject, and saw not the difficulty of believing in the presence of gemmules in the blood."[38] But he chided Galton for being "a little hasty" in pronouncing the epitaph for pangenesis, balancing this comment diplomatically with a nod to his cousin's "ingenuity and perseverance" in his experiments. Darwin concluded that pangenesis had not yet "received its death blow; though from presenting so many vulnerable points, its life is always in jeopardy; and this is my excuse for having said a few words in its defense."[39]

On May 4, Galton publicly apologized in *Nature*, blaming semantics for the misunderstanding. "I understood Mr Darwin to speak of blood when he used the phrases 'circulating freely,' and 'the steady circulation of fluids,' especially as the other words 'freely' and 'diffusion' encouraged the idea. But it now seems that by circulation he meant 'dispersion,' which is a totally different conception."[40] He quoted from the easily misinterpreted sentences, saying that he did "not much complain of having been sent on a false quest by ambiguous language, for I know how conscientious Mr Darwin is in all that he writes, how difficult it is to put thoughts into accurate speech, and, again, how words have conveyed false impressions on the simplest matters from the earliest times."[41] He ended his letter "Vive Pangenesis!" Galton was being gallant in return. There was a strange epilogue. The rabbit experiments continued for another year and a half and Darwin was very much involved![42] The results were negative as before, and a discouraged Darwin wrote Galton on November 8, 1872, that the rabbits "which you saw when here . . . are now ready to breed, or soon will be; do you want one more generation? If the next one is as true as all the others, it seems to me superfluous to go on trying."[43] Galton agreed that the experiments had "been carried on long enough. It would be a crowning point to them if your groom could get a prize at some show for those he has reared up so carefully, as it would attest their purity of breed. There is such a show, I believe, impending at the Crystal Palace."[43] What irony! Galton's transfused silver-greys were showing so little tendency to "mongrelize" that they should be upheld as examples of purity of breeding. In the 1875 edition of *Variation* Darwin paid heed to Galton's semantic difficulties and changed the wording of the offending sentences.

Meanwhile Galton had constructed his own theory of inheritance. At the end of a rabbit letter on May 28, 1872, he told Darwin that he had just corrected the proofs of a paper called "Blood Relationship," in which he tried "to define what kinship really is, between parents and their offspring. I will send a copy when I have one."[44] Assuming Darwin did try to peruse his cousin's manuscript,[45] he would soon have foundered as it verges on the incomprehensible.[46] Nevertheless, Darwin's influence on Galton's thinking was clearly recognizable. Darwin found it necessary to postulate the existence of two classes of gemmules. The first, like those sought in the rabbit experiments, were widely disseminated throughout the organism. These were subject to environmental modification, injury to reproductive organs, etc., resulting in variation. This was a process roughly equivalent to mutation, had not Darwin remarked that exposure to the changed conditions for several generations was necessary "in order that any modification thus acquired should appear in the offspring."[47] This introduced a temporal element in the acquisition of newly inherited characters, explaining why surgical alterations like circumcision, tail-docking in horses, and limb amputations were not heritable. Since gemmules were transmitted over many generations, the removal of a part posed no problem, "for gemmules formerly derived from the part are multiplied and transmitted from generation to generation."[48] But Darwin still had to account for the appearance of a characteristic seen in an earlier generation, a reversion.[49] Where did it come from? Darwin posited that some gemmules were dormant or latent and that their multiplication resulted in their visible manifestation as a reversion. From Darwin's reasoning three important notions emerged. First, one class of gemmules was widely dispersed and hereditarily modifiable over several generations. That is, variation could be acquired. Second, since gemmules reproduced and were transmitted over many generations, a mutilation was not heritable. Third, some gemmules were latent and only expressed under certain circumstances, explaining reversion.

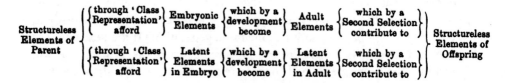

Fig. 13-1 The development of patent and latent elements from structureless elements according to Galton's scheme for inheritance in "On Blood-relationship," *Proc. of the Royal Society* 20 (1872): 394–402. Modified by Pearson, *Life*, II: 172.

Galton approved of Darwin's distinction between species of gemmules and also recognized two classes of elements, which he called "latent" and "patent." The patent elements were like Darwin's circulating gemmules and determined the character of the adult. The latent elements were equivalent to Darwin's latent gemmules and could explain reversion. As diagrammed by Galton, both kinds of elements differentiated from a common group of structureless elements in the fertilized ovum (Figure 13-1). A subset, the patent elements, were selected for development into embryonic elements. These in turn differentiated into adult elements while the residue of latent elements followed a parallel course, first as latent embryonic elements and then as latent adult elements. The two parallel pathways converged when a subset of patent and latent elements were selected to yield the structureless elements of the offspring. A key assumption in Galton's hypothesis was that the patent elements in the adult could be supplemented from the latent pool "because ancestral qualities indicated in early life frequently disappear and yield place to others."[50] But the reverse process, while it might occur at the embryonic stage, did not occur at the adult stage. By making information transfer a one-way street, Galton was trying to rule out acquired characteristics, for if patent elements could be environmentally modified and differentiated back into latent elements, acquired characteristics would be a reality. Unlike Darwin's pangenesis chapter, which gave many examples from nature with each chosen carefully to build the hypothesis, Galton failed to support his idea with appropriate examples. Nevertheless, "On Blood Relationship" represented an important step in the evolution of Galton's thinking as it meant that mankind could only be improved through selective breeding and not through environmental modifications, since these were not heritable.

On November 3, 1875, Galton responded to a letter from Darwin, who had heard he was "going to write on inheritance,"[51] informing his cousin that a new paper outlining his hypothesis was to be read before the Anthropological Institute the following Tuesday. He again assumed two types of elements, which he now called "germs."[52] The latent elements took center stage since "we must not look upon those germs that achieve development as the main sources of fertility; on the contrary, considering the far greater number of germs in the latent state, the influence of the former (patent elements) is relatively insignificant. Nay further, it is comparatively sterile, as the germ once fairly developed is passive; while that which remains latent continues to multiply."[53] One of his major conclusions was the "extremely small transmission of acquired modifications." And later he wrote that "I have, so far as the limits of a letter admit, made a clean breast of my audacity in theoretically differing from Pangenesis."[54] Darwin replied with a brief note the next day saying that "I can hardly form any opinion until I read your paper *in extenso*."[55]

As he often did, Galton had written two versions of "A Theory of Heredity," one for scientists in the *Journal of the Anthropological Institute* and the other for public consumption in the *Contemporary Review*. The latter paper was in proof when he wrote Darwin on November 5. He said he would send Darwin a copy as "I know you like to mark what you read, do not return it. I hope it will make my meaning more clear."[56] But why hadn't Galton sought his cousin's opinion earlier? He could easily have hopped on the train and gone to Down and discussed his ideas over lunch. Since Galton was so careful to keep his cousin informed on the progress of the rabbit experiments, one is led to conclude that he did not want or feel the need for Darwin's input. This was the third time he had published on the mechanism of heredity without soliciting his cousin's opinion. The first was in the *Proceedings* paper where he denounced pangenesis based on the negative results of the transfusion experiments. This was followed by "On Blood Relationship," which he seemingly wrote without consultation, an unfortunate decision in view of its incomprehensibility, and now the *Contemporary Review* paper. Perhaps because Galton was hammering away at various assumptions in Darwin's hypothesis, notably acquired characteristics, he wished to avoid wrangling with Darwin.

Like the *Fortnightly*, the *Contemporary Review*, founded in 1866, was a popular magazine where one could air a variety of opinions.[57] On the article's first page Galton confronted acquired characteristics head-on, writing that "the facts which a complete theory of heredity must account for may conveniently be divided into two groups; the one refers to those congenital peculiarities that were also congenital in one or more ancestors, the other to congenital peculiarities that were not congenital in any of the ancestors, but were acquired by one or more of them during their liftetime, through change in the conditions under which they lived; as of climate, food, disease, mutilation, or habit."[58] Shorn of confusing terminology and soaring flights of analogy Galton's message boiled down to this. Within the fertilized ovum is the sum total of germs or gemmules. This he termed the "stirp," which derives from the Latin, *stirpes*, a root. The fertilized ovum "receives nothing further from its parents, not even from its mother, than mere nutriment."[59] Effectively, the stirp was equivalent to what we would call the genome.

Galton then stated the four postulates upon which his theory was based. First "each of the enormous number of quasi-independent units the body is made up of, has a separate origin, or germ."[60] The modern equivalent would be that each gene specifies a different protein. Second, the germs in the stirp are much greater in number and variety than the structural units derived from them. Only a few germs actually developed and these were sterile. The notion that the developed germs are sterile is equivalent to saying the protein encoded by a gene does not contain the hereditary information necessary for its own transmission. Also embedded in this postulate is a vague glimmer of the concepts of dominance and recessiveness. That is, some genes are recessive and not expressed. Galton needed this postulate to account for variation and reversion. Third, the undifferentiated germs propagated themselves in the latent state and contributed to the stirp of the offspring. That is, genes, and not their products, were transmitted from one generation to the next. Fourth, the final structure, organization, and appearance of the adult organism depended "on the mutual affinities and repulsions of separate germs" within the stirp and during development. Positive and negative interactions of germs during differentiation can be visualized in terms of regulation of gene expression during development.

Next Galton addressed the question of sex. Unisexual organisms were rare, and bisexual organisms the norm. The reason unisexual systems were rare was that they tended to die out because "a deficiency of some of the structural elements, gradually sets in."[61] The modern analogy would be that deleterious recessive mutations begin to accumulate and become fixed among the progeny since they cannot be masked by outbreeding. Galton made this point beautifully a little later, writing that "when there are two parents, the chance deficiency in the contribution from either of them, of any particular species of germ, will be supplied by the other."[62] Thus, for most organisms, bisexuality is the result. He also recognized there must be some mechanism for reducing the size of the stirp at each generation. If the fertilized ovum contains the entire stirp from each parent, the progeny that arise will contain double the stirp of the two parents. So while "the stirp whence the child sprang can be only half the size of the combined stirps of his two parents, it follows that one half of his possible heritage must have been suppressed."[63] On theoretical grounds, Galton had predicted a process akin to the meiotic reduction divisions. But Galton, deeply steeped in evolutionary theory and unaware of the sophisticated mechanism that halves the number of chromosomes transmitted to sperm and egg at meiosis, assumed there was a struggle for existence between competing germs so only the fittest survived. He later returned to the heritability of acquired characteristics once more. He cited and dismissed several examples, concluding "that acquired modifications are barely, if at all, *inherited*, in the correct sense of the word."[64]

Galton had enunciated a form of the germ-line theory normally credited to the German biologist, August Weismann. Weismann acknowledged this in a letter to Galton dated February 23, 1889, writing that "It was Mr Herdman of Liverpool who—some years ago—directed my attention to this paper of yours.... I regret not to have known it before, as you have exposed in your paper an idea which is in one essential point nearly allied to the main idea contained in my theory of the continuity of the germ plasm."[65]

Having tortured himself over Galton's paper, Darwin wrote his cousin on November 7, 1875:

> I have read your essay with much curiosity and interest, but you probably have no idea how excessively difficult it is to understand. I cannot fully grasp, only here and there conjecture, what are the points on which we differ—I daresay this is chiefly due to muddle-headedness on my part, but I do not think wholly so. Your many terms, not defined "developed germs"— "fertile" and "sterile" germs (the word "germ" itself from association misleading to me), "stirp,"—"sept," "residue" etc. etc., quite confounded me. If I ask myself how you derive and where you place the innumerable gemmules contained within the spermatozoa formed by a male animal during its whole life I cannot answer myself. Unless you can make several parts clearer, I believe (although I hope I am altogether wrong) that very few will endeavour or succeed in fathoming your meaning.[66]

Darwin marked several passages in Galton's paper with numbers and enumerated his criticisms in the text. One dealt directly with their disagreement over the heritability of acquired characteristics. "If this implies that many parts are not modified by use and disuse during the life of the individual, I differ from you, as every year I come to attribute more and more to such agency."[67] Darwin was "very sorry to differ so much from you but I have thought that you would desire my open opinion."[68]

The correspondence continued for another couple of months. George Darwin was also keenly interested and sometimes Galton wrote to him. On December 18, 1875, Darwin wrote Galton that George had been explaining their differences to him. He also asked Galton whether he could answer a possible objection to his view concerning hybrid plants that are intermediate in many characters. "I cannot doubt that every unit of the hybrid is hybridised and sends forth hybridised gemmules. Here we have nothing to do with the reproductive organs."[69] Thus the act of hybridization caused these gemmules to acquire a new character just as prolonged exposure to a change in the environment might. Galton replied the next day with a solution to Darwin's objection.[70] He asked his cousin for simplicity's sake to consider a single character. Suppose the particular structure in the plant or animal being studied is

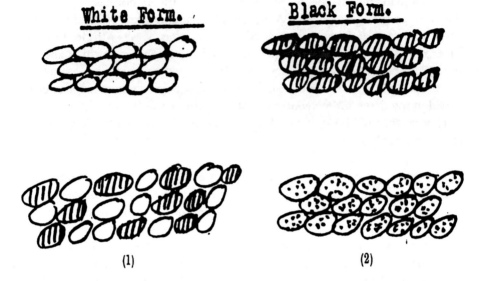

Fig. 13-2 Galton's explanation to Darwin of the intermediate character of hybrids. He imagines that some particular plant or animal structure consists of black cells in one parent and white cells in the other with the hybrid between them being exactly intermediate (i.e., grey). (1) The hybrid structure contains a mosaic of black and white cells. (2) The structural unit is not the cell, but instead there would be "an organic *molecule*" consisting of a group of gemmules of which black and white species are in statistically equal frequency so all cells appear grey. Modified from Karl Pearson, *Life*, II: 189–190.

black in one parent, white in the other, and grey in the hybrid. Galton envisioned two possibilities, which he diagrammed in his letter (Figure 13-2), neither of which required gemmule hybridization. First, the tissue being studied was a mixture of black and white cells "giving *on the whole* when less magnified a uniform grey tint."[71] Second, each cell had a uniform grey tint because black and white gemmules were, on statistical grounds, approximately equal. Then he began to derive the kind of distribution one would expect for a polygenic trait, that is a trait like height, which is determined by the incremental effects of a number of genes. "If there were 2 gemmules only, each of which might be either white or black, then in a large number of cases one-quarter would always be quite white, one quarter quite black, and one half would be grey. If there were 3 molecules, we should have four grades of color (1 quite white, 3 light grey, 3 dark grey, 1 quite black and so on according to the successive lines of 'Pascal's triangle')."[72] This statement reveals the direction in which Galton's thinking was going. He viewed the characters he was interested in, whether they be height or mental ability, as varying in a quantitative manner, for given enough determinants Pascal's Triangle will approach the normal distribution. Meanwhile, Galton revised his paper and published it in the *Journal of the Anthropological Institute*,[73] but in no way attempted to placate Darwin. Alfred Russel Wallace, who independently of Darwin had conceived of the theory of evolution through natural selection, read this version of the paper. He wrote Galton in March 1876, that "Your 'Theory of Heredity' seems to me most ingenious and a decided improvement over Darwin's, as it gets over some of the great difficulties of the cumbrousness of his Pangenesis."[74] On February 11, 1877, Darwin remarked in a short note to Galton "I shall never work on inheritance again."[75]

Galton had a workable theory of heredity that satisfied him, but it was not formulated in such a way that he could easily test it experimentally. Perhaps he did not care, for the rabbit experiments may have convinced him that the heritable gemmules could not be associated with the somatic tissues and by extension that any alterations they sustained (acquired characteristics) would not be transmitted to the progeny. In the prefatory chapter to the 1892 reprinting of *Hereditary Genius*, published ten years after Darwin's death, Galton wrote his epitaph on pangenesis. If the book were to be rewritten, its last chapter on Darwin's hypothesis would be revised and extended

> to deal with the evidence for and against the hereditary transmission of habits that were not inborn, but had been acquired through practice. Marvellous as is the power of the theory of pangenesis in bringing large classes of apparently different phenomena under a single law, serious objections have since arisen to its validity, and prevented its general acceptance. It would, for example, almost compel us to believe that the hereditary transmission of accidental mutilations and of acquired aptitudes would be the rule and not the exception. But leaving out of the question all theoretical reasons against this belief, such as those which I put forward myself many years ago, as well as the more cogent ones addressed by Weismann in late years,—putting these wholly aside, and appealing to experimental evidence, it is now certain that the tendency of acquired habits to be hereditarily transmitted is at most extremely small."[76]

In 1865, an Augustinian monk at the monastery in Brno named Gregor Mendel[77] summarized his experiments with the edible pea, *Pisum*, in one of the greatest scientific papers of all time, "Studies on Plant Hybridization."[78] There he described the segregation and assortment of what we have since come to call genes. Mendel's theory was far more comprehensible than either Darwin's or Galton's, but they were unaware of it at the time. In Galton's case this is not surprising, as he wasn't terribly good about keeping up on the literature, but Darwin was meticulous to a fault. Furthermore, the journal in which Mendel's paper was published was widely distributed in Europe, including Great Britain, and in the United States.[79] Nevertheless, only one contemporary biologist, Karl Naegeli, even seemed ready to correspond with Mendel about his work.[80] So his paper would gather dust until 1900 when it was rediscovered, supposedly independently, by three European scientists. Darwin was dead and Galton, aided and abetted by two younger colleagues, Karl Pearson and W. F. R. Weldon, was too old and too attached to his own theory of ancestral inheritance to regard Mendel's paper as of any great consequence. However, William Bateson, another disciple, became one of the great early champions of Mendelism.

FOURTEEN

Nature and Nurture

There is no escape from the conclusion that nature prevails enormously over nurture when the differences in nurture do not exceed what is commonly to be found among persons of the same rank of society and in the same country.

—Francis Galton, "The History of Twins,
as a Criterion of the Relative Powers
of Nature and Nurture"[1]

In December 1872, Galton received a prepublication copy of Alphonse de Candolle's *Histoire des Sciences et des Savants depuis Deux Siècles*. He had no prior contact with de Candolle, a highly regarded Swiss botanist, although he knew his family by reputation, having included it in *Hereditary Genius*.[2] De Candolle was stimulated to publish his *Histoire* after reading *Hereditary Genius*.[3] He was interested in the backgrounds of famous scientists, but his view of the relative importance of nature and nurture was the reverse of Galton's. His sample consisted of over 300 foreign members or associates of the Royal Society and the French and German Academies of Science. Since election as a foreign member was a rare distinction, de Candolle was certain these men had high reputation. His central argument was that eminent scientists were not randomly scattered by national origin and that their geographical distribution changed dramatically over time. For instance, Switzerland had produced over ten percent of the great scientists with less than one percent of the population of Europe. Eminent British and French scientists were twice as abundant as population size would predict, but Portugal and Spain together had half the number of distinguished scientists expected. By documenting changes in the frequencies of distinguished scientists in specific countries over time, de Candolle showed this irregular distribution did not reflect varied hereditary predisposition toward scientific ability in different populations. Thus, Holland produced eminent scientists at a rate only exceeded by Switzerland in the eighteenth century, but dropped below Britain, France, and Germany in the nineteenth.

De Candolle argued that these temporal changes reflected the degree to which science was being fostered in a given country at a specific time. He further pressed the environmentalist case by demonstrating the diverse national origin of Swiss scientists. The vast majority were not of native Swiss stock, but were descendants of refugee immigrants, principally French Huguenots like himself, with others from countries like Belgium, Italy, and Moravia. While refugee immigrants in other countries also produced eminent sons, it was de Candolle's impression that these men more often chose fields such as politics, law, or the humanities. So why did the environment of Switzerland guide talented immigrants toward science? De Candolle identified several factors. A temperate climate formed a suitable intellectual greenhouse while a hot one impaired the physical labor required in many forms of research. The demise of Latin as the dominant language of scholarly discourse meant that native speakers of English, French, or German were at an advantage. De Candolle also felt an authoritarian religious establishment was detrimental to science, so less dogmatic Protestant countries tended to produce a bigger crop of talented scientists than Catholic nations. Educational systems promoting free inquiry abetted production of able scientists and scientists most frequently originated in countries with high standards of living. This provided them with universities, library resources, and laboratory facilities compatible with productive rumination and experiment.

In spite of his strong bias toward nurture, de Candolle acknowledged that heredity also played a role, but he introduced heredity in a guise that allowed him to explain why gifted descendants of Huguenots such as himself, often became scientists in Switzerland, but lawyers or politicians in America. "Heredity, considered as a fact relative to the elementary faculties of the individual, and not to scientific specialties, will produce varied combinations, and will permit many young people to follow one career or another, one science or another, with the same probability of success."[4] Until he came to race de Candolle's analysis had a rather modern ring to it, but he argued that blacks, while physically strong, were not very bright. They lacked the intelligence to emigrate voluntarily or else the black race "would profit from its incontestable physical qualities, and would continue to invade the new world. Happily, the black is attached to the sun, and remains in the lands where his father lived."[5] Orientals were both bright and strong, and potentially serious rivals to Europeans, but their moral flaws, especially greediness, ensured they would never compete "as they have little courage and even less good faith."[6] This hereditarian notion of race, widespread in nineteenth-century Europe, provided an opening for the intellectual sparring that soon took place between Galton and de Candolle.

De Candolle's book irritated Galton, especially since he was clearly aiming at the central thesis of *Hereditary Genius*. He did not believe Galton had proved it nor had he "scrutinized the question in a specialized enough manner...."[7] His own accounts of scientific men were gathered differently from Galton's and he "employed completer biographical documents, drawn from French, English, and German works. I thus flatter myself to have penetrated farther into the heart of the question."[8] As Galton read the *Histoire* he wrote notes to himself highlighting passages he disagreed with. He disputed de Candolle's claim that his sample possessed more prominent scientists than Galton's. "I have taken much higher names, selected with great care, and studied their histories from all kinds of sources. His men are so unknown to fame that he can hardly learn anything about them."[9] Galton was being unjust, as de Candolle's list included prominent English scientists like Newton, Priestley, Halley, and Herschel. Galton thought he detected a technical flaw in de Candolle's tabular presentation of national scientific productivity rates. Since de Candolle's calculated productivity rates were based on raw population data, his analysis was biased against countries like Great Britain where the fraction of men too young to have yet achieved distinction was large enough to lower artificially the percentage of eminent individuals. He thought de Candolle should have analyzed only men of 50 or older. He also scoffed at de Candolle's claim that at conception the "momentary state" of the parents could influence the constitution of the offspring. Always on his guard for such claims of acquired characteristics, Galton scribbled "stuff" and "stuff and nonsense."[10]

But there were many parts of de Candolle's book he liked and he marked some passages "good" or "very good." He also realized he could use de Candolle's hereditarian views on race to undermine his overall argument and perhaps chuckled to himself when he translated this passage. "Obviously, Europeans and their descendants are the only ones who play a role in the sciences ... while the Asian, African, and indigenous American races have rested, to the contrary, completely outside the scientific movement."[11] Galton confided to himself that de Candolle admitted race was a fundamental deciding factor that "surpasses the others in importance," and penned a note to himself: "Quote the original—a complete concession."[12] After two weeks of studying the *Histoire*, Galton wrote de Candolle a long letter on December 27, 1872.[13] After opening with his usual Victorian decorum, he got down to business. "You say and imply that my views on hereditary genius are wrong and that you are going to correct them; well, I read on, and find to my astonishment that so far from correcting them you re-enunciate them.... I literally cannot see that your conclusions, so far as heredity is concerned, differ in any marked way from mine."[14] He cited specific examples from the *Histoire* concerning race, physical form, intellect, etc. that must have a hereditary basis, and continued with pique. "I feel the injustice you have done to me strongly, and one reason I did not write earlier was that I might hear the independent verdict of some scientific man that had read both books. This I have done, having seen Mr Darwin whose opinion confirms mine in every particular."[15] But Darwin had earlier written de Candolle saying he could not put the *Histoire* down as "I have hardly ever read anything more original and interesting than your treatment of the causes which favour the development of scientific men."[16]

But Galton's letter was frequently upbeat. He was pleased de Candolle pointed to the "chilling effect" religious authority could have on scientific inquiry and he applauded de Candolle's conclusions on the importance of a rigorously taught curriculum. However, he reproved de Candolle for not correcting his raw data for those populations having "a plethora of children and of persons too young to be academicians"[17] and chided him over his antique notions concerning defective children resulting from alcoholic conceptions. On rereading his letter Galton decided to end on a conciliatory note. Despite finding fault, he thought de Candolle had done a great service in writing the *Histoire*. His general impression of the book was very good and he would promote it in England.

De Candolle responded to his "Monsieur et honoré collègue" with an even longer letter. He understood the *Histoire* had caused Galton a mélange of impressions some agreeable and some disagreeable, but, he continued disarmingly, if "there has escaped me, in the 482 pages of my book, one sentence, one word, raising doubt about my respect for your impartiality, character, and talent for investigation, it absolutely could only have been contrary to my intentions. You have always sought the truth."[18] They were "remarkably in accord on the facts. We have the same ideas about race."[19] But de Candolle differed with Galton on their interpretation. "You habitually highlight, as the principal cause heredity. When you speak of other causes they are indicated accessorily" and the "very title of your work implies the idea of only studying heredity, its laws and consequences, or else you would have written: On the effect of heredity and other circumstances as to genius. Surely you have rendered true service to science, but your point has been essentially that of heredity."[20] De Candolle had hit the bullseye and Galton knew it. He had rightly taken Galton to task for ignoring the environment both in his book and in his letter and Galton would have to do something about it. But first there was the matter of damage control.

Galton reviewed de Candolle's book for the *Fortnightly*.[21] He chided de Candolle for accusing him of overstating "the influence of heredity, since the social causes, which he analyses in a most instructive manner, are much more important.... I am anxious to point out that the author contradicts himself, and that expressions continually escape from his pen at variance with his general conclusions."[22] Galton gave some examples one of which was de Candolle's assertion "that in the production of scientific men of the highest scientific rank, the influence of race was superior to all others." He also caught de Candolle in a little syllogism that mental qualities were connected with structure and since structure was inherited the former must be too. This positioned de Candolle exactly where he did not wish to be. So, tongue in cheek, Galton wrote that he considered "M. de Candolle as having been my ally against his will, notwithstanding all he may have said to the contrary."[23] He accused de Candolle of repeatedly trespassing "on hereditary questions, without, as it appears to me, any adequate basis of fact, since he has collected next to nothing about the relatives of the people upon whom all his statistics are founded."[24] Having let off steam, Galton's review became quite positive. He approvingly summarized de Candolle's analysis of the "the blighting effect of dogmatism upon scientific investigation"[25] in both Catholic and Protestant countries. He asked whether religion and science "could march in harmony" and launched into his own comparison of religious and scientific men.

In concluding, Galton gently reproved de Candolle. He did not believe the "acquired habit of drunkenness, which ruins the will and nerves of the parent"[26] was transmitted to the progeny. He had a simpler view more in tune with modern thinking. "The fluids in an habitual drunkard's body, and all the secretions, are tainted with alcohol; consequently the unborn child of such a woman must be an habitual drunkard also. The unfortunate infant takes its dram by diffusion, and is compulsorily intoxicated from its earliest existence. What wonder that its constitution is ruined, and that it is born with unstrung nerves, or idiotic or insane?"[27] But Galton had conceded the importance of the fetal environment in avoiding the trap of acquired characteristics.

Despite criticizing each other publicly, de Candolle and Galton entered into a long and warm correspondence, for they genuinely admired one another. They also influenced each other. In the 1885 edition of his *Histoire* de Candolle changed part of the original subtitle of his text from *sur la Sélection dans L'espèce Humaine* to *sur L'hérédite et la Sélection dans L'espèce Humaine*. Galton in turn was spurred to undertake a survey similar to de Candolle's, investigating the backgrounds of 180 eminent British scientists. He chose Royal Society members, but this was not enough as Society membership was merely a "pass examination" in vetting his thoroughbreds. His eminent men must have additional qualifications such as a prestigious medal, or have presided over a learned society or a section of the British Association.

To construct as complete a profile as possible Galton created the questionnaire, a novel technique at the time.[28] It was a monster of "seven huge quarto pages."[29] Naturally, he sent one to Darwin explaining that his inquiry was parallel to de Candolle's. Poor Darwin struggled manfully with the daunting questionnaire, finally writing his cousin that "I have filled up the answers as well as I could; but it is simply impossible for me to estimate the degrees."[30] Galton selected for "statistical treatment" the replies of more than 100 scientists and compiled his results in a rather inconsequential little volume with a major league title *English Men of Science: Their Nature and Nurture* (1874). He had used the expression earlier that year in the title of an address given at the Royal Institution.[31] Now it graced the cover of his book as the "phrase 'nature and nurture' is a convenient jingle of words, for it separates under two distinct heads the innumerable elements of which personality is composed. Nature is all that a man brings with himself into the world; nurture is every influence from without that affects him after his birth."[32] Nature represented "the latent faculties of growth of body and mind" that an infant possesses at birth. Nurture did not just refer to food, clothing, education, etc. but to the environment in which a person grows up "by which natural tendencies may be strengthened or thwarted, or wholly new ones implanted."[33] But nature should not be equated directly with heredity, since "natural gifts may or may not be hereditary."[34] Thus was Galton's most famous phrase born. Yet today surprisingly few people associate Galton's name with the phrase and even fewer with Shakespeare's *The Tempest* from whence Galton may have got it. In act IV, scene 1, Prospero, the Duke of Milan, is speaking to the spirit Ariel about the man-beast Caliban. "A devil, a born Devil, on whose nature nurture can never stick."

English Men of Science was only four chapters long. The first defined terms like "nature" and "nurture," described how the data were gathered, and summarized information on the geographical distribution of the scientists studied, and some statistics on their parents. It also presented pedigrees for 13 families in the now familiar manner adopted by Galton in *Hereditary Genius*. The remaining three chapters dealt at length with the mounds of data Galton had extracted from his questionnaires. In the second chapter, "Qualities," he presented qualitative measurements of characteristics like energy, perseverance, memory, and the like. And, of course, he wanted to know about parents as well as the eminent man himself. He also included extensive sample quotations in answer to his questions that, while serious, were occasionally amusing. This one came from a botanist, perhaps Hooker, regarding his memory. "Retentive for botanical names; rather deficient in other respects, especially as to persons."[35]

The third chapter was aimed directly at the heritability of scientific talent. He was elated, as 56 out of 91 scientists in his summary table believed their scientific aspirations were innate rather than environmentally determined. So Galton felt he had parried de Candolle's thrust, but he was gallant in victory and to be fair to his opponent, examined the contribution of environment including encouragement at home, and by friends, and the roles tutors and travel had played in the molding experience. He concluded these factors were also important, but added a clever hereditarian twist in summing up. He wrote that "a love of science might be largely extended by fostering, and not thwarting innate tendencies . . ."[36] The final chapter considered the role of education. Galton asked his subjects to summarize the good and bad points of their education and whether their health had been improved or had suffered, perhaps remembering his own experience and that of some of his friends. He particularly wanted to know if the education of his eminent men had been "conducive to, or restrictive of, habits of observation."[37] In his book, *Pioneers of Psychology*, Raymond Fancher remarks that, while Galton's analysis of his data was "naive," the real virtue of his book was "its demonstration that the statistical analysis of questionnaire data was a *potentially* valuable approach to psychological questions."[38]

Meanwhile, Galton had discovered a new way to assess the roles of nature and nurture in determining character, twin studies. He knew there were two classes of twins, those "closely alike in boyhood and youth" and those "who were exceedingly unlike in childhood."[39] These correspond to identical, or monozygotic twins and nonidentical, or dizygotic twins. The first derive from a single fertilized egg while the latter arise from two independently fertilized eggs. Hence, identical twins are genetically identical whereas nonidentical twins are no more similar than brothers and sisters. Galton clearly recognized this distinction when he wrote:

The word "twins" is a vague expression, which covers two very dissimilar events—the one corresponding to the progeny of animals that have usually more than one young one at a birth, each of which is derived from a separate ovum, while the other is due to the development of two germinal spots in the same ovum. In the latter case, they are enveloped in the same membrane, and all such twins are found invariably to be of the same sex.[40]

This distinction was critical for Galton's study. Identical twins did not differ in nature or nurture. To what degree would they diverge from one another in later life? Nonidentical twins were reared in the same environment, but differed genetically. How would identical nurture influence their similarities and differences? Armed with his new invention, the psychological questionnaire, Galton circularized persons whom he either knew or had heard about who were either twins themselves or closely related to twins. The last of his questions inquired whether these individuals knew of other pairs of twins. In this way he networked his way to 94 sets of twins, choosing to study those on which he had the most detailed information. Charles Ansell of the National Life Assurance Association also proved a great help to Galton, providing him with the names and addresses of 190 sets of twins.[41]

Galton classified his responses in different folders "girls alike, girls unlike, girls partly alike"[42] and the same for boys and wrote up his results. His paper was published in 1875 in *Fraser's Magazine*, a monthly periodical, under the title "The History of Twins, As a Criterion of the Relative Powers of Nature and Nurture."[43] He reprinted it "with revision, among the miscellanies" in the *Journal of the Anthropological Institute*.[44] *Fraser's* was a good choice, as the magazine commanded a wide readership. It was born in February 1830 with William Maginn at the editorial helm assisted by talented contributors like Carlyle and Thackeray.[45] Galton opened his article with an admission. He was keenly aware that his previous efforts to document the inheritance of "talent and character" by examining pedigrees of eminent men did not exclude the possibility that familial environment also played a role in achievement. A further criticism was that unpredictable events in the life of the eminent person were bound to be important in his success. Since these could not be quantified a critic might argue that his "statistics, however plausible at first sight, are really of very little use."[46] Twin studies were "wholly free from this objection."[47]

Galton analyzed 35 sets of identical twins and he did exactly what psychologists have done ever since. He collected anecdotes and recounted the more remarkable of these. Some dealt with the strikingly close physical similarity of the twins. "Two twins were fond of playing tricks, and complaints were frequently made; but the boys would never own which was the guilty one, and the complainants were never certain which of the two he was. One head master used to say he would never flog the innocent for the guilty, and another used to flog both."[48] Then there were cases where seemingly improbable events affected both twins. "Two twins at the age of twenty-three were attacked by toothache, and the same tooth had to be extracted in each case."[49] For mental likeness he turned to the dramatic tale told by a Parisian doctor named Moreau, who wrote of two twin brothers confined at his hospital for monomania. Not only were they physically very similar, but this was true "morally" as well. "They both consider themselves subject to imaginary persecutions; the same enemies have sworn their destruction, and employ the same means to effect it. Both have hallucinations of hearing."[50]

Galton was building a qualitative case for nature's importance in determining human behavior as psychologists have often done since, though they also use quantitative tools such as the IQ test that were unavailable to Galton. The emphasis is always on cases of striking similarities, sometimes bizarre, in behavioral traits between identical twins. Since similarities are easy to compare, but dissimilarities are not, behavioral differences between twins are rarely discussed, so nature seemingly reigns supreme. But the really crucial question for Galton was whether his identical twins diverged once they had flown the nest when their living environments would no longer be identical. He found that "in some cases the resemblance of body and mind had continued unaltered into old age, nothwithstanding very different conditions of life; and they showed in the other cases that the parents ascribed such dissimilarity as there was wholly, or almost wholly, to some form of illness."[51]

Galton also examined 20 sets of nonidentical twins. He breezed through these quickly with two or three examples only. "One parent says: 'They have had *exactly the same nurture* from their births up to the present time; they are both perfectly healthy and strong yet they are otherwise as dissimilar as two boys could be, physically and mentally, and in their emotional nature.'"[52] Galton was pleased with his findings as "there is no escape from the conclusion that nature prevails enormously over nurture."[53]

FIFTEEN

Sweet Peas and Anthropometrics

> My inquiries into hereditary genius ... were sufficiently advanced before the year 1865 to show the pressing necessity of obtaining a multitude of exact measurements relating to every measurable faculty of body or mind, for two generations at least, on which to theorise.
>
> —Francis Galton, *Memories*[1]

Following the publication of *Hereditary Genius*, Galton's attack on the human body and mind became two-pronged. The first prong involved the accumulation of quantitative data on easily measurable human physical parameters such as height coupled with the development of methods for their statistical analysis based principally on the properties of the normal distribution. Ideally, this information would be gathered in pedigree form so heritability of these characteristics could be analyzed. The observations obtained would serve as surrogate indicators of mental fitness, for Galton's Cambridge background had prompted him to believe that a healthy body means a healthy mind. The second prong of Galton's attack was on human personality and behavior. He wanted to develop methods for characterizing and quantifying human behavior. His desire in this regard led him into explorations as diverse as composite photography, psychology, and fingerprinting.

The ultimate purpose of these investigations was to establish methods for advancing the quality of the human stock and he laid out this agenda in 1873 in a *Fraser's* article entitled "Hereditary Improvement."[2] His bold goal was "to improve the race of man by a system which shall be perfectly in accordance with the moral sense of the present time."[3] He argued that civilization often acted "to spoil a race" and gave two examples. First, the transmission of wealth between generations interfered "with the salutary action of natural selection," because wealth "encouraged marriage on grounds quite independent of personal qualities."[4] Second, the centralizing tendency of English civilization attracted "the abler men to towns, where the discouragement of marriage is great, and where marriage"[5] was relatively unproductive of progeny. He analyzed census returns of 1,000 families of factory workers in Coventry and compared these to an equivalent number of families principally engaged in agricultural pursuits in small, neighboring parishes. While the two groups married with about equal frequency at similar ages, the workers in grimy Coventry had little more than half the adult grandchildren of those living in the idyllic English countryside. Galton attributed this discrepancy to two factors. Coventry families had fewer children and the fraction reaching adulthood was reduced. He blamed bad sanitary conditions and insufficient food in cities which "spoil . . . our breed" while favoring selection for "the classes of a coarser organisation."[6]

Galton argued that moral and intellectual qualities were closely interwoven amongst the nineteenth-century English and that "many of the wild instincts of our savage forefathers"[7] had been bred out. He invoked the normal distribution as a means for grading men according to "natural ability" hoping that the "average standard of a civilised race might be raised to the average standard"[8] of the best. However, while ability was normally distributed, the numbers of individuals in each category was not. The worst of the race were the most numerous while the best were the rarest. Hence, Galton's goal was to identify the "naturally gifted" and procure for them "such moderate social favour and preference . . . as would seem reasonable" in view of "their importance to the nation."[9] This would "bind them together"[10] and through intermarriage this intellectually and socially privileged caste would flourish.

To implement his proposal Galton asked that an appropriate British society undertake three tasks: first, to make "*continuous* enquiries into the facts of human heredity,"[11] meaning the collection of detailed pedigree data; second, to establish an information center on heredity for animal and plant breeders. Third, a separate unit would evaluate the information gathered by the first two. Galton imagined that within a few years pedigree data, accompanied by photographs and physical measurements, could be compiled for a thousand or more individuals in each region sampled. Schoolmasters, ministers, doctors, etc., would be asked to help in collecting the data. Eventually all schoolboys would be classified "in respect of their natural gifts, physical and mental together"[12] and inquiries would be made routinely "into the genealogies of those . . . who were hereditarily remarkable."[13] The most promising would be registered at their own local centers. They would be treated with more "respect and consideration than others whose parents were originally of the same social rank."[14] After a couple of generations, Galton reasoned "the selected race will have become a power"[15] and "the number of families of really good breed"[16] would have increased. With the passage of time these families "would multiply rapidly, while the non-gifted would begin to decay out of the land, whenever they were brought face to face in competition with them, just in the same way as inferior races always disappear before superior ones."[17] The inferior classes were to remain celibate, being treated "with all kindness" by their compatriots as long as they did so. But should they procreate the time would probably come "when such persons would be considered as enemies to the State, and to have forfeited all claims to kindness."[18] With these sentences Galton laid out his eugenic agenda, and it would change little over the years to come. Its goal was to identify and encourage the breeding of those perceived to have talent, while discouraging reproduction by the masses of people of inferior quality.

Although *Hereditary Improvement* laid out Galton's entire social agenda, there was much groundwork to be done. In 1874 the Council of the Anthropological Institute approved his proposal to collect measurements of heights and weights of students from selected schools.[19] To initiate his plan Galton enlisted the help of the Reverend Frederic Farrar, headmaster of the Marlborough School. Farrar, a willing accomplice, engaged the school medical officer, Walter Fergus, and G. F. Rodwell, the natural science master, to make the measurements. They dutifully estimated heights, weights, heads, chest girths, etc. of the 550 Marlborough School boys. Reviewing this work in 1991, J. M. Tanner remarked that Fergus and Rodwell "or perhaps Galton himself, it would have been characteristic of him—constructed a height measuring instrument far superior to anything in use at that time. It had a counterweighted headboard sliding between vertical guides, a design not reinvented (by R. H. Whitehouse and myself) until the 1950s and since then the standard."[20] Fergus and Rodwell reported their measurements in the *Journal of the Anthropological Institute*[21] with Galton's analysis following.[22] He concentrated on the data for the boys' heights, showing that they were normally distributed with a mean that increased with age.

To aid his analysis, Galton unveiled a new statistical tool. He wanted a metric that would allow him to arrange any set of measurements on a single statistical scale.[23-25] He discovered that if he graphed his data in a series of ranks according to the exponential function that describes the normal distribution, he obtained a graceful, sinuous curve that was concave at the bottom and convex at the top (Fig. 15-1). He christened it the "ogive," borrowing an architectural term with several meanings, one being an ogee moulding that has the same shape as his curve. The distribution could be divided into quartiles with the middlemost having a value of 0 (representing the average), with an individual in the upper quartile having a value of 1 (representing one probable error above the mean), and so forth. Stephen Stigler in his *History of Statistics* remarks that Galton's ogive, now called by the markedly less euphonious

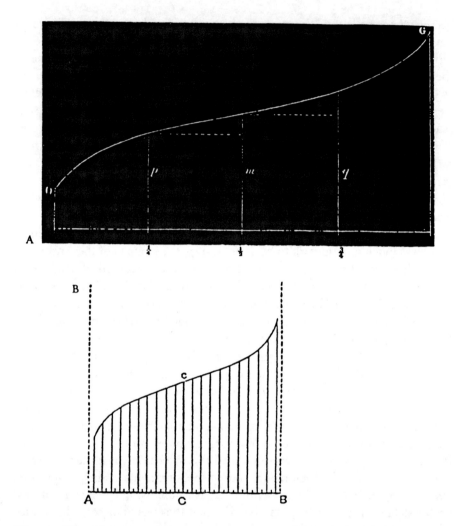

Fig. 15-1 Two of Galton's renditions of the ogive. A. The earlier, from 1875, shows the median and the quartiles. B. The later drawing, from 1883, shows spikes with heights representing 21 equally spaced ideal data values. From Stephen M. Stigler, *The History of Statistics*. Cambridge: Belknap Press of Harvard University Press, 1986, 270.

name "the inverse normal cumulative distribution function," would become the most used and abused method of scaling psychological tests.[26]

Meanwhile, data were pouring in on the comparative heights and weights of schoolboys. This enabled Galton to test his theory that growing up in a city was detrimental to one's health. He published his findings in the *Journal of the Anthropological Institute* in 1876.[27] He had gathered returns on "boys who were 14 on their last birthday, in two groups of public schools,"[28] restricting his analysis to boys of this age because his data were incomplete for younger or older boys. His hypothesis seemed vindicated, since country boys were "about 1 1/4 inch taller than those in the town group, and 7 lbs heavier,"[29] this height difference being due equally "to retardation and to total suppression of growth." In both cases heights were normally distributed. Galton admitted the town boys tended to make up the height differential as they grew older although never completely.

Curiously, Galton seemed unaware of the work being done at St. George's Hospital in London by Charles Roberts.[30] Roberts was one of a team of five doctors who examined and measured some 10,000 children of both sexes employed in textile factories or similar establishments. This survey, carried out between 1872 and 1873, was a follow-up on the Factory Commission Surveys of 1833 and 1837. Their purpose was to use the measurements to check supposed ages of children for entry into factory work and to monitor their health thereafter. Roberts results were published in the Parliamentary Papers in 1873, and analyzed by him the following year in the *St. George's Hospital Reports*. Roberts "had made an in-depth study of Quetelet's work . . . and laid as great a stress as Galton on the variation between different children of the same age."[31] After he learned of the Marlborough School study from Fergus, Roberts got in touch with Galton. He obtained permission to include the Anthropological Institute records in the *St. George's Hospital Reports* and later in his *Manual of Anthropometry*. Roberts joined Galton as a member of the Anthropometric Committee of the British Association in 1879, where they worked together in the early 1880s.

On the evening of February 9, 1877, Galton presented one of the famous Friday Evening Discourses in the theatre of the Royal Institution and titled it "Typical Laws of Heredity."[32] In this important lecture to scientists, persons of wide social interests, and the educated public he would illustrate the application of the normal distribution to quantitative data and unveil a new statistical concept, regression. He would also explain why he believed the return, or regression, of the progeny of individuals at the extremes of the normal distribution toward the mean of that distribution would act as a counterforce to natural selection.

The Royal Institution, located on Albemarle Street, was founded by Benjamin Thompson, a Massachusetts Royalist who escaped to England in 1776.[33] The front of the building was framed by an imposing row of elegant Corinthian columns for which the Temple of Antoninus in Rome served as a model. Over the years the distinguished scientist-lecturers appointed to the Royal Institution included Humphry Davy, Michael Faraday, and Galton's friend John Tyndall. By custom, the men in Galton's audience dressed in dark evening clothes and the women in brightly colored finery. They arranged themselves on ascending semicircular rows of benches in front of the podium.

Shortly before 9 P.M. the audience hushed as the president of the Royal Institution strode into the theatre to take his place in the front row. According to tradition, he would have had Louisa Galton on his arm, if she were well. Promptly at nine Galton was ushered in through a pair of swinging doors to stand behind a low horseshoe-shaped table on which scientific equipment could be arranged, since lecturers giving Friday Evening Discourses commonly illustrated specific points by experimental demonstration. Behind and above Galton were hung various charts that he would use to illustrate his talk, while behind him a lecturer's assistant hovered expectantly.

As the custom was for the speaker to start without introduction, Galton began immediately. The question he would address was why "two groups of persons selected at random from the same race, but belonging to different generations"[34] were so similar. Any statistical differences that existed were always "ascribed to differences in the general conditions of their lives; with these I am not concerned at present, but so far as regards the processes of heredity alone, the resemblance of consecutive generations is a fact common to all forms of life."[35] Galton intended to show how the normal distribution related to studies of inheritance, but first he needed to illustrate its application to a human characteristic. He called attention to a chart summarizing some of Quetelet's old data on the relative heights of men from America, France, and Belgium. He moved his pointer first down a column tabulating the number of American men falling in the different size classes. He then slid its bobbing tip down the adjacent column where the number expected in each category was calculated. He observed with evident satisfaction that "the close conformity between each of the pairs is very striking."[36]

Since the normal distribution could be applied to "the characteristics of all plants and animals," he intended to discover how the laws of heredity enabled "*successive* generations to maintain statistical identity."[37] He gestured to the assistant who brought forward a peculiar looking contraption some recognized as resembling one he had used in a lecture three years earlier. The original quincunx, shaped like half of an old-fashioned hot-water bottle with one side open and faced with glass, was an ingenious device he invented to illustrate the properties of the normal distribution. What looked like the bottle's top was a funnel through which lead shot was poured into a structure resembling the upper half of an hourglass. As the shot dropped through the narrow neck of the hourglass, the pellets cascaded through a series of rows of pins into individual bins at the bottom. Each row consisted of a series of arrays of five pins each arranged as a quincunx, that is with a pin at each of the four corners of a rectangle and the fifth in the middle. The uppermost row was a single quincunx with the rows successively expanding in length so that the final assembly of pins resembled an equilateral triangle.[38]

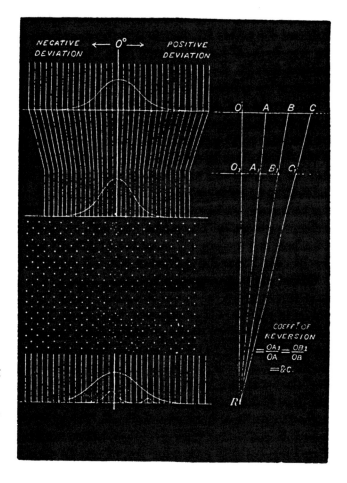

Fig. 15-2 Galton's 1877 published drawing showing the two-stage quincunx and the laws of reversion and family variability. From Francis Galton, *Nature* 5 (1877):493.

But the quincunx the assistant now placed beside Galton was a new model with two successive stages, which could be thought of as analogs of two successive generations (Fig. 15-2). Above the quincuncial arrays of pins, the funnel was divided into a series of individual slots. These were of two kinds. The uppermost row was vertical, with each slot blocked at the bottom by a trapdoor. Each of these fed into a second array of slots gently canted toward the center of the device that also had trapdoors. The quincunxes were arrayed below these slots and underneath them were slots in which the lead pellets would finally come to rest. Galton could easily create an array of normal distributions simply by choosing which slots to use and how many pellets to load per slot. He could achieve a similar goal by varying the number of quincuncial arrays and the distance between spikes in each quincunx. He had loaded the shot in the uppermost set of slots so that they formed a normal distribution.

He then opened the trapdoors on a slot in the middle and one on the right. In each case the pellets fell directly into the canted slot below. Then he opened the trapdoors of the two canted slots. As in a pinball machine, the shot bounced through the quincuncial arrays of spikes settling into the bins at the bottom of the quincunx.

The means and the variances of the resulting populations of lead pellets were different, as expected. Galton had often played with his quincunx in the months prior to his lecture and had cut out cardboard shapes to illustrate some of the distributions he obtained, demonstrating several of these for his audience. As he continued to lecture, he kept opening trapdoors in the quincunx and holding up cardboard shapes of normal distributions until he felt confident that his listeners understood what he was up to. He looked up and surveyed his audience expectantly. "I have now done with my description of the law. I know it has been tedious, but it is an extremely difficult topic to handle on an occasion like this. I trust the application of it will prove of more interest."[39]

Galton was ready to present his results. Since he was unable to test heritability in human beings, he turned to a model system just as he had when he used rabbits to test Darwin's pangenesis hypothesis. This time he selected sweet peas on the advice of Darwin and the botanist Joseph Hooker.[40] He cited three reasons. Sweet peas had little tendency to cross-fertilize; they were hardy and prolific; and seed weight did not vary with humidity.[41] His first experimental crop, planted at Kew in the spring of 1874, failed.[42] To avoid this outcome a second time Galton dispersed his seeds widely the next year, describing his method to his Royal Institution audience. "I weighed the seeds individually, by thousands, and treated them as a census officer would treat a large population. Then I selected with great pains several sets for planting. Each set contained seven little packets, and in each packet were ten seeds, precisely of the same weight."[43] The packets were lettered K, L, M, N, O, P, and Q, with K containing the heaviest, seeds, L the next heaviest and so forth down to packet Q, which had the lightest seeds. Galton sent sets of seeds to friends and acquaintances all over Great Britain from Nairn on the Moray Firth to Cornwall. Elaborate instructions for planting accompanied each set. Afterwards the beds were covered with underbrush to keep birds from digging up the seeds and the bag from which the seeds had come was impaled on a stick at the end of each bed. When the crop was coming to its end, the entire plants in each bed were uprooted, tied together, labeled, and sent to Galton. Darwin planted one set. By late September Darwin's vines were getting very unruly and he advised Galton to "come down and sleep here and see them. They are grown to a tremendous height and will be very difficult to separate."[44]

Galton summarized his results, having "obtained the more or less complete produce of . . . 490 carefully weighed seeds."[45] They gave him "two data, which were all that I required in order to understand the simplest form of descent

Parent seed	15	16	17	18	19	20	21
Mean Diameter of progeny seed	15.4	15.7	16.0	16.3	16.6	17.0	17.3

Table 15-1 Diameters of parent sweet pea seeds compared with the mean diameters of their progeny seeds in hundredths of an inch. Adapted from F. Galton, *Natural Inheritance* (1889), MacMillan and Co., London, p. 226.

and so got at the heart of the problem at once."[46] By simple descent Galton meant that, as far as he knew, sweet peas were self-fertilized. His discovery was that the "processes concerned in simple descent are those of Family Variability and Reversion."[47] Family variability referred to the degree of variation around the mean observed among progeny seeds, irrespective of whether they were large, small, or average in size. While the mean of the distribution shifted somewhat in different sets of progeny seeds depending on parental seed size, Galton discovered that the degree of variation around the mean was similar for all. He was "astonished to find the family variability of the produce of the little seeds to be equal to that of the big ones, but so it was, and I thankfully accept the fact for if it had been otherwise I cannot imagine from theoretical considerations, how the problem could be solved."[48] By reversion Galton meant "the tendency of that ideal mean type to depart from the parent type, 'reverting' towards"[49] the mean of the general population from which the seeds were selected (Table 15-1).

As he spoke he called attention to a diagram that he used in preparing his speech (Fig. 15-3). That diagram plotted the average diameter of progeny seeds on the Y axis against the average diameter of the parental seeds on the X axis. As Pearson later remarked, this was probably the first regression line ever to be computed.[50] Initially, Galton referred to the slope of the line as the "coefficient of reversion," but later he realized that this coefficient was not a hereditary property. Instead, it was a property of his own statistical manipulations so he changed the name to the coefficient of "regression."[51]

To illustrate reversion, Galton began playing with the quincunx again (Fig. 15-2). "I shall shortly open the trap-door on which the few representatives of the giant seeds rest," said Galton. "They will run downwards through an inclined chute and fall into another compartment nearer the centre than before. I shall repeat the process on a second compartment in the upper stage, and successively on all the others."[52] He opened the trap door in which the lead pellets representing the giant seed were gathered. They clattered down the chute and came to rest at a second trapdoor above the quincuncial arrays of spikes. He repeated the process for each slot. He had carefully varied the number of pellets in the slots at the top of the quincunx so they approximated

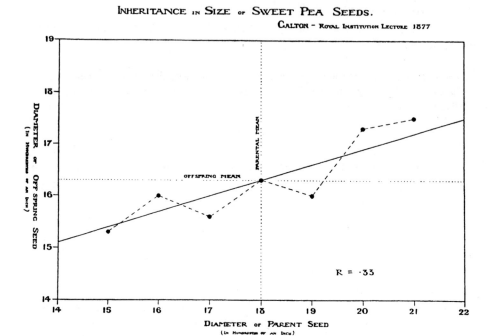

Fig. 15-3 The first regression line. From Karl Pearson, *Life* III: 4.

a normal distribution. As they came to rest at the bottom of the gently canted chutes they again formed a normal distribution, but this bell curve, while having the same mean as that above, was somewhat higher in the middle and more restricted in width. "It is obvious from this," said Galton, "that the process of reversion cooperates with the general law of deviation."[53] Now he turned to family variability, opening the trap door beneath the second stage chute containing the pellets representing the largest seeds. They bounced merrily back and forth among the quincuncial arrays of pins distributing themselves normally in the bins at the bottom with greater dispersion than before. As he repeated the process, with chute after chute, the bell curves obtained from each individual chute coalesced into a single normal distribution with a dispersion similar to that of the original population.

Galton thought he had discovered "a typical law" of heredity, whereby dispersion in the first generation was restricted by reversion in the next, following which dispersion occurred again. The alternation of dispersion and reversion would continue generation after generation "until the step by step process of dispersion has been overtaken and exactly checked by the growing antagonism of reversion."[54] For Galton, reversion was like an elastic spring. "Its tendency to recoil increases the more it is stretched, hence equilibrium must at length ensue between reversion and family variability, and therefore the scale of deviation of the lower heap (of pellets) must after many generations always become identical with that of the upper one."[55] Galton next applied his theory to sexual selection, a concept proposed and extensively developed by Darwin in his book *The Descent of Man and Selection in Relation to Sex* (1871),[56] and to natural selection. His discussion of sexual selection was difficult to follow, especially for an audience that had already been worked hard. But much more serious was the problem reversion to the mean posed for natural selection. How could evolution proceed in small increments as envisioned by Darwin if counteracted by reversion to the mean at every generation? This was a problem that would continue to bother Galton in years to come.

What had Galton's sweet pea experiments really shown? His notion of the laws of reversion and family variability working against one another to create an equilibrium was an oversimplification. As Pearson pointed out, the problem was experimental in nature.[57] First, Galton believed sweet peas invariably self-fertilize, which Pearson considered only partially true. What if a progeny plant grown from a large seed had been cross-fertilized by a small-seeded plant? The progeny seeds from the supposedly self-fertilized large-seeded plant would have been smaller than predicted. Second, he selected seed based on size, not on parentage. For example, his large seeds would have included the ordinary produce of large-seeded plants, occasional large seeds from plants yielding seeds of average size, and exceptional large seeds from small-seeded plants. Thus, large seeds from average and small-seeded plants might have been large because of the environment in which the parent plant was grown. Hence, Galton's results contained systematic errors so that reversion toward the ancestral mean was a necessary outcome. Nevertheless, he had made three important discoveries. Although flawed, Galton's data suggested that reversion in seed size was to some extent heritable. His law of family variability was equally important, for it showed that quantitative traits, such as seed size, are normally distributed in successive generations. Lastly he had invented a new statistical concept, regression analysis. But Galton was not interested in sweet pea seeds. "It was anthropological evidence that I desired, caring only for the seeds as means of throwing light on heredity in man."[58]

By this time in his life meetings, committees, and lectures were consuming much of Galton's time. While people often doodle to pass the time away, the numerically obsessed Galton counted fidgets.[59] Yawns were one example, but the sum total of yawns was insufficient as he wanted to know who yawned and how often so he devised an instrument to keep track he called a "registrator."[60] One can imagine Galton at a particularly turgid lecture. He is wearing a pair of white cotton gloves. A swarthy man to his left yawns. In a pocket on the palm side of Galton's glove is a small white rectangle the size of a calling card. He pushes his thumb inward toward the card and the needle point poking through the fabric of the glove and attached to a sliver of wood sewn inside the glove makes a tiny hole in the card. A tall elegantly dressed friend stifles another yawn. Galton makes another prick starting a new column. The swarthy man yawns again. Galton makes second prick in the first column.

Despite Galton's deep involvement in his scientific activities, there was many a year when personal difficulties intervened. Louisa noted in her diary for 1874 they were "uneasy from the very beginning about Mrs Galton. Frank went to see her early in Feby and she died Feby 12th aged 90. This coming so soon after my dear Mother made a sad blank, both homes gone."[61] Later that year Louisa broke a blood vessel and nearly died, but "all were so kind and good to me, and Frank especially."[62] However, her illness proved more severe than she realized and then Galton got sick in December. They "had a quiet dull Xmas, no going out and F. had to give up his promised lectures in Newcastle."[63] Galton's mood was probably not improved by Dr. Clarke's dietary prescriptions, which included restricting himself to a pint of claret at dinner, eliminating afternoon tea, avoiding hearty meals, and a variety of cakes, spices, and coffee. Louisa's illness kept her pretty much housebound in 1875, but they managed to get off for France and Switzerland in late May, staying until the British Association meeting intervened in August. In 1876 events moved along on a reasonably even keel and they left England on August 23 and "met Emma in Calais and with her and Mr Broderick travelled thro' Bavaria the Tyrol and to Venice and the Italian lakes. We returned Oct. 23rd much better and stronger and had a quiet domestic settlement to my unspeakable comfort."[64] But Louisa was not out of the woods and 1877 was "a year of illness for me but marked by so much kindness, love and affection, that in looking back, it seemed full of sweetness and to have brought me nearer to Frank and to the dear friends who solaced me."[65] Louisa's health had improved by 1878, but in March Galton got "rheumatic gout in his knee ... and was a prisoner for nearly three weeks."[66] And so it went, with Louisa recording their lives and her husband busily writing, making pictures, scurrying off to meetings, and calculating frequencies of fidgets.

Galton's many commitments must have proved a perpetual strain for Louisa, but the British Association meeting in August was a particular nuisance as it usually terminated their vacations. Galton was often obliged to hurry home in some official capacity in the Geography or Anthropology sections. Darwin's death in May 1882 cast gloom on the Galton household, but Louisa noted in her diary that they made a "summer ramble on the Rhine, in the Black Forest, Constance, and lastly Axenfels."[67] Despite bad weather she noted with relief that "it was such a boon not to be kept by a British Association Meeting this summer."[68] In the summer of 1884, Louisa had her heart set on a visit to the south of France, but an outbreak of cholera made this dangerous. They settled instead in the Lake District and Louisa, though disappointed, wrote she was quite willing to do so "in default of my pet scheme, which I had especially cherished, as there was no English British Association to spoil our holiday, it being held in Montreal."[69]

For several years Galton had set anthropometric data-gathering aside, perhaps because a vital component was missing. Unlike his sweet pea experiments his measurements on English schoolboys did not permit him to make intergenerational comparisons. He wanted pedigrees and he wanted them filled with useful data on weights, heights, medical histories, etc. Characteristically, he launched a multifaceted attack. In a *Fortnightly* article in March 1882, he called for the formation of anthropometric laboratories "where a man may . . . get himself and his children weighed, measured and rightly photographed, and have each of the their bodily faculties tested, by the best methods known to modern science."[70] But he did not intend merely to consider the usual anthropometric measurements (height, weight, chest girth, etc.). He wanted measurements of energy, which he defined as "the length of time during which a person is wont to work at full stretch, day by day."[71] He wanted measurements of arm strength, estimates of eye-muscle coordination, sensory stimulation, and persistence of impressions. He also wanted these data accompanied by proper medical histories and photographs. These measurements ought to be taken in laboratories scattered about the land.

While writing the *Fortnightly* article Galton was busy correcting the proofs of a new book, *Inquiries into Human Faculty and its Development*.[72] Some chapters were reprints, while others expanded on topics alluded to in earlier articles. He pulled together the results of his twin studies, his thoughts on anthropometrics and statistics, as well as psychometrics, psychology, race, and population. His purpose in writing the book was "to touch on various topics more or less connected with that of the cultivation of race, or, as we might call it, with 'eugenic' questions."[73] In a footnote Galton defined his new word. Eugenics, he wrote, deals with "questions bearing on what is termed in Greek, *eugenes*, namely, good in stock, hereditarily endowed with noble qualities."[74] And so Galton had brought into the vocabulary a word whose dark connotations have ever since been associated with his name.

The anonymous critique in the *Saturday Review* thought that Galton's collection of papers and essays made "very entertaining reading."[75] His book proved "that life need never be tedious to an observing person."[76] The reviewer joked that people were unlike dogs for whom life would be boring except when "fighting, feeding, or making love, were it not for fleas. These active little creatures . . . have the province of keeping dogs from feeling time lie too heavily on their hands."[77] But the reviewer was skeptical of Galton's nostrums for improving the human race through selective breeding, writing that "we expect little from our short-sighted race."[78] Faced with the diverse array of thoughts, scientific inquiries, and hypotheses embedded in Galton's book, the writer acknowledged, "the reviewer pants after him in vain."[79]

The critic for the *Spectator*[80] surefootedly picked his way through Galton's kaleidoscope of papers to identify three major themes in the book: (1) the great variation "which occurs in the psychological development of different individuals of the human race"; (2) the large extent to which this was due to heredity rather than environment; and (3) the duty "individually and socially, to use every lawful means to further"[81] the hereditary improvement of the human race. But a weak point was Galton's attempt "to impress his readers with the feasibility of *improving the breed* of our fellow-men."[82] The *Guardian* reported that Galton's main goal was the improvement of the human race "and this mighty *desideratum*" was considered from various angles.[83] Later the reviewer faulted Galton for naïveté, since the desire to influence the "future of humanity" had been the grail of great men ever since Socrates' time. It struck the critic "as almost ridiculous in a living author to put this forward as a new aim," especially when every "father capable of reflection has not only hoped, but has tried to do so."[84] And then the reviewer got to the nub of the problem. "When Mr. Galton passes from the speculative to the practical region, we find much not only to question, but to condemn. Who is to decide whether a man's issue is not likely to be well fitted 'to play their part as citizens?' Do not weak men have strong children, stupid ones wise, wicked good?—while, on the other hand, do we not find the weak emanating from the strong, and bad from good?"[85]

George Romanes, an acquaintance of Galton's and one of the brightest of the second generation of Darwinians, wrote the review for *Nature*. Galton had "no competitor in regard to the variety and versatility of his researches."[86] He had gathered together "in one series most of the investigations which he has published during the last ten years," which revealed themselves not "to be separately conceived," but throughout to be "united by the bond of common object," which was revealed by the book's title.[87] Romanes criticized Galton for trying to apply "the statistical method" to the question of prayer as "of doubtful validity," but he liked Galton's "well considered suggestions" that race could be improved by selection.[88]

The first edition of *Inquiries* did not sell particularly well, but enough copies were purchased to justify another edition in 1892. This, as well as later editions, omitted three earlier chapters including the abbreviated reprint of Galton's article on prayer. The other two, "Enthusiasm" and "Possibilities of Theocratic Intervention," had been equally jarring not only to the pious, but even closer to home with Sister Emma. While praising his chapters on twins and animal domestication she could not "help greatly deploring what you have said on prayer. Whatever may be your ideas, I cannot see any reason for publishing the fact to the World. It is a grave responsibility on your part. I do hope in some of the later editions many of your friends will persuade you to abstract that part from your volume, which in others ways is so much to be appreciated. Forgive me dearest Frank for saying this, you know how dearly I love you and how proud I am of your talents."[89] Sister Adèle's daughter Millicent Lethbridge was also offended by the prayer chapters, and so apparently was Louisa, three of whose brothers had taken Holy Orders, although she may not have said so directly to her husband.

Inquiries provided a useful road map to where Galton had been and where he was going, but data were what he craved, especially anthropometric measurements to which he could apply statistics. So he had a clever idea. Why not enlist the help of doctors and their families? He wrote up his plea in 1883 in another *Fortnightly* article entitled "Medical Family Registers."[90] To get the medics interested Galton would offer prizes up to a total of £500 to those doctors who best succeeded "in defining vividly, completely, and concisely the characters (medical and other) of the various members of their respective families, and in illustrating the presence or absence of hereditary influences."[91] Later he identified what are now recognized as two ethical dilemmas surrounding genetic disease. Should you inform yourself about whether you carry a genetic disease? Galton answered that it was "ignoble that a man should be such a coward as to hesitate"[92] to do so. If you find you have a genetic disease, should you tell other family members? "Parents may refrain from doing so through kind motives; but there is no real kindness in the end."[93]

The doctors didn't bite, but Galton, undaunted, widened the net. In early 1884 *Macmillan's* published a slim quarto pamphlet entitled *The Record of Family Faculties*.[94] According to the accompanying "fly-leaf," Galton was offering a total of £500 in prizes. The deadline for receipt of the completed family record was May 15, 1884. The record to be completed was 50 pages long and requested detailed pedigree information spanning four generations. A couple filling out Galton's record was asked to complete it for their children and for direct ancestors extending back through great grandparents. This 15-page section was the heart of the questionnaire and elicited detailed information on height, hair color, energy, mental powers, diseases, etc. In the rest of the record entries could be added on brothers, sisters, uncles, aunts, etc. This time Galton was successful. A total of 150 returns were received and he gave out 85 awards ranging from £5 to £7. In parallel with *The Record*, the *Life-History Album* was published with Galton as editor.[95] This was the work of a subcommittee on Life-History, chaired by Galton, whose parent was the Collective Investigation Committee of the British Medical Association. The idea was to present the *Life-History Album* to expectant parents or to parents with young children with the hope their progeny would later take over the *Album*. Once again Galton had invented a device for keeping detailed pedigree data.

Although 1883 was a whirlwind of activity for the ever-inquisitive Galton, it was also a year of tragedy.[96] In February Louisa's brother Montagu lost his wife Georgina. In late June Galton's good friend William Spottiswoode died and was buried in Westminster Abbey. But year's end brought an even more wrenching event when Galton's sister Adèle, dear Delly who had given him his first lessons, died on New Year's Eve. But for Adèle this was a blessing, as Louisa recorded. "She hoped she would not live to see the New Year and her wish was granted. She was weary with pain and suffering."[97]

The International Health Exhibition that opened on Thursday, May 8, 1884, in the Gardens of the Royal Horticultural Society in South Kensington was a major event and one Galton seized upon to establish an Anthropometric Laboratory.[98] It was in the tradition of the Great Exhibition of 1851, held in nearby Hyde Park, which, in many ways, had been responsible for the making of South Kensington as a citadel of museums and science.[98,99] The Brompton lanes, verdant and beckoning in summer, but dark and muddy by winter were gone, as was Lady Blessington's old mansion that had been razed and replaced by the towering circular structure of the Royal Albert Hall. The Royal Horticultural Society had moved from Chiswick to occupy the gardens at its rear, followed by the building of the South Kensington Museum, the precursor of the Victoria and Albert, and the National Art Training Schools. The Science Schools, the Indian Museum, the National Portrait Gallery, the New National History Section of the British Museum, and other institutions soon added to the hubbub of this intellectually charged part of London not far from Galton's home at 42 Rutland Gate.

The exhibition was designed to illustrate all manner of substances, fabrics, constructions, processes, and appliances related to food, dress, living space, and devoted to the management of schools, workshops, and factories as they were affected by sanitary conditions. The objects were arranged in five groups and carefully chosen to pique the curiosity. Depending on where one wandered one could see kitchens in which bread and pastry were being baked; figures displaying the history of the British national costume; textbooks and diagrams; toys and flower shows; and Chinese Courts with palanquins and lanterns. One could sample an ice-cream or a cherry cobbler or buy a sixpenny dinner in the restaurant of the Vegetarian Society, part of the colossal café where virtually any taste could be gratified. Unlike the Great Exhibition, the thirsty soul could purchase a variety of alcoholic beverages, some as exotic as an American mint julep or an arrack punch. The organizers deemed this essential because working class visitors to the Great Exhibition had simply brought stone bottles of beer with them or had piled into the pubs of nearby Knightsbridge, Kensington, and Brompton, making fortunes for their owners.

Brass bands played under the dazzling electric lighting that illuminated the Exhibition gardens and pavillions. The fountains, set out in the water garden and spurting to great heights with searchlights playing over them in the evening, were a masterpiece in pure artistry and ingenious engineering.[100] There were lectures and conferences and displays of all sorts of inventions and devices. Madame Eugénie Genty exhibited her new patented Health Busk, which enabled an indisposed lady to unclasp her corset instantaneously. In the South Gallery some 100 exhibitors displayed preserved foods such as fruit, vegetables, salmon in tins or smoked or salted or compressed, sausages, game, and army and navy rations. Many women paused before the glass display case of the International Fur Store, framed on either side by the antlered head of a great stag. Its centerpiece was a mannequin clad in a magnificent ermine robe. Cosmetics for milady were on view including perfumes and toilet soaps, with an especially handsome exhibit by the celebrated firm of A. and F. Pears of 38 Great Russell Street. In front of the display were portraits of four well-known ladies who endorsed Pears's products, including Mrs. Lillie Langtry, whose lovely profile was accompanied by these words. "Since using Pears' soap for the hands and complexion *I have discarded all others.*"

With its mobs of happy sightseers enjoying the multitude of objects arrayed to capture the eye and to whet the appetite, this venue was the perfect place for Galton to entice the curious into his Anthropometric Laboratory. Its original plan had been drawn up with the help of his friend George Croom-Robertson, the chair of Mental Philosophy and Logic at University College, London. In anticipation of its construction Galton circulated a pamphlet in the spring of 1883 stating that he wished "to compile a list of instruments suitable for the outfit of an Anthropometric Laboratory, especially those for testing and measuring the efficiency of the various mental and bodily powers."[101] He requested any useful information or suggestions as to instruments, methods of measurement, and so forth. Once the Anthropometric Laboratory was almost a reality, Galton wrote the leading psychologists in England to describe what he had in mind requesting "any special apparatus that you would allow me to exhibit in your name."[101] But there was little equipment available in England. The pioneering work on experimental psychology and sensory perception was being done in Germany by men like Herman Helmholtz and Wilhelm Wundt. In fact Galton had already established such a reputation for designing psychometric instruments that one eminent English psychologist wanted to visit his laboratory because "I expect you know a great deal more about the whole thing than I do."[102]

Galton's laboratory, six feet wide and 36 feet long, was fenced off from the adjacent gallery by open latticework so bypassers observing the proceedings could be induced to participate.[103] Once an onlooker was lured to the entrance he met a doorkeeper, who collected a three pence admission fee, and passed him on to the superintendent, Serjeant Williams.[104] The visitor received a frame containing a card with various entries on it for the measurements to be taken (Fig. 15-4) over which a thin transfer paper had been stretched with a piece of carbon paper in between.[105] At the end of the laboratory the participant was given the original card containing his vital statistics and ushered out the door with the superintendent keeping the carbon copy. In the evenings

5178 INTERNATIONAL HEALTH EXHIBITION, 1884.

ANTHROPOMETRIC LABORATORY,
Arranged by FRANCIS GALTON, F.R.S.

Sex M Colour of eyes Brown Date 5/9/84 Initials MXMS

EYESIGHT.
Greatest distance in inches, of reading "Diamond" type: right eye 34 | left eye 34

Colour sense, goodness: good

JUDGMENT OF EYE.
Error per cent. in dividing a line of 15 inches: in three parts 0 | in two parts ½

Error in degrees of estimating squareness: 0

HEARING.
Keenness can hardly be tested here owing to the noises and echoes.

Highest audible note: between 40,000 and 30,000 vibrations per second.

BREATHING POWER.
Greatest expiration in cubic inches: 204

SWIFTNESS
of blow of hand in feet per second: 19

STRENGTH
of squeeze in lbs. of right hand 70 | left 63 of pull in lbs. 67

SPAN OF ARMS
From finger tips of opposite hands: 5 feet 4.8 inches

HEIGHT
Sitting, measured from seat of chair: 2 feet 11.5 inches

Standing in shoes: 5 feet 5.8 inches
less height of heel: 1 inches
Height without shoes: 5 feet 4.8 inches

WEIGHT
in ordinary in-door clothing in lbs.: 124

Age last birthday?
Married or unmarried?
Birthplace? Edinburgh
Occupation? Clerk
Residence in town, suburb or country?

MR. FRANCIS GALTON'S ANTHROPOMETRIC LABORATORY

The Laboratory communicates with the "Western Gallery" in which the Scientific Collections of the South Kensington Museum are contained. The Western Gallery runs parallel to Queen's Gate, and is entered from the new Imperial Institute Road. Admission is free.

Date of Measurement. Day. Month. Year.	Initials.	Birthday. Day. Month. Year.	Eye Color.	Sex.	Single, Married, or Widowed?	Page of Register.
8 4 93	J.E.	14 11 72	Grey	m	S	4669

Head length, maximum	Head breadth, maximum	Height standing, one heels of shoes	Span of arms from opposite finger tips in front of chest.	Weight in ordinary clothing.	Strength of grasp. Right hand / Left hand.	Breath capacity.	Interval perceived across nape.	Color Sense.	Keenness of Eyesight. Diamond Numerals. read at inches. Right eye / Left eye.	Smallest Snellen's type read at 6 metres. Right Eye / Left Eye.
Inch. Tenths.	Inch. Tenths.	Inch. Tenths.	Inch. Tenths.	lbs.	lbs. lbs.	Cubic inches.	min.	? Normal.	No. No.	No. No.
7.69	6.05	67.5	72 9	161	112 108	230	8	Yes	25 25	5 5

Height sitting above seat of chair.	Length from elbow to finger tip. Left arm / Right arm	Length of middle finger	Keenness of hearing.	Highest audible note. (by whistle)	Reaction time. in hundredths of a second	Greatest speed of blow with fist.	Left Thumb.	Right Thumb.
Inch. Tenths.	In. tenths. In. tenths.	In. tenths.	? Normal.	Vibrations per second.	To Sight / To Sound	Feet per sec.		
35.5	18.5 18.75	4.6	Yes	19.000	19 13	broke		

One page of the Register is assigned to each person, in which his measurements at successive periods are entered in successive lines. A copy of these made at any specified date may be obtained on application by the person measured, or by his or her representative, at the cost of sixpence and postage.

Fig. 15-4 Record of anthropometric measurements. A. The form used in the original Anthropometric Laboratory. B. The form used following the move of the laboratory to the South Kensington Museum, which includes thumbprints. Reproduced courtesy of University College London Library Services from the Galton Archive, List No. 137.3.

Mr. Gammage, an optical instrument maker on the Brompton Road, made sure the various instruments were properly adjusted for the 90 or so subjects who would pour in the next day.

The examinee took the various tests in sequential fashion over about a half an hour, being measured with devices designed to detect "Keenness of Sight and of Hearing; Colour Sense, Judgment of Eye; Breathing Power, Reaction Time; Strength of Pull and of Squeeze; Force of Blow; Span of Arms; Height, both standing and sitting; and Weight."[106] However, Galton studiously avoided heads, a major reason being the difficulty of making accurate measurements on women "on account of their bonnets, and the bulk of their hair."[107] His instruments had to be strong, simple, and easy to understand since "the stupidity and wrong-headedness of many men and women" is "so great as to be scarcely credible."[108] For instance, the equipment used to measure the strength of a blow was a simple rod made of fir protruding out of a tube so that it could move freely. To the top of the rod a buffer was attached so the subject could smash his fist down on it without hurting himself. At the other end of the tube was a spring. A pointer attached to the rod allowed one to calculate the distance the rod travelled into the tube following the blow. However, one man punched the rod on the side instead of the top and broke it. Galton replaced it with "an oaken one, but this too was broken, and some wrists were sprained."[109] Alas, anthropometrics could be a dangerous game, but for most, the information they left with seemed worth the thruppence invested.

The Anthropometric Laboratory was very popular. The crowd was generally orderly although Serjeant Williams ushered out the occasional inebriate, and parents and children had to be kept separate since "the old did not like to be outdone by the young, and insisted on repeated trials."[110] By the time the International Health Exhibition closed in 1885, Galton had compiled data on 9,337 individuals, each measured in 17 different ways. The Anthropometric Laboratory was briefly put out to pasture on unoccupied land belonging to the Imperial Institute, but then moved to a room in the Science Galleries of the South Kensington Museum at Galton's request. There anthropometrics continued to flourish and, during the first three years alone, statistics were collected on 3,678 people, with repeated measurements being performed on a number of individuals. The form now included more parameters (Fig. 15-4). The head was no longer taboo and prints of the left and right thumbs were added as Galton was now interested in fingerprinting (chapter 17). One of his visitors was W. E. Gladstone, the Grand Old Man, and frequent prime minister. Gladstone "was amusingly insistent about the size of his head, saying that hatters often told him that he had an Aberdeenshire head—'a fact which you may be sure I do not forget to tell my Scotch constituents.'"[111] Gladstone asked Galton if he had "ever seen as large a head as mine?"[112] Galton replied undiplomatically, "Mr Gladstone, you are very unobservant!"[113]

As the mountains of data accumulated, Galton became engrossed in developing methods for their analysis. In a paper in the *Journal of the Anthropological Institute* in 1884, he made use of percentiles because they provided a convenient method of summarizing different data sets from many individuals.[114] Percentiling was a concept he introduced in an 1875 paper, but it was not until now that he coined the word.[115] One of the measurements Galton quantified was strength of squeeze in pounds. He reported his results with an attempt at humor, saying "very powerful women exist, but happily perhaps for the repose of the other sex, such gifted women are rare. Out of 1,567 adult females of various ages measured . . . the strongest could only exert a squeeze of 86 lbs., or about half that of a medium man. The population of England hardly contains enough material to form even a few regiments of efficient Amazons."[116]

Some wag at *Punch* picked this up and couldn't resist some fun at Galton's expense so this bit of doggerel appeared in April 15, 1884, issue.[117]

The Squeeze of 86

Maiden of the mighty muscles
   There recorded, you would be
Famous in all manly tussles,
   And its very clear to me,
That if in the dim hereafter
   Any husband should play tricks
You would with derisive laughter,
   Give a "Squeeze of 86."

Husbands be it sadly stated,
   Have been known their wives to whack,
You, unless you're over-rated,
   Could give endearments back.
Yours the task to try correction,
   Till your husband and your "chicks,"
Had a lively recollection
   Of your "Squeeze of 86."

In summing up the achievements of the Anthropometric Laboratory in 1924, Pearson noted that Galton had succeeded in collecting an immense amount of data, "which only forty years later is being adequately reduced."[118] In the process of analyzing his results Galton refined his concept of regression, invented the correlation coefficient, and proposed his ancestral law of heredity. In 1889 he published what was probably his most important and influential book, *Natural Inheritance*. There he summarized all of the statistical work he had done between 1877 and 1888. This book would become the foundation stone of modern biometrics.

SIXTEEN

Probing the Mind

> I thought it safer to proceed like the surveyor of a new country, and endeavour to fix in the first instance as truly as I could the position of several cardinal points.
>
> —Francis Galton,
> *Inquiries into Human Faculty
> and Its Development*[1]

From 1877 until 1885 composite photography and psychological studies occupied much of Galton's time. Both subjects might provide him a direct entree into understanding mental ability. As always, Galton wanted to apply quantitative techniques where possible and to determine heritability. Galton's adventures in composite photography relate directly to his budding interest in personal identification and ultimately in the use of fingerprinting as a means for distinguishing different individuals.

Galton's interest in mental phenomena was not new. Following his breakdown at Cambridge in 1843, he spent much of the summer in Germany with his sister Emma. While in Dresden he ventured across the Austrian border and learned mesmerism from an acquaintance there, the technique being forbidden in Saxony.[2] He "magnetised some eighty persons," subsequently deciding the procedure was "unwholesome" and never attempted it again.[3] Galton's curiosity about mesmerism was not surprising in view of the vogue it was enjoying in England at the time.[4] In his late twenties Galton had also delved into phrenology, popularized in Great Britain by the enterprising Scot George Combe,[5] and was examined by the chief phrenologist of the London Phrenological Institution.[6] Eventually phrenology lost its luster for him and he later wrote an acquaintance that "the localisation in quite modern times of the functions of the brain lends so far as I am aware no corroboration whatever, but quite the reverse to the divisions of the phrenologist."[7]

Galton also dabbled in spiritualism. His interest was probably whetted by his friend Sir William Crookes.[8] Crookes, a distinguished chemist, had convinced the well-known medium Daniel Dunglas Home to submit to a series of experiments designed to test his supernatural powers.[9] Home's stock-in-trade was mysterious rapping. His séances became famous and he added other inexplicable phenomena to his collection, including levitation and guitar-playing without hands. By the early 1860s, Home's ability to conjure up supernatural phenomena had caught the attention of scientists. Among his early converts were John Elliotson, the dynamic professor of practical medicine at University College London and renowned mesmerist, and Dr. Lockhart Robertson, editor of the *Journal of Mental Science*. Home submitted to the first series of experiments at Crookes's house in 1871. Crookes became convinced of Home's mystical abilities and published accounts in the *Quarterly Journal of Science* in 1871 and 1874. Galton became involved in these experiments. He wrote Darwin on March 28, 1872, describing the séances and commenting that "the absurdity on the one hand and the extraordinary character of the thing on the other, quite staggers me; wondering what I shall yet see and learn I remain at present quite passive with my eyes and ears open."[10] By the fall of 1872 Galton was becoming skeptical. He wrote Darwin that Crookes informed him that when Home was the medium "the experiments were far more successful" than otherwise. The spiritualism experiments continued until 1874, by which time the participants were pretty well convinced they were witnessing fraud.

In the summer of 1877 Galton found unexpectedly that he was to address the Anthropology Section of the British Association meeting at Plymouth.[11] Because he lacked time to prepare the usual formal remarks, Galton begged the audience's indulgence. He would confine himself to topics in which he "had been recently engaged," notably composite photography. If a certain group of individuals shared a particular mental trait and this was somehow reflected physically, the common features might be extracted by superimposing photographs of their faces upon one another. This should factor out the unique features and emphasize shared attributes, creating, as it were, a photographic mean or average. The idea that a composite photograph might reveal something about personality and the mind was not at all unreasonable given that Lavater's physiognomy still held a spell over Europe. Physiognomy supposed a direct relationship between a person's inner being and his or her outer physical appearance.

Galton decided to focus on criminals, a decision that owed much to a suggestion of Sir Edmund Du Cane, the director-general of Prisons.[12] With Du Cane's help Galton examined many thousands of photographs of criminals and procured copies of the ones that interested him. He requested that names not be attached to the pictures, but that they be classified in three groups according to the nature of the crime.[13] "The first group included murder, manslaughter, and burglary; the second group included felony and forgery; and the third group referred to sexual crimes." As Galton pored over the photographs sorting them, arranging them, and comparing them, he found that "certain natural classes began to appear, some of which [were] exceedingly well marked."[14] His feeling of excitement increased when he discovered "that the three groups of criminals contributed in very different proportions to the different physiognomic classes."[15]

In 1878 he read his first major paper on composite photography before the Anthropological Institute. He described his methods, remarking that "it was while endeavouring to elicit the principal criminal types by methods of optical superimposition of the portraits, such as I had frequently employed with maps and meteorological traces, that the idea of composite figures first occurred to me."[16] He overlaid photographs of eight different individuals (Fig. 16-1). The photos were all of the same size with the subjects posed similarly. The image in each picture was then exposed to the same photographic plate for ten seconds. Galton was pleased with his results for violent criminals as the uniquely "villainous" characteristics of each face had vanished, revealing "the common humanity that underlies them."[17] However, he cautioned that his composite photographic portrait had not identified the criminal, "but the man who is liable to fall into crime."[18] More dangerously, he mused over whether the composites might give "typical pictures of different races of men, if derived from a large number of individuals of those races taken at random."[19] Thus the reason for Galton's interest in composite photography becomes obvious. If physical features reflected specific behavioral or racial traits and if these were inherited, composite photography provided a rapid means of identifying these characteristics.

Following Galton's presentation Du Cane explained why he supplied Galton with the photographs. If criminals possessed "certain special types of features" and "certain personal peculiarities distinguish those who commit certain classes of crime; the tendency to crime is in those persons born or bred in them, and either they are incurable or the tendency can only be checked by taking them in hand at the earliest periods of their life."[20] Galton's procedure would help to establish this point, because if a distinguishing feature existed "it would come out in his mixed photographs in a clear line, whereas in those features which do not correspond the lines would be more or less blurred."[21]

But composite photography failed to reveal features typifying different groups of criminals and Galton admitted defeat in *Inquiries into Human Faculty and Development*, writing that the composites "produce faces of mean description, with no villainy written on them."[22] But this did not mean that composite photography might not be applicable in other ways. He wondered if composites of families could be used to forecast the physical appearance of

Fig. 16-1 Composite photographs made from portraits of criminals convicted of murder, manslaughter, or crimes of violence. From Karl Pearson, *Life* II: plate 28.

offspring of a proposed marriage. The composites produced took into account the degree of relatedness (e.g., father versus uncle) in terms of exposure time (e.g., longer versus shorter). Galton also practiced on ancient coins, producing composites of coins bearing the face of Alexander the Great or Cleopatra and so forth. Perhaps composite photography could be used to define a "physiognomy of disease," to identify heritable facial features, and perhaps to stereotype specific racial types. He collaborated with Dr. F. A. Mahomed of Guy's Hospital, London, and made composites of tuberculosis patients to see if they had anything in common. Together they photographed individuals, mostly outpatients, at Guy's Hospital, the Brompton Consumptive Hospital, and the Victoria Park Hospital for Diseases of the Chest.[23] For each patient, data were collected on age, duration of the disease, how advanced it was, time of onset, and whether there was evidence for heritability. Nontubercular patients served as controls. Mahomed and Galton concluded the tubercular physiognomy was no more common among people with the disease than among those suffering from other ailments. Perhaps the mentally diseased had characteristic physiognomic features. To find out, Galton photographed lunatics, but their features were so irregular they could not be blended. At the Hanwell asylum his photographer had an unsettling experience. The second patient to be photographed was insulted not to be first.[24] He considered himself a man of extraordinary importance, perhaps Alexander the Great, so "when the photographer had his head well under the velvet cloth, with his body bent, in the familiar attitude of photographers while focusing, Alexander the Great slid swiftly to his rear and administered a really good bite to the unprotected hinder end of the photographer."[25]

Galton also enlisted the help of Darwin's son Leonard, a lieutenant in the Royal Engineers and later president of the Eugenics Society, to obtain photos of officers and privates from which to make composites. He combined privates and officers and was pleased with his results. Although the composite was somewhat blurry it had "an expression of considerable vigour, resolution, intelligence and frankness."[26] Later Galton remarked that "this face and the qualities it connotes probably gives a clue to the direction in which the stock of the English race might most easily be improved."[27] He also published two papers in *The Photographic News* in April, 1885,[28] where he presented full face and profile composites of Jewish boys. He was elated and so was Pearson. As Pearson approvingly put it. "We all know the Jewish boy, and Galton's portraiture brings him before us in a way that only a great work of art could equal— scarcely excel, for the artist would only idealise from *one* model."[29] Galton felt he had demonstrated the utility of composite photography as an anthropometric tool and that it might be useful for racial characterization. He also believed that composite photography would reveal heritable physical features, so he urged providing standardized photographs, full face and profile, in keeping records of family histories. In a sense his hunch seems vindicated, for one of the projects being carried out today at the Galton Laboratory at University College, London, is aimed at understanding the genetics of human facial features.[30] These modern scientists use high-tech scanning equipment rather than composite photography, but to obtain the information they desire they use a method that Galton would have been comfortable with. "Volunteers will come for a 30 minute appointment for facial scanning and we will ask you to give us some information about your family."[31] Galton probably would have awarded monetary prizes as an inducement.

However, Galton sought a Rosetta Stone to the mind, and composite photography did not provide it. He needed some way to measure mental powers, but how was he to do it? The IQ test was still nearly a quarter century away, so how was he going to quantify the mind and relate this to heredity? He was, of course, thinking about anthropometrics, but what would his measurements tell him even if "the innate moral and intellectual faculties" were very "closely bound up with the physical ones"?[32] He began exploring his own mind and the minds of his friends searching for clues that might help him. His self-examination was designed to gain some idea "of the crowd of half-thoughts and faint imagery" that flit through a man's brain "and of the influence they exert upon his conscious life," he wrote in "Psychometric Facts" (1879), a popular article in *Nineteenth Century*, while adding "psychometric" to the lexicon of psychology.[33] In one of his first experiments, he walked from the Athenaeum Club along Pall Mall to St. James's Street, a distance of about 450 yards, keeping track of the ideas travelling through his consciousness as different objects caught his eye. He was "amazed at the amount of work" his brain had performed. His "mind had travelled" discursively "during that brief walk . . . through the experiences" of his entire life. It had "entered as an habitual guest into numberless localities that it had certainly never visited under the light of full consciousness for many years."[34] His "everyday brain work was incomparably more active, and . . . [his] ideas travelled far wider afield"[35] than he had previously thought possible.

Galton was now convinced that mental imagery was a useful tool to explore the human mind. But how was one to quantify these evanescent impressions passing so quickly through the mind's eye? He hit upon word association experiments. He would suddenly display a printed word and allow "a couple of ideas to successively present themselves, and then, by violent mental revulsion and sudden awakening of attention" he seized them "before they had faded," recording "them exactly as they were at the moment when they were surprised and grappled with. It was an attempt like that of Menelaus, in the Odyssey, to constrain the elusive form of Proteus."[36] Galton extracted 100 words from a dictionary all beginning with the letter "a," settling on about 75 for his experiments. Always the diligent counter, he found that once he displayed a word, the formation of two successive ideas associated with it took a little less than two and a quarter seconds. From this he calculated a rate of about 3,000 ideas per hour. In the space of a month he made 300 separate trials with a yield of 550 ideas in 660 seconds. But when corrected for recurrent ideas, the number dropped to 279. He estimated recurrences as a function of the word displayed, and classified ideas to the period of his life with which they were associated. He discovered that recurrent ideas tended to date "back to the period when I had not yet left college, at the age of twenty-two."[37] His nonrecurrent ideas fell into three groups. The most numerous were ones that gave "a vague sense of acting a part."[38] It was as if the actors were part of himself, but that he was simultaneously a spectator. The second group consisted of images "such as mental landscapes, sounds, tastes, &c."[39] The third was made up of verbal associations including "the mere names of persons or things, or bits of quotations in prose or verse."[40]

On Galton's mental video screen scenery and localities were most easily visualized, while faces were harder to formulate. Some ideas were remarkably vivid, but the background of mental imagery was constantly changing in "colour, tint, and pattern."[41] He was so impressed by the workings of his subconscious that he began to doubt the importance of consciousness "as a helpful supervisor, and to think that my best brain work is wholly independent of it."[42, 43] A more precise version of Galton's *Nineteenth Century* article was published in *Brain*,[44] to which Sigmund Freud subscribed,[45] but Freud never referred to Galton's paper nor did he credit Galton for suggesting the importance of unconscious mental processes.

Having been his own guinea pig Galton was ready to extend his studies to others. He began to read extensively in the emerging field of psychology, particularly about mental imagery.[46] He noted the work of the English psychologist Henry Maudsley, who stressed the importance of unconscious phenomena.[47] But he especially admired the research of the German psychologist Gustav Theodor Fechner.[48] He had read Fechner's *Elemente der Psychophysik* (1860) and lamented the absence of an English edition. It is not surprising that Fechner's work attracted Galton since Fechner, a former physicist, was seeking an "exact science of the functional relations or relations of dependency between body and mind."[49]

To collect data on mental imaging Galton designed a questionnaire, distributing several hundred copies between November 1879 and April 1880.[50] The first five questions were designed to test the respondent's ability to recall definite objects. "Suppose it is your breakfast-table as you sat down to it this morning—consider carefully the picture that rises before your mind's eye."[51] Galton wanted to know how bright the image was, how complete (i.e., details of the breakfast table), the degree of definition, and the colors of the objects (e.g., china, toast, mustard, meat). Questions six through 16 referred to definite kinds of images such as the details of scenery, the features of specific individuals, geographical descriptions, future chess moves, and mental figures associated with specific numerals. The last three questions related to the recall of tones of voice, smells, and tastes. Thus was born the "breakfast table questionnaire," a tool long-used by mental imagers.

Galton corraled his respondents through social networking. In November 1879, he spoke about his imagery studies before the membership of the Birmingham Philosophical Society, afterwards distributing copies of his questionnaire among the audience. This netted 11 responses. He coerced friends in the Royal Society, the Royal Institution, the Royal Geographical Society, and other learned societies to which he belonged to fill out the questionnaires. His friend George Croom Robertson, the editor of *Mind*, badgered some of his regular contributors into filling out questionnaires. Eventually, Galton received replies from 107 men and 180 women. The questionnaires were also sent to teachers who had responded to a published appeal. Galton summarized his results in "Statistics of Mental Imagery," published in *Mind* (1880).[52]

His goal was to analyze mental imagery statistically and determine the degree to which this quality was inherited (Fig 16-2). He culled through his questionnaires and decided to restrict his analysis to 100 adult men plus 172 boys from the Charterhouse School. He felt the Charterhouse data were his best for schoolboys, since the science master, Mr. Poole, obtained data from all of the boys in his classes after fully explaining the questionnaire's purpose. He did not analyze the data he had obtained from the 180 women although he made qualitative comments in the text. Initially, however, the scientists thought Galton had gone bonkers when he questioned them about mental imagery. "To my astonishment," the great majority "protested that mental imagery was unknown to them, and they looked on me as fanciful and fantastic in supposing that the words 'mental imagery' really expressed what I believed everybody supposed them to mean."[53] He likened the scientist to the "colour-blind man who has not discerned his defect."[54]

In contrast most other men, and even more women, as well as boys and girls, imaged vividly. They could often describe their images in detail. This made the inability of scientists to image even more puzzling. Galton ruminated over the problem, concluding that imaging had not really been lost in "the highest minds," but instead was "subordinated," being "ready for use on suitable occasions."[55] Having satisfied himself that scientists were not so unimaginative after all, Galton attempted to quantify mental imagery in terms of vividness, color, extent of field of mental view, and position of a mental image (e.g., corresponding to reality, in the head, just in front of the eyes). While he acknowledged it was premature to generalize, his results were sufficient to "give a fair knowledge of the variability of the visualising faculty in the English male sex."[56] Moreover he hoped that his statistical comparisons would

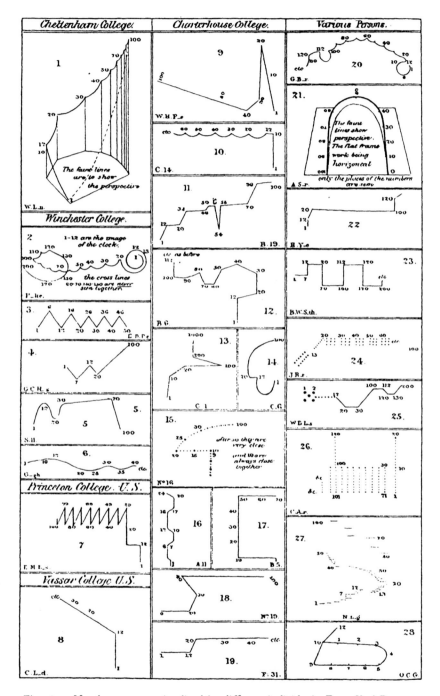

Fig. 16-2 Number patterns visualized by different individuals. From Karl Pearson, *Life*, II, plate 24.

"convince psychologists that the relative development of various mental qualities in different races admits of being pretty accurately defined."[57]

Despite Galton's own certainty of the significance of his results, many scientists had their doubts. The anatomist John Marshall returned his questionnaire with this comment. "I cannot place the slightest reliance on what a *person* says of the powers of seeing things with his mind's eye. . . . I believe that the replies you will get will be full of fallacies; and not of scientific value. . . ."[58] Galton pencilled in his notebook that "Marshall is obtuse" and very likely omitted him from his sample.[59] Other scientists were more polite, but just as skeptical. Sir James Crichton Browne, a distinguished physician, cautioned Galton that "the images which I see in my mind's eye, are never in any way comparable with the images which I see with my bodily eye."[60] He was concerned with "the enormous difficulty of accurately conducting such a process of mental analysis"[61] and worried about "the innumerable sources of error"[62] that would beset Galton's findings. The naturalist John Ball implied Galton was being a little naïve, saying "I think you will be led to recognise a great deal of complexity in many mental operations that are simply set down to the power of *visualising*—to use your convenient term."[63]

The visualization of numerals particularly intrigued Galton. In March 1879 Galton wrote to George Bidder junior, the son of a famous calculating prodigy, inquiring about his imaging abilities.[64] Bidder senior, as a boy gradually acquired the ability to perform arithmetically complex calculations in his own mind with great rapidity.[65] He eventually became a professional engineer, but was often called before parliamentary committees to deal with questions involving complex calculations. His son possessed his father's ability to visualize numbers and replied to Galton describing the novel way in which numerals appeared in a distinct spatial arrangement. Galton soon discovered that the ability to image such "number forms" was possessed by "about one man in 30, and one woman in fifteen."[66] While chairing a meeting of the Anthropological Institute on March 9, 1880, Galton summarized his findings on how people visualized numerals. He himself saw no "Form," but those who did arranged numbers in a variety of patterns in their mind's eye. There were "as many varieties as there are persons," but he drew attention to certain common features. Most of his respondents agreed that their "number forms" existed as far back as they could remember, predating the time they began to read. The figures frequently ran "to the left, and more often upwards than downwards."[67]

A lively discussion followed. Many of the commentators possessed the power to create mental number forms. Dr. Daniel Hack Tuke, a physician and coeditor of the *Journal of Mental Science*, remarked that "the grey matter of the visual centre of Mr Bidder and others who have given us their experience tonight" ought to contain "exquisitely adapted . . . cells possessing a receptive and retentive power to a superlative degree."[68] Colonel Yule remarked that he "had been visualising for a good deal more than 40 years, and but for their friend Mr. Galton he should never have become aware of the fact."[69]

Galton also wrote a popular article "Mental Imagery" for the *Fortnightly Review* (1880).[70] He did not detail his results, simply saying that he drew his conclusions "from no small amount of testimony."[71] He explained in lay language how he had organized his data using quartiles, with the two central quartiles containing "the broad middle class" while the first and last quartiles included "the exceptional cases." Thinking, perhaps, of George Bidder and his father, Galton remarked "that the visualising faculty is a natural gift, and, like all natural gifts, has a tendency to be inherited."[72] But it was "among uncivilised races that natural differences in the visualising faculty are most conspicuous."[73] He cited the marvellous abilities of Bushmen to draw cave paintings of men and animals and the ability of the Eskimo to draw maps. He mentioned the case of an Eskimo who drew from memory a map of 1,100 miles of Canadian arctic coastline he had skirted in a canoe. On comparing this "outline with the Admiralty chart of 1870 their accordance is remarkable."[74]

Galton presented his results in an address before the British Association in August 1880, and his speech was reviewed in the *Times*.[75] The reviewer, who had read the *Fortnightly* article, was quite critical. Having summarized Galton's main points, he took aim and fired. It was unfortunate "when a learned and skilful explorer, in a field of knowledge which is comparatively new to him ... omitted to make himself fully conversant with the labours of those who have previously trodden the same path."[76] He faulted Galton for departing from accepted terminology and for laboriously establishing "a good deal which has long formed part of the common stock of mental physiologists."[77] Then the reviewer got down to specifics. In his *Fortnightly* essay Galton had tried to contrast "sight" with "sight-memory." In the case of sight Galton wrote that the eye would record a flash of light and that the resulting "irritation" was propagated from the retina to the termini of the optic nerve and thence to the brain. Once the "irritation" reached the brain "it would be distributed in various directions, becoming confused with other waves of irritation proceeding from independent centres, lingering here and there longer than elsewhere and finally dying away."[78] Galton claimed that sight-memory of the flash involved a similar sequence of events, but they occurred in the reverse order.

Galton's critic disputed this description. What Galton called "a 'wave of irritation' is generally described as the conveyance of an impression; and the word irritation is hardly applicable to any stimulus which leads naturally to the performance of a natural function."[79] Galton's "notion of waves of irritation proceeding from independent centres"[80] was nonsense, according to current belief where "the received physiological doctrine is that the impressions are conveyed from peripheral organs."[81] His idea of reverse transmission of sight-memory from the brain outwards was pure nonsense in light of what "physiologists have been able to discover."[82] The reviewer lectured Galton for not citing the work of W. B. Carpenter, who had anticipated him in what was, "perhaps, the most interesting part of his work."[83] Galton, stung by the the *Times* critique, began penning a rebuttal, but then set it aside.[84,85]

Galton's ultimate impact in the field of experimental psychology was actually quite substantial. He is remembered not just for his pioneering work in mental imagery, but also for developing statistical tools like the correlation coefficient, which are widely used and, perhaps, abused in psychometrics today. Galton rates a chapter in Raymond Fancher's book *Pioneers of Psychology*.[86] Herrnstein and Murray tip their hereditarian hats to Galton in the introduction to *The Bell Curve* for trying to quantify individual differences in mental abilities and, naturally, for inventing the correlation coefficient.[87] An offering laid before Galton's intellectual altar is virtually obligatory in books on mental imaging. Alan Paivio in *Images in Mind* pays homage to Galton's famous "breakfast table" questionnaire.[88] In a volume of *Advances in Psychology* (1991) devoted to mental images[89] the following kinds of comments are to be found. Norman E. Wetherick remarks that Galton in 1883 "had already shown that introspectible mental contents vary markedly from person to person and it follows that theories asserting that mental contents of one kind or another play a crucial role in thinking can only be tested satisfactorily over a group of subjects."[90] John T. E. Richardson writing on gender differences points out "that the earliest formal research" on mental imagery was carried out by Galton, "who concluded from his inquiries that 'the power of visualising is higher in the female sex than in the male,' but his published accounts were based solely on the responses of men and boys."[91] Richardson continues that the most widely used technique for evaluating "the subjective vividness of experienced mental imagery is the Questionnaire Upon Mental Imagery (QMI)"[92] that was developed by Betts in 1909 on the basis of some of Galton's original questions. And Graham Dean and Peter E. Morris write that researchers investigating "mental imagery by subjective report . . . still tend to rely on a small set of overvalued questionnaires" whose content "is derived mainly from Galton's original study."[93] Overvalued they may be, but it is a tribute to Galton's insight that a questionnaire he designed over a hundred years ago is still at the root of studies designed to probe the nature of mental imagery today.

In his authoritative *History of Experimental Psychology*, Edwin Boring long ago contrasted Galton with Wilhelm Wundt, probably the most famous psychologist of the day:

> It is impossible to get the entire difference expressed in a phrase, but the important thing that Galton lacked was Wundt's professionalism. As a professional psychologist Wundt always bore upon himself the weight of his past, of the logic of his systematic commitments and of his philosophical

predilections. He could only work within the shell of what he made psychology to be. Galton was free. He had no major commitments. He was not a psychologist nor an anthropologist nor anything at any time except what his vivid interests made him. He had the advantage of competence without the limitation of being an expert.[94]

In the fall of 1886 James McKean Cattell, a young American psychologist, arrived in Cambridge fresh from Wundt's Leipzig laboratory where he had just received his Ph.D.[95] Students from all over the world were flocking to Wundt's laboratory to study experimental psychology, but Cattell proved so capable that he eventually became Wundt's first assistant and prize pupil.[96] Cattell was originally interested in reaction times, a problem he became involved in while a fellow in philosophy at Johns Hopkins University in 1883.[97] But Wundt was more focussed on the general features of the mind than in the specifics of reaction times, so Cattell set his studies aside until he could complete his Ph.D. on a topic more acceptable to Wundt.[98] Meanwhile he was reading Galton's papers, writing his parents gaily in January 1884, that he had picked up Galton's new word "psychometry" from the title of his *Nineteenth Century* article.[99] What appealed to Cattell was Galton's attempt to quantify the speed and number of new ideas arising in his word-association experiments. Like Galton, Cattell sought to attach numbers to mental processes.

During the summer of 1884 Cattell travelled from Leipzig to the United States and returned by way of England where his parents were staying.[100] He especially wanted to visit Galton's Anthropometric Laboratory and was probably dumbstruck by the experience. For a psychologist interested in reaction times and in quantifying mental processes, Galton's elaborate instrumentation must have been a revelation. But Cattell still had to finish his doctoral work, so it was back to Leipzig where he was gaining an ever higher appreciation for his own aptitudes and an increasing disdain for Wundt's. At the end of May 1886, German doctorate in hand, Cattell arrived in London where he was greeted at his West End hotel with a note from Galton[101] asking him to have lunch that same day.[102] Afterwards, he took Cattell to a meeting of the Anthropological Institute.[103] Cattell soon secured an appointment at Cambridge where he arrived in the fall of 1886.[104] Only about 75 minutes by rail, London was a constant attraction to Cantabrigians including Cattell, who now fell under Galton's influence and would later call him "the greatest man I have ever known."[105] Galton, similarly impressed by Cattell, had begun quoting from Cattell's letters before the Anthropological Institute even before Cattell had left Wundt's laboratory.[106]

Cattell's life at Cambridge was intellectually fulfilling. There were interesting colleagues to talk to, research papers to craft, and book reviews to write. By the spring of 1888 Cattell was at work on a book he intended to title *Prac-*

tical Psychology.[107] In May he sent the first chapter to Galton for his comments and enclosed a letter outlining his plans for the rest of the volume. He assured Galton that some of his pet topics, "association of ideas, mental imagery etc." would be included and asked Galton to contribute a chapter on physical anthropometry, which Galton agreed to do. He even tried to induce Galton to be his coauthor, but Galton demurred, a wise decision as the book was never completed.[108] While at Cambridge Cattell established an anthropometric laboratory similar to Galton's.

Cattell retained his enthusiasm for the anthropometric approach upon returning to the United States in 1888 to teach first at the University of Pennsylvania, and subsequently moving to Columbia in 1890 as a psychology professor.[109] His anthropometric measurements, which he called "mental tests," bore strong similarities to Galton's and included such items as strength of squeeze, a pain sensitivity assay, and reaction time for response to a sound. The tests had a strong sensory and physiological bias, consistent with Galton's hypothesis that such measurements could serve as surrogate indicators of mental ability. "Mental testing" was taken up enthusiastically by investigators in several countries, but before long it became evident something was amiss, as the tests failed to predict mental accomplishment. The death knell was sounded by Clark Wissler, one of Cattell's graduate students, who obtained mental test scores and academic records for more than 300 Columbia and Barnard students and showed, ironically using correlation analysis, that the mental tests exhibited virtually no tendency to correlate with academic achievement. Wissler's dissertation effectively ended Cattell's mental testing program. He turned his attention away from psychology to scientific administration and in the early 1900s became editor of *Science*, one of the most important journals in the discipline, a position he held for decades thereafter.[110]

Well before Galton began his psychological studies, John Stuart Mill had ventured into the murky waters of mental ability in the name of nurture.[111] He was educated under the wing of his father, James Mill. This rather peculiar schooling had a profound effect on Mill, who came to believe that virtually anybody could be taught anything given the proper environment. Mill began to develop a psychology rooted in the associationism of John Locke, who assumed that the human mind at birth was in a blank state, a tabula rasa. Onto this tabula rasa, Mill believed, ideas and impressions would enter the consciousness of an individual and become interconnected or *associated* with one another. But Mill wondered whether association was an adequate theory. To what extent were the contents of one's consciousness the result of association and experience as opposed to inborn ideas or responses, and to what degree did these factors explain individual differences? In developing his answer Mill did not deny the role of innate factors in determining one's mental constitution. After all, animals have instinctive behavior. However, he argued that even these innate instincts were subject to marked modification or suppression through learning.

Across the English Channel Mill's ideas had a profound effect on a young French psychologist named Alfred Binet[112] who would later referred to Mill as "my only master in psychology."[113] In 1883 Binet began work at the Salpêtrière Hospital in Paris, whose director, Jean Martin Charcot, was profoundly interested in hypnosis and hysteria. At Salpêtrière Binet learned how to study individual cases. This case-study approach helped him to appreciate the individuality and complexity of real people in contrast to Galton and Cattell, who preferred to assess the responses of large numbers of people using specific anthropometric measurements and analyze the results statistically. While at Salpêtrière Binet's interest in associationistic psychology broadened and deepened, but he also began to recognize that the theory was too passive. The mind had a penchant for asserting itself actively and *attention* was the most important process for accomplishing this end. His two little girls became his experimental subjects. At home he would try out tests and puzzles on them, several of which assessed reaction times and were derived initially from those devised by Galton and Cattell.

Binet then moved to the Sorbonne as an assistant to Henri Beaunis, a physiologist, where he continued with his case-study approach. At Charcot's suggestion he carried out case studies on two men, Inaudi, a Piedmontese, and a Greek named Diamandi, who were reported to be extraordinarily gifted calculators.[114] He compared their capacities for auditory and visual imagery and delved into their family histories. He concluded that Inaudi and Diamandi were not predisposed by heredity to calculate with such facility, but rather they had developed their precocity through intensive practice. He also examined chess players who could play several games simultaneously while blindfolded, expecting them to exhibit vivid mental imaging. Surprisingly, this was not the case, leading Binet to the notion that there must be something that he called "imageless thought."

Binet collected his observations in a book titled *Psychologie des Grands Calculaterus et Joueurs d'Échecs* (1894). Galton reviewed the book for *Nature*.[115] He was quite complimentary, but directed most of his attention toward the cases of Inaudi and Diamendi, perhaps because of his own experience with the calculating prodigy George Bidder and his equally talented son. Binet believed that Inaudi and Diamandi had learned their calculating skills, but Galton was convinced that Bidder *fils* had inherited his from his talented father. Since he assumed the same was probably true of Inaudi and Diamandi he came up with a novel explanation. "Two mental peculiarities have to concur in the making of a calculating boy; the one is a special capacity for mental calculation, and the other is a passion to exercise it."[116] Galton argued that the capacity for mental calculation might be heritable, as in the case of the Bidders, but if not used it would not be apparent. With this argument about hereditary predisposition Galton had neatly given nurture a secondary role and opened the possibility that nature was at the root of the remarkable calculating prowess possessed by Inaudi and Diamandi. He had also predicted the existence of what we today call susceptibility genes.

By 1899, Theodore Simon, a young physician, was carrying out doctoral research with Binet.[117] In 1904 Binet was appointed to a commission by the French government, the purpose of which was to investigate the incidence of mental deficiency among French schoolchildren, as universal education laws had recently been enacted. Those with subnormal intelligence were qualitatively classified by a tradition of sorts into three groups: seriously mentally deficient (*idiots*), moderately mentally deficient (*imbeciles*), and weakly mentally deficient (*débiles*), but, while there was agreement that these general categories existed, there was great confusion in matching specific individuals to the appropriate category. At first the task of classifying the mentally deficient unambiguously seemed impossible until Binet realized that age might be the key. Both subnormal and normal children might be able to pass the same set of tests, but the age at which normal children did so should be younger. In 1905 Binet and Simon published their first "Test" of intelligence. This test and its various modifications were eventually incorporated into what we know as the IQ test. So Binet had succeeded where Galton and Cattell had failed, but the irony was that in creating a quantitative measure that might in some way be related to intelligence, he had fashioned a tool that has led to countless arguments ever since on the heritability of intelligence.

SEVENTEEN

Fingerprints

Galton laid the foundations on which are based the fingerprint systems employed today by police forces throughout the world.

—D. C. Browne and A. Brock,
*Fingerprints: Fifty Years
of Scientific Crime Detection*[1]

Francis Galton's interest in personal identification led him from composite photography, to the comparison of facial features (a life-long interest), and finally to fingerprints. On May 25, 1888, Galton presented a Friday Evening Discourse at the Royal Institution entitled "Personal Identification and Description"[2] that dwelt on these interests. Standing before his audience resplendent in their evening finery, he began: "It is strange that we should not have acquired more power of describing form and personal features than we actually possess."[3] He illustrated the problem with a pair of contour lines that failed to superimpose (Fig. 17-1). His goal was to determine the *least discernible difference* between two such lines. This could be done by drawing other contour lines between them until one finally superimposed with one of the original lines. By giving each of the intermediate contours a number, the "grades of unlikeness" between the two original contours could be determined. Realizing he was separating a continuum into discrete units, Galton imposed specific limits. The contours, now defined as human profiles (Fig. 17-1), might be sorted into classes, "gradually building up a well-selected standard collection."[4] Later he returned to this subject, demonstrating his mechanical selector, a kind of analog of the punchcard system, which one could use to enter quantitative comparisons of facial profiles.

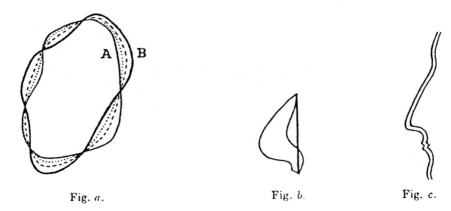

Fig. a. Fig. b. Fig. c.

Fig. 17-1 Measurement of resemblance. A. Two irregular contour lines A and B are drawn, that surround the same area, but which differ in shape. Other contour lines drawn through the intervals between A and B will resemble one or the other more closely until they actually superimpose with either A or B allowing one to determine what Galton refers to as the least discernible difference. What Galton tries to do in parts B and C of the figure is to show how this methodology can be applied to the human face. B. shows the extreme case whereas C shows two profiles that are closely similar. Galton proposed using this method for the classification of different facial profiles. From Francis Galton, "Personal Identification and Description I," *Nature* 38 (1888): 173–177.

In discussing methods for distinguishing individuals, Galton introduced a system devised by M. Alphonse Bertillon, chief of the Service of Judicial Identity in Paris. For criminal identification Bertillon used precise measurements of height, limb length, appendage length, head width, etc., coupled with photographs of the head taken in profile and full face.[5] Once his data for different individuals were recorded on cards, he divided them into three groups based on head length: large, medium, and small (Fig. 17-2). Using head width these three groups were subdivided into three more, making nine groups (3^2). These nine groups were subdivided based on left middle finger length, yielding 27 groups (3^3). These were divided again into three using left little finger length giving 81 groups (3^4). A cabinet with eighty-one drawers, with nine horizontal and nine vertical rows, could thus accommodate the vital statistics of many thousands of individuals with a balanced distribution of the cards through the drawers.

Galton identified a couple of potential shortcomings. First, by its arbitrary use of large, medium, and small, Bertillon's classification system divided a continuously varying character such as length into three discrete, discontinuous groups. His second criticism foreshadowed one of his most important intellectual contributions, the concept of correlation (see chapter 18). He cautioned that "bodily measurements are so dependent on one another that

Length of the Head	Width of the Head	Length of Middle Finger	Length of Little Finger	Drawer No.
α – 18.3	α – 15.2	α – 10.9	α – 8.0	1
			8.1 – 8.3	2
			8.4 – ω	3
		11.0 – 11.4	α – 8.5	4
			8.6 – 8.8	5
			8.9 – ω	6
		11.5 – ω	α – 8.9	7
			9.0 – 9.2	8
			9.3 – ω	9
	15.3 – 15.7	α – 11.0	α – 8.2	10
			8.3 – 8.5	11
			8.6 – ω	12
		11.1 – 11.4	α – 8.5	13
			8.6 – 8.8	14
			8.9 – ω	15
		11.5 – ω	α – 9.0	16
			9.1 – 9.3	17
			9.4 – ω	18
	15.8 – ω	α – 11.1	α – 8.2	19
			8.3 – 8.5	20
			8.6 – ω	21
		11.2 – 11.6	α – 8.7	22
			8.8 – 8.9	23
			9.0 – ω	24
		11.7 – ω	α – 9.1	25
			9.2 – 9.4	26
			9.5 – ω	27

α = lower limit   ω = upper limit

Fig. 17-2 Bertillon's system of classification of measurements of different individuals shown for one of three possible head lengths. The remaining subdivisions eventually led to a total of 27 categories and if the same is done for the other two head lengths he recognized all individuals can be classified into a total of 81 categories. From H. T. F. Rhodes, *Alphonse Bertillon: Father of Scientific Detection*. New York: Abelard Schuman, 1956, 92.

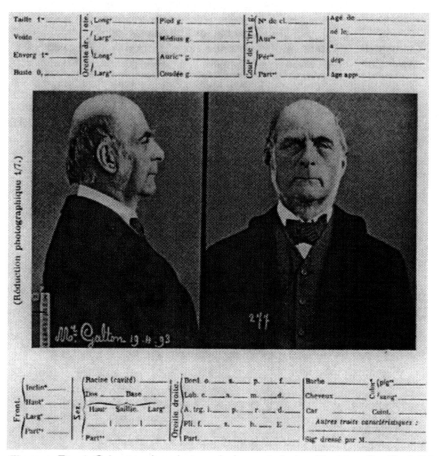

Fig. 17-3 Francis Galton, aged 71, photographed as a criminal on his visit to Bertillon's Criminal Identification Laboratory in Paris, 1893. From Karl Pearson, *Life*, II, plate 52.

we cannot afford to neglect small distinctions. Thus long feet and long middle-fingers usually go together."[6] In the summer of 1893, Galton visited Bertillon in Paris to be photographed and have his measurements taken by the methods of Bertillonage (Fig. 17-3).

Near the end of his lecture Galton dropped his bombshell. "Perhaps the most beautiful and characteristic of all superficial marks are the small furrows with the intervening ridges and their pores that are disposed in a singularly complex yet even order on the under surfaces of the hands and the feet."[7] He summarized the short history of fingerprinting. In 1823 Johann Evangelist Purkinje, professor of Anatomy and Physiology at the University of Breslau, wrote a long Latin thesis calling attention to the diversity existing among fingerprint patterns. He also suggested a system for fingerprint classification, but his proposal aroused little interest and was soon forgotten. Galton briefly referred to several more recent studies and then made a Freudian slip. He credited Sir William Herschel for developing the first systematic method for personal identification by fingerprinting. Herschel "described his method fully in *Nature* in 1880 (vol. xxiii, p. 76), which should be referred to by the reader; also a paper by Mr. Faulds in the next volume."[8]

Henry Faulds, a young Scottish doctor, must have been furious upon reading these words in *Nature*, for then as now priority mattered and Faulds's letter to *Nature* preceded Herschel's, a point Herschel himself acknowledged.[9] In 1878 Faulds, who was attached to the Tsukiji Hospital in Tokyo, became interested in fingermarks on fragments of prehistoric pottery found on the beaches of Tokyo Bay made before the pottery hardened. Since the impressions were poorly preserved, he turned to the fingerprints of monkeys, noticing their similarity to human fingerprints. From monkeys he progressed to Japanese citizens, taking careful note of the spirals, loops, and whorls found in fingerprints. His method for taking fingerprints involved covering a slate or sheet of tin with printer's ink. The subject would press his fingertips down onto the ink-covered surface and transfer the impression to damp paper.

On February 16, 1880, Faulds wrote Darwin an enthusiastic letter saying he was "an ardent student" of Darwin's writings and hoped he could "venture to address" Darwin "on a subject of interest."[10] He observed that the furrows of the hand "form singular and intricate patterns which vary in detail with each individual but may be classed according to their leading lines without much difficulty."[11] He described his work with the pottery fingerprints and suggested a comparative study of finger marks in "lemuroids" might provide useful insights on "man's origin. I hope for this and have bethought myself of your powerful aid—A word or two would set observers working everywhere."[12] Faulds summarized the various advantages of fingerprinting, especially for criminal identification, showing the outline of a right hand with thumb, finger, and palmprint. He referred to this as one of his "filled up" forms and would make similar forms available to anyone wishing them.

On April 7 Darwin forwarded the letter to Galton with an accompanying note remarking that the "enclosed letter and circular may perhaps be of interest to you, as it relates to a queer subject. You will perhaps say: hang his impudence. But seriously the letter might be worth taking some day to the Anthropolog. Inst. for the chance of some one caring about it."[13] Darwin said he had notified Faulds he had forwarded his letter to Galton. Galton promptly responded that he would take Faulds's letter to the Anthropological Institute, noting that he had "got several thumb impressions a couple of years ago . . . but failed, perhaps for want of sufficiently minute observation, to make out any *large* number of differences."[14] The Anthropological Institute did not publish Faulds's letter and that letter together with Darwin's cover letter were returned to Galton in 1894.[15]

Meanwhile Sir William Herschel of the Bengal Civil Service had begun using fingerprints for an unrelated reason.[16] In 1858 Herschel had a road-metalling contractor at Jungipur on the upper reaches of the Hooghly River named Rajyadhar Konai sign his contract in the usual way by pressing his hand, which had been dipped in oil ink, upon its surface. Looking at the whorls and ridges that Konai's handprint left on the document, Herschel realized that fingerprints might prove extremely useful for personal identification. Two years later Herschel, now magistrate at Nuddea near Calcutta, began employing the method to detect fraud. Many Indian citizens receiving government pensions were illiterate and unable to sign receipts indicating payment had been made. Because of this, fraud by impersonation began to cause difficulties. To counter the problem Herschel introduced the first systematic use of fingerprints in the 1860s. Upon receipt of payment illiterate pensioners were required to leave one or two finger impressions in place of a signature. Although Herschel was employing fingerprints to prevent crime, he recognized their potential for criminal identification. In 1877 he wrote to the Inspector of Jails in Bengal suggesting the use of fingerprinting during convict registration, but nothing was done to implement his proposal.

Having heard nothing from Galton, Faulds summarized his findings on the potential uses of fingerprinting in an October 28, 1880, letter to *Nature*.[17] As he had written Darwin, he felt fingerprints might prove useful for comparative purposes, adding that he had passed from examining human fingerprints to those of monkeys and "they presented very close analogies to those of human beings."[18] Since he had little hope of following up this work in his current position, he suggested that "others more favourably situated" should study lemurs . . . as an additional means of throwing light on their interesting genetic relations."[19] He described some of the characteristic features of fingerprints and listed several applications, but it was Faulds's fifth and last suggestion that caught the attention of most readers. "When bloody finger-marks or impressions on clay, glass, &c., exist, they may lead to the scientific identification of criminals."[20] He cited two examples from his own experience. Then Faulds made a remark that Galton must surely have filed away in his memory for future reference. Perhaps fingerprints would be useful for forensic identification since if "the hands only of some mutilated victim were found . . . heredity might enable an expert to determine the relatives with considerable probability in many cases, and with absolute precision in some."[21]

Now in Oxford, Herschel saw Faulds's letter and responded in *Nature* on November 25, 1880.[22] He acknowledged Faulds's report, adding that he had employed fingerprints for over 20 years in the identification Indian pensioners and criminals. He also observed that fingerprints remained unaltered with time, a crucial point if they were to be used for identification purposes. He suggested that fingerprinting might curb desertion from the army. When a new recruit enlisted, three sets of fingerprints should be taken "one to stay with the regiment, one to go to the Horse Guards, and one to the police at Scotland Yard."[23]

Both letters addressed a sensational case of deception that had caught the British public's imagination. In April 1854 Roger Tichborne was lost at sea and presumed drowned when the ship *Bella* disappeared en route from Rio de Janeiro to Kingston, Jamaica, and New York.[24] He was heir to the estates of his father Sir James Francis Tichborne and his uncle Sir Edward Doughty. Several years after Sir James's death in June 1862, a man claiming to be Roger Tichborne, but variously identified as Arthur Orton and Thomas Castro, appeared in Paris and then London. Roger Tichborne's mother was taken in, and settled an allowance of £1,000 per year on her presumptive son. This comfortable arrangement lasted for only 15 months, for she passed away in March 1868. By 1869 the presumptive Roger Tichborne was declared bankrupt, but some friends hit upon an ingenious scheme that allowed him to raise the funds necessary to hire the legal talent required for him to lay claim to the inheritance. Two trials followed, both of which Orton alias Castro lost. After the second trial the judge began the court's sentence thus. "Thomas Castro, otherwise called Arthur Orton, otherwise called Roger Charles Doughty Tichborne, Baronet, after a trial of unexampled duration, you have been convicted by the jury of the several perjuries charged in the counts of this indictment, and which were truly described by your Counsel as 'crimes as black and foul as justice ever raised her sword to strike.' "[25]

Neither Bertillonage nor fingerprinting was available at the time of the Tichborne trial and, as Herschel pointed out in *Nature*, if "there existed such a thing as a finger-mark of Roger Tichborne, the whole Orton imposture would have been exposed to the full satisfaction of the jury in a single sitting by requiring Orton to make his own mark by comparison."[26] Galton recognized how precipitately the Tichborne trial would have been altered had anthropometric evidence been available. In his Presidential Address to the Anthropological Institute in January, 1888, he remarked that "it is impossible not to allude to the Tichborne trial, and the enormous waste of money, effort, and anxiety which might have been spared, had Roger Tichborne passed through an anthropometric laboratory before he went abroad."[27]

On November 27, 1890, Galton read a major paper on fingerprinting before the Royal Society.[28] In recent years he had obtained thumbprints from 2,500 visitors to the Anthropometric Laboratory (Fig. 15-4B). His countless hours spent analyzing photographic enlargements of the thumbprints had permitted him to elaborate a classification system for sorting the ridges, arches, and interspaces into patterns (Fig. 17-4). The ridges running under the thumbnail, defined as the "primaries" or arches, were followed by an "interspace" below which they tended to run horizontally. It was the patterns of whorls and loops

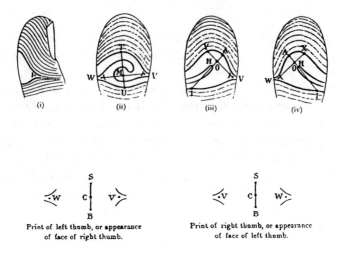

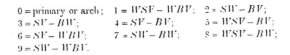

0 = primary or arch; 1 = $WSV - WBV$; 2 = $SW - BV$;
3 = $SV - BW$; 4 = $SV - BV$; 5 = $WSV - BV$;
6 = $SV - WBV$; 7 = $SW - BW$; 8 = $WSV - BW$;
9 = $SW - WBV$.

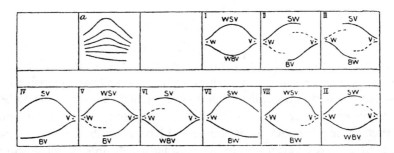

Fig. 17-4 Galton's method of fingerprint analysis illustrated for the thumb. TOP PANEL. (i) The ridges on the upper end of the thumb below the fingernail are referred to as the primaries. The unmarked area between the upper ridges and the lower ridges is referred to as the interspace. Note the triangle that forms at one end of the interspace where the primaries and the lower ridges come together. (ii)-(iv) show that some thumbs have two triangles while others have either a triangle on the outside or the inside of the thumb. The whorls seen in the interspace in (ii)-(iv) are what Galton referred to as the nucleus. MIDDLE PANEL. This indicates the horizonal (w-v) and vertical (s-b) coordinates Galton used to define the ridges circumscribing the interspace. BOTTOM TWO PANELS: The ridges fall into nine patterns plus those cases (a) where the interspace is missing, giving a total of ten patterns. These can be subdivided by examining the pattern of ridges in the nuclei and the characteristics of the minutiae (e.g., when two ridges split into three or coalesce into one). From Francis Galton, "The Patterns in Thumb and Finger Marks," *Philosophical Trans. of the Royal Society of London, Series B* 82 (1891): Figs. 1–7.

in the interspace on which Galton focused. At the point where the arches below the thumbnail diverged from the lower row of ridges to form the interspace, Galton noted a short perpendicular line that characteristically formed a tiny triangle together with the ridge above and below the interspace (Fig. 17-4, top, i). In many thumbprints there were two such triangles (Fig. 17-4, top, ii). He connected these by a baseline to two "cardinal points," which he referred to as V (on the outside of the thumb) and W (facing the rest of the hand). Other thumbprints lacked either the V or W triangle (Fig. 17-4, top, iii, iv). By using V and W as the horizontal anchors and drawing a second, vertical line S-B through the most central part of the pattern, Galton was able to break down the "divergent lines" that bounded the interspace (Fig. 17-4, middle). These yielded "nine, and only nine, possible variations" (Fig. 17-4, bottom) which produced a total of ten patterns when combined with the one instance in 30 where the interspace was missing (Fig 17-4, bottom *a*). By comparing photographic enlargements of 1,000 thumbprints, Galton arranged them into 25 main divisions that were sorted to correspond with the ten classes already described. Obviously such a system could not be used for personal identification if the variability stopped there, since individuals would often have identical fingerprints. This was not the case, however, since each thumbprint had additional, unique features that related not to the bounding ridges of the interspace, but to the structure of the whorls and loops within the interspace, which Galton called "nuclei." There were also the minutiae that formed "minor patterns of their own quite distinct from the larger patterns." For instance, two ridges running in parallel might be replaced by three. Hence, Galton was systematically developing a method for comparing human thumbprints that could be used to reveal the uniquely descriptive features possessed by each individual.

He could undertake this painstaking work with confidence only because of Herschel's key observation that fingerprints did not vary over time. To emphasize this point Herschel provided Galton with paired sets of prints taken from eight individuals at intervals ranging from nine to 31 years. Galton was delighted by the "absolute and most extraordinary coincidence between the details of each of the two impressions of the same finger and of the same person." Never did an old ridge disappear or a new ridge arise. Later Galton used his classification system to try and quantify ridges and other parameters of his fingerprints. The results were tabulated as frequency distributions and the six distributions he obtained were combined "by an artifice" to obtain a single "ogive" curve that formed "a quasi-normal series." The ever-admiring Karl Pearson was too good a statistician to buy this. "Personally I do not see why it is needful to show accordance, quasi or otherwise, with the normal law of error," wrote Pearson following a careful analysis of Galton's paper.[29] From conversations with Galton, Pearson thought his mentor probably realized he was forcing his data to fit the cherished curve. Near the end of his paper Galton mused that fingerprints might be heritable, but he lacked sufficient data to test this hypothesis. Galton's paper was an intellectual tour de force, a breakthrough whose consequences for criminal identification have been of lasting importance. In a much shorter paper published later that year, Galton proposed a method for indexing fingerprints "after the fashion of a dictionary, and on the same general principle as that devised by A. Bertillon" for anthropometric measurements.[30]

As usual when Galton was onto something he wrote an article for public consumption, "Identification by Finger-Tips" published in *Nineteeth Century* (1891).[31] Fingerprints represented a "visible token of identity" for each individual and fingerprinting could be "easily applied to show either (1) that a man is the person he professes to be, or (2) that he is not the person whom he is suspected to be, or (3) that he is not included among the persons whose names and tokens are to be found in any given register."[32] Borrowing extensively from Herschel's *Nature* letter, he described the potential importance of fingerprinting in criminal investigations, convicting army deserters, as an adjunct to the passport system, and "in our tropical settlements, where the individual members of the swarms of dark and yellow-skinned races are mostly unable to sign their names and are otherwise hardly distinguishable by Europeans, and ... are grossly addicted to personation and other varieties of fraudulent practice."[33] Galton suggested that British emigrants should be fingerprinted as fingerprints would prove useful if the prodigal son later returned "in proving claims to kinship and property."[34] Conjuring up Roger Tichborne's impersonator, Galton warned that "some alien scoundrel from foreign parts may assert himself to be the long-lost rightful claimant to an estate held in previous security by others on the supposition of his decease."[35] Lastly, he argued that fingerprints might be useful in the identification of accident victims "such as bodies washed up after a wreck, or other ghastly contents of a Morgue."[36]

Galton crowed that he was the first to provide evidence for the lifelong persistence of fingerprints, for classifying the complexities of fingerprints, and for devising consistent methods for taking fingerprints. Bertillon did "not use fingerprints in connection with his system of anthropologic identification" and Herschel "was the only person who has used the method on a large scale ... during the tenure of magistracy in Bengal."[37] He acknowledged Herschel's help without which he "could not have planted" his "first step," while ignoring Faulds as usual. He looked

> forward to a time when every convict shall have prints taken of his fingers by the prison photographer, at the beginning and end of his imprisonment, and a register made of them; when recruits for either service shall go through an analogous process, when the index-number of the hands shall usually be inserted in advertisements for persons who are lost or who cannot

be identified, and when every youth who is about to leave his home for a long residence abroad, shall obtain prints of his fingers at the same time the portrait is photographed for his friends to retain as mementos.[38]

Galton published additional articles and three books on fingerprinting, the most important being his 1892 monograph, *Finger Prints*.[39] This masterful work of 13 chapters began with the history of fingerprinting, gave detailed descriptions of the various features of fingerprints, their classification, their persistence, evidential value, peculiarities, indexing, inheritance, and their relationship to race and class. For once Galton was careful to note the contribution of Faulds as well as Herschel. The book also set forth detailed instructions for those who wished to apply the technique, as well as recommending equipment and materials such as rollers, dyes, and the best methods for rendering accurate fingerprint impressions photographically.

In chapter 7, "Evidential Value," Galton dealt with the critical question of the probable uniqueness of individual fingerprints.[40] To determine uniqueness, he devised an ingenious test. To break a single fingerprint into separate components he asked what would happen if he dropped a small square at random on a fingerprint hiding the portion that lay beneath. What was the liklihood that an experienced analyst would successfully reconstruct the ridges of the hidden portion correctly with a probability of 1/2? Based on 75 trials he estimated that a six-ridge square could be reconstructed with a probability of 1/3, but he chose to overestimate the probability of a match as 1/2, meaning that he was underestimating the probability that two fingerprints were unique. A full fingerprint consisted of 24 six-ridged squares. Now he made the key assumption that the six-ridge squares could be treated "as independent units, each of which is equally liable to fall into one or other of two alternative classes, when the surrounding conditions are alone known."[41] Since the probability of guessing the correct contours of the fingerprint under any square with full knowledge of the surrounding fingerscape was 1/2, the probability of successfully reconstructing an entire fingerprint of 24 six-ridged squares was $1/2^{24}$. Galton recognized that he might be "overlooking correlations between variables" and falsely assume their independence "with the result that inflated estimates" were made that must "be proportionately reduced," but he felt that in this case there was "little room for such an error."[42]

Now Galton was ready to calculate the critical number. He estimated the chance he would have correctly determined "the general course of the ridges adjacent to each square" as 1/24 and the number of ridges entering and leaving each square as $1/2^8$. Both numbers were assumed to be gross overestimates. He could now estimate the probability that a randomly selected fingerprint would match a specified one as $1/2^{24} \times 1/2^8 \times 1/2^4 = 1/2^{36}$, "or 1 in about sixty-four thousand millions. The inference is, that as the number of the human race is reckoned at about sixteen thousand millions, it is a smaller chance than 1 to 4 that the print of a single finger of any given person would be exactly like that of the same finger of any other member of the human race."[43] Later he corrected this figure for the population to 1.6 billion, which would give odds of 1 to 39.[44] Galton, had, through a clever calculation, provided ammunition for the use of fingerprints as unique signatures for personal identification. In DNA fingerprinting today two sorts of results may obtain. The fingerprint of the suspect does not match the fingerprint at the crime scene, in which case he is excluded. But if the fingerprint does match, an inclusive test, it is of crucial importance to establish that the probability of finding another matching fingerprint in the population as a whole is infinitesimal before concluding the match is meaningful. Galton's calculation represents the first attempt to establish a standard for an inclusive fingerprint test.

Galton dealt with many other subjects, including the taxing problem of establishing a workable system for indexing fingerprints, but two topics, heritability and race, are especially germane. Since he knew that even close relatives were distinguishable by their minutiae, he focused on gross patterns. He compared 105 pairs of siblings, using his arch-loop-whorl classification, and found that the probability of obtaining correlated patterns between siblings (e.g., arch-arch) seemed higher than expected on a random basis. He repeated the comparisons with 150 pairs of siblings using a finer classification. Since he had no way to determine statistical significance, he assumed that the positive deviation from randomness indicated that there was a "decided tendency to hereditary transmission."[45] This conclusion was bolstered by his observations on prints from three fingers of the right hand in twins whose origin, identical or fraternal, was not distinguished. In the first set of 17 pairs he found 19 of 51 fingerprints that gave the same pattern for the same fingers of both twins, 13 that gave partial agreement, and 19 that disagreed. From these results, together with those from a second series of twins, Galton concluded that "there cannot be the slightest doubt as to the strong tendency to resemblance in the finger patterns of twins."[46,47]

To determine whether racial differences existed in fingerprints, Galton employed a method he had used successfully before. He contacted headmasters of schools, this time in London, Cardiff, and Niger, who supplied him with fingerprints of their pupils. This enabled him to compare the fingerprints of English, Welsh, Jewish, and African schoolchildren. Although some differences appeared significant, Galton's firm conclusion was that "there is no *peculiar* pattern which characterises persons of the above races."[48] But he made one telling remark. "Still, whether it be from pure fancy on my part, or from some real peculiarity, the general aspect of the Negro print strikes me as characteristic. The width of the ridges seems more uniform, their intervals more regular, and their courses more parallel than with us. In short, they give an idea of greater simplicity, due to causes that I have not yet succeeded in submitting to the test of measurement."[49] So Galton thought blacks' fingerprints were less complex, suggesting that in his mind their prints like they themselves were less sophisticated than those of other races. Nevertheless, he was uncompromising that a person's intelligence was not reflected in his fingerprints. "I have prints of eminent thinkers and of eminent statesmen that can be matched by those of congenital idiots. No indications of temperament, character, or ability can be found in finger marks, so far as I have been able to discover."[50]

Finger Prints was widely and favorably reviewed. In its "Books of the Week," the *Times* included this admiring comment: "It is needless to say that the whole subject is handled with that rare patience and thoroughness in investigation, and that keen but cautious acumen in interpretation which are characteristic of all Mr. Galton's work."[51] The reviewer for the *St. James's Gazette* was convinced that fingerprints would have important forensic applications.[52] "Let us assume . . . that Mr. Galton is right . . . and that the results were available for and recognized as legal evidence: we should never hear another Tichborne case."[53] The reviewer believed fingerprinting so important that he presciently suggested Galton was "within a measurable distance of a Royal Commission to ascertain how much can be safely adopted for the purpose of administering justice."[54] The critic for the *Perthshire Advertiser* thought "that for the purposes of identification nothing" could "equal the evidence afforded by [fingerprints]."[55] "A capital title for a detective story, *Fingerprints*," commented the *National Observer*, tongue-in-cheek, adding its praise to the accolades Galton's new work had received.[56] *The Scotsman* remarked eruditely that the "old professions of Palmistry and Cheiromancy quite missed their mark. The human hand—to say nothing of the human foot—is a palimpsest scored over with curious knowledge."[57] The reviewer for the *British Medical Journal* declared that "I will be astonished if Mr Conan Doyle, the creator of Sherlock Holmes, does not work something up by Mr Francis Galton's unique book on thumbmarks."[58] The critique in the *Saturday Review* concluded that with "regard to anthropometry Mr. Galton has made out his case in favour of finger prints over measurements as being both quicker and easier to take and more accurate in reproduction for the sake of comparison. The two together might well defy all possible frauds."[59]

Other publications followed in quick succession. A booklet, *Decipherment of Blurred Finger Prints* (1893), showed how to prepare fingerprint evidence from badly impressed prints. *Physical Index to 100 Persons Based on their Measures and Finger Prints*, issued privately in 1894, was Galton's second attempt at setting up a fingerprint index. In 1895 MacMillan published *Finger Print Directories*, Galton's final major work on the subject. Its main purpose was to provide a means for indexing the fingerprints of several hundred thousand individuals. Pearson in 1930 painfully worked through his classification system.

Despite Galton's "most imposing battery of additional suffixes"[60] and "its cumbrous character," he was generally positive, writing that if a student of finger-prints asked him how to index his material "I could still not refer him to anything better than Galton's Finger Print Directories of more than thirty years ago!"[61]

The potential of fingerprinting became so widely appreciated that in October 1893, Herbert Asquith, home secretary in the last Gladstone government, appointed a committee chaired by C. E. Troup of the Home Office, whose other members were M. L. Macnaghten, chief constable of the Criminal Investigation Division, Scotland Yard, and Major Arthur Griffiths, inspector of Prisons.[62] Their charge was to inquire into (1) the method of registering and identifying habitual criminals then in use in England; (2) the "Anthropometric System" of classified registration and identification being used in France; and (3) identification by means of fingerprints. They were to report whether either method or both would be useful supplements to the existing techniques of criminal identification. The Troup Committee paid several visits to Galton's Anthropometric Laboratory, saw the methods being used to take fingerprints, and was impressed by the relative ease with which Galton picked out from his cabinet a set of prints of an individual whose prints had just been provided. In their report they wrote that a "visit to Mr. Galton's laboratory is indispensable in order to appreciate the accuracy and clearness with which fingerprints can be taken and the real simplicity of the method."[63] The committee was particularly impressed by two qualities of fingerprints. These adapted them:

> for use in deciding questions of identity. In each individual they retain their peculiarities, as it would appear, absolutely unchangeable throughout life, and in different individuals they show an infinite variety of forms and peculiarities. Both these qualities have formed the subject of special investigation by Mr. Galton, and having carefully examined his data, we think his conclusions may be entirely accepted.[64]

But as Galton pointed out to the committee, his system of indexing was still incomplete and fraught with difficulties.

The Troup Committee laid down three main conditions in deciding what system should be adopted:

> (1) The descriptions, measurements or marks, which are the basis of the system, must be such as can be taken readily and with sufficient accuracy by prison warders or police officers of ordinary intelligence. (2) The classification of the descriptions must be such that on the arrest of an old offender who gives a false name his record may be found readily and with certainty.

(3) When the case has been found among the classified descriptions, it is desirable that convincing evidence of identity be afforded.[65]

The committee felt Galton's methods admirably fulfilled the criteria imposed by the first and third conditions, but that indexing of fingerprints should be by Bertillonage. In fact, the committee "were so much impressed by the excellence of Mr Galton's system in completely answering these conditions that they would have been glad if, going beyond Mr Galton's own suggestion, they could have adopted his system as the sole basis of identification."[66]

In March 1894, the conclusions of the Troup Committee were widely aired in the press. The reporter for the *Standard* wrote that the committee had recommended a system "which borrows M. Bertillon's method of classification and at the same time embodies the practical results of Mr. Galton's investigations."[67] the *Yorkshire Post* compared Bertillonage and fingerprinting, concluding that "Galton's system would seem to be a more sure means of proving identity, though it could not be carried out except with the aid of skilled experts."[68] The *Daily Chronicle* was supremely upbeat. "There can be no doubt that the Bertillon-Galton combination will render a wrong identification practically impossible, and add a new terror to crime."[69] But, while lavish in their praise, the *Evening Standard* and *Morning Post* worried over the classification dilemma. The *Evening Standard* mused over the problem posed by a library composed of "not less than a hundred thousand minute descriptions, and as many finger stamps" for "how on earth can the police refer to a single one on short notice."[70] But, the paper noted, Bertillon did have a workable system and "this is where genius comes in."[71] The *Morning Post* echoed similar sentiments.[72]

The one man distinctly unhappy with the acclaim Galton was receiving was Dr. Henry Faulds, who undoubtedly felt he was ill-used by Galton. He had written to Darwin about fingerprinting in the spring of 1880, but Darwin forwarded his letter to Galton. Faulds's scientific antennae would have been acute enough to recognize that his letter to the great Darwin had been shunted to a highly regarded, but less illustrious scientist. That was all well and good, but nothing came of it. Instead, Faulds had to act as his own scientific advocate by communicating his ideas about fingerprinting to *Nature*.

In 1885 Faulds returned to England where he practiced in the Potteries in Kensington as a police surgeon. In 1886 and 1888 he had interviews with officials at Scotland Yard, in particular with Inspector J. B. Tunbridge, concerning the use of fingerprints as a means of identification.[73] He even offered to establish a small laboratory free of charge to test the feasibility of his method. Tunbridge filed a report in 1887 indicating he thought the fingerprint method was accurate, but did not see how it could be applied in practice. The next year Galton's Royal Institution speech was reprinted in *Nature*, wrongly attributing priority for fingerprinting to Herschel. This undoubtedly incensed Faulds, who had a ferocious temper. Then, in 1891, Galton published his first major paper on fingerprint analysis, crediting Herschel for providing proof of the permanency of fingerprints. There was nary a mention of Faulds. Although Galton briefly recognized Faulds in *Finger Prints* while detailing the history of fingerprinting, Herschel again played the major role with Galton acknowledging his help. In fact, Galton was so grateful for Herschel's aid that he later dedicated *Finger Print Directories* to him.[74] Faulds must also have been stung badly when he became aware of the Troup Committee report. Galton was given full credit for developing the fingerprinting method and it was clear the committee thought fingerprinting had great potential for criminal identification. What an irony this must have been for poor Faulds. Not only had he not convinced Scotland Yard to adopt the method, but his archrival had done so with relative ease and was getting full credit for his idea.

Faulds, craving some modicum of recognition, published a letter in *Nature* on October 4, 1894.[75] He recognized Galton's acknowledgement of his work in *Finger Prints*, but faulted him for misspelling his name and especially referencing his first *Nature* letter as 1881 rather than 1880, effectively making it appear Herschel's letter was published first. He quoted from his 1880 letter as evidence he was the first to suggest using fingerprints for criminal identification. "As priority of publication is generally held to count for something, and as I know nothing of Sir W. Herschel's studies, nor ever heard of anyone in India who did"[76] it was important to learn more about Herschel's semi-official report to the Inspector of Jails, which he stumbled on in Galton's book. He briefly recounted his experiences with fingerprinting, including Scotland Yard's interest in the method. At the end of his letter, Faulds returned to the question of priority again, saying he had "not the slightest wish to diminish the credit that may be due to Sir W. Herschel. What I wish to point out is that his claim ought to be brought out a little more clearly than has yet been done, either by himself or by Mr. Galton. What precisely did he do, and when?"[77]

Herschel replied graciously in late November in *Nature*.[78] He acknowledged that as far as he knew Faulds's 1880 letter was "the first notice ... of the value of finger-prints for the purpose of identification."[79] He explained that he had "chanced upon" fingerprinting in 1858 and the way in which he had "followed it up afterwards" had been stated on his authority by Galton "at whose disposal I gladly placed all my materials on his request."[80] His semi-official report to the Inspector-General of Jails in Bengal had elicited what seemed a discouraging reply, but he attributed his own reaction to a "very depressed state of health at the time. The position into which the subject has now been lifted is therefore wholly due to Mr. Galton through his large development of the study, and his exquisite and costly methods of demonstrating in print the many new and important conclusions he has reached."[81] So there it was. Faulds could have his priority, but it was Galton who had really brought fingerprinting to the point where it could used in forensic analysis.

Faulds was not pleased by Herschel's trivializing of his priority and he retaliated in 1905 with *Guide to Finger-Print Identification*. He again raised his various grievances about lack of recognition while emphasizing the importance of his work. He belittled Herschel's contributions while excusing his own ignorance of them.[82] He dismissed "Mr Galton who frequently acts as a graceful chorus to Sir William."[83] His chronological bibliography on fingerprinting started with his own paper in 1880 and proceeded almost yearly until 1890, skipping nimbly to 1894 and avoiding all reference to Galton's papers and books. He misrepresented the Troup Committee's conclusions published in the Parliamentary Blue Book devoted to criminal identification. According to Faulds, "Mr Galton's own system, afterwards expounded in a work [i.e., his *Finger Prints* of 1892] abounding in grave errors and set forth in a way which the Blue Book of 1894 characterises."[84]

Now it was the 83-year-old Galton's turn to be incensed and he evened the score in his scathing review of Faulds's book in *Nature* in 1905.[85] He briefly mentioned Faulds's 1880 *Nature* letter, writing that his suggestions for using fingerprints for legal purposes "fell flat" presumably because "he supported them by no convincing proofs of three elementary propositions on which the suitability of finger-prints for legal purposes depends."[86] These were evidence for permanence; proper documentation rather "than opinions based on mere inspection, of the vast variety in the minute details of those markings, and finally . . . that a large collection could be classified with sufficient precision to enable the officials in charge of it to find out speedily whether a duplicate of any set of prints that might be submitted to them did or did not exist in any collection. Dr. Faulds had no part in establishing any one of these most important preliminaries."[87] While Faulds contributed "the first *printed* communication on the subject, it appeared years after the first public and *official use* of finger-prints had been made by Sir William Herschel in India, to whom the credit of originality that Dr. Faulds desires to monopolise is far more justly due."[88] Galton's meaning was clear. Faulds's scientific and practical contributions to fingerprinting were negligible in comparison to his own and Herschel's.[89]

Despite the obvious utility of fingerprinting, a disaster occurred on June 16, 1896, that returned Bertillonage to the limelight.[90] Shortly before midnight in a pummelling rain, the 3663-ton steamship *Drummond Castle*, bound from South Africa to London, was driven onto the Pierres Vertes, a rock formation at the southern entrance of Fromveur Sound at the western tip of Brittany. The ship was running 12 miles off course because the visibility was so poor that the lookout could not see the beam from the Ushant lighthouse. Captain Pierce and his third officer, Brown, were on the bridge, while virtually all of the passengers and many of the crew were below. The captain apparently thought the bulkheads would hold, allowing sufficient time to muster everyone and lower the lifeboats, but the ship foundered rapidly, leaving the passengers and most of the ship's company trapped beneath the deck hatches as the ship slid below the waves.

The sinking of the *Drummond Castle* with the loss of nearly 250 lives and but three survivors was one of the greatest British marine disasters of the late nineteenth century. By June 19, bodies began to wash ashore. Diligent searching of the rocky coasts of the islands of Ushant and Molène and of the adjacent coast of Brittany led to recovery of 40 bodies by day's end on June 20. How were they to be identified? On the evening of June 19, Alphonse Bertillon received urgent official instructions to proceed to Brest and Ushant to help in the identification of the bodies. Bertillon first visited Ploudalmézeau 16 miles west of Brest, where seven bodies were housed in a temporary morgue upon improvised biers each with a crucifix and candles. He had designed special equipment to photograph the corpses by his usual method, full face and profile. Two bodies carried enough information to permit immediate identification. Bertillon and his assistants had to work fast and hard because there were many more bodies at Ushant. By June 27, Donald Currie and Company of Fenchurch St., London, the owners of the *Drummond Castle*, were able to report that 53 bodies had been recovered and 27 "official descriptions" received, presumably based on Bertillon's reports and photographs. Ten had already been positively identified. Further victims were probably identified later although the record is unclear on this point. Bertillon's efforts were recognized by the British government on the anniversary of the sinking, when the British ambassador conferred on him the *Drummond Castle* Medal.

In 1899 the Belper Committee was appointed by the government for yet another assessment of methods of criminal identification.[91] Galton and Dr. J. G. Garson, scientific advisor to the Convict Supervision Office of Scotland Yard, were called to testify. Garson, a skilled craniologist interested in variations in size and shape of the skull,[92] was inclined toward Bertillonage. Fingerprints played a relatively minor role in this inquiry although Galton's strong advocacy brought them into tentative use at Scotland Yard. However, it fell to Sir Edward Henry to develop a workable system of fingerprint classification.[93,94] He joined the Indian civil service and was posted to Bengal as assistant magistrate-collector, eventually rising to Inspector-General of Police in Bengal in 1891. While in Bengal Henry found Herschel's fingerprint system still in use for its original purpose. In October 1894 he visited Galton's laboratory, where Galton demonstrated his fingerprint methodology. Henry told Galton he would collect as many fingerprints as Galton wanted upon his return to India and a lengthy correspondence ensued between them. While in Bengal, Henry developed the method of fingerprint classification that would soon come into general use. His system was adopted by the Indian Government in 1897, which, three years later, commissioned the printing of Henry's *Classification and Uses of Fingerprints*, a work that has subsequently gone through many editions. Henry's success at indexing fingerprints tipped the scales for the Belper Committee. Its members unanimously voted in favor of adopting the fingerprint system as modified by Henry as the primary method of criminal identification in Great Britain. On May 31, 1901, Henry took up his appointment as assistant commissioner in charge of the Criminal Investigation Department at Scotland Yard, and the Central Fingerprint Branch was born that July.

The Deptford Murders represented the first high-profile case in which fingerprint evidence was introduced in court in Great Britain.[95] An elderly couple named Farrow lived over an oil-and-color business that Farrow managed at 34 High St., Deptford. On the morning of March 27, 1905, Farrow was found in his shop with his head battered in and Mrs. Farrow was found in the bedroom on the second floor with similar injuries. She lived four more days without regaining consciousness. In the kitchen the black stocking masks used by the assailants were discovered. The cash box had been broken into, but the thieves probably escaped with little money given the Farrows' parsimonious nature. The murderers' knowledge of the Farrows habits and the brutal slaying of the elderly couple suggested the killers were local residents. Suspicion soon fell on two brothers, Alfred and Albert Stratton, who were known to be of unsavory character. One of the brothers was apprehended at a pub six days later and the other was detained the following day.

The Fingerprint Branch had been awaiting an arrest, for they had found a thumbprint on the cash-box tray, which had been thrown aside during the robbery. Two potential suspects, Mr. Chapman, the owner of the business, and the shop boy, had already been ruled out. Their fingerprints were on the box, but they, together with a young detective-sergeant, had pushed the box out of the way to make way for the stretcher bearers who came to carry the dying Mrs. Farrow away. The thumbprint was found to be identical to that of Alfred Stratton. The case was tried at the Old Bailey in May before a judge named Channell, who was not wholly convinced that fingerprints were infallible. The prosecution was conducted by Richard Muir, soon to become senior treasury counsel. At Henry's insistence Muir had already been thoroughly coached in the fingerprint system by a Detective-Sergeant Collins in connection with a lesser trial. Muir proceeded confidently, calling Collins to the witness stand. Collins stated that there were already 80,000 to 90,000 sets of fingerprints on file at Scotland Yard and these were equivalent to 800,000 to 900,000 impressions of digits. Furthermore, he had found 11 points of agreement between the print on the cash box and Alfred Stratton's thumb and would have been satisfied with only four. Mr Justice Channell could only agree that resemblance between Alfred Stratton's thumbprint and that on the cash box was remarkable, but he went on to express his belief that the jury was unlikely to act on such evidence alone. Nevertheless, both Strattons were found guilty and hanged.

EIGHTEEN

The Birth of Biometrics

> Galton's line of analysis led ultimately to the concept of correlation, which is a measurement of how closely any two series vary relative to one another whether it be size of parent and child, rainfall and crops, inflation and interest rates, or the stock prices of General Motors and Biogen.
>
> —P. L. Bernstein,
> *Against the Gods: The Remarkable Story of Risk*[1]

While personal identification and psychology represented one prong of Galton's research agenda, the second involved the detailed analysis of the data he had obtained from his Anthropometric Laboratory and the development of statistical methods for its treatment. On the basis of this research Galton in 1888 completed his two most influential scientific works: a book, *Natural Inheritance*, and a paper in the *Proceedings of the Royal Society* describing the concept of correlation.[2] Louisa remarked in her diary that while they enjoyed glorious weather at Vichy in early September 1888, "Frank was busy about his Book, 'Natural Inheritance' correcting Proofs and I led a quiet life, as suited the cure, taking the waters and we did so enjoy the fresh morning air."[3] That slim volume pulled together in one place all of Galton's thoughts on heredity plus the results of applying his biometric methods to the copious data from the Anthropometric Laboratory.

Some of the most significant findings Galton analyzed in *Natural Inheritance* were presented earlier on September 10, 1885, in his presidential address to the Anthropology Section of the British Association meeting in Aberdeen.[4,5] Height was an obvious metric to examine as it was easily quantified.

Since Galton was intent on gathering heritability data, he had succeeded in obtaining partial pedigrees of many of the visitors passing through the Anthropometric Laboratory. This permitted him to compare stature between parents and children. Hence, he could ask whether regression toward the mean applied in people as well as in sweet peas. However, the anthropometric data posed problems he had avoided in sweet peas. Each child had two parents, whereas self-fertilization was thought to be the modus operandi of the sweet pea (chapter 15). Consequently, Galton faced three potential difficulties. First, parental heights could not be compared directly since the average height of women is less than for men. Second, sexual selection might cause a preference between potential mates for partners of similar or contrasting height. Finally, the parental difference in height might influence their offspring, so it might be necessary to determine the contribution of each parent.

Galton solved his first problem by multiplying each woman's height by a correction factor of 1.08, which he determined as appropriate from his data. The second problem was one he had wrestled with in *English Men of Science*. There he argued on statistical grounds that he had no reason to believe there was sexual selection for height among the couples he had analyzed. The third problem presented more of a sticky wicket. Galton wanted to replace the individual heights of the two parents with their average, with mothers scaled up by 1.08. But was this notion of the "mid-parent" biologically valid? Perhaps the heights of progeny did not depend only on the height of the "mid-parent," but also on the difference between parental heights. When Galton grouped 525 children according to the difference in the heights of their parents, he later reported in *Natural Inheritance* that this had no effect on progeny heights.[6] But he was already confident enough about the answer in 1885 for him to ask his central question: How is height inherited from mid-parent to child?

Having justified the use of the mid-parent, Galton could treat his data much as he had for sweet peas. He could tabulate or plot the height of the mid-parent versus the height of the adult child. Furthermore, he had a complete sample of anthropometric data while for sweet peas he had taken equal numbers of seeds of seven different sizes and measured the sizes of the progeny seeds. He had learned that the sizes of the progeny seeds "regressed" to the mean. However, he had imposed an artificial constraint on a normal distribution of parental seed sizes by interposing artificial selection for seven discrete classes. To analyze his anthropometric data he constructed a highly informative table (Table 18-1).[7] On the left was a vertical scale in inches of the mid-parent height. At the top was a horizontal scale for height of adult children. The table itself summarized stature data for 928 adult children. By reading across the horizontal rows Galton could determine the median height of adult children of a specific mid-parent. To calculate his medians he used varying numbers of classes. For example, for children having mid-parents 70.5

NUMBER OF ADULT CHILDREN OF VARIOUS STATURES BORN OF 205 MID-PARENTS OF VARIOUS STATURES.
(All Female heights have been multiplied by 1·08).

Heights of the Mid-parents in inches.	Heights of the Adult Children.													Total Number of		Medians.
	Below	62·2	63·2	64·2	65·2	66·2	67·2	68·2	69·2	70·2	71·2	72·2	73·2	Above	Adult Children.	Mid-parents.
Above	..	..	..	..	..	..	..	..	..	..	1	3	..	4	5	..
72·5	..	..	..	..	..	..	1	2	1	2	7	2	..	19	6	72·2
71·5	..	..	..	..	1	3	4	3	5	10	4	9	2	43	11	69·9
70·5	1	..	1	..	1	1	3	12	18	14	7	4	3	68	22	69·5
69·5	..	..	1	16	4	17	27	20	33	25	20	11	5	183	41	68·9
68·5	1	..	7	11	16	25	31	34	48	21	18	4	3	219	49	68·2
67·5	..	3	5	14	15	36	38	28	38	19	11	4	..	211	33	67·6
66·5	..	3	3	5	2	17	17	14	13	4	..	..	..	78	20	67·2
65·5	1	..	9	5	7	11	11	7	7	5	2	1	..	66	12	66·7
64·5	1	.1	4	4	1	5	5	..	2	..	..	..	..	23	5	65·8
Below	1	..	2	4	1	2	2	1	1	..	..	..	..	14	1	..
Totals	5	7	32	59	48	117	138	120	167	99	64	41	17	14	928	205	..
Medians	..	..	66·3	67·8	67·0	67·7	67·9	68·3	68·5	69·0	69·0	70·0	..	..	..	..	..

NOTE.—In calculating the Medians, the entries have been taken as referring to the middle of the squares in which they stand. The reason why the headings run 62·2, 63·2, &c., instead of 62·5, 63·5, &c., is that the observations are unequally distributed between 62 and 63, 63 and 64, &c., there being a strong bias in favour of integral inches. After careful consideration, I concluded that the headings, as adopted, best satisfied the conditions. This inequality was not apparent in the case of the Mid-parents.

Table 18-1 Number of Adult Children of Various Statures Born of 205 Mid-Parents of Various Statures

inches tall, Galton normalized the data for adult children of 69.2 and 70.2 inches yielding a median of 69.5 inches, which gave him 31 children above the median and 37 below.[8] For a mid-parent size of 69.5 inches, he used children of 68.2 and 69.2 inches, which gave a median of 68.9 inches with 98 children of greater stature and 85 that were smaller. He was probably delighted with his creation for by simple inspection of the medians on the right he could see that average height of the adult children regressed toward the mean. The columns in the table allowed him to compute the median size of mid-parents for each size class of adult children.[9]

Next Galton plotted the median heights for adult children against the heights of their mid-parents (Fig. 18-1). The mid-parents, by definition, fell on a straight line as they were numbers selected to differ by one inch each (see Table 18-1). Galton's audience would have noticed that the slope of the regression line for the mid-parents was steeper than that for the adult children, dictating wider deviations from the mean (M) for the mid-parents. At the point where the two regression lines intersected, a horizontal line was drawn that crossed the left hand ordinate at M. By computing the ratio of the deviation of heights of the mid-parents (A-M) to the children (C-M) he calculated the value for children to be 2/3 that of the mid-parents. This little pictorial device neatly illustrated that the children were regressing in size toward the mean. This simple diagram with its intersecting lines was of great significance to

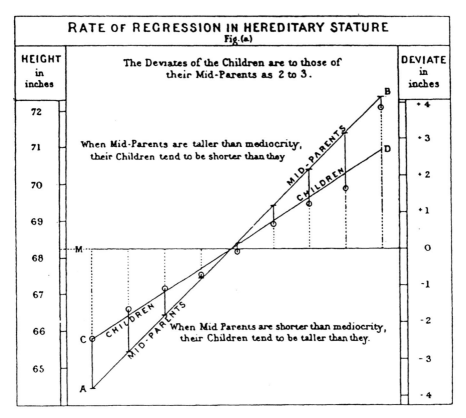

Fig. 18-1 Regression lines for heights of children and their mid-parents (see text for definition). The mean of the population (M) is indicated by the horizontal line drawn through the intersection point of the two regression lines and extrapolates through the left-hand ordinate at 68 1/4 inches. The ratio of the distance M-C to M-A is approximately 2 to 3. That is, the children are regressing to the mean with regard to the mid-parents. Francis Galton, "Regression towards Mediocrity in Hereditary Stature," *J. of the Royal Anthropological Institute* 5 (1886): plate IX.

Galton for it generalized regression to the mean from sweet peas to human beings. "If this remarkable law had been based only on experiments on the diameters of the seeds . . . it might well have been distrusted until confirmed by other inquiries."[10]

Regression to the mean seemed to pose a serious obstacle to Darwin's theory of natural selection because it implied that evolution could not proceed by small, incremental steps. These would be inexorably reversed by the built-in hereditary mechanism of regression to the mean. Galton may not yet have recognized this point or, if he did, he was not prepared to raise it in the absence of a plausible alternative evolutionary mechanism. However, in *Natural Inheritance* he would propose that evolution must proceed in disjunct steps, jumps, or saltations (see also chapter 20). He also recognized that stature was "not a simple element, but a sum of the accumulated lengths or thicknesses of more than a hundred bodily parts."[11] So he was on the threshold of taking the next step. If stature is complex and the sum of many individual elements, then perhaps the dimensions of these elements correlated in contributing to stature.

As Galton gazed at his stature data (Table 18-1), he recognized that they exhibited a beautiful symmetry. In each row summarizing the heights of adult children with the height of the mid-parent held constant, the individuals appeared to be distributed normally. Normal distributions were similarly apparent in the columns where the height of the adult child was held constant, but the mid-parent varied. Galton "found it hard at first to catch the full significance of the entries in the table which had curious relations that were very interesting to investigate."[12] To aid his analysis, Galton " 'smoothed' the entries by writing at each intersection of a row with a column the sum of entries in the four adjacent squares."[13] Smoothing simplified the table considerably.[14] Smoothing, as Galton called it, is a process every scientist is familiar with. It aids in the interpretation of data and is guided to some degree by the scientist's intuition about what they may mean. However, smoothing can be carried too far. This happens when a scientist tries to make data fit a hypothesis to which they are not properly suited. In such cases the scientist is not attempting to falsify or fudge his results, but is simply too enamored of his own idea. Galton would later be guilty of this when he tried to apply regression to the mean to artistic faculty[15] and tuberculosis.[16]

Galton was probably staring intently at this smoothed table one morning on the Kentish coast "while waiting at a roadside station near Ramsgate for a train."[17] As he scrutinized this "small diagram" in his notebook he was struck by the fact "that the lines of equal frequency ran in concentric ellipses. The cases were too few for certainty, but my eye, being accustomed to such things, satisfied me that I was approaching the solution. More careful drawing strongly corroborated the first impression."[18] So in the diagram which he likely showed his Aberdeen audience next, an illustrative ellipse touched or came close to frequencies of three or four (Fig. 18-2). Galton observed that the common center of the ellipses "lay at the intersection of the vertical and horizontal lines, that corresponded to 68 1/4 inches. Their axes were similarly inclined." So the common center for the ellipses was at the predicted mean for midparents and adult children (Fig. 18-1). The normal distribution of the numbers in Galton's smoothed table of rows and columns (Fig. 18-2) was even more striking than in his original table (Table 18-1). He drew a line that he called the "Locus of horizontal tangential points," which fell between the two largest numbers in most of the rows. This was bisected by a second line, the "Locus of vertical tangential points," which fell between the largest numbers in the columns. The fact that he could plot straight lines in both cases con-

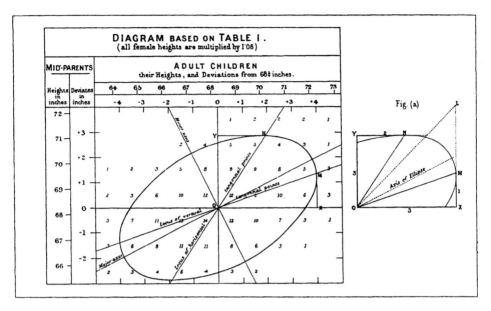

Fig. 18-2 Francis Galton's geometric analysis, his "smoothed" data for numbers of adult children of different heights (rows, and columns, see Table 18-1 for the original data). Galton's use of ellipses to estimate variance is illustrated in the figure by a single ellipse that follows a course between 3 and 4 individuals per height class. The line labelled "Locus of horizontal tangential points" is drawn between those height classes having the largest number of individuals in each row. The line labelled "The locus of vertical tangential points" is drawn between those height classes having the largest numbers in each column. Note that both lines are approximations and, while they fit most rows and columns using these criteria, they do not fit all. These two lines intersect at the mean height for adult children (68 1/4 inches) In the inset (Fig. (a)) Galton has drawn tangents to his sample ellipse. The ratio of the length of the horizontal tangent (Y-N) to the ordinate (Y-O) is 2/3 while the ratio of the vertical tangent (M-X) to its ordinate (X-O is 1/3). See text for discussion. *J. of the Royal Anthropological Institute* 5 (1886): plate X.

firmed that the distributions were normal. Had these values been variable with regard to size the lines could have meandered back and forth like crooked pathways. "The points where each ellipse in succession was touched by a horizontal tangent, lay in a straight line inclined to the vertical in the ratio of 2/3; those where they were touched by a vertical tangent lay in a straight line inclined to the horizonal in the ratio of 1/3."[19] That is, the ratio of the length of the horizontal tangent (Y-N) to the ordinate (Y-O) was 2/3 while the ratio of the length of the vertical tangent (M-X) to its ordinate (X-O) was 1/3 (Fig. 18-2, inset Fig. (a)). Galton's concentric ellipses, nested within each other like a family of Russian dolls, provided a geometrical solution to the variances of his normal distributions, each reassuringly yielding the predicted mean of 68 1/4 inches.[20]

But Galton worried that his empirical solution lacked mathematical rigor. All "the formulae of Conic Sections having long since gone out of"[21] his head, he rushed over to the Royal Institution, "to read them up. Professor, now Sir James Dewar, came in, and probably noticing signs of despair in my face, asked me what I was about; then said, 'Why do you bother over this? My brother-in-law, J. Hamilton Dickson of Peterhouse, loves problems and wants new ones. Send it to him."[22] Galton did so with relief. After all, as a Cambridge passman instead of an exalted wrangler, he was always somewhat worried about his mathematical prowess. Besides he wanted the analysis to be unbiased and "disentangled from all reference to heredity."[23] Dickson quickly resolved the problem and Galton was delighted. "I never felt such a glow of loyalty and respect towards the sovereignty and magnificent sway of mathematical analysis as when his answer reached me, confirming, by purely mathematical reasoning, my various and laborious statistical conclusions with far more minuteness than I had dared to hope, for the original data ran somewhat roughly, and I had to smooth them with tender caution."[24] To Galton it was obvious "that the law of error holds throughout the investigation with sufficient precision to be of real service, and that the various results of my statistics are not casual and disconnected determinations, but strictly interdependent."[25]

The concept of regression to the mean has had far-ranging ramifications and applications in many areas. In his absorbing book, *Against the Gods: The Remarkable Story of Risk*, Peter Bernstein wrote that regression "to the mean motivates almost every variety of risk-taking and forecasting. It is at the root of homilies like 'what goes up must come down,' 'Pride goeth before a fall,' and 'From shirtsleeves to shirtsleeves in three generations.' " And later, "It is what J. P. Morgan meant when he observed that 'the market will fluctuate.' It is the credo to which so-called contrarian investors pay obeisance: when they say that a certain stock is 'overvalued' or 'undervalued,' they mean that fear or greed has encouraged the crowd to drive the stock's price away from an intrinsic value to which it is certain to return."[26]

Sitting in the audience listening with rapt attention was Francis Ysidro Edgeworth. Edgeworth, who admired Galton's statistical work, was also a speaker at the British Association meeting. His curious middle name derived from his mother, the former Señorita Rosa Florentina Ercoles, who had met his father, Francis Beaufort Edgeworth, on the steps of the British Museum at the age of 16.[27-29] While reading for the bar, Edgeworth, a Balliol graduate, undertook a rigorous self-study program in mathematics. Although called to the bar, he obtained a position as lecturer in Logic at King's College, London, in 1880. In 1881 he published *Mathematical Psychics: An Essay on the Application of Mathematics to the Moral Sciences*, which extended his earlier mathematical treatment of ethics to the more tractable subject of economics. This work brought him wide recognition and received approving reviews from two of the preeminent economists of the day, Alfred Marshall and W. Stanley Jevons, although Jevons complained that Edgeworth's prose was very difficult. Galton, who had read Jevons's review, disagreed. He wrote Edgeworth on October 28, 1881, to congratulate him on his "powerful work of Math. Physics, and especially those parts of it that claim the right of Mathematics to deal even with the loosest quantitative data."[30] Edgeworth and Galton developed an extensive correspondence that could only have served to reinforce their mutual interest in statistics.

In 1885, the year of Galton's Aberdeen address, Edgeworth was also very active, reading four important papers including one at Aberdeen. Another, given in June before an international gathering honoring the jubilee of the Royal Statistical Society, acknowledged his debt to Galton's 1875 paper, "The Statistics of Intercomparison."[31] He even borrowed an illustrative example from that paper. The analogy was to a garden with fruit trees. Fruit gathered from the garden as a whole represented the total population corresponding to Quetelet's normal distribution in the broad sense. Hidden within this were smaller normal distributions of fruit picked from particular trees. Edgeworth then succeeded in taking Galton's conceptual description of the problem as visualized via the quincunx and provided it with a proper statistical underpinning.[32] In a sense, Edgeworth was Galton's first disciple. He refined and modified some of Galton's statistical techniques and brought them to bear in the social sciences.[33]

On New Year's day, 1886, Galton submitted "Family Likeness in Stature" for publication in the *Proceedings of the Royal Society*.[34] Although studded with confusing quartiles, grades, and ogives,[35] this paper represented Galton's next step toward the concept of correlation, for he considered not only mid-parents and children, but brothers of men of various heights. This eliminated the generational component, allowing Galton to begin thinking about correlated variations in stature among contemporaneous relatives.[36] But what about correlation itself? There are two stories of how Galton came to the idea of correlation. Unfortunately, the more romantic is probably wrong, but it is worth telling because it reveals how scientific insights sometimes occur. In the summer of 1889 the Galtons stayed for a fortnight in the vicinity of Naworth Castle.[37] Built during the reign of Edward III, this russet-pink stronghold, surmounted by two towers, lies in wild Cumberland countryside, north of Carlisle.[38] Beyond the castle rises the bulk of Hadrian's Wall. Galton wrote in his autobiography that "the circumstances under which I first clearly grasped the important generalisation that the laws of Heredity were solely concerned with deviations expressed in statistical units, are vividly recalled to my memory. It was in the grounds of Naworth Castle, where an invitation had been given to ramble freely. A temporary shower drove me to seek refuge in a reddish recess in the rock by the side of the pathway. There the idea flashed across me, and I forgot everything else for a moment in my great delight."[39]

Having quoted this passage, Pearson wrote "that recess deserves a commemorative tablet as the birthplace of the true conception of correlation."[40] The trouble is that, in quoting the above passage, Pearson references a footnote containing an abstract from Louisa Galton's record for 1889, indicating that the Galtons stayed in the vicinity of Naworth Castle in late August. Galton's paper on correlations was submitted to the *Proceedings of the Royal Society* on December 5, 1888, for publication on December 20, a turnaround time equivalent to the speed of light by modern scientific publishing standards. So what does the memorable passage quoted above refer to? Perhaps the Galtons paid an earlier visit to Naworth Castle, because the sentences in question are embedded in a discussion of the normal curve, statistical scale, and the laws of heredity. The sweet pea experiments leading to the concept of regression to the mean come next, followed by correlation. Possibly while waiting for the rain to stop in his rocky crevice, Galton was beginning to grasp the concept of regression. This slip by Pearson is surprising, for his immense biography of Francis Galton is nothing if not meticulous.

Galton told the true story of the discovery of correlation in an American review.[41] It is more plausible, if lacking in romance. After *Natural Inheritance* went to press, Galton was using his anthropometric data to plot forearm length against height one day when he noticed that the problem was intrinsically the same as that of kinship. He summarized these data in one of the tables in his paper on "Co-relations and their Measurements, chiefly from Anthropometric Data."[2,42] Galton extended correlation to other physical parameters such as head breadth versus head length, head length versus height, etc. He also determined the first set of correlation coefficients, using the now familar symbol r. Most of his correlation coefficients were pleasingly high, between 0.7 and 0.9. Galton's pathbreaking memoir of 1888 on correlation, together with his greatest scientific book, *Natural Inheritance*, not only were the stimuli that activated his first disciples, but would form the cornerstone of a new science, biometrics.

With the notable exception of correlation, *Natural Inheritance* pulled together in one place much of Galton's work on heredity, anthropometrics, and statistics. That was his intent as he wrote early in chapter 1. "I have long been engaged upon certain problems that lie at the base of the science of heredity, and . . . have published technical memoirs concerning them. . . . This volume contains the more important of the results, set forth in an orderly way, with more completeness than has hitherto been possible, together with a large amount of new matter."[43] The book followed a logical progression from heredity, to a description of statistical methods (frequency distributions and normal variation), to Galton's anthropometric data, to the statistical analysis of the data.

The sections covering hereditary processes were completely theoretical. There was no mechanistic basis for any statement, probably because there were as yet no obvious cellular structures whose behavior could be correlated with hereditary units. Galton knew about chromosomes and their segregation as he cited an excellent review by John McKendrick on contemporary cell biology, but the significance of these "chromatin filaments" was still unknown.[44] Nevertheless, Galton succeeded in elaborating what may be described as the second best theory of heredity. He came close to deducing several of the fundamental genetic truths arrived at by Mendel. His theory of particulate inheritance with its emphasis on quantitative variation in numbers of different genetic elements actually fitted the metric characters, like stature, he analyzed. His main reason for discussing "the chief processes in heredity" was to present them in a way "that best justifies the methods of investigation to be employed."[45] So his theoretical analysis was designed to be as consistent as possible with his experimental observations.

Galton began by distinguishing between "Natural and Acquired Peculiarities." He would only consider the former characters as they were "noticeable in every direction." To justify ignoring "Acquired Peculiarities" he made an astute analogy using nonidentical twins. Natural characters "are nowhere so remarkable as in those twins who have been dissimilar in features and disposition from their earliest years, though brought into the world under the same conditions and subsequently nurtured in almost an identical manner."[46] He also anticipated the chromosome theory of inheritance, hypothesizing that human beings were built from a plethora of "minute particles of whose nature we know nothing . . . which are usually transmitted in aggregates."[47] In dealing with the problem of family likeness and individual variation, he used a dreamy analogy that began with "those miniature gardens, self-made and self-sown, that may be seen in crevices or other receptacles for drifted earth, on the otherwise bare faces of quarries and cliffs." He meant that these gardens exhibited vegetational differences as two parents possessed genetic differences. The parents became two floating islands in "a desolate sea" each with its own unique bodily and mental features. The two islands (parents) anchored near each other in the proximity of several more islands (their children) devoid of vegetation. Seeds from the parental islands "gradually make their way to the islets through the agency of winds, currents, and birds."[48] The "vegetation" growing up on the islands "represents the features of the several children." Galton reasoned that the seeds would not distribute equally between the desert islands (children) so, while there would be many similarities between the newly acquired flora of these islands, there would also be differences. Like his grandfather Erasmus Darwin, Francis Galton enjoyed imagery. But, as a prominent Italian once said, for his countrymen the shortest distance between two points is the arabesque, and so it often seems with Galton's analogies.

Next Galton discussed latent characteristics. Once again the parents and children were islands, but this time a few long dormant seeds "find their way" to the islands represented by the children. Often these seeds sprouted into plants whose existence was obscured, "being hidden and half smothered by rivals; but whenever these seeds happened to find their way to any one of the islets while those of their rivals did not, they would sprout freely and assert themselves."[49] What Galton wished to explain with this bit of exposition was the observation that some traits appear to skip generations in their expression. With more attention to precision and less attempted metaphor he might have stumbled on the concepts of dominance and recessiveness, for he was nearly there. Next he distinguished characters that seemed to blend from those that were mutually exclusive. To illustrate blending characters he chose skin color variation among the progeny of matings between blacks and whites. Building on the idea of particulate inheritance he argued that skin color variation could be explained "as a fine mosaic too minute for its elements to be distinguished in a general view."[50] This was an idea he had expressed many years earlier in a letter to Darwin (chapter 13). Thus hue depended on the ratio of different particle types. However, Galton regarded eye colors as representing mutually exclusive characters. He came close to defining discontinuously varying traits when he observed that children of parents of dark and light eye color usually do not have intermediate or blended tints to their irises, but rather resemble one parent or the other.[51]

He returned once more to acquired characteristics, dismissing their importance by using an alcoholic mother as an example. She had normal children when she was sober and neurotic children when on the bottle. These neuroses were not newly acquired hereditary characteristics, wrote Galton, because the fetus became "alcoholised" and the child grew up under the care of a tipsy mom. So he conveniently raised nurture instead of an acquired character as the cause for neurosis. Even today the relative roles of nature and nurture in provoking alcoholism remain a subject of debate.

In the next chapter, "Organic Stability," Galton wrestled with the problem of evolution in the face of regression to the mean. How were continuously varying traits like height going to increase or decrease by small incremental steps if the process was foiled by regression to the mean at each generation? Galton's hypothesis of organic stability began with the notion of a type, a typical form of an organism. This was what he referred to as a stable variety. They were limited in number and had subtypes. As in the case of the quincunx, he built a toy by way of illustration. This time it was a polygon whose many faces were uneven in length (Fig. 18-3).

From his description we can imagine Galton proceeding as follows. Perhaps he is seated at a writing table in the dining-room at 42 Rutland Gate facing the front window. On the walls around him are prints of his friends—Darwin, Hooker, Spencer, Spottiswoode, etc. Now in his mid-sixties, he is dressed in a dark worsted suit with a fine herringbone pattern complete with

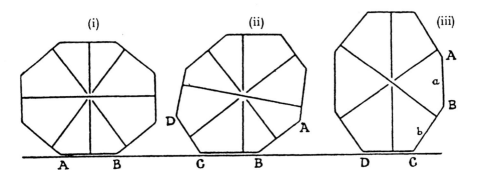

Fig. 18-3 Galton's use of a polygon with slightly asymmetric surfaces to illustrate his hypothesis of organic stability. In the left diagram the polygon rests on face A-B. This is a stable, symmetric configuration from which it will not budge without a forceful nudge. Such a nudge causes the polygon to come to rest on face B-C. This is an asymmetric, unstable face and a gentle shove will bring the polygon back to rest in its original stable position on face A-B. A harder shove in the opposite direction will, however, bring the polygon to rest on face C-D. This is a stable, symmetric position once again. In short, the polygon has now achieved what Galton refers to as a "new system of stability." From Francis Galton, *Natural Inheritance*. London: Macmillan, 1889, 27.

vest. He is wearing a white shirt and a bowtie, blue with tiny white polka dots. His hairline has marched relentlessly back across his head so he only sports some white thatch hanging over his ears set off by white sidewhiskers. His lips are pursed in the familiar V and his piercing blue eyes, hooded by bushy brows, stare intently at the polygon before him, which is resting on one of its longer sides, a stable position. He pokes the polygon gently. It tilts toward one of the shorter sides and then rocks back to its original position. It regresses to the mean. He pushes a little harder and the polygon tilts over onto the shorter face. This is a subtype. He easily nudges the polygon back to its original position, regression to the mean. Then he flicks the polygon back on the short face again and jolts it sharply in a new direction. It lands on another long side, a new position of stability. A "sport" has occurred that produces such a marked change that the new type created is "capable of becoming the origin of a new race with very little assistance on the part of natural selection."[52] This sort of major discontinuous change is what we would refer to as a saltation today. In deference to Darwin, Galton argued gamely that a new type could also arise "without any large single stride, but through a fortunate and rapid succession of many small ones."[53] But he was obviously concerned about this since subtypes could revert. So, while Galton acknowledged the possibility of evolution by incremental steps, his own mental polygon sat solidly in a position of stability that dictated large, discontinous changes as being the stuff of evolution.

At the end of *Natural Inheritance* Galton returned to heredity again (chapter 11), differentiating between two sorts of genetic elements, personal and latent. Personal elements were transmitted by both parents. They were found in various proportions in their progeny while latent elements popped up unexpectedly. They represented some ancestral character. They were the dormant seeds that suddenly sprout. But Galton made an important deduction. Each parent contributed no more than half of his or her latent plus personal elements to each offspring for "if every variety contributed its representative, each child would on the average contain actually or potentially twice the variety and twice the number of elements (whatever they may be) that were possessed at the same stage of its life by either of its parents, four times that of any one of its grandparents, 1024 times as many as any one of its ancestors in the 10th degree, and so on, which is absurd."[54] Galton had predicted on theoretical grounds the existence of the process of reduction division, meiosis, which ensures that a human egg or sperm has half the number of chromosomes present in a somatic cell (see also chapter 13). This obviates the geometric increase in genetic elements at fertilization that would obtain at each generation in its absence. On theoretical grounds, Galton had edged close to developing a whole set of important genetic concepts that would emerge early in the twentieth century. But unlike Mendel, he did not know how to test his model and there were no physical structures within the cell on which he could hang his personal and latent elements so far as he knew. He was acutely aware of the limitations of his theory. "I have largely used metaphor and illustration to explain the facts, wishing to avoid entanglements with theory as far as possible, inasmuch as no complete theory has yet been propounded that meets with general acceptance."[55]

Galton surely felt he was on firmer ground when he showed his readers how to apply statistical tools to his anthropometric data. He illustrated how percentiling (ogives) and the normal distribution could be applied to measurements of strength of pull (chapter 4). His results were presented lucidly in descriptive, graphical, and tabular form. The mean, the median, the mid-parent, and the predictive value of percentiling were discussed. Next (chapter 5) he considered the properties of the normal distribution in more detail, discussing deviations from the mean, the probable error (a term he disliked), and the probability integral. The quincunx was briefly trundled out as a mechanical analog to illustrate the interconvertibility of the normal curve and the frequency distribution, the ogive. Having enumerated his statistical tools, Galton reviewed his methods for collecting and recording anthropometric data (chapter 6). His aim was always to obtain records from at least two successive generations for each family. Sweet peas came in for an encore.

In the next four chapters Galton summarized his analysis of four different kinds of traits. He began with a lengthy discussion of the data on stature (chapter 7) because it was "an excellent subject for statistics."[56-59] The concepts of regression to "mediocrity" between generations and fraternal regression were introduced together with the Galtonian ellipses of scatter about the regression lines.[60,61] He also raised a subject he had touched upon in his earlier writings. He wanted to calculate the contribution of different ancestors in a kinship to the traits he was quantifying. His approach was to multiply regression coefficients. If the regression of the son on the father was 1/3 and that of the father on his father was 1/3, Galton reasoned the regression of the son on one of his grandfathers would be 1/3 x 1/3 or 1/9.[62] Continuing with this vein of logic, he tried to determine the separate contribution of each ancestor to an individual. There was more reasoning and multiplication of regression coefficients, leading him to conclude that the mid-parental contribution to the offspring was somewhere between 4/9 and 6/11, which conveniently "smoothed" to 1/2. Hence, the influence of each parent would be 1/4 and each grandparent 1/16 and so forth. Galton was beginning to formulate what would later be named Galton's Law of Ancestral Heredity by Karl Pearson.[63,64]

At the end of his chapter on stature, Galton reported on what must be the first directional selection experiment. In the absence of human data he again turned to a genetic model, in this case the Purple Thorn Moth (*Selenia illustraria*). The moth's attraction was that it was double-brooded and could be reared inexpensively in a small amount of space. Galton proposed setting up three lines with respect to wing-length. Long-winged males and females mated to each other would constitute one line, short-winged moths would constitute the second, and medium-winged moths would serve as the control.[65] Moths from each generation would then be mounted and measured. At the sixth generation, selection would be relaxed and specimens of medium wing-length from each line would be bred until all trace of long and short wing-length had disappeared from lines selected for these traits. The data Galton obtained could be used to test his ideas concerning "organic stability" and the validity of his notions about ancestral inheritance. An entomologist, Mr. Frederic Merrifield, would do the actual experiments and an assistant took care of photography. Environmental variables like temperature and feeding were to be held constant. Wing-lengths would be classified by percentiling (ogives) and regression coefficients calculated.

Given that these directional selection experiments in the Purple Thorn Moth were proposed by Galton over a century ago, their design is remarkably well thought-out. Thus, it is sad to report that, as often happens, biology did not cooperate with lofty expectation. The spring and autumn broods were dimorphous, with males being larger in one brood and females in the other.[66] The fertility of large and small moths proved less than those of average size. In attempting to increase the number of broods per year to satisfy Galton's thirst for data, Merrifield raised the temperature. This resulted not only in an overall increase in wing-length, but the "giant" and "dwarf" lines became sterile and Merrifield had to start all over again with average-sized moths. Food supplies (leaves) began to run low, and Merrifield tried to coax the caterpillars to moult over shorter intervals so they would pupate sooner. Naturally, these environmental changes were reflected in variations in size (and wing-length) of the adults, so that proper statistical treatment of the data became impossible. A frustrated Galton gave up in despair.

From stature Galton moved to eye color (chapter 8).[67] He made an important distinction, writing that "parents of different Statures usually transmit a blended heritage to their children, but parents of different Eye-colours usually transmit an alternative heritage."[68] In modern-day terminology what he meant was that eye color was a discontinously varying character (e.g., you are blue-eyed or brown-eyed), but stature varied continuously. He classified eye color into eight grades and plotted his data by generation, with children being generation I and great-grandparents representing generation IV (Fig. 18-4). The discontinuity between blue and brown eyes persisted over all four generations, but there were some intermediates, which he called hazel, so he grouped his results into three categories: light, hazel, and dark, and attempted to calculate ancestral contribution to eye color. To do this he used a line of reasoning he had applied to his stature data, employing various arithematic circumlocutions to determine the ancestral contribution for parents, grandparents, etc., to the frequency of children with light eye colors. He was pleased with the apparent fit of the calculated and observed frequencies, writing that a "mere glance" at the tabulated data revealed "how surprisingly accurate the predictions are, and therefore how true the basis of the calculations must be."[69] But he was piling one assumption on another to build a precariously teetering ediface. Pearson dutifully deconstructed Galton's airy castle block by block, scratched his head and observed that it "is certainly remarkable that the predictions should be even as accurate as they are—and they are indeed not perfect—considering the contradictory assumptions on which they are based."[70]

From eye color, Galton turned to artistic ability. Did the inheritance of the artistic faculty follow "a similar law to that" governing stature and eye color?[71] Galton recognized four classes of aptitudes: music alone; drawing alone; music and drawing; and "those about whose artistic capacities a discreet silence was observed."[72] He found that among the mobs of people who had gaily tramped through the Anthropometric Laboratory, only about one-third had artistic tendencies. He calculated the probabilities of an artistic individual marrying a like-minded spouse or marrying an unartistic person, and of two of the latter "barbarians wedding each other." The calculated and observed frequencies agreed pretty well, but he detected "some slight disinclination to marry within the same caste."[73] His rationale was that a

> man of highly artistic temperament must look upon those who are deficient in it, as barbarians. . . . On the other hand, every quiet unmusical man (per-

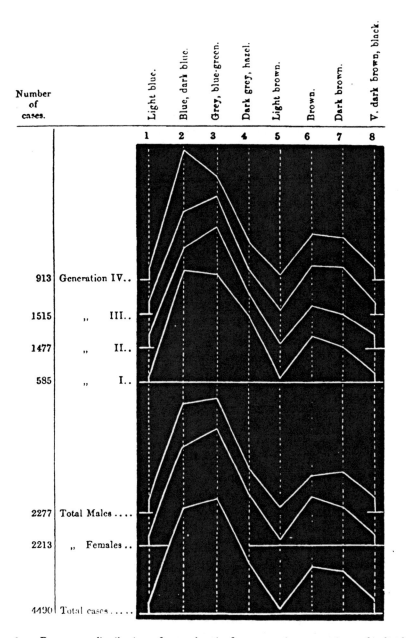

Fig. 18-4 Percentage distribution of eye colors in four successive generations of individuals. The overall results suggest relatively little change of eye color distribution as a function of generation. From Francis Galton, *Natural Inheritance*. London: Macmillan, 1889, 143.

haps Galton was thinking of himself here) must shrink a little from the idea of wedding himself to a grand piano in constant action, with its vocal and peculiar social accompaniments; but he might anticipate great pleasure in having a wife of a moderately artistic temperament who would give colour and variety to his prosaic life.[74]

Pearson reworked Galton's data and found that "the tendency for like to marry like is increased at the expense of the unlike marriages.[75] He thought that Galton's contrary observation represented a case in which his personal views constrained the data unnaturally. "I can well picture what torture to him it would have been to be wedded to 'a grand piano in constant action.' While always exhibiting the best of old-fashioned courtesy to women, he had, when I first knew him, little belief in their intellectual strength; just as he held, that while women gifted with great physical strength existed, it was well for the repose of the other sex that they were rare."[76] Galton concluded from his tortured calculations that "the same law of Regression . . . which governs the inheritance both of Stature and Eye-colour, applies equally to the Artistic Faculty."[77] Pearson, who patiently worked through Galton's numbers using more powerful statistical tools, concluded that Galton's data did not support his conclusion.[78]

Lastly, Galton tackled the heritability of disease (chapter 10). His constituency at the Anthropometric Laboratory had suffered from a wide variety of ailments, but he recognized that his data were too fragmentary to reach any conclusions. Nevertheless, he took a stab at "consumption" (tuberculosis) because he stated that one in every six or seven persons in England died from the disease. He investigated "fraternities," which he hoped would vary according to the normal curve in the degree of consumptive "taint." This was not to be. "They make a distinctly double-humped curve, whose outline is no more like the normal curve than the back of a Bactrian camel is to that of an Arabian camel."[79] From this he concluded that kinships separated into two groups: highly susceptible or highly resistant to consumption. He was ambivalent on heritability, first saying that consumption was acquired and referring to the role of diseased mothers in infecting their infants. Later he reconsidered and assumed that resistance or immunity might be normally distributed in the population and used regression analysis to assay the tendency to contract consumption. He clearly recognized the practical importance of obtaining accurate information on the heritability of different diseases. "The knowledge of the officers of Insurance Companies as to the average value of unsound lives is by the confession of many of them far from being as exact as desirable . . . . Considering the enormous money value concerned, it would seem well worth the while of the higher class of those offices to combine in order to obtain a collection of completed cases for at least two generations, or better still for three."[80] Genetic disease is, of course, an important concern for insurers today.

Natural Inheritance was published in February 1889. Louisa Galton remarked in her diary that the book had attracted "but small notice, it is beyond the range of most minds."[81] But she was off the mark, for the book was widely reviewed. Professor Patrick Geddes in the *Scottish Leader*[82] observed that Galton nearly "had the subject of heredity to himself."[83] It was curious that "most eminent biologists, from Huxley downwards, despite their strong Darwinian faith and advocacy have" largely "stuck to the pre-Darwinian problems, instead of attempting the solution of the far more important post-Darwinian ones."[84] Geddes contrasted *Hereditary Genius* with *Natural Inheritance* and made the acute observation that Galton was formerly interested "with the conscious pride of an intellectual patrician, himself sprung from the mighty races of Darwin and Wedgwood, in compiling a sort of spiritual peerage; now he insists not only upon the fraternity, but even that it be a large one."[85] For Geddes "the harder and later book" in the end really was the "tortoise" that "outstripped the hare."[86] The critic for the *Spectator* wrote that on "the subject of statistics Mr. Galton writes with an enthusiasm well warranted by the results of his long and costly investigations in a field of inquiry that has hitherto possessed but little attraction for the scientific mind."[87] The reviewer remarked on the book's difficulty even "for those competent to understand it" for it would require their "careful and concentrated study."[88] But it "is well worth the time and trouble needed to master it. It lays the foundations of what one day will be a great science, one that will not merely satisfy scientific curiosity, but will be eminently useful to society."[89] This was a prescient remark, for with the publication of *Natural Inheritance* were born the fields of biometrics and social statistics. As the American educator, John Dewey, wrote in his appreciative review, "It is to be hoped that statisticians working in other fields, as the industrial and monetary, will acquaint themselves with Galton's development of new methods, and see how far they can be applied to their own fields."[90]

One enthusiast for social statistics who briefly corresponded with Galton was Florence Nightingale.[91] She had, of course, achieved fame first for her herculean efforts on behalf of British soldiers wounded during the Crimean War. After the war she began her crusade to improve sanitation in the army. She also struck up a friendship with Galton's cousin Douglas shortly after the end of the Crimean War. Captain Galton, a brilliant young Royal Engineer, was the army's leading expert on barrack construction, ventilation, heating, water supply, and drainage.[92] His interests and expertise dovetailed neatly with Nightingale's.

Because of her desire to obtain proper social statistics, Douglas Galton made contact with his cousin Francis Galton on Nightingale's behalf. On

February 7, 1891, she wrote Galton presenting him with an elaborate agenda for areas in which proper social statistics were needed.[93] She hoped that a professorship or readership could be obtained at Oxford in statistics. She was not so concerned about hygiene or sanitary work because statistics in these areas were already closely studied in England. Rather she wanted statistics on the success of the elementary education program then in place; on whether legal punishments acted as appropriate deterrents; on the success of workhouses; and whether the population of Britain's greatest colonial possession, India, was growing richer or poorer under the Crown's hand. In each case she posed a series of questions she felt would benefit from number-crunching and statistical analysis. But what she was ultimately leading up to was "teaching how to use these statistics in order to legislate for and administer our national life with more precision and experience."[94] The Government needed to make use of "the statistics which it has in administering and legislating,"[95] and politicians, the great majority university-educated, needed to understand the "practical application" of statistics.

Galton heartily agreed, writing Nightingale that "obtaining a supply of men well versed in the appropriate methods of statistics, who shall apply them to the social problems of the day, seems to me a *most worthy* one, and well deserving of *great effort*."[96] He considered the pitfalls inherent in obtaining accurate social statistics and then took up Nightingale's suggestion of an endowed professorship or readership in statistics. He thought such a person would be isolated at Oxford and suggested the Royal Institution as an alternative, as the able academics there would "stimulate on the one hand and . . . curb the vagaries of the inquirer in the other."[97] The correspondence proceeded vigorously until it became apparent that the £4,000 endowment Galton had suggested for the chair was more than Nightingale's trustees felt she could afford. "It does not seem to me that the £100 annually for 7 years would as your trustees are inclined to think be equivalent in the end to an endowment that would produce that annual sum."[98] Galton then asked Nightingale to select two or three subjects that seemed particularly appropriate for presentation at the Demographic Congress that summer, but it was clear his enthusiasm for the project was substantially dampened by Nightingale's inability to endow a professorship. The correspondence began to trail off, ending in June 1891.

NINETEEN

Galton's Disciples

> It may be said that [*Natural Inheritance*] created Galton's school; it induced Weldon, Edgeworth and the present biographer to study correlation and in doing so to see its immense importance for many fields of inquiry.
>
> —Karl Pearson[1]

Galton had always worked largely alone. If he needed help he hired men like Serjeant Randall of the Anthropometric Laboratory or asked friends to aid him, as in the case of the sweet pea plantings. Since he had no professional position, academic or otherwise, he so far had no intellectual disciples. Edgeworth came closest as he was greatly influenced by Galton's writings and corresponded with him extensively. However, with the publication of *Natural Inheritance* and his paper on correlation, Galton soon acquired two new acolytes who complemented each other perfectly (Table 19-1). One was Galton's devoted first biographer, Karl Pearson (Fig. 19-1), who added mathematical rigor to his intuitive and graphical approaches to statistical questions. The second was a zoologist, Raphael Weldon (Fig. 19-2), who applied Galton's statistical methods to his extensive quantitative data on shrimps and crabs. Pearson and Weldon became good friends and intellectual collaborators while both held positions at University College, London. This chapter is the story of how this relationship came about.

Imagine a February morning in 1863. The skies are dark and lowering. Fierce gales blow rain and sleet against the windows of a London house, rattling the windowpanes. William Pearson, a barrister and Queen's Counsel, is standing ramrod straight eating breakfast from a high table.[2] He is a man of grim demeanour, a stern disciplinarian. His son Carl, aged six, watches as his

Table 19-1 Pedigree of Francis Galton's Intellectual Disciples

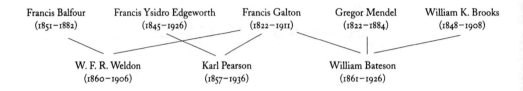

perpetually downtrodden mother Fanny silently begins to clear away the breakfast dishes. A carriage parked outside stands ready to whisk William Pearson off to the Inner Temple. Carl's brother Arthur, two years older, has received a paternal reprimand for some perceived infraction while sister Amy, two years younger than Carl, has conveniently absented herself from view in the pantry. Of his father Pearson wrote, "During the legal terms, winter and summer, he was up at 4 a.m. to read his briefs and prepare his speeches for Court. Home at 7 p.m., dinner followed and bed at 9 p.m. Only in the vacations did we really see him; then he was shooting, fishing, sailing with a like energy which astonished me even as an active boy."[3] Carl would follow obediently in his father's footsteps as the elder Pearson worked his way through the high grass at the river's edge until he came to a quiet pool where a trout might lurk. He would snake a fly across the promising stretch of water while his son watched. Carl was not permitted to cast if there were likely to be fish about.

Arthur Pearson was sent off to Rugby, but Carl was educated at home until he was nine, when he entered the University of London College School. He remained there for seven years, but was withdrawn in 1873 for reasons of health, spending the next year with a private tutor. Arthur briefly practiced law, but was made rich by one of his father's clients for taking his surname and, as Arthur Beilby Pearson-Gee, set about enjoying life. Meanwhile Carl convinced his father "to let him go up to Cambridge and work as a 'beast' under Routh"[4] along with another young man, Josh Conway. Edward John Routh was the greatest of the Cambridge maths coaches, surpassing even his mentor William Hopkins who had tutored Galton. Between 1862 and 1882 he had an unbroken string of 22 senior wranglers.[5] Routh wasted no time with his two young beasts. "You have a year," he said, "before entering college; we will devote it to reading subjects not of first-class importance for the Tripos,"[6] whereupon he began to lecture on the theory of elasticity.

Josh Conway couldn't keep up and dropped out after a term, leaving Pearson as Routh's only student. Routh's coaching paid off and in 1875 Pearson went up to King's College, Cambridge, on a mathematics scholarship. At Cambridge, Pearson studied with some of the great mathematicians and

Galton's Disciples

Fig. 19-1 Karl Pearson from a pencil drawing made in 1924. From E. S. Pearson, *Karl Pearson: An Appreciation of Some Aspects of his Life and Work.* Cambridge: Cambridge University Press, 1938.

Fig. 19-2 Walter Frank Raphael Weldon. From Karl Pearson, "Walter Frank Raphael Weldon, 1860–1906, A Memoir," *Biometrika* 5 (190): 1–52.

physical scientists of the era. They included Routh, Arthur Cayley, a leading mathematician who had earlier coached Galton, and the physicist J. Clerk Maxwell. The community at King's was small, just 30 to 40 in number, and Pearson learned from dons and fellow undergraduates alike. Oscar Browning introduced him to Rousseau, Goethe, and Italy, and he became fast friends with the university librarian, Henry Bradshaw. He read Dante in the original, and he began to explore Spinoza's writings as his interest in philosophy grew. Not content only to read, Pearson wrote reviews of two books on Spinoza plus one on the twelfth-century Jewish sage of Cordoba, Maimonides, for the *The Cambridge Review* and another on Spinoza for *Mind*. He sniggered privately about Cambridge undergraduates who wished "to gain social stamp, but not to learn/ While teachers only teach to earn."[7]

In 1879 Pearson won honors in the Mathematics Tripos as Third Wrangler. He also took the Smith's Prize Examination, held on four successive days at the houses of his four examiners, Stokes, Clerk-Maxwell, Cayley, and Todhunter. His performance was insufficient for him to receive the award, but his answer to a question of Todhunter's was so apt that the don bound it up with his unfinished manuscript on the *History of the Theory of Elasticity*, commenting that "this proof is better than De St Venant's."[8] Later Pearson was appointed by the Syndics of the Cambridge University Press to edit and complete Todhunter's history. Pearson rebelled against compulsory religious indoctrination. One day he confronted his tutor saying "I am not going to attend any more divinity lectures."[9] The bewildered tutor countered that "it is an inexorable rule of the College." Pearson fired back "then I am going to another college." His composure regained, the tutor intended to put an end to this nonsense. "It is unheard of," he said, "We shall not give you a bene decessit."[10] But the fleet Pearson was one step ahead of the don. "Mr Ferrers of Caius has already agreed to accept without a bene decessit." The tutor was having none of this. "This matter must come before the College Council," he intoned. That is what happened, but Pearson fought on and mandatory divinity lectures were eventually abolished. Despite his hatred of compulsion, Pearson had thought deeply about religion and several years later wrote, and anonymously published, *The Trinity; a Nineteenth Century Passion Play*.

Pearson's mathematical prowess caught the attention of the authorities and the Fellowship of King's College was conferred on him on April 5, 1880. This assured him of financial independence and complete freedom from duties of any sort. He departed for the Universities of Berlin and Heidelberg, where he attended lectures on philosophy, Roman law, biology, and physics. But most of all Pearson's year abroad got him interested in German culture and literature. Pearson now began to spell his Christian name with a K rather than a C. Although this may have been because he was so enamored of things German, J. B. S. Haldane raised a more intriguing possibility in a centenary lecture honoring Pearson's birth.[11] He conjectured that the C to K change may have been in homage to Karl Marx.

After his year in Germany, Pearson took rooms in the Inner Temple and was called to the bar in 1881. But Germany was his passion and in 1880 he published *The New Werther* under the nom de plume of Loki. In it a young man, Arthur, wanders through Germany writing letters to his intended, Ethel. These letters form the substance of the book. As Egon Pearson points out in his biography of his father, Arthur's letters to Ethel contained bits of the philosophy that Pearson would later espouse in *The Ethic of Freethought* and *The Grammar of Science*. For instance, Arthur asks Ethel in one letter whether she has "ever attempted to conceive all there is in the world worth knowing.... The giants of literature, the mysteries of many-dimensional space, the attempts of Boltzmann and Crookes to penetrate Nature's very laboratory, the Kantian theory of the universe, and the latest discoveries in embryology, with their wonderful tales of the development of life — what an immensity beyond our grasp."[12] This is not your typical love letter, but presumably Ethel was not your average girl.

By 1882 Pearson was lecturing at the South Place Institute on "German Social Life and Thought" up until 1500 A.D. He followed this with lectures in a university extension course on the Reformation in Germany, Humanism in Germany, and so forth. The four hundredth anniversary of Luther's birth was celebrated in 1883. In the *Athenaeum*, Pearson ridiculed a Luther exhibition in the Grenville Library of the British Museum as "a slur on English Scholarship." The rebuttal noted that "there is but a step between hyper-criticism and hyper-nonsense."[13] Bradshaw reprimanded his headstrong young friend with a firm letter:

> I have not the slightest wish to defend the Museum ignorance. But ... when a man who might by his own deeper knowledge help to make such an exhibition very much more interesting and instructive wastes his energies in writing to the *Athenaeum* as you do, it naturally produces the impression that his main object is to let the world see how much more he knows of the subject than the idiots to whose care he says these treasures are entrusted. Those who know you know also that that is not the object you have in view, but it is a pardonable inference for ordinary people to draw.[14]

In *The Ethic of Freethought* (1888) Pearson assembled many of the letters, articles, and reviews he had written on Luther.

Pearson was an active socialist, lecturing on Sundays to revolutionary clubs on the two leaders of German socialism, Ferdinand Lassalle and Karl Marx, and contributing hymns to the *Socialist Song Book*. But he was more a socialist in the abstract and, as an intellectual snob, believed that social progress would inevitably favor those who worked mainly with their brains rather than their hands. Pearson's interest in socialism bloomed at a time when this credo had gained considerable popularity in England.[15] His circle included Beatrice and Sidney Webb, Marx's daughter Eleanor, George Bernard Shaw, Havelock Ellis, and the South African novelist Olive Schreiner, an enthusiastic proponent of women's rights. Ellis and Schreiner became involved, but never intimately as Ellis wrote that "she swiftly realised that I was not fitted to to play the part in such a relationship."[16] "Tussy" Marx lived openly with a scientist, Edward Aveling, in what she termed a "free marriage," later bowing slightly to convention by styling herself in a recognizably modern vein, Marx-Aveling.

Association with such radicals and freethinkers led Pearson to participate in the rededication of the Men and Women's Club in 1885 to focus on the relations between the sexes.[17] A precursor club of the same name had discussed a broad range of topics including art, theatre, and the influence of science on modern thought. The reborn club's purpose was solemnly proclaimed to be "for the free and unreserved discussion of all matters in any way connected with the mutual position and relation of men and women."[18] Club membership was limited to 20 with ten of each sex participating. However, equality in numbers failed to translate into unrestricted participation of both sexes, as the women tended to defer to the men, particularly Pearson. Members included Pearson's Cambridge friends, Robert Parker and Ralph Thicknesses, Marx-Aveling's doctor Bryan Donkin, Schreiner, the Sharpe sisters, and Annie Besant, the birth-control pioneer, socialist, and theosophist. Elizabeth Sharpe, a freethinker married to the solicitor Henry Cobb, introduced Pearson to her sisters, one of whom he would marry.

Pearson's inaugural address to the club was on "The Woman Question." Other subjects discussed at these fraught meetings ranged from prostitution to "preventive checks." Pearson's absorption with "The Woman Question" probably derived partly from observing his mother's miserable marriage. He urged careful scientific study of the problem by impartial minds. "Is there like or unlike inheritance by male and female children of their parents' intellectual capacity?"[19] Is it not likely, Pearson later queried, "that in the future the best women will be too highly developed to submit to childbearing; in other words, the continuation of the species will be left to the coarser and less intellectual of its members?"[20] Galton would have loved it. Here was a protoeugenic concern voiced independently of his own.

Briefly estranged from Havelock Ellis, Olive Schreiner felt her heart go aflutter as the handsome Pearson, tall, slender, with curly blond hair, and a fine, angular lower jaw, solemnly preached on "The Woman Question." Pearson's lectures were published in *The Ethic of Free Thought* and she reviewed them for the *Pall Mall Gazette*. She was most complimentary, but the *Glasgow Herald* was not.[21] The reviewer concluded with this barb: "Mr Pearson would nationalise land and nationalise capital; he at present stands alone in proposing to nationalise women also."[22]

Now approaching 30, Pearson seems never to have been seriously involved with a woman, but in 1886 he formed an intellectual relationship with Schreiner, two years his senior.[23] Schreiner's love life was indeed complex. She was still in touch with Ellis and Donkin had proposed to her, but she was enamored of Pearson. Initially, her letters to Pearson were vivacious and lighthearted.[24] She tried vigorously to suppress her feelings, pretending to play the role of the disinterested researcher, signing one of her letters "Your man-friend OS."[25] As the year wore on, Pearson must have been less diligent about responding, so on December 11 she pleaded that he write her. "My man friend write to me. . . . My man-friend some day when your spiritual life is burning low and dim I will put out my hand and help you if you will help me now."[26] A couple of days later Schreiner fell seriously ill and the rejected suitor Donkin nobly begged Pearson to visit her. Pearson apparently dropped round and then wrote Schreiner implying that her passion for him extended to other parts of his anatomy than his head. She vehemently denied any such primal urge. "If [Donkin] told you I loved you with sex-love it was only a mistake on his part. You will forgive him. I do."[27] Later Pearson wrote that she had the "greatest mind of any woman he had met," but he worried over how to behave in the presence of a woman "who has a sexual passion for him which he does not reciprocate."[28]

The reason was that Pearson had his eye on Maria Sharpe. She presented the same problem for Pearson that Pearson had posed for Schreiner, for it was her mind rather than her body for which she wished to be valued. But Pearson was persuasive because he was believable. He recognized that the comparative lack of intellectual development in women was partially the result of inadequate opportunity, the nostrum for which was socialism since this would eliminate their economic enslavement. He also drew a clear distinction between sexual intercourse and childbearing for he felt that the right of all to decent working conditions required limitation on population growth. At the same time he argued to Sharpe that the primary reason for having sex was to express "the closest form of friendship between a man and a woman."[29] Slowly Pearson wore Sharpe down, but even so his proposal of marriage in 1889 triggered a nervous breakdown. After another six months, Sharpe pulled herself together and married Pearson. Evidently they were unsuccessful at decoupling sex from fertilization for she bore two daughters and a son, all of whom received Germanic names.

Despite his diverse adventures in German history and culture, philosophy, and socialism, Pearson remained true to science. He was appointed Goldsmid Professor of Applied Mathematics and Mechanics at University College, London, in 1884, a position he held until 1911, when he was named the first

Galton Professor of Eugenics. As Goldsmid Professor, Pearson lectured to classes of 40 or 50 engineering students. Apparently he was good at his job, for his colleague Sir William Ramsay once remarked to him: "You and I, you know, Pearson, are the only men who can hold big classes in complete silence in the College."[30] In the fall of 1890 Pearson applied for the position of Gresham Lecturer in Geometry at Gresham College, an appointment he held simultaneously with the Goldsmid professorship from 1891 to 1894. Founded by Sir Thomas Gresham at the end of the sixteenth century, Gresham College was organized like a medieval university with a professor for each of seven subjects: divinity, astronomy, geometry, music, law, physic (medicine), and rhetoric. But even at its founding, this rigid division of knowledge was dying. Hence, the first professors of geometry regarded their mandate as the application of mathematics to all branches of physical science. Like his predecessors, Pearson interpreted geometry broadly to include "courses of lectures on the elements of the exact sciences, on the geometry of motion, on graphical statistics, on the theory of probability and insurance."[31]

Devoted as he was to teaching, Pearson found time for his scholarly pursuits. He finished a manuscript begun by his penultimate predecessor in the Goldsmid chair, William Kingdon Clifford, entitled *The Common Sense of the Exact Sciences*, following which he completed and edited Todhunter's *History of the Theory of Elasticity*. In 1887 he published *Die Fronica*, a historical study that focused on the development of the Veronica legend and the history of the Veronica portraits of Christ. Written in German, the book was dedicated to his old friend and mentor, Henry Bradshaw. Like Edgeworth, Pearson read *Natural Inheritance* shortly after its publication. He was so intrigued by the book that he volunteered to lecture on it to the Men and Women's Club and prepared a 25 page précis, which he delivered on March 11, 1889.[32] While his presentation was perceptive and lucid, Pearson did not yet see the general applications of Galton's statistical methods. He apparently dismissed *Natural Inheritance* from his mind, for Galton is unmentioned in his book *The Grammar of Science* (1892).[33] Although Pearson discussed heredity in that book, it was only in terms of Darwin's hypothesis of pangenesis and Weismann's theory of the germplasm. Correlation was only considered in the nonstatistical sense. Galton was also missing from Pearson's Gresham lectures in March and April, 1891, on which *The Grammar of Science* was based, as well as the next two sets of lectures on the "Laws of Chance" given in the fall of 1892 and the winter of 1893.[34] But he appeared briefly with Weldon, in the Gresham lecture series beginning in November 1893, on "The Geometry of Chance."

So Pearson apparently forgot the past when he rose to speak at a dinner given in his honor at University College on April 23, 1934, shortly after his retirement. "After Bradshaw came Francis Galton," said Pearson. "In 1889 he published his *Natural Inheritance*. In the Introduction to that book he writes:

'This part of the inquiry may be said to run along a road on a high level, that affords wide views in unexpected directions, and from which easy descents may be made to totally different goals to those we have now to reach.'"[35] Pearson then spoke passionately about his mentor, perhaps slightly influenced by a glass of fine claret:

> "Road on a high level," "wide views in unexpected directions," "easy descents to totally different goals"—here was a field for an adventurous roamer! I felt like a buccaneer of Drake's days—one of the order of men 'not quite pirates, but with decidedly piratical tendencies,' as the dictionary has it! I interpreted that sentence of Galton to mean that there was a category broader than causation, namely correlation, of which causation was only the limit, and that this new conception of correlation brought psychology, anthropology, medicine and sociology in large part into the field of mathematical treatment. It was Galton who first freed me from the prejudice that sound mathematics could only be applied to natural phenomena under the category of causation. Here for the first time was a possibility, I will not say a certainty, of reaching knowledge—as valid as physical knowledge was then thought to be—in the field of living forms and above all in the field of human conduct.[36]

A finer tribute could not be given by a protégé to his mentor. The only trouble is that it probably did not happen quite that way.[37] Edgeworth rather than Galton may have been central to Pearson's intellectual development. The two were in correspondence at least as early as April 1891, but only Edgeworth's letters survive. Edgeworth began by enticing Pearson into sending an article to the *Economic Journal*.[38] Seemingly Pearson penned a scathing critique, which Edgeworth tried to tone down. "We of the *British Economic* do not lay ourselves out for controversy. The method of rebuttal and rejoinder does not seem particularly suited to our subject.... Hence I would rather that you omitted the second part of your title 'a rejoinder to Mr C.' and if possible consent to soften passages relating specially to him rather than generally to the subject."[39]

Edgeworth corresponded extensively with both Galton and Pearson, so it is surprising Edgeworth is only mentioned twice in Pearson's massive and meticulous biography of Galton. He is mentioned directly in the quotation at the beginning of this chapter and in a letter of Galton's to W. F. Sheppard.[40] One reason may be the nasty little contretemps that Pearson and Edgeworth got into over skew curves in the early 1890s. With respect to the normal distribution, these distributions are asymmetric, with one or the other tail of the curve extending further out than expected. Such distributions were recognized by Quetelet himself, and in 1879 Galton introduced a method for dealing with asymmetry.[41] Interest in these distributions was aroused in 1887 by John Venn's letter to *Nature*, which argued that the theoretical assumption that there was only one normal distribution was incorrect. Worried about the same problem, Edgeworth had considered methods for fitting asymmetric binomial distributions to asymmetric frequency data in a paper published a year earlier in the *Philosophical Magazine*. Stimulated by Venn's letter he wrote to *Nature* himself. In 1887 he prepared a longer paper on the subject for the *Philosophical Magazine*. Pearson's first statistical publication, an October 1893 letter to *Nature*, was a lineal descendant from these contributions. In it he boasted that he had developed a better technique for fitting skewed distributions by purposely ignoring Edgeworth's method. The full account of Pearson's analysis appeared in 1894 and 1895 in two enormous articles in the *Philosophical Transactions of the Royal Society*.[42]

In the fall of 1893, before Pearson made his initial presentation to the Royal Society, Edgeworth sent him a paper developing his own approach to the analysis of skew curves. Pearson was upset so Edgeworth agreed to hold up his own paper in an act of courtesy rarely seen in science. On June 21, 1894, Edgeworth finally read his long paper before the Royal Society. Galton and George Darwin reviewed the paper favorably, but the major mathematical critique took almost a year to arrive and the review was negative, so the paper was rejected. Suspecting Pearson of foul play, Edgeworth failed to call his attention to Erastus De Forest's earlier manuscript on skew curves until it was too late for Pearson to cite the work in his article. As Pearson wrote to George Udny Yule in August 1895, "I saw Edgeworth and he told me with some glee that an American had in 1884 reached my skew curve of type III! So he has and quite nicely: see *Nature* this week."[43] There Pearson ate a little crow and acknowledged De Forest's priority.

Then Edgeworth took Pearson's paper to task in print.[44] Pearson was stung and apparently blasted Edgeworth in a letter, for Edgeworth replied "You bring so many charges against me—(1) misinterpretations, (2) mathematical error, (3) logical fallacy and (4) unjustifiable tone. My withers are not wrung equally by all these."[45] Eventually Pearson and Edgeworth were on good terms once more, but there is reason to suspect that Pearson felt that his methods were far superior to Edgeworth's.[46] A revealing passage is to be found in Pearson's 1920 paper on the history of correlation. On reexamining Edgeworth's paper 25 years after its publication, Pearson wrote that "he harnessed imperfect mathematical analysis to a jolting car and drove it into an Irish bog on his road, and that it was doubtful analysis not errors of printing which led to his obscure conclusions."[47] Perhaps Pearson's diminishing regard for Edgeworth's work explains why he ignored Edgeworth's relationship with Galton. But perhaps he was jealous of Edgeworth's lengthy correspondence with Galton concerning the statistical analysis of the anthropometric data. In one letter Edgeworth wrote Galton that his own mathematical extension of the theory of correlation "bristles with partial differential equations much like thistles."[48]

Whatever the derivation of Pearson's statistical roots, he came to regard Galton as his great intellectual mentor as the final decade of the nineteenth century progressed. This was also true of Walter Frank Raphael Weldon, the son of Walter and Anne Weldon.[49] After a brief stint in his father's business, Weldon senior moved to London where he landed a job as a reporter with the *Dial and Morning Star*. From 1860 to 1864 he edited *Weldon's Register of Facts and Occurrences relating to Literature, the Sciences and the Arts*, for which he had numerous distinguished contributors. Like many Victorians, he was a man of broad accomplishments, and his knowledge of chemistry led him to develop a method for regenerating the manganese peroxide used to manufacture chlorine. This brought the Weldon family comparative wealth and fame, and Weldon senior a fellowship in the Royal Society in 1882.

In contrast to Pearson's mother, Anne Weldon exhibited marked strength of character, helping her husband to cope early on when he was struggling. She also made sure that the two Weldon boys, Dante, and especially Raphael, toed the mark. At age 13, Raphael was sent to a boarding school at Cavarsham. After nearly three years he left and matriculated at University College, London, in 1876, at age 16 following a few months of private study. There he took courses in Greek, English, Latin, French, and pure mathematics. He also attended Daniel Oliver's lectures in botany and Ray Lankester's on zoology. Lankester apparently ignited Weldon's interest in the subject. A year later he transferred to King's College to prepare for a medical career by taking additional science courses, but stayed only two years and departed for St. John's College, Cambridge, to study zoology.

While studying for the Natural Sciences Tripos, Weldon came under the influence of the gifted young morphologist Francis Balfour, but the strain of reading for the Tripos took its toll. He began to suffer from insomnia and later had a breakdown. Nevertheless, he recovered sufficiently to start plugging away at the Tripos once again in 1881. Balfour arranged for a scholarship to be awarded to Weldon, recognizing that he was very talented. Teetering on the brink of nervous exhaustion again, Weldon took a three-month break in the south of France. Upon arriving he learned that his brother Dante, a student at Peterhouse, had just died of apoplexy. But the hot Provençal sun, the hillside vineyards, the lavender fields, the vistas of jagged and bleached mountains, and the hilltop towns of clustered houses with stuccoed walls and red tile roofs worked their magic, and a rejuvenated Weldon returned from France to take a first class in the Tripos. Meanwhile those who were nearest and dearest to him were departing rapidly. Broken-hearted at the loss of her son Dante, his mother died the same year, and Weldon's mentor Balfour was killed in an alpine accident in the summer of 1882. His father followed soon after in 1885.

Having successfully navigated the shoals and reefs of the Tripos, Weldon was off to the Zoological Station in Naples to begin research on the developmental biology and anatomy of marine organisms. He returned to Cambridge in September 1882, and was invited to demonstrate for Adam Sedgwick, who had inherited Balfour's courses. As a demonstrator, Weldon would have been expected to aid Sedgwick's students in a practical sense much as a laboratory instructor does today. In the spring of 1883 Weldon married Florence Tebb, a Girton graduate, who would become his lifelong research associate. There were more trips to Naples, the Bahamas, and to Guernsey to collect and study marine creatures. Weldon was now publishing a steady flow of papers on the development and anatomy of these organisms, focusing ever more on the Crustacea. In 1884 he was appointed university lecturer in Invertebrate Morphology at St. John's.

By the fall of 1888, the buildings of the Marine Biological Laboratory in Plymouth were nearly completed and visits there now replaced those to Guernsey. Weldon's interests were turning slowly from morphology and anatomy to racial variation within species. He began a large-scale analysis of variation in carapace length of adult females of the common shrimp *Crangon vulgaris*. He had samples of 400 individuals from Plymouth Sound in Cornwall and 300 from Southport on the coast of the Irish Sea in Lancashire. Mr. W. H. Shrubsole had kindly measured another 300 from Sheerness near the mouth of Thames. Weldon was also reading *Natural Inheritance* and got in touch with Galton soon thereafter. His first paper on variability in carapace length in *C. vulgaris* was received for publication in the *Proceedings of the Royal Society* on March 20, 1890.[50] There he effusively acknowledged Galton's input. "My ignorance of statistical methods was so great that, without Mr. Galton's constant help, given by letter at the expenditure of a very great amount of time and trouble, this paper would never have been written."[51] Weldon should also have acknowledged his wife's contribution. She had helped with the laborious calculations, the marvellous Brunsviga mechanical calculator having yet to arrive on the scene. Weldon analyzed his data principally by the percentiling (i.e., ogive) method, with medians and quartiles being determined "in accordance with Mr. Galton's notation."[52]

Weldon observed "that not only does the average size of the carapace differ in different local varieties, but the range of deviation from that average differs also."[53] Nevertheless, carapace length in all three populations of *C. vulgaris* was normally distributed. He added a teaser at the end, writing that he had "attempted to apply to the organs measured the test of *correlation* given by Mr. Galton . . . and the result seems to show that the degree of correlation between two organs is constant in all the races examined."[54] But this result was "so important to the general theory of heredity" that Weldon decided "to postpone a discussion of it until a larger body of evidence has been collected."[55] Hence, using a time-honored scientific caveat, Weldon was getting his conclusions on correlation in print without actually presenting the data. Galton was an enthusiastic referee for his paper, the first of thousands to follow in the next century on quantitatively varying characters in different races and species of plants and animals.

In May 1890, Weldon wrote Galton apologizing for the long delay in preparing the *Crangon* correlation curves. He was working on them in the evenings and hoped Galton did not mind as he did not want to "do arithmetic by daylight, because I want very much to do some anatomical work."[56] These results were eventually sent for Galton's perusal and in 1892, Weldon, now a Fellow of the Royal Society himself, submitted his second *Crangon* paper for publication in the *Proceedings*.[57] He had attempted "to apply Mr. Galton's method to the measurement of the correlation between four organs of the common shrimp."[58] "Organ" was a bit of a misnomer as Weldon was measuring parameters like carapace length, length of the sixth abdominal tergum, etc. His sample numbers were large, ranging from 300 to 1,000 adult female shrimp. They came from the three English locations previously sampled, plus Roscoff in Brittany and Helder in North Holland.

Not only was Weldon's paper a cornucopia of correlation coefficients, summarized in 22 tables, but he had calculated means and variations around the mean. The results were a triumph. There were clear variations in average length of the organs measured in the different shrimp populations and in the probable error of the distributions, but they were not reflected in the correlation coefficients. This was highly encouraging, for it meant that correlations between different organs were going to hold despite variations in average size for different populations. Galton must have been delighted. Weldon was too, writing that eventually the "study of those relations which remain constant through large groups of species would give an idea, attainable at present in no other way, of the functional correlations between various organs which have led to the establishment of the great sub-divisions of the animal kingdom."[59] Weldon's mathematical progress between the first and second *Crangon* papers caused Pearson to remark in "Notes on the History of Correlation" that the "pupil . . . was soon to outdistance the master in his width of theoretical knowledge."[60]

The third paper in the series appeared in the *Proceedings* in 1893.[61] By now Weldon and Pearson were close collaborators. In this paper Weldon reported on correlated variations in the shore crab, *Carcinus moenas*. He examined two sets of 1,000 adult females, with one sample coming from Plymouth and the other from the Bay of Naples. He measured eleven different "organs" (e.g., breadth of carapace, length of meropodite of the right chela, etc.) and used one measurement, length of carapace, to normalize the rest of his data, taking the total length of the carapace as 1,000 (Fig. 19-3 top, line A-B). Thus, the mean value for the right antero-lateral margin of the carapace came out to be

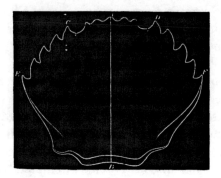

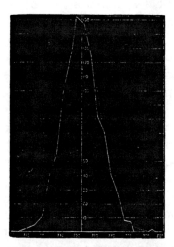 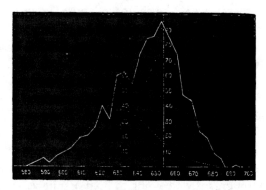

Fig. 19-3 Variation in two different measurements made by Weldon on the carapace of the crab *Carcinus moenas*. *Top*. Diagram of the carapace of *C. moenas*. The length of the carapace was measured as the distance A-B and and was used to normalize other measurements using a value for this distance of 1,000. *Bottom left*. Distribution of the normalized frequency of occurrence of different lengths of the right antero-lateral margin (solid line, distance A-F in Fig. 19-3 top) of the carapace for 999 female crabs from the Bay of Naples compared to a theoretical normal distribution (dotted line). The abscissa represents thousandths of a carapace length and the ordinate represents numbers of individuals. All parameters save the one shown on the bottom right exhibited similar, unimodal distributions. *Bottom right*. Distributions of frontal breadths (Fig. 19-3 top, distance C-D) for the same crabs (solid line). The three dotted lines are theoretical distributions for these data. Each ordinate of the upper dotted curve is the sum of the corresponding ordinates of the two component curves. That is frontal breadth exhibits a bimodal distribution in the Bay of Naples population. The abscissa represents thousandths of a carapace length and the ordinate numbers of individuals. From W. F. R. Weldon, "On Certain Correlated Variations in *Carcinus moenas*," *Proc. of the Royal Society* 54 (1893): 318–333.

752 thousandths of a carapace length. Although there seems not yet to have been a statistical test for goodness of fit other than the eyeball, Weldon calculated the expected bell curves and compared them to the observed distributions and found the agreement to be very good.

All of the measurements distributed normally (Fig. 19-3, bottom left) save one. The frontal breadths of the carapaces of the Naples sample showed an asymmetric distribution (Fig. 19-3, bottom right). Weldon thought the explanation might be "the presence . . . of two races of individuals clustered symmetrically about separate mean magnitudes."[62] Pearson came to the rescue and tested this supposition, showing that Weldon was right—"female *Carcinus moenas* is slightly dimorphic in Naples with respect to its frontal breadth."[63] Weldon again calculated a wealth of correlation coefficients, reaching the same conclusion as before. His results did "not demonstrate a difference between the value of Galton's function for a given pair of organs in Naples and the corresponding value in Plymouth."[64] Near the end of his paper Weldon remarked that it "cannot be too strongly urged that the problem of animal evolution is essentially a statistical one."[65] Galton must have been pleased.

Pearson and Weldon were initially drawn together not by science, but by university politics, a subject most academics will recognize as an enormously time-consuming and often low-stakes process. But in this case the cause that united the two scientists had more merit than usual. It involved the reorganization of the University of London, which was mainly an examining body for a collection of independent colleges held together in a loose confederation.[67] Since 1884 reform of the University of London had been in the wind and "association followed on association, royal commission on royal commission."[68] After several upheavals, the professoriate at University College had won representation on the governing body. Ray Lankester was one of the standard bearers, but his appointment to the Linacre chair of comparative anatomy at Oxford in 1891 took some steam out of the movement. With his boundless energy and enthusiasm, Weldon helped to fill the gap. The rebels found they were in a fight with their own College authorities as well as with interfering outsiders. They wrote a letter to the *Times* that provoked growls from the authorities about dismissing them from their chairs.

The main sticking point was a proposal to unite King's and University College in a what the professors regarded as a second-rate duplicate of London University to be named Albert University. While this may have pleased the Queen, it provoked the wrath of the professors and one evening Weldon, with one or two compatriots, bustled around gathering signatures on a petition to block the Albert University charter, which was to be discussed the next day in Parliament. With great alacrity Weldon and his coconspirators managed to have the widely signed petition in the hands of every member of the House of Commons the next day. The Albert University charter was dead. The insurgent professors now went on the offensive. They formed an "Association for Promoting a Professional University of London." They proposed the "creation of a homogeneous academic body with power to absorb, not to federate existing institutions of academic rank," a professorial university as opposed to "a collegiate or federal university."[69]

The Association got the support of scientific and literary giants and trotted out T. H. Huxley as its president. But Huxley stood for compromise. This was too much for the fiery Pearson, who resigned as secretary of the Association, so the more level-headed Weldon took his place. Pearson was determined to send the *Times* an open letter to Huxley, but Weldon tried to talk him out of it, fearing reprisals. Undeterred, the headstrong Pearson prevailed and the letter was published on December 3, 1892. The Executive Committee delivered its report at a general meeting of the Association on December 21 and Huxley spoke strongly for its adoption. Pearson continued to push and got an amendment passed stating that "the Association trusts that its Executive Committee will persevere in its efforts to establish as far as possibly may be a professorial as distinguished from a federal university."[70] This was carried.

The dispute dragged on with Huxley opting for compromise and Pearson pushing him with amendments. On January 23, 1893, Huxley presented his own vision of a teaching university; a vague motion to prepare a scheme to be submitted to the Association. Pearson, seconded by Lankester, tightened up the charge to the Executive Committee to prepare a report "in general accordance with the proposals adopted by the Association."[71] Hence, the Executive Committee was instructed to develop a plan for an integrated university into which the separate colleges would be absorbed and in which professors would not only teach, but exercise governance. Pearson's amendment passed. During the contretempts Weldon, to whom Huxley was a hero, wrote Pearson a strongly worded letter cautioning him to mollify his attacks upon the older man. With Huxley's death in 1895 the Association for all intents and purposes closed up shop and Weldon succeeded Huxley in 1896 as Crown nominee on the University Senate.

Weldon, a genial man of strong convictions with a great dark walrus mustache and thinning hair, had all hallmarks of an able academic administrator. "An impulsive loveable man going to the heart of any subject immediately," said one of his colleagues, "and always speaking up with great feeling for what he thought right."[72] But Weldon delighted in research and it was a nuisance for him to be pulled away from his work to deal with one tempest in a teapot or other. It seems plausible that then as now he occasionally had to preen the ruffled feathers of this or that academic prima donna while gently defusing some madcap scheme. In Pearson's words, Weldon " 'played the game,' threw firmly and well the lance for the cause he thought right, and went his way."[73] And, as often as not, that way would take him back to his beloved Brunsviga calculator with its characteristic grinding sound, made more pronounced in Weldon's case because he habitually forgot to oil it.

Weldon and Pearson became very close friends. Before they both lectured they lunched together. As they ate they argued, raised problems, and suggested solutions, often working them out on the back of the menu or using pieces of bread. "Weldon, always luminous, full of suggestions, teeming with vigour and apparent health, gave such an impression to the onlookers of the urgency and importance of his topic that he was rarely, if ever, reprimanded for talking 'shop.'"[74] Considering that Pearson had temporarily dismissed *Natural Inheritance* while Weldon embraced Galton's statistical methods in his first *Crangon* paper, it seems likely that Weldon lit the Galtonian fire in Pearson. Perhaps it was at one of these luncheons. But now both Pearson and Weldon, firmly committed to each other, were to form a powerful triumvirate together with their aging master. They would try to make statistical sense out of heredity and evolution.

TWENTY

Evolution by Jumps

> From Darwin's gradualism to the saltationism of the Mendelians, continuous and discontinuous change seemed fundamentally distinct. Galton only made explicit a difficulty that everyone seems to have felt.
>
> —J. Maynard-Smith,
> *Galton and Evolutionary Theory*[1]

Galton genuinely disagreed with Darwin over whether evolution occurred in small incremental steps. In *Natural Inheritance* he acknowledged that whenever intermediates were sought "between widely divergent varieties ... a long and orderly series can usually be made out, each member of which differs in an almost imperceptible degree from adjacent specimens."[2] But he discounted these intermediates as "unstable varieties" whose descendants eventually reverted. Instead, he argued that "sports" occurred frequently and were genetically stable, implying that these were the stuff of evolution.

For the young British biologist William Bateson (Fig. 20-1), Galton was preaching to the converted.[3] The son of William Henry Bateson, the master of St. John's College, Cambridge, he attended the Rugby School and matriculated at St. John's in 1879. By his own admission "mathematics were my difficulty. Being destined for Cambridge I was specially coached, but failed. Coached once more I passed, having wasted, not one, but several hundred hours on that study."[4] Weldon was Bateson's closest friend at Cambridge and, like Weldon, he came under Balfour's influence. Balfour encouraged Bateson to work on a strange marine organism called *Balanoglossus* that was perceived to be allied to the vertebrates and was abundant in Chesapeake Bay.[5] Weldon

Fig. 20-1 William Bateson. Reproduced courtesy of the John Innes Foundation Historical Collections, John Innes Centre, Norwich, England.

helped Bateson make contact with W. K. Brooks at Johns Hopkins University. Shy, retiring, and gentle, Brooks was a descriptive evolutionary morphologist. He did most of his research at the Chesapeake Zoological Laboratory, a movable marine station established each summer between 1878 and 1906 somewhere on the shores of the bay.[6] Bateson visited Brooks in 1883 and 1884 when the laboratory was located at Hampton, Virginia. Brooks seemed happy to have Bateson work on *Balanoglossus* and Bateson later wrote that Brooks handed over "to a young stranger one of the prizes which in this age of more highly developed patriotism, most teachers would keep for themselves and their own students."[7]

After the long, hot summer days with humidity so palpable that it turned the sky milky and the great afternoon thunderheads reared up over the shimmering vastness of the bay, Bateson and Brooks engaged in long hours of animated discussion deep into the thick night. Brooks would lie on his bed in shirtsleeves, his forehead covered with prespiration. Bateson sat nearby, the back of his shirt soaked through. There is little doubt that during those steamy evenings there was much talk of discontinuous evolution. Brooks had just completed his book *The Law of Heredity: A Study of the Cause of Variation and the Origin of Living Organisms*. There he proposed a new theory of heredity, designed to replace pangenesis, that permitted discontinuous evolution by jumps.[8] In a key chapter, "Saltatory Evolution," he cited arguments of Huxley,

Galton, and Mivart in support of this mechanism. He gave examples of new races formed in sudden jumps or saltations to illustrate that "the evolution of organisms may . . . be a much more rapid process than Darwin believes."[9] Bateson was convinced and pinned his colors to the masthead of discontinous variation. This positioned him uniquely to embrace Mendel's principles upon their rediscovery in 1900.

Much later, Beatrice Bateson drew a verbal portrait of her husband.[10] He was a man intolerant of slowness in others who could accomplish the work of many days in one. But he enjoyed socializing, when his apparent reserve would suddenly dissipate into mirth. He was absent-minded, often mislaying his notebooks, forceps, scissors, pipe, and glasses. "Of his clothes he was as reckless as a school-boy. He was capable of going up to London in old 'garden' flannels, darned across the knee, or (in the other extreme), he might be found kneeling on the gritty garden path, in a brand-new 'town' suit, recording some batch of seedlings."[11] He also had a passion for fine art. He once remarked in a letter to his sister Anna, then visiting Dresden, that "I am glad you are not overwhelmed by the Gallery. But I felt one could trust as good a Blake-ite as yourself with Rembrandt and Corregio."[12]

Although Bateson's papers on *Balanoglossus* were well-regarded,[13] he came to view this work as trifling. In 1885 he was elected a Fellow of St. John's College. In the spring of 1886 he set out for Central Asia hoping to study variation in specific animals in response to the degree of salinity of the Aral Sea and nearby lakes whose waters varied from fresh to salty. He camped nearby at Kazalinsk, spending 18 months there through searing summer and freezing winter looking out over the vast stony red expanses of the Kyzyl Kum desert. There were shells everywhere along the shores of the Aral Sea, and the dry, salt-bottomed lakes of Jaksi Klich, Jaman Klich, and Shumish Kul. Strong south winds blew for days across the sea driving the water for hundreds of feet over the beach at Sary Cheganak. When the winds subsided, cockles (*Cardium edule*) were stranded all over. These were the animals Bateson chose to study as he tried to probe correlations between environmental differences and morphological alterations. He returned from Russia in the autumn of 1887, was elected to the Balfour Studentship, and was soon off to northern Egypt to collect cockles from lakes of varying salinity in the vicinity of Alexandria.

He wrote up his results and titled his paper "On some variations of *Cardium edule* apparently correlated to the conditions of life." He shipped it off for publication in the *Philosophical Transactions of the Royal Society*[14] and sent Galton a copy. Galton's correlation paper had been published the previous year and he chided Bateson for using correlation improperly in his title. Bateson replied that he had "a sort of idea" that he had misused the word, "but allowed it to stand through negligence."[15] It was too late to undo the damage, as the proofs of the paper had been returned. Soon thereafter he became fully aware of the power of Galton's work when he read *Natural Inheritance*. He later recalled "the thrill of pleasure" with which he had "first read *Hereditary Genius* and the earlier chapters of *Natural Inheritance*."[16] The chapters in *Natural Inheritance* are those summarizing Galton's hereditary theories where he first hypothesized discontinuous evolution via "sports."

In 1891 Ray Lankester became the Linacre Professor of Comparative Anatomy at Oxford, a position Bateson applied for just in case the electors denied the distinguished Lankester's petition. Weldon moved into Lankester's old slot as Jodrell Professor at University College, leaving his position as demonstrator under Sedgwick open at Cambridge. But Bateson failed to profit from this game of academic musical chairs, as Sedgwick believed he had gone "too far afield" and his "things are a 'fancy subject.'"[17] He retained his Johnian fellowship and was elected to the Stewardship of St. John's College in 1892. For the next 15 years these positions would relieve him of personal and financial anxiety.[18]

Bateson began publishing a steady stream of papers on subjects as diverse as the sense organs and perceptions of fishes to the nature of supernumerary appendages in insects, but his heart was in the study of discontinuous variation. He "travelled ceaselessly to see for himself alleged cases of abnormality and variation; he endeavoured, as far as possible, to examine every specimen, and verify every statement."[19] In 1891 he made his first major pronouncement on discontinuous variation in the *Journal of Linnean Society*.[20] There he documented a series of instances of discontinuous variation in the floral symmetry of different plants. For example, in *Streptocarpus rexii* normal flowers have five petals with the upper lip possessing two and the lower lip three. But symmetric variants arise with four petals of equal size forming a cruciform structure. Other abnormal flowers have seven petals. He dismissed reversion to an ancestral type as explaining the variants because several distinct types frequently occurred (e.g., four- and seven-petalled *Streptocarpus* flowers), not all of which could be revertants. Although Darwin had recognized that certain characters vary discontinuously, Bateson's object was to show that such abrupt changes were of wide occurrence. He acknowledged the existence of continuously varying characters, citing Galton's anthropometric studies of size in humans and Weldon's *Crangon* measurements as examples. But if there was still "little evidence that species may arise" by discontinuous variation, there was even less "that new forms" arose by the action of natural selection on continuously varying traits.[21] So Bateson threw down the gauntlet. The charge for those seeking to understand the origin of species was to recognize the existence of these two causes of variation and then to determine their role in the speciation process.

In 1892 Bateson published a paper describing size variation in the hornlike processes seen in males of certain beetles and in the terminal forceps in earwigs.[22] In so-called "high" lines these structures were more highly developed than in "low" lines (Fig. 20-2, top right). The high and low earwig lines were dimorphic and exhibited two nonoverlapping, essentially normal, distributions of forceps length (Fig. 20-2, top left). Bateson was pleased because the two normal distributions appeared to represent positions of stability as defined by Galton. The same was true of the Javanese beetle, *Xylotrupes gideon*. Males from high lines possess two horns that come together much like a lobster claw with which they capture females prior to mating (Fig. 20-2, bottom right). Bateson measured the lower or cephalic horn. The measurements again fell into two nonoverlapping, approximately normal distributions (Fig. 20-2, bottom left), while measurements of the length of the elytra (wing covers) exhibited a single normal distribution. Excited about the dimorphism in earwigs and *Xylotrupes*, Bateson wondered whether they might represent the beginning of a division into two new species. Hence, he was discounting the continuous variation observed around the two peaks of the bimodal distributions as being of great evolutionary consequence. What mattered was the discontinuity, the fact that there were two modes in the first place. Bateson consulted Galton prior to the paper's publication, thanking him for his advice.[23]

By now Bateson was so wound up about variation that he "ransacked museums, libraries, and private collections; he attended every kind of 'show' mixing freely with gardeners, shepherds and drovers, learning all they had to teach him."[24] He made good use of this information in his monograph, *Materials for the Study of Variation* (1894).[25] His thesis was straightforward. The naturalist's duty was to codify the facts concerning variation to rid biology of "the burden of contradictory assumptions by which it is now oppressed."[26] Since variation was the stuff of evolution, it was crucial to understand its nature. Even if Darwin had not provided an answer "we shall not honour Darwin's memory the less; for whatever may be the part which shall be finally assigned to Natural Selection, it will always be remembered that it was through Darwin's work that men saw for the first time that the problem is one which man may reasonably solve."[27]

Bateson's conundrum was that members of a species were similar, but distinct from those belonging to another. Although transitional forms between related species were sometimes recognizable, in most cases none were detected so "the forms of living things do . . . most certainly form a discontinuous rather than a continuous series."[28] Since this was true at present, there was no reason to think that it had ever been otherwise so evolution must be the story of discontinous changes, of saltations, of jumps from one species to the next. Bateson accepted the Doctrine or Theory of Descent, in its assertion "that all living things are genetically connected,"[29] but environments blended into one another "to form a continuous series, whereas the Specific Forms of life which are subject to them on the whole form a Discontinuous Series."[30] Since all theories of evolution started from the premise that the various forms of life

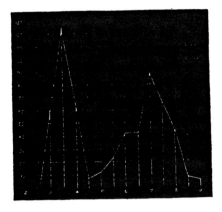
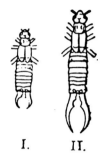
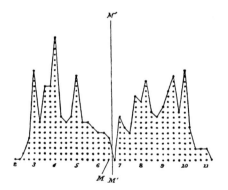
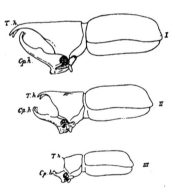

Fig. 20-2 Dimorphism in earwigs (*Forficula auricularia*) and the Javanese beetle (*Xylotrupes gideon*). *Top right.* Low (I) and high (II) forceps males of the common earwig. *Top left.* Curve showing frequency of occurrence of forceps of various lengths in male earwigs. Ordinate gives numbers of individuals and abcissa gives length of forceps in millimeters. *Bottom right.* Diagrams of *Xylotrupes gideon* males seen from the side. Legs not shown. High (I), medium (II), and low (III) males are illustrated. *T.h.*, thoracic horn; *Cp.h.*, cephalic horn. *Bottom left.* Curves of the frequency of various lengths of the cephalic horns in *X. gideon* males. M', mean value. Ordinates show numbers of individuals; abscissae show relative approximate size (shortest cephalic horn 0.4 cm and longest 2.4 cm.). There are hardly any individuals at the mean as the population is dimorphic. From William Bateson and H. H. Brindley, "On some cases of Variation in Secondary Sexual Characters statistically examined," *Proc. of the Zoological Society* (1893): 585–594.

were related to each other and that their diversity was the result of variation, "variation, in fact, *is* Evolution."[31] The critical question for evolutionary theory was "the degree of continuity with which the process of Evolution occurs."[32] He summed up as follows: "The first question which the Study of Variation may be expected to answer relates to the origin of that Discontinuity of which Species is the objective expression. Such Discontinuity is not in the environment; may it not, then, be in the living thing itself."[33]

In his long introduction to *Materials* Bateson referred admiringly to the chapter on "Organic Stability" in *Natural Inheritance*.[34] While Galton's use of the normal distribution to analyze size variation in human beings was appropriate, Bateson argued that if two populations existed, one tall and the other short with intermediates being rare, a normal distribution could be fitted to each. They would be dimorphic like the beetle *Xylotrupes* or the earwigs. He later cited Galton again. "To employ the metaphor which Galton has used so well . . . we are concerned with the question of the positions of Organic Stability; and in so far as the intermediate forms are not or have not been positions of Organic Stability, in so far is the Variation discontinuous."[35]

Despite its intimidating size and density, 886 examples of discontinuous variation, Galton was so enthusiastic about *Materials* that he published an article in *Mind* entitled "Discontinuity in Evolution."[36] There he reviewed organic stability and considered how "the centre of a race may be changed." At one time its position was A, but much later it had switched to position B. Did this switch occur via a long succession of tiny steps, each so small as to be imperceptible, but large in aggregate, or were there abrupt changes? Galton felt it was specious to argue that finding intermediates between A and B was evidence that evolution occurred in small steps because these apparent intermediates merely represented extremes of two normal distributions. Their progeny would regress toward the mean in the next generation coming to resemble more closely the original race be it A or B.

Galton cited three explanations for the differences between races A and B, but the one he favored was organic stability. No variation could "establish itself unless it be of the character of a sport, that is, by a leap from one position of organic stability to another, or as we may phrase it through *'transilient'* variation."[37] He was "unable to conceive the possibility of evolutionary progress except by transiliencies."[38] Perhaps Galton used "transilient" in place of Bateson's term "discontinuous" since he felt the term actually described the process of evolution as he envisioned it. A transiliency is a saltatory change, a jump or a leap, from one state to another, one race to a new race, one species to a new species. He had aired these views recently "in various publications," but "seemed to have spoken to empty air."[39] Hence, he was delighted when he "read Mr. Bateson's work bearing the happy phrase in its title of 'discontinuous variation.'"[40] Bateson sent Huxley a copy of his book. Huxley approved. He replied that he was "inclined to advocate the possibility of considerable 'saltus' on the part of Dame Nature in her variations. I always took the same view, much to Mr. Darwin's disgust, and we used to debate it."[41]

In contrast, Weldon's review in *Nature*[42] was negative. While he congratulated Bateson and hoped that he would not "rest content with his already great achievement, but will proceed with his promised second volume,"[43] he shredded Bateson's ideas in the second two-thirds of his article. To be sure Bateson had done a nice job of assembling a great many facts, but his interpretation of them, the most precious of things to the scientist, was fundamentally flawed, reflecting a lack of familiarity with the "history" of his chosen subject. He disputed Bateson's contention that discontinuity was not environmental, saying that he referred only to the continuity of the physical environment. However, Darwin and Wallace had argued that "the most important part of the environment against which a species has to contend consists of other living things."[44] Furthermore, environments do not form a continuum, but over geological time are discontinuous. "These preliminary arguments in favour of Mr. Bateson's main contention therefore fail . . . when applied to any part of the process of evolution of which we can know anything."[45] One suspects Bateson was furious on reading Weldon's review. He could not have expected such vituperation from his erstwhile friend, especially having presented so many observations that he thought he had carefully knit together in support of his thesis.

Even more devastating was Alfred Russel Wallace's two-part article in the *Fortnightly Review*.[46] Wallace had, of course, formulated the principle of evolution by natural selection independently of Darwin.[47] While Wallace probably read, or at least skimmed, Bateson's book with mounting concern, Galton's highly supportive article in *Mind* likely alarmed him even more. Bateson was just getting started, but Galton was an eminent scientist. With Darwin dead it was up to the venerable Wallace to defend the faith he cofounded. "The effect of Darwin's work," wrote Wallace, "can only be compared to that of Newton's *Principia*. Both writers defined and clearly demonstrated a hitherto unrecognized law of nature, and both were able to apply the law to the explanation of phenomena and the solution of problems which had baffled all previous writers."[48]

But a reaction had developed. Natural selection was threatened not only because Lamarck's theories were being reinstated in America and England as having equal merit, but "some influential writers" were "introducing the conception of there being definite positions of *organic stability*, quite independent of utility and therefore of natural selection."[49] These positions were attained by discontinuous changes. Bateson had recently "advocated these views in an important work on variation."[50] Hence, Wallace decided to speak out, for he believed that such views were "wholly erroneous."[51] They constituted "a backward step in the study of evolution."[52] Those variations important for evolution were not necessarily "infinitesimal, or even as small as they are constantly asserted to be."[53] Most species exhibited extensive variability and the struggle for existence favored only the fittest individuals. It was also intermittent since there were long intervals when the environment was benign with adverse "meteorological" conditions intervening only occasionally. Wallace used examples to show how combinations of characters led to adaptation "and if we assume that these several characteristics are positions of 'organic stability,' acquired through accidental variation, we have to ask why several kinds of variation occurred together."[54]

Materials went far beyond enumerating "interesting and little-known facts" concerning discontinuous variation. It was aimed at "discrediting the views held by most Darwinians" in favor of a new theory based on the plethora of observations summarized in the book.[55] The problem revolved around "sports." Darwin rejected their evolutionary significance, but to Bateson they were the stuff of evolution. Wallace's blood pressure probably soared when he read Bateson's "Concluding Reflexions," for Wallace quoted several inflammatory statements including this one. "The existence of Discontinuity in Variation is therefore a final proof that the accepted hypothesis is inadequate."[56] Wallace attacked Bateson's main point, that species form a discontinuous series in a continuous environment. He failed to appreciate that even in a single locality extreme environmental variability existed and "nothing can be more abrupt than the change often due to diversity of soil, a sharp line dividing a pine or heather-clad moor from calcareous hills."[57]

With his pen practically smoking, Wallace dismissed the several hundred pages of Bateson's book, addressing meristic variations as catalogs of "malformations or monstrosities which are entirely without any direct bearing on the problem of the 'origin of species.' "[58] Bateson had mixed malformations together with more normal variants under the heading of discontinuous variations and faulted Darwin for ignoring them. But, in so doing he had "failed to grasp the essential features which characterise at least ninety-nine per cent. of existing species, which are, slight differences from their allies in size, form, proportions, or colour of the various parts or organs, with corresponding differences of function and habits."[59] Bateson, Wallace implied, had emphasized the exception rather than the rule, giving a distorted picture of how evolution proceeded.

Having dismasted Bateson's frigate, Wallace trained his broadsides on Galton. He seized immediately on the key problem. Galton was so focused on regression to the mean that, although he admitted there was such a thing as natural selection, he reasoned as if it did not exist. In Galton's view evolution took place by leaps or saltations, but he had missed the essential point. Natural selection was a force so powerful that

It destroys ninety-nine per cent. of the bad and less beneficial variations, and preserves about the one per cent. of those which are extremely favourable. With such an amount of selection how can there be any possible 'regression backwards towards the typical centre' when any change in the environment demands an advance in some special direction beyond it as the only means of preserving the race from extinction?[60]

Wallace proceeded to give Galton, always so enamored of numbers, a lesson in arithmetic. Suppose you have an animal that lives for ten years and produces five pairs of young each year on average. If none died within the first five years, there would be 6,480 pairs, far too many pairs for the environment to support. If selection was now interposed so only one pair survived each year to breed, after ten years the original pair would be replaced by 512 pairs. This was still too many so Wallace supposed that only one fiftieth of the progeny survived and estimated the original population would only expand 2.5-fold over ten years, which seemed more reasonable. Then he came to his main point. The power of natural selection was such that when a typical center, as represented by a normally distributed function, ceased to be the most advantageous, a new bell curve with its own unique center would be formed. "There could not possibly be regression from the new typical centre unless the inevitable survival of the fittest in a rapidly increasing population can be got rid of."[61]

Wallace next dissected Galton's theory of organic stability. What were these variations of Galton's that formed races and ultimately new species? Did they arise independent of environment and if so how did they come into harmony with the environment? Discontinuous variants were rare in the first place. Few of them had "the alleged character of 'stability,'" and they only altered a single part or organ. Adaptation did not involve the modification of a single character, but rather the correlated alteration of groups of characters. Even supposedly stable variants would be subjected to natural selection and could survive only if they were beneficial or at least neutral in effect. If a new variety was among the fittest one or two percent "it does not need this purely imaginary quality of 'organic stability' in order to survive; if it is *not* among this small body of the most fit . . . then . . . it will certainly *not* survive."[62] Hence, "organic stability" was a meaningless concept except in the sense of adaptation to the environment in response to natural selection.

Curiously, Galton had discussed the influence of natural selection near the end of his 1891 paper on fingerprints.[63] He remarked that different classes of patterns were distinguishable from each other much as were different genera of plants and animals. However, natural selection was "wholly inoperative in respect to individual varieties of patterns, and unable to exercise the slightest check upon their vagaries."[64] Using a rather rickety line of reasoning he concluded from the fingerprints that "natural selection has no monopoly influence in forming genera," but that internal conditions alone were sufficient.[65] Exactly how he made the transition from fingerprint classification in human beings, a single species, to distinct genera of plants and animals, often comprising many species, is obscure. However, at the end of the paper, he stated, with more clarity than usual, how he thought the process of evolution proceeded. "A change of type is effected, as I conceive, by a succession of sports or small changes of typical centre, each being in its turn favoured and established by natural selection to the exclusion of its competitors."[66] This is really not a bad description of the way we think that natural selection proceeds today, substituting mutation for sport.

Wallace ignored Galton's last statement and pounced. He correctly accused Galton not only of using terms vaguely, but of comparing apples and oranges. Galton's fallacious analogy between classes of similar fingerprints and biological genera "depends on applying the terms of classification in systematic biology to groups of single objects which have no real relation with the genera and species of the naturalist."[67] Galton himself believed that fingerprint patterns were at best only slightly heritable while heritability was the very essence of the distinguishing features characterizing species and genera. Wallace's logic was devastating, his analysis impeccable, and near the end of his article he tried to diagnose why these two bright scientists had been so badly misled concerning the workings of the evolutionary process. He concluded that they had both looked too narrowly at "one set of factors, while overlooking others which are both more general and more fundamental."[68] He listed these for the edification of Bateson, Galton, or any others who might be led astray. Because they had not recognized these factors they had "completely failed to make any real advance towards a more complete solution of the Origin of Species than has been reached by Darwin and his successors."[69] Beatrice Bateson glumly noted that "Will's misgivings as to the fate of his work were justified. The book was not a success—the Professors and lecturers of the day did not introduce their students to it."[70] The annual arrival of the publisher's account "was a dismal event," the book was remaindered, and the second volume never written.

Weldon's review of *Materials* marked the beginning of his falling out with Bateson. But the *Cineraria* controversy, which should have been no more than a tempest in a teapot, opened the crevasse between Bateson and Weldon into a yawning void. The episode began innocently enough at a meeting of the Royal Society on February 28, 1895. Weldon presented a paper on the frequency of specific abnormalities affecting two dimensions of the shore crab in the Plymouth Sound population. He suspected their frequencies might vary as a function of age. Bateson, recently elected to the Royal Society, was in the audience as was the distinguished botanist Sir William Thiselton-Dyer, the director of Kew Gardens. Afterwards Dyer stood up and made a long statement on the stability of the mean that seemed to exist for certain species, but not others that yielded rapidly to changed conditions. He congratulated Weldon who, he thought, had demonstrated the selective destruction of female shore crabs that deviated from the mean before they became adults. Weldon had also noted that, while there were normally five teeth along the dentary margin, only four could be found in about one percent of the individuals. Although sporting might account for some of these cases, Weldon felt this was unlikely. Dyer concurred saying he believed that evolution proceeded through the agency of small variations and not via sports. In illustration he placed a wild *Cineraria* with its panicle of purple starry flowers on the table in front of him near a horticultural variety with electric blue flowers. The different flower color of the cultivated form, he said, was achieved by human selection and was accomplished by the gradual accumulation of small variations as far as was known.

Dyer elaborated in a letter to *Nature* on March 14, 1895.[71] On April 25, Bateson responded with a rebuttal. He emphasized with many examples the importance he felt sports had played in the origin of new horticultural varieties of *Cineraria*.[72] Bateson concluded that Dyer's statement that modern *Cinerarias* "evolved from the wild *C. cruenta* 'by the gradual accumulation of small variations'" was misleading as it neglected "two the chief factors in the evolution of the *Cineraria*, namely, hybridisation and subsequent 'sporting.'"[73] A furious Dyer grabbed his pen and sent off a stinging counterblast, published on May 2.[74] Having huddled with his staff Dyer examined authentic specimens of the species Bateson regarded as sharing in the origin of modern *Cineraria*. After a brief technical discussion, since a longer one like Bateson's "would necessarily take up a good deal of space, and would not be very interesting to readers of *Nature*,"[75] Dyer concluded that Bateson was talking through his hat. In commenting on *C. lanata*, of which there were numerous examples at Kew, Dyer admonished Bateson that had he examined these specimens he would probably regret "for the sake of his reputation as a naturalist, that he committed himself to print on a subject on which he evidently possesses little objective knowledge."[76]

Bateson shot back angrily on May 9.[77] Dyer's rebuttal was meaningless, for he had "made two objections to Mr Dyer's account of the history of the Cineraria; the careful reader will observe that his letter meets neither."[78] His purpose had been to prevent Dyer's dismissal of sports from being accepted without additional proof that cultivated *Cinerarias* had been produced by the gradual accumulation of small variations. Bateson was confident that he had shown the importance of sports in the origin of these plants so the hybrid origin of cultivated *Cineraria*, the point stressed by Dyer, was of secondary interest. Weldon, watching from the sidelines, chimed in "having consulted the principal authorities cited by Mr. Bateson in *Nature* of April 25."[79] His May 16 letter accused Bateson of omitting "from his account of these records some passages which materially weaken his case."[80] Weldon reviewed the papers cited by Bateson, presenting interpretations that he felt largely compromised the case for the origin of cultivated *Cineraria*, via hybridization and sporting. His letter ended with a reprimand. "As to the actual pedigree of the modern varieties, I am not qualified to express an opinion. All I wish to show is that the documents relied upon by Mr. Bateson do not demonstrate the correctness of his views; and that his emphatic statements are simply evidence of want of care in consulting and quoting the authorities referred to."[81]

Bateson concluded that Weldon had initiated a personal attack on him and arranged to meet him on May 21. He thought he understood Weldon to say that Dyer was bluffing in his counterargument. In his notes on the conversation Bateson remarked that "Weldon's position in writing is therefore that of the accomplice who creates a diversion to help a charlatan. I cannot at all understand his motives, or how he can bring himself to play this part."[82] Three days later Weldon wrote Bateson saying that "you accuse me of attacking your personal character; and when I disclaim this, you charge me with a dishonest defense of some one else."[83] Weldon had only discussed "what appeared to me to be facts, relating to a question of scientific importance."[84] He ended stiffly that if "you insist upon regarding any opposition to your opinions concerning such matters as a personal attack upon yourself, I may regret your attitude, but I can do nothing to change it."[85]

The acrimonious letters to *Nature* flew back and forth until June 6 when both Dyer and Weldon independently had a go at Bateson.[86] The inevitable *Cineraria* facts had been trotted out repeatedly for the readers of *Nature* who by now were probably chuckling over the invective being hurled back and forth and ignoring the convoluted arguments. Dyer ended on a preachy note. Weldon washed his hands of the subject, having shown that "Bateson's original evidence does in fact bear the interpretation I put upon it."[87] *Nature* agreed and published no further letters.[88]

The controversy over the roles of continuous and discontinuous variation in evolution next surfaced in a different setting. Galton was impressed by Weldon's papers on variation and correlation in shrimps and crabs, especially by his discovery of dimorphic forms of shore crabs in the Bay of Naples. Bubbling over with new ideas to test, Weldon had recruited a little army of enthusiastic assistants to aid him, but was short of funds. Galton apparently was toying with the notion of forming a Royal Society Committee to sponsor investigations like Weldon's. Such a committee might also help in giving them direction. This idea was likely stimulated by earlier correspondence between Galton and Alfred Russel Wallace. On February 3, 1891, Wallace wrote Galton urging the formation of an experimental farm or institute to sponsor research on disputed points in the theory of evolution. He was particularly vexed by two problems: (1) "Whether individually acquired external characters are inherited, and thus form an important factor in the evolution of species,—or whether as you & Weismann argue, and many of us now believe, they are not so, & we are thus left to depend almost wholly on variation & natural selection" and (2) "What is the amount and character of the sterility that arises when closely allied but permanently distinct species are crossed, and then 'hybrid' offspring bred together."[89] Wallace thought both questions could be answered if systematic experiments were conducted over a period of ten to 20 years. But how could such long-term experiments be funded? Wallace proposed that a British Association Committee sponsor the research with aid from the Royal Society.

Galton responded diplomatically that he doubted there were enough experiments or experimenters to justify such an elaborate proposal. Wallace replied with a short cover letter and a detailed list of possible experiments. Galton talked the matter over with Dyer and Herbert Spencer and wrote Wallace. While "I hate destructive criticism—for it is so easy to raise objections—& want constructive criticism . . . I see serious difficulty without any considerable gain."[90] He presented a detailed critique of Wallace's suggested experiments, worrying over possible ambiguous outcomes, numbers of animals involved, cost, etc. Wallace replied that he thought Galton's objections could be easily answered, but the matter became dormant. Then, probably because Weldon's results seemed so promising, the idea of a steering committee to sponsor quantitative research on evolution was brought up on December 9, 1893, at an informal meeting at the Savile Club attended by Galton, Weldon, and others who were willing to assist with the project.[91] It was proposed to the Council of the Royal Society that a "Committee for conducting Statistical Inquiries into the Measurable Characteristics of Plants and Animals" be formed. The committee was appointed on January 18, 1894, with Galton as chairman, Weldon as secretary, and several distinguished scientists as members.

The committee's first report was read at the February 28, 1895, meeting of the Royal Society where Dyer had plunked down his *Cineraria* plants. The first part summarized Weldon's research on shore crabs discussed earlier.[92] He had found that the frontal breadth of the carapace was normally distributed in both adults and juveniles, but exhibited a wider deviation in juveniles. Hence, he concluded that the deviant juveniles must be destroyed selectively. In the second part of his report he considered the theory of natural selection and the more recent view, supported by "various writers," that large deviations were important in in the evolutionary process.[93] While not denying "the possible effect of occasional 'sports' in exceptional cases,"[94] Weldon had demonstrated evolution in action via the accumulation of small deviations. His shore crab results contained "all the information necessary for a knowledge of the direction and rate of evolution."[95] His measurements of the frontal breadth of the carapace suggested selective elimination of juveniles at the extremes of the normal distribution of carapace breadths during maturation.

Not so fast, thought Bateson. Weldon's method was invalid because he failed to measure crabs at the same stage of molting.[96] Since the frontal breadth length changed at each molt, Bateson thought Weldon should not have pooled the sample of 7,000 juveniles. He wrote four letters to Galton, as committee chairman, explaining his criticism of Weldon's work and offered to print them for distribution to the committee members. Galton passed Bateson's letters on to Weldon who became incensed. Communiques whizzed back and forth with the two antagonists thrusting and parrying about the report itself and the aims of the committee. Galton was in the middle trying to supply balm. He was immensely supportive of Weldon's groundbreaking quantitative studies on variation, but he was also impressed by Bateson's arguments for discontinuous evolution. In Bateson he sensed a kindred, if prickly, spirit. What better way to terminate the endless bickering between Weldon and Bateson than to appoint Bateson to his "Statistical Inquiries" committee? On November 17, 1896, an exasperated Galton wrote Weldon that it would be most helpful "if Bateson were made a member of our Committee, but I know you feel that in other ways it might not be advisable."[97]

This suggestion would not have pleased either Weldon or Pearson, who would soon be a member of the committee himself. As Pearson later wrote, "Bateson had absolutely no sympathy with the statistical treatment of biological problems, the very work for which the Committee had been appointed."[98] Galton eventually prevailed and in December wrote Bateson that the committee would soon meet and that Weldon and he were "desirous that you should join us. Would it be agreeable to you that we should suggest your name?"[99] Bateson declined, for "I am not convinced that the present lines of inquiry of the Committee are fruitful and I do not think it is likely that the results will be at all proportionate to the labour expended."[100] This was a nasty little stab at Weldon's painstakingly accumulated shore crab measurements, but Galton soldiered on, presumably in the face of objections from Weldon and Pearson. Within a month he had convinced Bateson that the committee would change its direction. In late January, 1897, Bateson, along with Lankester, Dyer, and six others joined the committee, which met on February 11. Its name was changed to the Evolution Committee of the Royal Society, reflecting the admission of biologists with divergent and nonquantitative backgrounds to its membership. The next day a disgruntled Pearson wrote Galton that he "felt sadly out of place in such a gathering of biologists, and little capable of expressing opinions, which would only have hurt their feelings and not have been productive of real good. I always succeed in creating hostility without getting others to see my views; infelicity of expression is I expect to blame. To you I mean to speak them out even at the risk of vexing you."[101] Galton's method seemed the "right one—a Committee to undertake experiments of a definite statistical character. But your actual Committee is quite a wrong one. It might be a good Committee to press the public with subscription lists; but it is, I believe, a hopeless one to devise experiments which will solve in the only effective way these problems."[102]

Meanwhile a radical and unexpected change was about to occur in Galton's life. In the summer of 1896 the Galtons made their annual excursion to the continent. Now 75, Galton must have been relieved to get away from the all of the friction being generated by his younger colleagues, especially as the previous year had not been particularly kind. Louisa noted in her diary that Galton had a bad cold and cough early on and "fell ill on the 19th June with Gastric Catarrh like he had before, but much milder and it might have been nothing, had he not gone to Kew with Temperature 102 so with guests and nursing I was nearly done for."[103] Their physician, Dr. Chepmell, advised a course of hot baths at Wildbad where Galton, always restless, wrote his one substantial paper of the year on the possibility of establishing visible signal contact with Mars.[104] The planet had made a near approach to earth four years earlier causing furious speculation in the popular press about the existence of life on its red surface. Galton left Wildbad feeling fit and the couple toured in Switzerland and Germany, not returning home until September. But they could not shake off bad news. Their good friend Emily Gurney expired, followed on their wedding anniversary by Sir William Grove. Then Dr. Chepmell moved out of London. Despite these setbacks, Louisa was fairly upbeat in her diary entry for 1896, ending wistfully that "surely do good things come to us and pass from us, and I try to be thankful enough for the innumerable blessings we have had even with the pain of feeling them gone. So ends our year, not an eventful one, but a calmly happy one ending with a merry Xmsas at Spencers', the young folk full of life and ambition."[105]

In the winter of 1897 the Galtons visited Bournemouth to consult with Dr. Chepmell about an asthmatic cough that was plaguing Galton.[106] He recommended a visit to Cauterets or Royat for the cure the following summer. Louisa's health was up and down. She attended a tea party on May 7 and felt well enough to go with Galton to the Athenaeum on June 21, Jubilee Day. But on June 26 Galton observed a Naval Review without her as she was not feeling strong enough to accompany him. On July 14 they departed for France, staying in Boulogne the first night, but then experienced an interminable wait at Gare de Lyon in Paris for the overnight train south. They settled in the Auvergne at Royat. On July 24 they took a short trip to the Puy de Dôme where Galton scaled the peak with some friends while Louisa had a luncheon set out in the garden at a nearby auberge. She was happy, animated, and acted the part of the gracious hostess perfectly. On August 3 they all came down with diarrhea, probably the result of those rich French cream sauces, but soon recovered. Louisa was fine on August 8 and getting ready for a new excursion the next day. But she became sick again with diarrhea and her condition steadily worsened, she was vomiting continuously, and died, just before 3 A.M., on August 12.

At 5 A.M. a distraught Francis Galton was writing his sister Emma about Louisa's death in order to catch the morning post to England. Everything had happened so fast. Only a few days earlier they had been planning a tour of the Dauphiné mountains when Louisa fell ill. "I had a nurse to sit up through the night, who awoke me at 2 1/2 A.M. when dear Louie was dying. She passed away so imperceptibly that I could not tell when, within several minutes. Dying is often so easy!"[107] After briefly mentioning the burial plans, Galton's anguish at the yawning void in his life came through painfully. "I cannot yet realise my loss. The sense of it will come only too distressfully soon, when I reach my desolate home. Please tell the brothers and sisters. I am too tired to write much, having had long nursing hours." After telling Emma how kind his French hosts had been, Galton ended ever so softly. "Dear Louisa, she lies looking peaceful but worn, in the room next to where I am writing, with a door between. I have much to be thankful for in having had her society and love for so long. I know how you loved her and will sympathise with me. God bless you."[108] A couple of days later a sleepless Galton wrote Emma again. He had pulled himself together a bit, but the wound was still deep. In the middle of the letter was a passage directed more to his dead wife than to his sister. "Dearest Louisa—I have very much to be grateful for, but our long-continued wedded life must anyhow have come to an end before long. We have had our day, but I did not expect to be the survivor."[109]

TWENTY-ONE

The Mendelians Trump the Biometricians

> Galton was claimed as a father by both Mendelians (for his stress on discontinuous variation) and biometricians (for his introduction of statistics to biological studies), and his relations to both sides of the turn-of-the-century controversy hold a unique significance.
>
> —Robert de Marais, *The Double-Edged Effect of Sir Francis Galton: A Search for the Motives in the Biometrician-Mendelian Debate*[1]

Francis Galton's Ancestral Law of Heredity (1897) had its roots in the pedigrees he had analyzed for *Hereditary Genius*. The underlying assumption was that the closest male relatives of the eminent man were the most likely to be eminent, with eminence diluting over hereditary distance. The ancestral model seemed to accommodate continuous variation satisfactorily and would be adopted by both Pearson and Weldon. However, for Bateson discontinuities were what counted in heredity and evolution, and this belief found strong reinforcement in Galton's hypothesis of organic stability. The rediscovery of Mendel's paper in 1900 at once provided the key for the hereditary cipher and unlocked in Bateson one of the first great geneticists. This chapter chronicles the ensuing controversy.

Using anthropometric data, Galton had tried to estimate the ancestral contribution to progeny over several generations, but this had proved difficult.[2,3] Then he stumbled on a goldmine. Sir Everett Millais, eldest son of the Pre-Raphaelite painter Sir John Everett Millais, had scrupulously recorded the characteristics of a large pedigree of Basset hounds in *The Basset Hound Club Rules and Studbook* kept from 1874 to 1896. Bassets are either white with blotches ranging from red to yellow (lemon and white) or have additional black mark-

Fig. 21-1 Galton's Law of Ancestral Heredity. Male members in the pedigree are represented by even numbers in the white squares and female members by odd numbers in the black squares. Parents (2,3) contribute half of the heritage of their progeny, grandparents (4,5,6,7) a quarter and so forth. From Francis Galton, "A Diagram of Heredity," *Nature* 57 (1898): 293.

ings (tricolor). Hence, Galton had only two color varieties to contend with and a large sample size. In June 1897, he read his Basset paper before the Royal Society and unveiled his ancestral law.[4] In its simplest form his law is easy to comprehend (Fig. 21-1). He contemplated a continous series with parents contributing one half (0.5) the heritage of their offspring; grandparents one quarter $(0.5)^2$, the eight great-grandparents one-eighth, etc. The whole series $(0.5) + (0.5)^2 + (0.5)^3$ ... sums to 1 accounting for the total heritage.[5] He was pleased as the predicted and observed frequencies of tricolor Basset progeny agreed well. It "is hardly necessary to insist on the importance of possessing a correct law of heredity. Vast sums of money are spent in rearing pedigree stock of the most varied kinds, as horses, cattle, sheep, pigs, dogs, and other animals, besides flowers and fruits.... A correct law of heredity would also be of service in discussing actuarial problems relating to hereditary longevity and disease, and it might throw light on many questions connected with the theory of evolution."[6]

Pearson found Galton's paper intriguing and decided to "represent his views in my own language."[7] He dedicated the manuscript to Galton and sent it to him as his 1898 New Year's day greeting. His modified law would predict "the values of all the correlation coefficients of heredity" and form a "fundamental principle of heredity from which all the numerical data of inheritance can in future be deduced."[8] Galton was overjoyed with his "most cherished New Year greeting. It delights me beyond measure to find that you are harmonizing what seemed disjointed, and cutting out and replacing the rotten planks of my propositions."[9] Pearson read his paper before the Royal Society on January 27, 1898. It was decorated with sophisticated derivations and festooned with elegant equations going far beyond anything Galton imagined. While Galton had treated the two Basset color patterns as discontinuously varying characters, Pearson's conception of the law was consistent with continuously varying characters that could be subjected to classical Darwinian selection. Pleased with his handiwork, Pearson ended with a self-congratulatory sentence. "If Darwinian selection be natural selection combined with heredity, then the single statement which embraces the whole field of heredity must prove almost as epoch-making to the biologist as the law of gravitation to the astronomer."[10]

While Pearson's mathematics undoubtedly nonplussed Bateson, he grasped the basic concept of ancestral inheritance from Galton's Basset article. He summarized Galton's law and its application in a paper read before the Royal Horticultural Society on May 8, 1900,[11] where he also presented a revolutionary new concept of heredity he had learned about from Professor Hugo de Vries in Holland.[12] As Beatrice Bateson wrote, her husband's "delight and pleasure on his first introduction to Mendel's work were greater than I can describe.... He was fortified with renewed faith in the largeness of his research; he found in it new interest, new possibilities, and drew from it new inspiration."[13] He communicated his enthusiasm to Galton on August 8, 1900, suggesting that he look up Mendel's paper "in case you may miss it. Mendel's work seems to me one of the most remarkable investigations yet made on heredity, and it is extraordinary that it should have got forgotten."[14] But Galton either did not take advantage of this suggestion or failed to appreciate the significance of Mendel's hypothesis.

For Bateson, the ancestral law, Pearson's modifications included, was inadequate because "*it does not directly attempt to give any account of the distribution of the heritage among the gametes* of any one individual."[15] The ancestral law dealt with populations and continuous variation, but Mendel's laws applied to discontinuously varying traits in individual progeny sets. Mendel's approach was direct. His results were easy to understand for the nonmathematical Bateson, who probably threw up his hands at Pearson's paper as pages of equations and derivations danced dizzily before his eyes. The split between Pearson and Bateson soon widened further. Pearson submitted a gargantuan memoir to the Royal Society on a new theory for which he coined the term "homotyposis." He read an abstract describing his theory on November 15, 1900. He estimated that among the offspring of a single parent, variability for a given character was reduced only slightly with respect to the race as a whole, even when selection for ancestry was interposed for many generations.[16] Hence, he wanted to (1) determine the ratio of individual to racial variability, and (2) understand how individual variability was related to racial inheritance. He hypothesized that each individual elaborates "undifferentiated like organs" such as blood corpuscles, fish scales, or flower petals, which may or may not undergo further specialization. The variability of the individual's undifferentiated like organs should be somewhat less "than that of all like organs of the race."[17] Pearson assumed that progeny inherit parental "germs" and that the degree of resemblance between parents and offspring reflected the "variability of the sperms and the ova which may be fairly considered as undifferentiated like organs."[18] He argued that "if we can throw back the resemblance of the offspring of the same parents upon the resemblance between the undifferentiated like organs of the individual, we shall have largely simplified the whole problem of inheritance."[19] For undifferentiated like organs elaborated by the same individual, Pearson coined the term *homotype*.

Bateson, sitting in the audience, reacted negatively. A pained Pearson wrote Galton that Bateson "came to the R.S. at the reading and said there was nothing in the paper—that it was a fundamental error to suppose that number had any real existence in living forms. That this criticism did not apply to this memoir only but to all my work, that all variability was differentiation, etc. etc."[20] What likely happened can be reconstructed from Bateson's critique of Pearson's paper.[21] Bateson saw Pearson, the mathematical wizard, wave his magic wand and dress homotyposis in the glittering raiments of elegant equations. But for Bateson this emperor had no clothes because the underlying biology was faulty. Bateson knew biology and Pearson did not, despite examining hundreds of leaves from beeches, chestnuts, and hollies, seed vessels from poppies, etc. For Bateson the foundation of Pearson's hypothetical edifice, the undifferentiated like organ, was cracked and with intellectual probing broke asunder, bringing the whole structure toppling down. Pearson's hypothesis depended "on the assumption that there is an absolute distinction between differentiation and varying among repeated parts, and its solubility depends on the assumption that this distinction can be perceived. The proviso that such a distinction is to be observed stultifies the whole inquiry."[22]

The editors sent Pearson's manuscript to Bateson who rejected it while identifying himself as a referee. Worse, they disseminated his negative remarks to the other reviewers, an inappropriate action that insured they would be biased against the manuscript. Pearson complained to Bateson and to the Royal Society about the shabby trick they had played on him. He groused to Galton that "if the R.S. people send my papers to Bateson, one cannot hope to get them printed. It is a practical notice to quit. This notice applies not only to my work, but to most work on similar statistical lines. It seems needful that there should be some organ for publication of this sort of work and talking it over with Weldon, he drew up the prospectus, I gave a name."[23] Thus the distinguished journal *Biometrika* was conceived as a Bateson-avoidance mechanism. Pearson's memoir was finally published late in 1901.

Biometrika debuted in 1902 with Pearson, Weldon, and the American Charles Davenport as editors. Davenport, an assistant professor at the University of Chicago, was an ardent follower of Galton and Pearson who introduced Pearson's techniques to America in his book, *Statistical Methods*. He would manage the review of manuscripts submitted by American scientists. Also listed on *Biometrika*'s masthead was Galton, whose advice the editors would seek. The first issue set forth the journal's high aspirations. *Biometrika* was intended to "serve as a means not only of collecting under one title biological data of a kind not systematically collected or published in any other periodical, but also of spreading a knowledge of such statistical theory as may be requisite for their scientific treatment."[24] The editorial credited Galton with initiating "the recent development of statistical theory, dealing with biological data."[25] The editors wanted to bring together biologists, statisticians, and mathematicians in a seamless union, because "the danger will no doubt arise in this new branch of science that — exactly as in some branches of physics — mathematics may tend to diverge too widely from Nature. . . . As Mr Francis Galton said a few years ago, for these new problems we want a scientific firm with a biologist and a mathematician as acting partners and a logician as a consulting partner."[26]

The editors gallantly refused to fortify their journal against a Batesonian blitzkrieg. "Extensions, corrections, criticisms of the results published in our pages we shall heartily welcome whatever their source. We expect to give stalwart blows as well as to receive them. All we shall demand in this respect is the chivalry which is needful in scientific controversy, which while combating error does not discourage honest scientific endeavour."[27] This seemed magnanimous especially since Bateson's dark forces had enveloped the Evolution Committee of the Royal Society from which Galton, Pearson, and Weldon had all resigned. Pearson wrote that their "capture of the Committee was skilful and entirely successful."[28] The editorial was followed with a short essay on biometry by Galton.[29]

Meanwhile Galton's personal circumstances took a turn for the better. Within two years of Louisa's death, a solution to his loneliness emerged in the form of Evelyne "Eva" Biggs, the granddaughter of his late sister Lucy (Table 2-1). Eva Biggs, her thoughtful face crowned with a mass of upswept black hair, was in her mid-thirties. In March 1899, she sailed off with her great uncle for a two-month holiday in Spain and France accompanied by an old and half-blind courier named Eschbach, who proved so capable that Galton half joking wrote that he was "a capital lady's-maid for Eva."[30] From Granada, Eva wrote to Galton's sister Emma that her "Uncle Frank" was "really the perfect person to travel with, because he never fusses or gets impatient *or* grumbles if we are kept waiting ever so long for food and luggage!"[31] He was back at 42 Rutland Gate by early May and already missing Eva. "I want her father to lend her to me a good deal, and she wants to come," Galton wrote his sisters Emma and Bessy, for "if she could make my house a good deal her home and be with me again when abroad, it would help me a great deal."[32]

That summer they visited Royat and toured Switzerland, returning to England in September so Galton could deliver a peculiar little paper entitled "The Median Estimate" at the British Association meeting in Dover.[33] He spent the autumn at home, but by December he was off with Eva to Egypt. They steamed up the Nile all the way to Wadi Halfa on the Sudanese border before returning. On the way back, they visited a site being excavated by the prominent Egyptologist Professor W. M. Flinders Petrie, an old acquaintance of Galton's. Their relationship had commenced in 1880 when Flinders Petrie responded to an appeal in *Nature* [34] for cases in support of Galton's research on the mental visualization of numbers and calculations (chapter 16). Flinders Petrie was a master at mental arithmetic, being able to create the equivalent of a cerebral slide rule. Galton's notions of eugenics, in turn, profoundly influenced Flinders Petrie, who advocated encouraging the "best stocks" to breed by means of grants and privileges, while penalizing the "lower class of the unfits" with compulsory work and by encouraging their women to undergo voluntary sterilization.[35] At the end of February Eva and Galton returned to England. They were now very close and Eva's father agreed to let her reside at 42 Rutland Gate. An enthusiastic Galton wrote her in October 29, 1900, "My dear 'chattell' Eva. I am delighted that you are now to be altogether transferred to me and to take charge of my household henceforth. You weren't transferred *quite* as a 'chattell' (I don't know how many *t*'s and *l*'s there are in the word) as I have said in my letter to your father 'if she acquiesces . . .' "[36] More properly Eva Biggs became Galton's chatelaine and remained so until his death in 1911, after which she married Guy Ellis. She apparently relished her role, as she wrote Galton's great-great-nephew, Hesketh Pearson, nine years after Galton's death: "He was particularly amusing in repartee. When he and Mary Coleridge met, they were most witty together, and one was simply astonished at the fund of wit and learning of each. They were the best of friends. But when Dr. Lilias Hamilton came to the house, the repartee was like lightning; no one spoke but these two, and the company laughed all the time without stopping. It was pure fun, and hardly any dinner was eaten—even I forgot my food, which is rare!"[37]

Hesketh Pearson, a 19-year-old businessman in the City, first met Galton as a sparkling octogenarian.[38] He seemed quite short, an impression accentuated by his stooped walk. His deep blue eyes were surmounted by prominent brows set on either side of a well-chiseled nose. To Pearson he appeared simple and unaffected and spoke in a soft voice with a smooth, sweet quality. He was unfailingly courteous and brought out the best in everyone by gently probing for their chief interests and making these the topic of conversation. "I could never find a subject that Galton was not willing and eager to discuss—from golf to Egyptology," Pearson wrote, "and he always managed to throw new light on matters upon which one liked to believe oneself an expert."[39]

Pearson observed Galton's penchant for quantification at a lecture on eugenics. Owing to his age Galton wrote it out and had a surrogate read it for him. Afterwards, he was asked to make a few remarks. He walked slowly up to the podium and said: "I have often observed that when people are interested in a discourse, the movements of their hands or legs are roughly two in every minute. When they are bored this number may be multiplied by four, or, at moments of excessive ennui, five. It gave me real pleasure to perceive that you were even absorbed in my paper. Your movements have averaged only one to the minute."[40] Pearson also found Galton highly practical. In illustration, he referred to his unpublished research for a "Beauty-map" of the British Isles.[41] Galton used a thimble with a spike on the end to prick holes in a piece of paper. One counter was in his left-hand pocket and the other in his right. A girl who got a prick on the right passed muster, but the unfortunate young woman who got one on the left failed. Galton found London was blessed with the highest number of attractive women, but poor Aberdeen had the fewest.[42]

Like a newly forged sword, a scientific theory must be beaten into shape through searching criticism and honed to perfection by thoughtful rebuttal. Mendel's theory of inheritance was no different. Volume one of *Biometrika* featured a detailed critique by Weldon.[43] It began by contrasting *blended* inheritance, by which he meant Galton's ancestral law, with *particulate* or *alternative* inheritance, that is Mendel's theory. He summarized the main points in Mendel's theory, beginning with dominance and recessiveness of alternative traits. For instance, yellow seed leaves (cotyledons) are dominant to green, so in crosses of green by yellow the first filial generation (F_1) had yellow cotyledons. To explain this, Mendel assumed that each germ cell or gamete, pollen grain or egg cell had one genetic alternative, or allele, for cotyledon color. Hence, the F_1 progeny had an allele from each parent, and the yellow color specified by the dominant allele was expressed. In the second generation (F_2) the recessive trait appeared again in one quarter of the progeny. This fit prediction exactly since by random combination one quarter of the F_2 progeny should have two yellow determinants, one half should possess one of each, and one quarter, two green determinants. That is, the *genotypic* ratio among progeny should be 1:2:1, but the *phenotypic* ratio should be 3:1. This was Mendel's law of segregation.

Weldon tabulated Mendel's results illustrating dominance for all seven pairs of segregating characters he studied, which Weldon thought "no one who reads his paper will find the slightest difficulty in accepting."[44] With modern statistical tools at his fingertips, Weldon was the first, but by no means the last, scientist to marvel at how closely Mendel's actual results for segregation in the F_2 corresponded with theory.[45] He then summarized Mendel's second law, positing that the seven pairs of characters studied assorted independently of one another (i.e., in all possible combinations). Having reviewed Mendel's results, Weldon asked whether his laws applied generally. "It is almost a matter of common knowledge that they do not hold for all characters, even in Peas, and Mendel does not suggest that they do."[46] The ancestral law, as transmogrified by Pearson, predicted that each individual and each trait contained ancestral contributions beyond the immediate parents with this variation being distributed continuously. If the pairs of alternative traits upon which Mendel based his theory were simply waystations in a continuum, the underpinnings of his theory would collapse.

Weldon began with seed shape. Mendel recognized variability existed, but this was subsumed within the two alternative states postulated for each character and posed no problem, for round and wrinkled seeds were easily scored in crosses. But Weldon emphasized this variability. "A race with 'round smooth' seeds, for example, does not produce seeds which are exactly alike" [so] "both the category 'round and smooth' and the category 'wrinkled and irregular' include a considerable range of varieties."[47] If seed shape varied so much, how could one be sure only two alternatives existed? Weldon applied this logic to the other characters and concluded that all work based on Mendel's hypothesis was vitiated by neglecting ancestry. Weldon's paper angered Bateson, who exchanged letters with Pearson that culminated in an attempt to win Pearson over. "I respect you as an honest man and perhaps the ablest and hardest worker I have met, and I am determined not to take a quarrel with you if I can help it. . . ."[48] He added that for Weldon and himself it was too late. Pearson replied that Bateson seemed unaware "that Weldon has been for many years past one of my closest and most valued friends; that I do not readily make friends, and that when I say a man is my friend [I] am prepared to do for him and to accept from him anything that one human being can or will do for another."[49]

The die was cast. Bateson was quick to refute Weldon's arguments in a little book called *Mendel's Principles of Heredity: A Defence*.[50] Since Weldon was a highly regarded senior professor, "there was the danger—almost the certainty" that junior people entering the field would regard his critique of Mendelian theory as definitive "and look elsewhere for lines of work."[51] Bateson provided translations of Mendel's papers and mounted a 104-page critique of Weldon's paper. He pulled no punches. Weldon's criticism was "baseless and for the most part irrelevant."[52] He dissected each piece of evidence Weldon used to marginalize Mendelian theory. He began with "alternative inheritance," the notion that each gamete carries only one member of a pair of alleles. The coat-color variations in Bassets that Galton used to exemplify his ancestral law were no more than cases of alternative inheritance. Weldon had referred to Mendel's "Law of Dominance." Bateson acknowledged that dominance was important, but observed that Mendel had proposed no such law. Dominance depended "on the specific nature of the varieties and individuals used, sometimes probably on the influence of external conditions and on other factors we cannot now discuss."[53] Weldon probably thought Bateson was waffling, but Bateson knew there were going to be exceptions to absolute dominance and recessiveness. He recognized that Mendel should be interpreted elastically just so long as a few cardinal rules such as gametic purity and allelic segregation were observed. He concluded his book tongue-in-cheek, quoting Weldon's declaration that he meant not to belittle Mendel, but only to call attention to facts that suggested fruitful avenues of inquiry. Bateson proposed to assist him, for he believed "that unaided he is—to borrow Horace Walpole's phrase—about as likely to light a fire with a wet dish-clout as to kindle interest in Mendel's discoveries by his tempered appreciation."[54] Although Bateson's attack may have "deeply pained Weldon"[55] the truth was that Weldon and Pearson could give as good as they got, and the insults had really begun with Weldon's review of Bateson's *Materials* years earlier.

Next Pearson resurrected homotyposis.[56] His line of attack was clear-cut. Unable to cope with Pearson's sophisticated equations, Bateson fell back on definitions. "The contrast between the old and the new methods of dealing with biological conceptions has recently been emphasised by my memoir on Homotyposis, and of Mr. Bateson's criticism of it."[57] So the mathematically illiterate Bateson was old-fashioned to boot. Pearson argued that "the whole problem of evolution is a problem of vital statistics—a problem of longevity, of fertility, of health, and of disease, and it is as impossible for the evolutionist to proceed without statistics, as it would be for the Registrar-General to discuss the national mortality without enumeration of the population, a classification of deaths, and a knowledge of statistical theory. Yet this . . . is precisely what the school of biologists represented by Mr. Bateson are attempting to do."[58] Pearson, "with a considerable sense of gravity," took "up the gauntlet thrown down by Mr. Bateson."[59] He had hoped to continue his work unimpeded "leaving the old school of biologists rigidly alone," but Bateson, "by a brilliant but logomachic attack," had sidestepped Pearson's biometrics and appealed to the significance of words. Two could play at that game so Pearson's rebuttal of Bateson concentrated initially on "an analysis of terms." Homotyposis was "the resemblance of certain like parts." It was a correlation to which a numerical value could be attached so Pearson did not have to quibble over definitions. His method could assign precise quantities, correlation coefficients. Pearson lectured Bateson about correlation and regression because throughout "all Mr Bateson's writings, as well as in his criticism of my paper, there runs a hopelessly confused notion of ... regression."[60] Pearson threw in Bateson's ally de Vries for good measure, whose notion of regression was "equally obscure." Bateson was free to use terms like "variation," "regression," and "correlation" as he liked *except on an occasion when he is attacking a biometric memoir*" since he lacked even "preliminary biometric training."[61]

Having exposed Bateson's soft underbelly on biometric matters, Pearson, who had painfully waded through *Materials* to understand Bateson's usage of terms, emerged "not much wiser," he analyzed Bateson's discussion of discontinuous variation concluding that Bateson had actually given three distinct definitions of discontinuity.[62] If Bateson himself didn't understand what he meant by "discontinuity," how could he use it in criticizing homotyposis? Having skimmed over the hundreds of examples of discontinuous variation in *Materials*, Pearson concluded that if Bateson wished to use discontinuous variation to attack the problem of evolution "he must go further than forming a useful catalogue of museum and collectors' deviations from 'type.' "[63] Within 20 years biometric methods would not "have to justify themselves to a non-mathematical biological world; mathematical knowledge will soon be as much a part of the biologist's equipment as to-day of the physicist's."[64]

Bateson fired back with "The Facts of Heredity in the Light of Mendel's Discovery,"[65] a section of a Report to the Evolution Committee of the Royal Society written with Miss E. R. Saunders. It briefly elaborated Mendel's principles and terms and generalized his findings to other organisms, including humans, since alkaptonuria, an affliction that results in the accumulation of a dark substance in the urine, had been shown to behave like a Mendelian recessive by Sir Archibald Garrod. Later Bateson considered a series of cases of "Compound Allelomorphs." This led him to a momentary insight whose significance would only be proved by R. A. Fisher in 1918. In a compound allelomorph a given phenotype can be broken up into more than two elements among the progeny because there are several possible genetic variants (alleles) at a genetic locus A_1, A_2, A_3, etc. Bateson observed that for a characteristic like stature, which varies continuously, there must be "more than one pair of possible allelomorphs .... If there were even so few as, say, four or five pairs of possible allelomorphs, the various homo- and hetero-zygous combinations might, on seriation, give so near an approach to a continuous curve, that the purity of the elements would be unsuspected, and their detection practically impossible."[66] Bateson had deduced the concept of a continuously varying, polygenic character by extrapolating the definition of a complex allelomorph. The key distinction between Mendel's and Galton's laws of inheritance was for "each allelomorphic pair of characters we now see that only four kinds of zygotes can exist, the pure forms of each character, and the two reciprocal heterozygotes. On Galton's view the number of kinds is indefinite."[67] However, Galton's law might describe "particular groups of cases which are in fact Mendelian, in the sense . . . that there may be purity of gametes in respect to allelomorphic characters."[68] The paper ended with a footnote urging the adoption of a uniform description for each generation and suggested the classic terms P for parental, F_1 for the first filial generation, and F_2 for the second.

The brushfires of Mendelism were breaking out everywhere, and Weldon's strategy for stamping them out was containment and counteroffensive. He replied to Bateson's report with another *Biometrika* paper whose title stressed the "Ambiguity of Mendel's Categories."[69] He again argued that Mendel defined character differences arbitrarily and trotted out his own Miss Saunders, her initials being C. B. instead of E. R. Bateson and Saunders had mentioned the presence and absence of hairs on Campion (*Lychnis*) leaves as an example of a simple pair of genetic alternatives. Weldon and Saunders counted hair numbers per square centimeter of leaves at similar stages of development in several "races" of *Lychnis*, observing wide variations. Weldon argued that the arbitrary adoption of the category 'hairy' concealed the nature of variation within the races examined. Hence, his adversaries' statements were "utterly inadequate, either as a description of their own experiments, or as a demonstration of Mendel's or of any other laws."[70, 71] Weldon viewed the absence of hairs as one extreme in a continuum, but Bateson was focusing on their presence or absence, not on their number.

In a separate *Biometrika* paper[72] Weldon engaged de Vries. His plant hybridization experiments antedated the rediscovery of Mendel's paper and were undertaken to test his version of pangenesis and elaborated in his book *Intracellular Pangenesis* (1889).[73] His pangenes had two fundamental properties: they varied in number, and assorted in different combinations as the result of hybridization. During division they occasionally produced altered pangenes that became active when sufficiently numerous. De Vries assumed that the first property could explain small individual differences of the type emphasized by Darwin, while the second accounted for the sudden appearance of new, discontinuous variations. To test his theory, de Vries began hybridization experiments, carrying out many of them with a species of evening primrose, *Oenothera lamarckiana*. He observed many variants that he thought demonstrated the predicted mutability of pangenes.[74] He presented these findings in the first volume of his monumental *Die Mutationstheorie* (1901), arguing that selection alone could not explain the origin of new species and proposing a new theory of evolution by mutation. He sent an advance copy to Bateson with a note saying: "I have now the pleasure of offering you my work on the origin of species, as discontinuous as you could hope it."[75] Anticipating strenuous objections from Darwinians and biometricians, he wrote "there must be no discontinuity between us, not even in the use of the word."[76]

Weldon's critique of de Vries's mutation theory rested on two arguments.[77] The first involved environmental variables. How could de Vries be sure that some of his mutations were not irrelevant environmental modifications? He next attacked de Vries's understanding of regression. De Vries had supposed that "the focus of regression" was fixed. This was simply one of two limiting cases postulated by Pearson, the other being that the focus of regression changes at every generation. Both Galton and Pearson had shown that the focus for regression at each generation "is its own mean." Since de Vries provided no evidence that the focus of regression was fixed, "the statements concerning the focus of regression on which the whole theory of the instability of varieties depends, are erroneous."[78] Both Bateson and de Vries failed to understand that the "relation between the phenomenon of 'regression' and the stability of specific mean character through a series of generations" was, as "a little knowledge of the statistical theory of regression will show to be wholly imaginary."[79]

Meanwhile Weldon teamed up with his student A. D. Darbishire for a data-gathering offensive of his own, enlisting a new weapon—the mouse. In his earlier attempt to refute Mendel's hypothesis, Weldon had mostly used examples from peas and other plants, but he cited "one case among animals" as fitting the ancestrian model.[80] When "the ordinary European albino mouse is paired with the piebald Japanese 'dancing' mouse, the offspring are either like wild mice in colour, or almost completely black."[81] Weldon used the reappearance of the wild color pattern to argue against dominance and for the cropping up of the ancestral character, not realizing that distinct pairs of genes were probably involved in determining albino and piebald. In his *Defence*, Bateson easily countered Weldon, pointing out that when these "reversionary" mice were interbred, parental types and wild types were obtained once again, so the wild pattern was merely dominant to black or albino. This was a "compound allelomorph" where the character of one of the original parental varieties was "split up" among the progeny. Bateson understood that such compound allelomorphs resulted because of the interaction of the products of more than one gene and that the different alleles of these genes could segregate and assort amongst the progeny.

Despite Bateson's critique, Weldon felt that breeding experiments with mice would support the ancestrian interpretation and he set Darbishire to work. Darbishire rushed out a preliminary report in volume two of *Biometrika*.[82] Although he only had first generation hybrids between albino and

Japanese waltzing mice, he believed the hybrids disproved dominance since the progeny were not uniform in coat color, but exhibited one of four different color patterns. Bateson was most interested and wrote Darbishire about his results. Although Darbishire replied that he was "absolutely unbiased about Mendel," he later admitted that this was not true at the time.[83] Bateson soon recognized that something was amiss. He wrote Darbishire pointing out that he failed to mention that the progeny, unlike their pink-eyed parents, had dark eyes and that none were waltzers, suggesting that waltzing and pink eyes were recessive traits.

Darbishire's second paper presented results for first-generation progeny of additional crosses, which he again viewed as inconsistent with Mendelian theory as the mice were not uniform in coat color.[84] However, he noted that these mice all had black eyes. He also reported on the second-generation progeny of his original crosses, observing that pink-eyed albinos and waltzers reappeared. He admitted that their appearance was "in possible accordance" with Mendelian segregation, but that production of black-eyed mice by pink-eyed parents in the first generation was not. Bateson responded in a letter to *Nature*[85] that touched off another nasty little public correspondence with Weldon reminiscent of the *Cineraria* dust-up. He focused on whether the mice had pink or dark eyes and whether their coats were albino or exhibited color, dismissing coat color variability as "being too complex for consideration in a few lines."[86] He hypothesized that pink-eyed albinos possessed G gametes while waltzers possessed G' gametes. The GG' progeny formed by these gametes would be dark-eyed mice with some color in their coats. Crossing these mice should yield a Mendelian ratio of 1 GG (pink-eyed albino) : 2GG' (dark-eyed with coat color) :1 G'G' (pink-eyed with coat color) progeny. This ratio fit Darbishire's results well. However, Bateson failed to explain why a cross of two pink-eyed mice should yield dark-eyed progeny. Since albinos are pink-eyed, he must have assumed that the trait for pink eye was genetically distinct in the mice with coat color. When albinos and waltzers were crossed, their progeny were black-eyed because each parent carried the normal allele for the pink-eyed mutation present in the other parent. The presence of coat color was dominant to its absence. But he never stated these points explicitly.

Weldon's reply in the April 2 *Nature*[87] indicated bafflement about how a cross of two pink-eyed mice could yield a dark-eyed mouse. He reprimanded Bateson for ignoring the fact that Darbishire's progeny mice had coat colors different from their parents and grandparents (a few even assumed a new lilac hue) even though Bateson had made clear his decision to focus on the presence of color or its absence. Bateson's rejoinder on April 23[88] faulted Weldon for focusing on coat color variation. Although he resignedly wrote that debating these points "with one who doubts the Mendelian nature of the phenomena . . . is like discussing the perturbations of Uranus with a philosopher who denies that the planets have orbits."[89] Bateson unwisely ventured into the coat color morass, a problem he had actually thought about extensively. The "lilac" mice resulted because of the "partial disintegration of characters commonly witnessed when a compound colour is crossed with an albino,"[90] meaning that color is probably determined by several genes and alleles. Furthermore, the diversity of coat colors "in the first crosses" pointed "to heterogeneity among the gametes of one or both 'pure' races."[91] The color patterns seen in the heterozygotes probably depended on factors "other than the visible colours of the parents, and having an independent distribution amongst their gametes."[92]

Bateson did not mean the gametes were heterogeneous for the presence or absence of coat color, but only with regard to the coat color variations, but he gave Weldon an opening and he struck on April 30.[93] Intentionally or not, he began by misinterpreting Bateson's original letter. Bateson had stated that his hypothesis was meant to explain color versus its absence, but Weldon again brought up color variations in the progeny of the albino-waltzer crosses, arguing that they could not have been produced by just two kinds of gametes. He then turned to Bateson's April 23 letter and accused him of abandoning "his first formula" and arguing that in the parental crosses there were more than two gamete types fusing, with the gametes of one or both being heterogeneous. But Bateson had not abandoned his "first formula," which explained the segregation of coat color or its absence. He had only added a "compound allelomorphic" explanation of variation in pattern and shade in colored mice. Not wishing to tackle this problem with a "brief treatment," Weldon turned to Bateson's claim that "coat-colour is split into simpler elements when the hybrids form gametes."[94] This was not what Bateson had said. He had supposed that color pattern probably depended on factors other than those whose products were visible in the parents. But Weldon argued that, while Bateson's original hypothesis of two gametic types was "not contradicted by the facts," his new hypothesis should yield a lower frequency of albinos than observed, since there would now be "different gametic elements for the black and for the yellow mice."[95] Hence, Bateson's two Mendelian predictions were "mutually contradictory; with which of them is the inheritance of coat-colour 'in punctilious agreement?'" This was patently unfair. Bateson had used "in punctilious agreement" to refer to the segregation of pink and dark eyes, coat color versus albinism, and waltzing versus normal, all of which exhibited Mendelian inheritance.

The letters to *Nature* climaxed on May 14[96] with Bateson reiterating his original position that there were two classes of gametes: G, albino coat, and G', colored coat. He explained coat color variations in terms of various classes of G' gametes such as aG', bG', cG', etc. The coat colors observed depended on the allelic combination so that aG'aG', aG'bG', bG'cG', etc., individuals would all have different colors. Bateson treated each of these variants as if they were "compound allelomorphs" at a single genetic locus. Weldon's rejoinder[96] chided Bateson for masking the departure from Mendelian expectation "by the simple device of calling the whole series of different colour-bearing gametes by the same name G'."[97] Bateson had modified "first one and then another of Mendel's statements" so that "his name is made to shelter almost any hypothesis, and almost any experimental test is evaded."[98] Having had the last word, Weldon did "not propose to continue this correspondence."[99]

Despite the acrimony and confusion, Weldon was fighting a losing battle while Bateson was extending the frontiers of Mendelian logic, explaining one new observation after another. Unsurprisingly, his next paper was a sober analysis of coat color genetics in mice and rats.[100] He began with a problem every geneticist faces at one time or another, nomenclature. To deal effectively with the inheritance of coat-color variation, he needed a system for classifying all variants. He recognized 13, cautiously pointing out that it was essential to compare specimens of similar age and molt stage. Having defined the ground rules, Bateson began with the simplest case, albinism, and described Cuénot's results showing that the grey color of the wild mouse was dominant over albino. He cited other papers establishing that albinism was recessive in rats, guinea pigs, and rabbits. Colored animals, he continued, often yielded some albino progeny, but the converse was never true, with a single exception. Dr. Carter Blake had reported that albino mice produced only albino progeny except in one case where "a pair of albinos produced some brown-and-white, some plum, some grey, and some albinos."[101] Now Bateson demonstrated his true genius as a geneticist. Recognizing that these were early days in the science, he wrote that "we should be cautious in declaring the result impossible, for in Mendelian experiments the observer must be on the look-out for the appearance of a character, elsewhere a definite dominant, *as the consequence of crossing two dissimilar recessives.*"[102] Hence, a mutation in either of two genes might cause albinism so when a strain carrying a mutation in one gene was mated to one with a mutation in the other, their progeny would be colored, since each parent carried the dominant normal allele for the albino mutation it did not possess. Why Bateson did not lay out a parallel explanation when describing the black-eyed mice Darbishire observed on crossing waltzers and albinos is puzzling.

Having shown that albino coat color was recessive, Bateson reviewed in Mendelian terms the results obtained by various investigators including Darbishire on heritability of coat-color patterns, clearly delineating those cases in which the findings had so far resisted easy explanation. He concluded that the "majority of the observations are in accord with the Mendelian hypothesis in a simple form."[103] Meanwhile, Darbishire's third report appeared.[104] Darbishire's paper was in tandem with one from Weldon entitled "Mr Bateson's Revision of Mendel's Theory of Heredity."[105] Weldon accorded Bateson five revisions of Mendelian theory. The first concerned the nature of gametes in hybrids. Bateson had correctly asserted that for specific characters among the progeny of hybrids those which were pure for either the dominant or recessive trait were identical to the pure parents. Weldon said Bateson was alone in holding this view. He attacked Bateson's ideas about dominance, arguing that, while Mendel had always observed dominance in hybrids, Bateson claimed that dominance was not essential and thus had effectively committed a breach of faith with his hero. Weldon dealt similarly with each of Bateson's other supposed transgressions including the nature of "compound allelomorphs." The confusion in these early days of Mendelian genetics was understandable, as new phenomena were always being uncovered. Bateson usually could work out a Mendelian interpretation, but this often required modification of the original hypothesis. Understanding but not believing Mendel's interpretation of heredity, Weldon was acutely sensitive to each new Batesonian modification. He was anxious at every opportunity to hoist Bateson by his own petard. None of this was very helpful for the development of Mendelian theory.

Darbishire's magnum opus appeared in *Biometrika* in 1904.[106] The paper overflowed with 54 correlation tables between offspring and parents with regard to whiteness of coat, waltzing, etc. Darbishire's aim was to determine whether increasing the amount of albino ancestry in hybrid mice expanded the frequency of albinos among their progeny. If so, he believed he could eliminate the essential Mendelian concept of gametic purity. In his first set of crosses (i) all four grandparents were hybrids (H) derived by crossing waltzers and albinos. They were bred ([H x H] x [H x H]), their hybrid progeny selected, crossed, and the proportion of albino and colored mice among their progeny (the grandchildren) examined. The second cross (ii) was identical except one of the grandparents was an albino ([H x H] x [H x A]). In the third cross (iii) two of the grandparents were albinos ([H x A] x [H x A]). The number of albino progeny varied from about 11 percent in the first cross to 25 percent in the third. Darbishire purred with delight that "the doctrine of gametic purity asserts that a series of individuals having any of the pedigrees above represented should contain 25 per cent. albinos and 75 per cent. of coloured mice. The law of ancestral inheritance proclaims that the percentage of albinos will be greater in (ii) than in (i) and greater in (iii) than in (ii): which is exactly what we find to be the case."[107]

Soon after publication of his manuscript, Bateson began corresponding with Darbishire, asking critical questions and pointing out discrepancies within his paper and between it and his other three.[108] Bateson convinced Darbishire that the waltzers failed to appear in Mendelian ratios because their viability was reduced. Darbishire began to recant. In a paper delivered on March 15, 1904, he admitted that among the offspring of hybrids the Mendelian expectation was 25 percent waltzing mice and "this is very roughly what happens."[109] Weldon and Pearson were irritated, putting Darbishire in a delicate position as he depended on Weldon for recommendation letters. Next Bateson discovered that Darbishire had made an elementary error in his analysis. Although his grandparent mice in the first cross were heterozygous for the normal allele for coat color (A) and the recessive albino allele (a), their supposed hybrid progeny were of two types that he failed to distinguish. Because of Mendelian segregation, two-thirds were hybrid (A/a), but one third were homozygous (A/A) for the normal coat-color allele. By increasing the number of albino grandparents to one in cross ii and two in cross iii, he simultaneously decreased the proportion of homozygous (A/A) to hybrid (A/a) progeny while increasing the proportion of albino (a/a) progeny at the next generation. His results matched Mendelian expectations nicely.[110] Not only was Darbishire guilty of this analytical error, but Bateson cast doubts on his record-keeping in a letter dated May 22, 1904. Darbishire now was in the unenviable position of having antagonized Pearson and Weldon, while Bateson undermined his scientific findings. His whole scientific career hung by a thread.

Darbishire wrote Bateson pleading for a secret meeting at which he could put his records in order. He hoped Bateson would keep his discoveries under wraps and that "you will do your best to get me out of the position I am in as soon as possible and I pray you not to mention this letter to anyone."[111] Darbishire's main worry was that "to have my records discredited would be heartbreaking and render it useless and a waste of time for me to go on with the costly experiments I am carrying on now." Bateson retorted that it "will, I think, be obvious to you on reflexion, that any communication between us which is to serve as a basis of discussion must be of a public nature."[112] But Bateson did not openly pillory Darbishire, for Darbishire was about to defect to the Mendelians. His conversion was engineered not only by Bateson, but independently by a brash young Mendelian named W. E. Castle across the Atlantic at Harvard.

Castle, studying the inheritance of coat color in small mammals like rats, mice, and guinea pigs, published a paper on the inheritance of albinism in 1903, soon after the appearance of Darbishire's first article.[113] At that point Darbishire only had the results from the first generation hybrids between albino mice and waltzers. Castle recognized that "Darbishire's observations, when rightly interpreted, afford strong evidence" for Mendelian inheritance, predicting that his "premature conclusion, that . . . 'albinism is not recessive,' will undoubtedly be abandoned by him when he has reared from them a second generation of hybrids."[114] Castle's prophesy was fulfilled in Darbishire's second paper, so Castle added a footnote to his paper that began smugly, "this prediction has been fulfilled sooner than we expected."[115]

In the midst of the fracas George Udny Yule stepped calmly in. Yule studied engineering at University College, after which he spent a year in Bonn examining the properties of electric waves with Hertz.[116] In 1893 he returned to

London where Pearson, now professor of Applied Mathematics, offered him a demonstratorship, having been impressed by his undergraduate performance. But Yule's salary was so meager that even a promotion to assistant professor did not help. In 1899 he left his position for secretarial work in the Department of Technology of the City and Guilds of London Institute. This paid him a living wage while allowing him to continue a less formal relationship with University College.

Yule's main interest was in applying statistical methods to the social sciences, but Bateson's *A Defence* apparently both incensed him and aroused his curiosity about Mendel's laws. In 1902 he published a two-part article in *The New Phytologist* entitled "Mendel's Laws and their Probable Relations to Intra-racial Heredity."[117] After taking Bateson to task for treating Weldon shabbily, Yule got down to business. Unlike Bateson or the biometricians, he seemed able to examine the merits of the ancestral and Mendelian laws without obvious prejudice. The first part of his article was a clear and simple development of the ancestral law while the second began with Mendel's hypothesis, which Yule acknowledged was "ingenious and remarkable" and possibly of "far-reaching importance."[118] Yule wished to determine whether the theoretical predictions of the two hypotheses could be conformed assuming random breeding between a population homozygous for a dominant trait (A/A) and one homozygous for its recessive allele (a/a). The first-generation progeny expressed the dominant allele as they were heterozygous (A/a), but progeny in the following generations would segregate in a 3:1 ratio expressing the dominant to the recessive allele. This fitted both the Mendelian and ancestrian predictions, so Yule concluded that Mendel's law was a special case of the ancestral law.

Later Yule ran into a roadblock, being unable to get the Mendelian and ancestral laws to agree in predicting the probability of recessive parents yielding recessive progeny in a randomly breeding population. Hence, he dropped the notion of complete dominance and assumed the environment also affected the gametes. When he manipulated this model with these new variables, he was able to make the two laws fit the same set of data. While the validity of Yule's environmental assumption is questionable, his assumption of lack of dominance enabled him to apply Mendel's laws to continuously varying characters like stature. He concluded that if variations in the contribution of the hereditary units "take place by discrete steps only (which is unproven), discontinuous variation must merge insensibly into continuous variation simply owing to the compound nature of the majority of characters with which one deals."[119] Unfortunately, like Bateson a year earlier, Yule failed to pursue his insight that continuously varying traits could be explained on a Mendelian basis.

The rancorous tempest churned up between Bateson and the biometricians peaked at the Cambridge meeting of the British Association that began

Thursday, August 18, 1904.[120] Section D, Zoology, commenced that morning with Bateson's presidential address before a large audience in the new Sedgwick Museum of Geology.[121] He intended to avoid throwing brickbats at the biometricians, but expressed surprise that "assimilating the new knowledge" imparted by Mendel's discovery had "proved so difficult.... Had a discovery comparable in magnitude with that of Mendel been announced in physics or in chemistry, it would at once have been repeated and extended in every great scientific school throughout the world."[122] He spoke enthusiastically about the new science while toying with a possible name. "The breeding-pen is to us what the test-tube is to the chemist—an instrument whereby we examine the nature of our organisms and determine empirically what for brevity I may call their genetic properties."[123] Two years later Bateson formally proposed genetics as the name for his science in a book review in *Nature*.[124]

The fireworks began on Friday, which was devoted to papers and discussions on heredity.[125] Bateson's collaborator Miss E. R. Saunders led off with a paper on Mendelian inheritance in stocks, followed by Darbishire who described his crosses of albino and waltzing mice. Under the gimlet eyes of Pearson, Weldon, and Bateson, Darbishire sought the middle ground. Some of his results seemed to conform to Mendelian interpretation while others fitted the ancestral law. Darbishire's waffling was followed by Mr. C. C. Hurst who interpreted his results on coat-color inheritance in crosses between inbred albino Angora rabbits and Belgian hares in a straightforward Mendelian manner. Weldon began the discussion by referring to Miss Saunders's description of one of Mendel's experiments. Next he raised a point he had made before. Bateson had mentioned that Campion plants had either hairy or smooth leaves, a character that behaved in a Mendelian manner. It was a minor point, one of many instances Bateson had given of Mendelian traits, but Weldon used it to skewer Bateson for his "looseness of Mendelian descriptions."[126] He again fastened on the question of hair-number variability rather than the presence or absence of hairs. How could one be sure that the absence of hairs was not just part of a continuum? "Questions of this kind, which were vital to the Mendelian hypothesis, could not be answered without the adoption of finer methods of description and observation than any of which Mendelians at present condescended."[127]

Bateson did not answer immediately. Instead, the afternoon session began with his student, R. C. Punnett, describing some experiments demonstrating Mendelian inheritance in chickens, following which Professor Minot from Harvard reviewed his breeding experiments with guinea pigs. Then Bateson rose to rebut Weldon. "They were to be congratulated that a clear issue was before them," for he was going to speak "plainly"; "To the Mendelians it appeared that the hypotheses of the ancestrians were disposed of, and that the voluminous works based on those hypotheses had no scientific value."[128] Weldon "had passed very lightly over the critical fact" that Mendelian segregants were pure for the traits they expressed and ancestrians could not deal with purity of type. Weldon could not explain away purity of type as reversion for "such reversion was Mendelian segregation by another name."[129] Bateson scoffed at the antediluvian Weldon:

> Without doubt ... disputants in the past had maintained the flatness of the earth before applauding crowds, much as Professor Weldon had to-day upheld the view of the Ancestrians. The paths of the heavenly bodies had been harmonised with the theory of the flat earth, as some of the facts of heredity had been with the law of Ancestry; but as the theory of gravitation had brought together great ranges of facts into one coherent whole, so had Mendelian theory begun to co-ordinate the facts of heredity, till then utterly incoherent and apparently contradictory.[130]

For reasons of decorum the *Times* account of Bateson's remarks was not quite as plain-spoken as was the ardent Mendelian. After a visit from Weldon and Pearson, Galton wrote his niece Millicent Lethbridge on September 6 that the the *Times* reporter had omitted "some rather savage phrases of Bateson."[131]

Now it was Pearson's turn.[132] He attempted to play the mediator, saying that "the great revolution which Francis Galton introduced into biological study was purely a difference of method. He taught biologists to look at the subject exactly." The Mendelians presented figures without really showing they fitted theory. He had elaborated the most complete "Mendelian system" yet worked out and it led to general principles like those proposed by Galton. By contrast the Mendelians' theories could not be grasped. But in truth it was Pearson's theory that eluded all but the most mathematically sophisticated while the simplicity of Mendel's laws with their great predictive power was rapidly drawing new adherents. The biometricians were fighting a desperate rearguard action that they would soon lose. Pearson ended by appealing for an armistice. "No hostile criticism of the Mendelians should appear in the journal of the Biometricians for the next four years if Mr. Bateson could see his way to going on quietly with his own work. This controversy could only be settled by investigation, not by disputation."[133] Pearson would not renounce his own opinions, but would withdraw from the fray, as "he did not wish to see time and energy wasted on it."[134] The confrontation was over, but the Reverend T. R. Stubbing, who had relished the show, hoped that Bateson and Weldon "would not accept the suggestion of an armistice."[135] Stubbing egged the two combatants on in continued conflict "from which the world could only gain light."[136] Professor Hickson, the chairman, closed the discussion, emphasizing the great significance of the debate especially for those biologists who were still "sitting on the fence."[137]

The British Association meeting was the ancestrians' last stand. The Mendelian juggernaut was now gathering disciples right and left. Weldon had planned most of Darbishire's experiments and prepared the correlation tables,[138] but under Bateson's tutelage and Castle's assault Weldon's puppet cut his strings and became a confirmed Mendelian. Much to Pearson's disgust, Davenport defected too. He would soon become the most powerful advocate of eugenics in the United States. W. S. Sutton proposed the chromosome theory of heredity in a paper in 1902 and brought it out in full flower in 1903.[139] The obvious problem of too few chromosomes for too many genes was soon solved with the discovery of genetic linkage.[140] Curiously, Bateson fell into his own intellectual trap by formulating a hypothesis to explain his genetic linkage results, which did not fit Sutton's chromosome theory. Like the ancestrians before him, his enchantment with his own theory prevented him from recognizing the truth. And poor Weldon, a superb scientist despite his flawed view of heredity, died unexpectedly in his prime in 1906 of an illness that began as a cold and turned into pneumonia. After his death, Pearson retreated behind the walls of *Biometrika* to continue his pioneering work in statistics and would, in a few years' time, become the first Galton Professor of Eugenics at University College. So by the time of the first International Congress of Eugenics in 1912, the year after Francis Galton's death, the pedigree and Mendelian segregation carried the day. Yet strangely, at the level of the whole genome, the ancestral law has appeal, for each of us has a genomic contribution of one half from each parent, and they, in turn, from each grandparent and so forth.

TWENTY-TWO

The Triumph of the Pedigree
Eugenics

> I take Eugenics very seriously, feeling that its principles ought to become one of the dominant motives in a civilised nation, much as if they were one of its religious tenets.
>
> —F. Galton[1]

The opening decade of the twentieth century found the educated classes in England primed to welcome Galton's eugenics. There were two main reasons for this. First, there was an overriding concern about biological degeneration in the country.[2,3] Statistical data indicated that the birth rate in England was declining, with this process being far more pronounced in the upper and middle classes than among the lower classes. Second, the battle between the Church of England and the Darwinians was mostly over and evolution by natural selection had achieved widespread acceptance. Hence, it seemed logical to many that the quality of the British population as a whole could be improved by reducing the reproductive rate of those perceived as less fit while increasing the propagation of those of good stock. This notion soon became popular not only in England, but in much of Europe and the United States. It required only that men like Galton and Pearson popularize eugenics and justify their conclusions with apparently sound scientific arguments.

On October 29, 1901, Galton chose to address eugenic issues when he delivered the second Huxley lecture at the Royal Anthropological Institute. This was a timely decision especially in view of Charles Booth's recent sociological studies. While chairing the Booth steamship company that he cofounded with his brother Alfred,[4] Booth had undertaken a massive investigation of the welfare of working men. He was aided by a team of enthusiastic young investigators including Beatrice Potter, later Mrs. Sidney Webb.[5] His major contribution, the *Life and Labour of the People of London*, was a gargantuan work of 17 volumes published from 1891 to 1903. It ranked people from A to H. A's were the "lowest class of occasional labourers, loafers and semi-criminals" while H's were of the wealthy, upper middle class. The 13,600 London streets surveyed were color-coded. Those populated by A's were black, those inhabited by H's were yellow while classes in between had their own colors and mixed streets were purple.

Galton made extensive use of Booth's data in his lecture. With some caveats, he argued that the distribution of talent could be predicted from the bell curve. Galton's five grades below the mean (v-r) had subnormal talent while the five (R-V) above had greater than average talent. To apply expectation to observation, Galton chose the population of East London, possibly because Booth's bar graphs already hinted at a normal distribution.[6] He conformed Booth's A-H categories to his own, "smoothed" the raw data, and found the numbers fitted expectation. Next he turned to children. If the fittest children (his class V or above) could be identified, it would be a "bargain for the nation to buy them at the rate of many hundred or some thousands per head."[7] In contrast the "worth of an average baby born to the wife of an Essex labourer" using his "cost of maintenance" and lifetime earnings had been calculated as five pounds by an eminent statistician. Would regression to the mean (Fig. 22-1) pose a problem for talented couples? Galton thought not. In a sample of 10,000 "while 35 V-class parents suffice to produce 6 sons of the V class, it takes 2500 R-class fathers to produce 3 of them."[8] Since there was a strong tendency to intermarry within one's class, there was "a marked effect in the richness of brain power of the more cultured families."[9] The converse applied at the other end of the societal scale, and Booth came galloping to Galton's aid with an apt quote. "Their life is the life of savages. . . . From them come the battered figures who slouch through the streets and play the beggar or bully. They render no useful service, they create no wealth; more often they destroy it."[10] Booth admitted that those who washed "the mud may find some gems in it."[11] But he gloomily concluded that while it "is much to be hoped that this class may become less hereditary in its character; there appears to be no doubt that it is now hereditary to a very considerable extent."[12]

But what about the best and brightest? Booth's classification did not differentiate well between the uppermost classes, so Galton suggested using scholastic and athletic competitions for young men at universities. A committee would select winners by interviewing each prospect and considering "all favourable points in the family histories of the candidates, giving appropriate hereditary weight to each."[13] The "opportunities for selecting women in this way" were fewer since women students were rarer and nothing was "known of

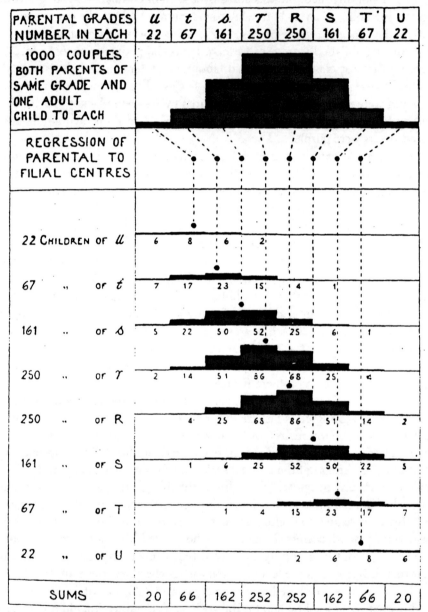

Fig. 22-1 Galton's standard scheme of descent according to which, even with regression to the mean, the more talented classes beget talented progeny while the less talented produce progeny whose ability is somewhat improved, but not to the same extent. Note that the figure does not include members of the most extreme V and v classes discussed in the text. From Francis Galton, "The Possible Improvement of the Human Breed under the Existing Conditions of Law and Sentiment," *Nature* 64 (1901): 659–665.

their athletic proficiency."[14] Since there was no easy way to identify the right mate for a high-class male, he suggested a rigorous medical examination stressing "hereditary family qualities, including those of fertility and prepotency."[15] The fortunate men and women who passed muster would receive diplomas marking them as particularly well endowed. However, if the diplomas were to be meaningful, it was critically important to determine the "correlation between youthful promise and performance in mature life."[16] He chided the "vast army of highly educated persons who are connected with the present huge system of competitive examinations" for neglecting this correlation as "gross and unpardonable."[17]

Augmentation of the favored stock was "far more important than . . . repressing the reproductivity of the worst,"[18] although the two processes were complementary. Talented individuals should be encouraged to marry early, and to rear their children "healthily." Galton suggested providing dowries for elite women, especially those of modest circumstance, to give security and induce early reproduction. Honors would be awarded to those Stakhanovite couples producing the most babies because "an enthusiasm to improve the race is so noble in its aim that it might well give rise to the sense of a religious obligation."[19] A potential stumbling block was "the tendency among cultured women to delay or even to abstain from marriage" to preserve their freedom.[20] Early marriage would enhance the yield of desirable progeny by shortening the generation time and saving "from barreness the earlier child-bearing period of the woman."[21] A healthy environment would lead to increased fertility and decreased child mortality, and the children born were more likely to become parents of healthy stock. Finally, a hereditary predisposition to high fertility was a requirement for a diploma. The grail Galton sought was "one of the highest man can accomplish" since "the faculties of future generations will necessarily be distributed according to the laws of heredity."[22] In "no nation is a high human breed more necessary than to our own, for we plant our stock all over the world and lay the foundation of the dispositions and capacities of future millions of the human race."[23,24]

Galton's appeal for eugenics was buttressed by his reputation as a scientist, which was being burnished to ever higher luster. On December 1, 1902, he received the Darwin Medal of the Royal Society and was cited by Sir William Huggins "for his numerous contributions to the exact study of heredity and variation contained in Hereditary Genius, Natural Inheritance and other writings."[25] But what delighted the old poll man even more was a telegram he received while on the Riviera with Eva from his brother-in-law Montagu Butler, master of Trinity College. It appointed him an honorary fellow. He jubilantly wrote his sister Emma, now 91, that she, Bessy, and Erasmus would not only "be glad to hear of the Darwin Medal," but also that he had been elected an Honorary Fellow of Trinity College: "This is the sort of recognition

I value most highly.... I seem to owe almost everything to Cambridge. The high tone of thought, the thoroughness of its work, and the very high level of ability, gave me an ideal which I never lost."[26] Early in 1903 Galton learned from Emma that his brother Darwin had died at 89. From Rome he wished his late sister Adelé's daughter, Millicent Lethbridge, "hearty New Year's wishes," wistfully recalling that on "Twelfth day just 50 years ago I first made the acquaintance of Louisa and of her family party."[27] Darwin's death was a shock. "Darwin used to have a terror of death.... Now he is initiated into the secret and has passed the veil."[28]

Another major opportunity to promote eugenics emerged when the newly formed Sociological Society met on April 18, 1904, in the School of Economics and Political Science at the University of London with Galton as a featured speaker.[29] The *Westminster Gazette* applauded its formation, remarking that while "sociology has won a distinct and important place among the academic studies of France, Germany, and the United States... Great Britain has lagged behind."[30] Galton's address, "Eugenics: Its Definition, Scope and Aims,"[31] was intended to influence his audience and ultimately to shape national policy. At 82 years of age he was ever more aware of his own mortality and the need to act with dispatch. He believed that such "a learned and active" society should promote eugenics and laid out a five-point plan. First, it should disseminate knowledge of the laws of heredity and promote their further study. Second, inquiry was needed to determine the extent to which different social classes had contributed to the population over time, as there was "strong reason" to believe that "national rise and decline is closely connected to this influence."[32] A major problem was the tendency of "high civilisations to check fertility in the upper classes, through numerous causes."[33] Third, records should be compiled to establish the circumstances under which "large and thriving families" of the intellectually adept "have most frequently originated; in other words, the *conditions* of Eugenics."[34] Fourth, although the "passion of love seems so overpowering that it may be thought folly to try to direct its course,"[35] social pressure would ensure few eugenically unsuitable marriages would be made. Lastly, persistence was required to realize a eugenic program in Great Britain. Eugenics must pass through three stages for this to happen. First, its importance had to be recognized by the academic community to secure the requisite professional endorsement. Second, eugenics must be seen as having such significance that it required practical schemes for implementation. Finally, eugenics would "be introduced into the national conscience, like a new religion" [ensuring] "that humanity shall be represented by the fittest races. What nature does blindly, slowly, and ruthlessly, man may do providently, quickly, and kindly."[36]

Pearson, who chaired the session, opened the discussion with fulsome praise.[37] He was taken with Galton's "boyish hopefulness" [in a man who] "is mentally half my age."[38] Given his "eternal youth, elasticity of mind, and his keen insight,"[39] Galton could be crucial in solving one of "the most vital of our national problems,"[40] which was to ensure that the "next generation of Englishmen" [was] "mentally and physically equal to the past generation,"[41] which had provided "the great Victorian statesmen, writers and men of science."[42] Other discussants were more dubious. Dr. Maudsley, concerned about the assumption that "talent and character" were largely heritable, used Shakespeare as an example.[43] "He was born of parents not distinguished from their parents"[44] and of his five brothers "none distinguished themselves in any way."[45] From his experience as a physician he could cite many instances "in which one member of a family, born of the same parents and brought up in the same surroundings, has risen to extraordinary prominence" [while] "another has suffered from a mental disorder."[46] He cautioned against being overzealous in laying down rules for human breeding. Dr. Mercier endorsed Maudsley's comments while reiterating Galton's observation that little was known of hereditary laws.[47]

Weldon jumped up to rebut Maudsley and Mercier.[48] Even though eugenicists might not be able "to account for the production of a Shakespeare ... we are certainly able to tabulate a scheme of inheritance which will indicate with very fair accuracy the percentage of cases in which children of exceptional ability result from a particular type of marriage."[49] But H. G. Wells got to the heart of the problem.[50] In *Hereditary Genius* and elsewhere Galton had picked eminent judges, etc., and determined their eminent relatives. This approach ignored

> the consideration of social advantage, of what Americans call the "pull" that follows any striking success. The fact that the sons and nephews of a distinguished judge or great scientific man are themselves eminent judges or successful scientific men, may after all be far more due to a special knowledge of the channels of professional advancement than to any distinctive family gift. I must confess that much of Dr. Galton's classical work in this direction seems to me to be premature.[51]

He also urged caution because of the recent rediscovery of Mendel's principles, whose ramifications Bateson was just beginning to explore. He was dissatisfied with Galton's implied suggestion that criminals should not procreate, since "a large proportion of our present-day criminals are the brightest and boldest members of families living under impossible conditions, and that in many desirable qualities the average criminal is above the average of the law-abiding poor, and probably of the average respectable person."[52]

Written commentaries were appended to Galton's published speech. George Bernard Shaw was enthusiastic. "I agree with the paper," wrote Shaw, "and go so far as to say that there is now no reasonable excuse for refusing to face the fact that nothing but a eugenic religion can save our civilisation from the fate that has overtaken all previous civilisations."[53] Shaw continued "that we never hesitate to carry out the negative side of eugenics with considerable zest, both on the scaffold and on the battlefield."[54]

In August Emma died, but despite these ever more frequent reminders of his own mortality, the buoyant Galton was soon at work on a new scheme to promote eugenics. On October 10, 1904, he wrote Sir Arthur Rücker, principal of the University of London, that he wished to support "the *exact* study of what may be called *National Eugenics*," meaning "the influences that are socially controllable, on which the *status* of the nation depends."[55] If a satisfactory scheme could be developed, he would donate £1,500 to cover a three-year period and would consider a permanent endowment of £500 per year to support an investigator, "a clerk," and provide expenses if the trial was a success. A scheme was drawn up by several prominent individuals including the principal, Galton, and Pearson, and accepted by the University Senate. A fellowship committee that included Galton and Pearson recommended Mr. Edgar Schuster, a student of Weldon's at Oxford, who had done good biometric work. London University provided rooms at 50 Gower Street, which, at Galton's request, were designated the "Eugenics Record Office." The *Pall Mall Gazette* nodded approvingly, congratulating the Sociological Society and the University of London for enabling "Galton to develop and further promulgate his new study of *eugenics*."[56]

Galton was again on stage before the Sociological Society on Valentine's day, 1905, and this time he gave two presentations. "Restrictions on Marriage" classified different marital customs giving examples from various ethnic groups.[57] Galton's point was that free choice was often restricted, but that marital rules could change over time, frequently in connection with a specific religious belief. He concluded that "limitations to freedom of marriage might, under the pressure of worthy motives, be hereafter enacted for Eugenic and other purposes."[58] His second paper, "Studies in National Eugenics," outlined those topics he thought required careful investigation as eugenics moved onto the national stage.[59] There were seven points to his manifesto. Pedigrees would provide information on "the average quality of the offspring of married couples, from their personal and ancestral data."[60] Next he wanted to know the extent to which state institutions were already effecting eugenic solutions. Public opinion was beginning to favor prolonged segregation of habitual criminals partly to prevent them from "producing low class offspring."[61] Since assisting institutions for the feebleminded might "eventually promote their marriage and the production of offspring like themselves,"[62] there was merit in giving large-scale aid only to those who were mentally gifted. Third, marriage restrictions needed investigation. Next human heredity required exploration. Fifth, the vast amount of published material relevant to eugenics should be hunted down and catalogued. Sixth, eugenicists should cooperate and encourage others to contribute data. Finally, eugenic certificates could be used to attest to such qualities as "constitution, physique and intellect."[63]

A lively discussion followed, most of which was positive. Dr. F. W. Mott, pathologist to the London County Council Asylums, supported Galton's contention that "improvement of the stock" could be effected by segregating the unfit to check their reproduction.[64] The government should establish registry offices where a couple anticipating marriage filled out a form that detailed marriage plans and also served as a bill of health. This would be useful to children of good heritage in "obtaining life insurance policies at a more reasonable rate; also in obtaining municipal and government employment."[65] Mott had anticipated the current debate on providing genetic information to insurers and employers. Dr. Alice Drysdale Vickery brought a unique viewpoint to the male-dominated discussion.[66] She felt the success of any attempt to improve the human race depended on three factors. The first was "the economic independence of women, so as to render possible the exercise of selection, on the lines of natural attraction, founded on mental, moral, social, physical and artistic sympathies, both on the feminine and masculine side."[67] Second, girls and boys needed to understand "their future responsibilities as citizens of the world, as co-partners in the regulation of its institutions, and as progenitors of the future race."[68] Third, childbirth should be restricted intelligently in "proportion to the requirements of the community" to ensure "the efficient development of future citizens."[69] In Dr. Vickery's opinion the "present economic dependence of women upon men was detrimental"[70] to their physical, intellectual, and moral growth. "It falsified and distorted her views of life, and, as a consequence, her sense of duty. It was above all prejudicial to the interests of the coming generation, for it tended to diminish the free play and adequate development of those maternal instincts on which the rearing of children mainly depended."[71]

Written comments included a letter from Havelock Ellis rejecting Galton's analogy between animal breeding and human eugenics. Animals were bred for specific purposes "by a superior race of animals, not by themselves. . . . It is important to breed, let us say, good sociologists; that, indeed, goes without saying. But can we be sure that, when bred, they will rise up and bless us?"[72] It would require "a race of supermen" to "successfully breed human varieties and keep them strictly chained up in their stalls." Dr. Max Nordau raised a question that should have caused eugenicists to pause.[73] "It is clear that we cannot apply the principles of artificial breeding to man. . . . There is no recognised standard of physical and intellectual perfection. Do you want inches? In that case, you would have to exclude Frederick the Great and Napoleon I, who were undersized; Thiers, who was almost a dwarf; and the Japanese as a nation, as they are considerably below the average of some European races."[74] Despite such occasional skeptics most of the discussants bought Galton's view, and the juggernaut of eugenics rolled forward, no doubt helped by publication of Galton's speeches in the *Times*.

Galton had long thought certain diseases might run in families, an idea supported by the well-known actuary W. Palin Elderton.[75] Elderton felt that life insurance offices should be helpful in studying the heritability of disease since when an insured person died, the death certificate was filed in the office together with the original insurance papers. These papers stated "the causes of death of parents, brothers, and sisters and their ages at death, or their ages if they were alive when the assurance was taken out."[76] Such information would be valuable in studying the relationship between inheritance and disease, as actuarial experience indicated "that environment operates merely as a modifying factor after heredity has done its work."[77] Galton wrote Elderton in January 1905, requesting help in "obtaining Eugenic data from Insurance offices" and framing a series of questions. Elderton suggested that Galton contact the Institute of Actuaries requesting that they distribute an appropriate circular to insurance agencies. Galton wrote the president and Council of the Institute of Actuaries hoping to convince them such data would serve their own self interest, but his plea apparently fell on deaf ears.[78]

Shortly after the Sociological Society meeting of February 1905, Galton and Eva were nestled in at Bordighera on the Italian Riviera. They returned to England in early May where Galton was greeted by an invitation from W. A. Herdman, general secretary of the British Association, to accept its presidency.[79] George Darwin wrote that he had put him up and "speaker after speaker endorsed what I have said."[80] He need not preside at Council meetings because of his deafness, but Galton replied that, while flattered and appreciative of Darwin's desire to ease the workload, "the fatal fact remains that I am not strong enough even under these alleviations."[81] Although physically frail, Galton's mental powers seemed as sharp as ever and he was hard at work on his new book, *Noteworthy Families*. With the aid of Schuster, the Galton Research Fellow, Galton canvassed Fellows of the Royal Society mentioned in the yearbook for 1904 seeking evidence for the heritability of high ability.[82] *Noteworthy Families* was a minor book. Its main conclusion was that noteworthiness diminished rapidly "as the distance of kinship to the F.R.S. increases,"[83] supporting "the great fact upon which Eugenics is based, that able fathers produce able children in a much larger proportion than the generality."[84] As always, Galton ignored the possibility that noteworthy men influenced their sons' careers.

By mid-November Galton and Eva were esconced in the south of France. He described Wellington's successful assault on Bayonne in a letter to Milly, noting that he was "wrong in saying that bayonets got their name from Bayonne. It was from a neighboring village Bayonnette."[85] Meanwhile he was corresponding extensively with Schuster over pedigrees and their notation.[85,86] In early January 1906, a telegram from his nephew Edward Wheler Galton informed him of the death of his only surviving sister Bessie at age 98. "It is the last link to my own boyhood, for Erasmus was at sea, etc., and knew little about me then."[87] Galton's party left Biarritz shortly thereafter for St. Jean de Luz in the lower Pyrenees, and later in the month moved to the Hôtel de la Rhune in nearby Ascain. Eva loved Ascain. "This is a duck of a place, so very simple and picturesque."[88] The ever-inquisitive Galton wrote Milly on February 16 that he had embarked on yet another project. He was developing a method for measuring visual resemblance. As someone approaches the "*general markings* of the face are seen" and as the person moves closer one "sees *individual features* clearly."[89] Each grade of resemblance was connected with a critical distance, but simple distance was an insufficient measure since one needed to know both distance and size so "the unit is the *angle*" and "size at a distance is expressed as the *angular size;* the distance and area by the *angular area*."[90] He explained to his perhaps nonplussed niece how the calculations were made using the angle that "is approximately that subtended by the disc of the sun (paled by a cloud)." Later that year he published a letter detailing his method in *Nature*.[91] In another letter to Milly he wrote that the Hôtel de la Rhune was "liable to inrushes of noisy French, who go up the Rhune (3000 ft) and have grand dinner, sleep here and return to Biarritz, etc. on the morrow. One noisy party of six men in two motors appeared here three days ago. They drank like Britons and sang the Marseillaise like Frenchmen and danced in rhythm to the chatter of the motors in the *place* in front of the hotel."[92] Galton and Eva returned home on April 7 to receive from Pearson the shocking news of Weldon's untimely death. Galton was one of two or three donors who contributed generously to Weldon's memorial. It included a biennial Weldon medal for the best biometric memoir of the previous year. Pearson, stewing over *Biometrika*'s future, wrote Galton for suggestions. He reacted modestly to Pearson's suggestion that the *Biometrika* title page should now state that it had been founded in 1901 by Pearson, Weldon, and Galton. "You must not give so much prominence to me. Why not keep to the existing formula and say: 'Founded in 1901 by Professor K. Pearson and W. F. R. Weldon in *consultation* with Francis Galton.'"[93]

During the spring of 1906 Galton began designing eugenic certificates.[94] Although never used, they provide insights into his thinking. A panel of judges would evaluate the certificates and pronounce the awardee and "his near kinsmen . . . to be distinctly superior in Eugenic Gifts to the majority of those in similar position."[95] Eligibility was restricted to educated men between 23 and 30, since younger men had yet to prove "their powers," while at a later age "memories of the youthful achievements of their kinsfolk"[96] would be hard to substantiate. These men must have easily verifiable qualifications bolstered by the results of "numerous competitive trials"[97] to document athletic and intellectual ability. Women were excluded, an odd proposition since Galton's ancestral law predicted that they contributed equally with their spouses to the "nature" of their progeny. Galton airily brushed this contradiction aside, saying that it "would require a special discussion."[98] In October Pearson wrote that Schuster had returned to Oxford to study real biology thus escaping the drudgery of Galton's pedigrees.[99]

Despite Schuster's resignation, Galton wrote Pearson that he would continue funding the fellowship and broached a new subject. Upon his death he would consider endowing a professorship with £30,000.[100] By late November negotiations to establish the Galton Professorship were underway with the University of London.[101] Galton and Pearson were also planning to convert the Eugenics Records Office into the Eugenics Laboratory.[102] On December 22, 1906, Pearson wrote Galton outlining an elaborate scheme for the Francis Galton Laboratory for the Study of National Eugenics.[103] It would be supervised by Pearson and employ a Francis Galton Fellow in National Eugenics, a Francis Galton Scholar, and a person with computational skills. Oversight would be provided by an advisory committee and the laboratory would serve as a center for publication and dissemination of information about "National Eugenics." The negotiations with the university went well and in early 1907 Pearson was installed as director, David Heron, a skilled mathematician well-known to Pearson, as Francis Galton Fellow, Ethel Elderton, W. Palin Elderton's sister, was promoted from clerk to Francis Galton Scholar. Miss Amy Barrington joined the laboratory part-time to carry out computations.

On March 2, 1907, Pearson wrote Galton that he was giving the Boyle lecture at Oxford on May 19 on "The Scope and Importance to the State of the Science of National Eugenics."[104] Galton replied that he was invited to give the Herbert Spencer lecture at Oxford on June 5, but had declined because of his health.[105] But on April 16 he wrote Pearson that the vice-chancellor had assured him that his lecture could be read in his presence or absence, so Galton agreed to go ahead. Pearson worried that he might "unwittingly have taken your subject from you."[106] Could Galton consider him "your John the Baptist, making the way straight?"[107] He would send Galton a typed version of his lecture and eliminate any overlaps. Galton was amused that at Oxford they would "both be proclaiming Eugenics as one of the large progeny of the University of London! Really the study is gaining an academic *status*!"[108]

On May 25 Galton sent Pearson the text of his lecture and Pearson replied the next day.[109] He liked the beginning on the history of eugenics and the ending that reviewed its underlying philosophy, but he worried about the middle with its dense exposition on probability. Galton's text would be fine "if you were teaching the teacher to teach," but his condensed object lessons might confound his audience as "your *child*, not your teacher."[110] Galton's lecture was read by his nephew

Arthur, as Galton had recently fallen after getting out of bed, lying helplessly on the floor through the night with his "ribs etc. loudly" proclaiming "in their language of feeling" until the house staff got up.[111] Late in November he was approached by Methuen to write his autobiography and soon was hard at work on *Memories of My Life*.[112] He wrote Pearson that "all my life from 5 years to 85 is beginning to seem to me 'present,' like a picture on the wall."[113] But in truth the book was strongly biased temporally, with the first part of his life through his meteorological studies taking up over two-thirds of the text. The remainder of the book dealt topically and briefly with subjects like anthropometrics, heredity, and biometrics. Perhaps this was intentional, as Galton's Namibian exploits made a better story for the average reader than the discovery of correlation and regression. But perhaps his early memories were the most vivid.

Galton tackled eugenics in the last chapter of *Memories*, covering many now-familiar themes.[114] He concluded with a succinct statement of the philosophy that had guided him ever since reading the *Origin of Species*. Since man "was gifted with pity and other kindly findings," it was within his power "to replace Natural Selection by other processes that are more merciful and not less effective."[115] "Natural Selection rests upon excessive production and wholesale destruction; Eugenics on bringing no more individuals into the world than can be properly cared for, and only those of the best stock."[116] This must have seemed a potent, powerful, and desirable objective in those heady days of social Darwinism. Since natural selection no longer reigned supreme in man's evolution, human beings needed to take charge of their own destiny. *Memories* was the first really popular book Galton had written since the *Art of Travel*, and he got it to press in only six months. The first edition of 750 copies quickly sold out and a second printing was soon on the way.[117] In a letter to Milly a pleased Galton wrote that the "book continues to be reviewed very favourably. *The Times* had a careful review in its *Literary Supplement* last Friday."[118] The book's widespread popularity resulted in a third edition in 1909.

Meanwhile the dragon's teeth Galton had sown and patiently tended had grown into a small army of professional and lay devotees to eugenics. Pearson wrote in late June 1907, that Galton "would be amused to know how general now is the use of your word *Eugenics*! I hear most respectable middle class matrons saying, if their children are weakly, 'Ah, that was not a eugenic marriage!'"[119] In late March the next year Pearson wrote that Lord Rosebery, the Chancellor of the Exchequer, had visited the University of London to open a new wing. Pearson showed him around the Eugenics Laboratory and Rosebery queried "Now how do you pronounce that word? I shall call it Eughennics."[120] Later Rosebery gave a speech in which he referred to "Eughennics" and paid Galton and the Eugenics Laboratory "quite a pretty compliment."[121]

An energetic social activist and born organizer, the young widow Sybil Gotto had also developed an enthusiasm for eugenical reform.[122] She was introduced to Montague Crackanthorpe, a friend and neighbor of Galton's in Rutland Gate. By happy coincidence Galton had written Crackanthorpe in December 1906 asking him whether "the time was not ripe for some association of capable men who are really interested in Eugenics."[123] Crackanthorpe introduced Mrs. Gotto to Galton and she laid out her proposal to form a society to educate the public about eugenics. Galton responded enthusiastically. Before long, Sybil Gotto and Galton had begun to interest their friends in the proposed society, among whom was Lady Emily Lutyens, wife of the distinguished architect, Sir Edwin Lutyens. This little band of enthusiasts made an overture to the Committee of the Moral Education League, of which Lady Emily was a member, and presented a proposal, presumably drafted by Sybil Gotto, at a meeting at Caxton Hall on November 15, 1907. Mrs. Gotto used persuasion and flattery in her presentation. She hoped that the new association might have sufficient clout to gain the essential backing of the medical establishment. After discussion, the participants decided that the Eugenics Education Society should branch off as an entity distinct from the Moral Education League. Its credo, drawn up by Mrs. Gotto, was to eliminate the "conspiracy of silence" that enveloped "the subject of birth and parenthood" in children's education; to heighten public opinion on questions of morality; and to "strengthen public opinion against unhealthy marriages, and a wilful propagation of an unhealthy and suffering race."[124]

The Eugenics Education Society next formed a link with the Society for the Study of Inebriety, which embraced both the medical and legal aspects of alcoholism. Its first president, Sir James Crichton-Browne, was also a vice-president of the Society for the Study of Inebriety. Other influential members of the latter society included F. W. Mott, director of the Laboratory and Pathologist to the London County Asylums, and F. W. Archdall-Reid, a surgeon, both of whom had heard Galton's lectures to the Sociological Society. In the spring of 1908, Crackanthorpe persuaded Galton to serve as honorary president of the Eugenics Education Society and to present a paper on eugenics.[125] This little speech, read to a small assemblage, was one of the clearest expositions of eugenic philosophy Galton had yet given. It enjoyed wide popular recognition for it was published in full in the *Westminster Gazette*.

Meanwhile, Galton's work had stimulated scientists elsewhere in Europe and in the United States. In 1903, Wilhelm Schallmeyer took the first prize of 10,000 marks in the Krupp competition awarded for the best manuscript explaining the political and social meaning of Darwin's theory.[126] His winning entry, a book entitled *Vererbung und Auslese im Lebenslauf der Völker* (Heredity and Selection in the Life History of Nations), would become the standard German work in eugenics. Alfred Ploetz launched the principal German journal in eugenics, *Archiv für Rassen- und Gesellschaftsbiologie*, in 1904, and founded the Race Hygiene Society in Berlin in 1905. In France, Georges

Vacher de Lapouge, a librarian in Rennes with a collection of 25,000 beetles, who also studied human anatomy and amassed skulls, was drawn to Galton's work.[127] In a *cours libre* he gave in 1886 at Montpellier on "Anthropology and Political Science," Lapouge promised his students that he would end the course by revealing "the theory of Mr. Galton on eugenics—the laws which regulate the production, the conservation and the propagation of superior families, the heart of any race—and which can, by a wise selection, permit the substitution of a superior humanity in the future for the humanity of today."[128] Lapouge advocated an idea that would later become a reality, the use of the sperm bank for artificial insemination. He wrote enthusiastically about the prospect of "Minerva replacing Eros, a single (male) reproducer in a good state of health would suffice to assure 200,000 births annually."[129]

French scientists also came into eugenics from another direction, puericulture. The word, coined in 1858 by Alfred Caron, a Paris physician, referred to the notion that improvement of the species could be effected by enhancing the health of newborns. Adolphe Pinard, a Parisian obstetrician, became an advocate. Puericulture soon became intertwined with the idea that nature might also play a significant role. In an 1899 paper Pinard called for puericulture before procreation with a focus on the role of heredity in determining the well-being of the new child. Conversion of puericulture before procreation into full-blown eugenics of necessity depended on Frenchmen more familiar with the work of Galton and other eugenicists in Europe and the United States. One of these was Lucien March, who had worked with Pinard and was head of the Statistique Générale. March, Pinard, and others eventually formed the French Eugenics Society in 1912.

In the United States eugenics flourished under the guiding hand of Charles Davenport. He founded a station for the experimental study of evolution at Cold Spring Harbor in 1904, having persuaded the Carnegie Institution of Washington to fund it.[130] The small staff he recruited carried out research on variation, hybridization, and natural selection. Much of the work was of good quality and dealt with biometric or Mendelian analyses using poultry, canaries, and animals. Castle's work on coat-color variation in animals was done partly at Cold Spring Harbor. But Davenport was also interested in human heredity and eugenics and began to accumulate pedigree data on traits that seemed to show Mendelian behavior such as brachydactyly, hemophilia, and Huntington's chorea. Davenport convinced Mary Harriman, a social activist, of the significance of the work being done by his group. Her mother had just taken over the reins of her late husband's vast railroad fortune and with her support Davenport's Eugenics Record Office was established on 75 acres she purchased up the hill from the Cold Spring Harbor experimental station. Davenport's workers used pedigrees almost exclusively in analyzing their results and proposed Mendelian inheritance for various purported human genetic traits, some real, but others imaginary. American eugenicists soon had their own periodical, the *Journal of Heredity*, which intermixed straightforward papers on genetics with articles, editorials, and book reviews of eugenic interest.

So eugenics was gaining acceptance throughout the Western world while the aging Galton accumulated honors, acclaim, and recognition. On July 1, 1908, he participated in the Darwin-Wallace Celebration of the Linnean Society of London.[131] It was the last gathering of the old guard of evolutionary biology. Darwin and Huxley were gone, but Galton, Haeckel, Hooker, Lankester, Strassburger, Wallace, and Weismann were present. Each was to receive a medal. The ailing Galton leaned on Pearson's arm as he walked slowly to the podium. Some "wag on the Linnean Executive" seated Pearson next to Bateson. Determined to disappoint the wag, Pearson planned to greet Bateson politely, if not warmly. But Bateson sat down sideways in his chair with his back to Pearson and remained so during the whole ceremony. In awarding Galton's medal, Dr. Dukinfield H. Scott, president of the Linnean Society, remarked that it was Galton "who first showed the way by which exact measurement could be applied to the problems of evolution and heredity, and indicated their laws must be susceptible of proof."[132] Afterwards a worried Pearson escorted the thoroughly fatigued Galton home safely. But Galton was soon in good form again. "That Eugenics Education Society seems really promising" he wrote to Pearson.[133] "The prospectus has been re-worded and members are coming in. Mrs Gotto is marvellous in her energy."[134]

Galton addressed the Eugenics Education Society on October 14, 1908. His subject was the formation of local associations for promoting eugenics.[135] Establishment of "any general system of constructive eugenics" would depend on the "efforts of local associations acting in close harmony with a central society" like the Eugenics Education Society. Aided by the central society, each association would be self-governing and would provide lectures on eugenics designed to arouse a wide interest in the subject. Associations would seek the cooperation of upstanding representatives of the community. Persons would be classified by "physique, ability, and character" in that order, with an inferiority in one category outweighing superiority in the other two. Galton drew the curtain aside revealing a utopian world where "family histories would become familiar topics, the existence of good stocks would be discovered, and many persons of 'worth' would be appreciated and made acquainted with each other who were formerly known only to a very restricted circle."[136] They would be aided in finding suitable appointments by "local sympathisers with eugenic principles."[137] If the local societies did no more than this, they would have succeeded. Once "public opinion in favour of eugenics" had taken hold and eugenics was accepted "as a quasi-religion" the result would "be manifested in sundry and very effective modes of action."[138] Galton expected local eugenic action to take numerous directions including "the accumulation of considerable funds to start young couples of 'worthy' qualities in their married life."[138] The local associations by fostering formation of circles of individuals of "worth," by molding public opinion, and by making subsidies available to worthy young marrieds would automatically ensure that the most fit wed each other and felt financially comfortable with vigorous procreation.

The *Eugenics Review*, the journal of the Eugenics Education Society, commenced publication with 1909–1910 volume. Galton, the honorary president, wrote a foreword emphasizing that the journal disclaimed any rivalry "with the more technical publications issued ... from the Eugenics Laboratory," but proposed "to supplement them."[139] By distinguishing between the highly technical memoirs that Pearson preferred to pen, and articles laymen associated with the Eugenics Education might write, Galton intended a preemptive strike. The exuberant Sybil Gotto was buzzing around the Eugenics Laboratory requesting that Pearson allow recent lectures given to the Society by Miss Elderton and Heron to be published in the *Eugenics Review*. An upset Pearson had written Galton that he hoped Mrs. Gotto would not think him "churlish" for refusing.[140] Miss Elderton's research was incomplete and Heron planned to publish his findings in the Galton Laboratory Memoirs. Galton replied that it would "never do to allow the Eugenics Education Society to anticipate and utilise the Eugenics Laboratory publications."[141] He would explain the problem to Mrs. Gotto and had written "a brief send-off" to the *Eugenics Review* making the distinction between the two kinds of publications explicit. A relieved Pearson replied that if the "youthful efforts" of the Eugenics Laboratory had been "mixed up in any way with the work of Havelock Ellis, Slaughter or Saleeby, we should kill all chance of founding Eugenics as an academic discipline."[142]

Pearson's concern was genuine. He was planning the *Treasury of Human Inheritance*, which became a multivolume series published as Eugenics Laboratory Memoirs from 1909 to 1933. The help of the medical community was crucial and "medical men," he confided to Galton, were "coming in and giving us splendid material ... often confidential and personal histories. But Saleeby and others on the Eugenics Education Society's Council are red rags to the medical bull, and if it were thought we were linked up with them we should be left severely alone."[143] He tried to be tactful with Mrs. Gotto, informing her that the position of the Eugenics Laboratory was "one of sympathy but independent action"[144] with the Eugenics Education Society. If so, he failed to convince Mrs. Gotto, who probably thought he was being snobbish, and a fissure opened between the Eugenics Laboratory and the Eugenics Education Society.

Shortly thereafter, Galton wrote Pearson a note adding the news, almost as an afterthought, that his sole remaining sibling, Erasmus, had died at the age of 94.[145] Pearson replied sympathetically and Galton remarked that Erasmus had been cremated. "It is strange that a living human being should be so quickly reduced to four handfuls of ash, and that scattered over the soil of a garden."[146] Galton soon regained his sense of humor. Perhaps thinking of the ordeal of Prometheus he wrote Pearson in March that the "demon lumbago has planted beak and claws into my loins and sent me helplessly to bed."[147] But soon he was better.

In April there was another dust-up with Sybil Gotto.[148] She informed Heron that she had found two appropriately qualified individuals interested in doing eugenics research, so Heron suggested sending them to the Eugenics Laboratory. She ruled this out, saying that they should work under the direction of the Society and requested that Pearson give her copies of the forms being used by the Eugenics Laboratory to record pedigree information. Pearson agreed, on the understanding that they were to be returned to the Eugenics Laboratory upon completion, but this was not what Mrs. Gotto had in mind. If the two eugenicists were going to work for the Society, the pedigrees they collected would be for its use. Pearson already had been burned by investigators in Scotland and the United States who had borrowed forms "on the excuse that they were going to return them to us and then using them to collect facts for themselves! That does not seem quite playing the game!"[149] The first issue of the *Eugenics Review* appeared in the spring of 1909. Pearson wrote Galton that, while he liked his foreword, "the text is a little 'thin' and the statements a bit misleading."[150] He thought the journal could succeed if it emulated its German counterpart the *Archiv für Rassenbiologie*. Galton, agreeing that the *Eugenics Review* was "rather feeble," thought it would probably "mend." In June, while Eva was away, Galton received a confidential letter from the British prime minister, H. H. Asquith, informing him that he was to "receive the honour of Knighthood on his Majesty's approaching birthday."[151] He joked in a letter to Eva, "I have to live until November 9 and then shall blossom."[152] To Pearson he wrote a "precious bad *knight* I should make now, with all my infirmities."[153]

The Eugenics and Biometrics Laboratories were in high gear, pouring out memoirs with support not only from Galton, but from the Worshipful Company of Drapers, one of the old chartered companies of the City of London. It had sponsored Pearson's research since 1903.[154] There were ongoing investigations on albinism in man and other animals, tuberculosis, insanity, and alcoholism, with each leading to several massive memoirs. Both Eldertons, David Heron, and others were intensely involved. Pearson, still suffering from the slings and arrows of his old scourge Bateson, had now provoked prominent members of the Eugenics Education Society with a memoir on the heritability of alcoholism, coauthored by Ethel Elderton. They had analyzed a set of data from Edinburgh and Manchester to see whether parental alcoholism had any marked influence on the mental capacity and physical characteristics of offspring.[155] Their main conclusion was that there was no obvious relationship "between the intelligence, physique or disease of the offspring and parental alcoholism.... The balance turns as often in favour of the alcoholic as the non-alcoholic parentage." They were careful to add that they did "not attribute this to the alcohol, but to certain physical and possibly mental characters which appear to be associated with alcohol."[156] Nevertheless, they strongly suspected that alcoholism itself had a hereditary basis and if so, "the problem of those who are fighting alcoholism is one with the fundamental problem of eugenics."[157] Any "unjustifiable statistics" that zealous temperance reformers produced must be supplanted by "real knowledge" in place of "energetic but untrained philanthropy in dictating the lines of feasible social reform."[158]

The study was summarized in the *Times* on May 21.[159] The authors had waved the red sash before the Councils of the Eugenics Education Society and the Society for the Study of Inebriety. Crackanthorpe, the chairman of the Eugenics Education Society, charged horns down in a letter to the *Times*. "To those," wrote Crackanthorpe, "who are familiar with the methods of eugenic . . . research the Report causes no surprise at all. It simply confirms their belief that, serviceable as biometry is in its proper sphere, it has its limitations, and that a complex problem such as that of the relation of parental alcoholism to offspring is quite beyond its ken."[160] Crackanthorpe trampled roughshod not only on Pearson, but on some of Galton's most important convictions, writing that "the biometrical method is based on the 'law of averages' which again is based on the 'theory of probabilities,' which again is based on mathematical calculations of a highly abstract order."[161] He implied that biometrics was so much hocus pocus and supplied "no practical guide to the individual."[162] And here he caught the essence of the problem. While biometrics was hard to understand for the average member of the Eugenics Education Society, Mendel's straightforward pedigrees were easy to grasp even when grossly misapplied, as they often were, to complex social phenomena involving "feeblemindedness," criminality, disease, and poverty. Then Crackanthorpe took a shot that surely incensed his elderly neighbor in Rutland Gate, writing "that some of the new technical phraseology used by the biometricians" was "rather repelling—notably, their coefficient of correlation."

Galton wrote a firm, but circumspect rejoinder to the *Times*.[163] Crackanthorpe's letter implied that biometric conclusions were a house of cards, but Galton felt this portrayal "inaccurate." He briefly summarized his understanding of biometric methods, explaining their great analytical power, and disputed Crackanthorpe's assertion that the relationship between parental alcoholism and progeny was "quite beyond the ken" of biometry. The controversy simmered throughout 1910 with the Temperance Press vigorously stirring the noisome brew. In an article in the *British Journal of Inebriety*, Saleeby hinted that a crevasse had opened between the Eugenics Education Society and the Eugenics Laboratory. Galton's exasperation with Saleeby boiled over when Saleeby, signing himself X, like a persistent mosquito succeeded in irritating Pearson in a series of articles attacking the Eugenics Laboratory and its memoirs.[164] He wrote to the journal, "that an antagonism exists between at least one member of the Society, namely Dr Saleeby, and the Laboratory is absolutely shown in this article."[165]

But he was in a difficult position, shielding the biometricians on one hand while stroking the laymen interested in eugenic questions on the other, his shock troops in practical implementation. The supercharged rhetoric on alcoholism had confused the Eugenics Education Society with the Eugenics Laboratory in the public mind, so Galton tried make the distinction clear in a November 3, 1910, letter to the *Times*. "Permit me, as the founder of the one and the honorary president of the other, to say that there is no other connection between them. Their spheres of action are different, and ought to be mutually beneficial."[166] The Laboratory's purpose was to permit highly trained experts to gather and analyze masses of data pertinent to eugenics while the raison d'être of the Society was "to popularize results that have been laboriously reached elsewhere and to arouse the enthusiasm of the public."[167] If only the Council of the Eugenics Education Society had felt the same way. In October 1910, Galton learned that he had been recommended for the Copley Medal of the Royal Society. To Pearson he wrote, "people die so fast that I can find only five other living Englishmen ... How *age* counts!"[168] Sir George Darwin received the medal on Galton's behalf on November 30. In his presentation, the president managed to summarize, in a single paragraph, many of Galton's achievements beginning with geography and ending with eugenics.[169]

Earlier that year the irrepressible Galton, physically failing but mentally sharp, embarked on a novel, *Kantsaywhere*, about a eugenic utopia.[170] *Kantsaywhere* was supposedly extracted from the journal of the late I. Donoghue, a professor of vital statistics, who arrives in Kantsaywhere, a colony governed by a Council, perhaps like one of Galton's local eugenic councils. He meets Miss Augusta Allfancy, to whom he takes a fancy. She is about to take her honors examination at the Eugenics College of Kantsaywhere and Donoghue decides to stand for the highest degree he can obtain. The eugenically correct structure of Kantsaywhere owes its existence to Mr. Neverwas, who left his property to its Council so that the income could "be employed in improving the stock of the place especially of its human breed."[171] The Eugenics College granted "diplomas for heritable gifts, physical and mental."[172] It encouraged those graduating with high distinction to marry early by offering them "appropriate awards of various social and material advantages to relieve the cost of nurturing their children."[173] Mr. Neverwas specified that none of the estate's income should be "spent on the support of the naturally feeble."[174]

Donoghue arrived in the nick of time to take the pass examination, the preliminary to the honors examination for which Augusta Allfancy was a candidate. The pass examination vetted the candidate genetically. Failures were segregated in labor colonies "under conditions that were not onerous"[175] except that they had to work hard and remain celibate. Galton, always aware of the normal distribution, had to do something about the people in the middle. These individuals received a "second-class certificate," meaning they could propagate "with reservations."[176] Augusta's brother Tom was Donoghue's examination sponsor. He assured Donoghue that his personal capabilities should rate well, but his *ancestral* claims were another matter. While satisfactory for the pass examination, they were "insufficiently authenticated" for the honors examiners. This was a continuing problem with immigrants as their pedigree data tended to be fragmentary.

Augusta passed her examination with flying colors and was elected a Probationer in the College. Donoghue's pass examination began with rigorous physical tests following which the Examiners questioned him closely about the papers he had presented. They liked his responses and "smilingly gave" him "a first-class P. G.—Passed in Genetics—degree," following which he was fingerprinted "for future identification if necessary."[177] The next day he stood for the annual honors examination. The examination had four divisions. "The first is mainly anthropometric, the second is aesthetic and literary, the third is medical, and the fourth is ancestral."[178] He handed in his Pass Certificate, and was fingerprinted again. Like most immigrants he did not do as well on the ancestral part of the examination as on the rest. Donoghue (like Galton) did "not know much in detail about the examination for girls. It was carried out by women examiners who had taken medical degrees elsewhere,"[179] but was thorough.

Augusta engaged in a round of parties designed to insure that she met each of her male probationer counterparts under conditions "which approached merrymaking and banished diffidence."[180] Ideally, an older male probationer was united with a younger woman "say about 22 years of age, which admits more than four generations being produced each century."[181] Later Donoghue learned that those who failed the pass examination did not always end up in the labor colony, but were given incentives to emigrate and subjected to "surveillance and annoyance" if they refused. Propagation by the unfit was "looked upon by the inhabitants of Kantsaywhere as a crime to the State."[182] Donoghue did well enough on the honors exam to engender a marked improvement in his reception. He "was begged to accompany my host's family to half a score of different places to which they were invited."[183] He observed that the Kantsaywhere women were similar to those in Guido Reni's "Apollo and the Hours preceded by Aurora." They had "the same massive forms, short of heaviness, and seem promising mothers of a noble race."[184] They were, however, garbed in keeping with Victorian mores, with their dresses "more decorously buttoned."[185] The Kantsaywhere men were "well built, practised both in military drill and in athletics, very courteous, but with a resolute look that suggests fighting qualities of a higher order. Both sexes are true to themselves, the women being thoroughly feminine, and I may add, mammalian, and the men being as thoroughly virile."[186] Kantsaywhere was Galton's eugenic utopia and this fanciful, but unpublished, novel expressed more clearly than any dry scientific paper or popular article what he hoped eugenics would achieve.

Kantsaywhere was completed by late fall and Galton, who had let a house in Haslemere for the winter, asked Methuen over for tea on December 4, as he hoped to interest the publisher in the manuscript, but Methuen was not taken with the manuscript and rejected it. On December 28 Galton wrote Milly saying, "*Kantsaywhere* must be smothered or be superseded. It has been an amusement and it has cleared my thoughts to write it. So now let it go to 'Wont-say-where.'"[187] Pearson knew nothing of *Kantsaywhere* when he visited Galton that same glorious late December day.[188,189] Eva Biggs greeted him and requested that, should Galton mention the novel, Pearson must persuade him not to publish it because, she confided, the love scenes seemed completely unrealistic. She was clearly worried that *Kantsaywhere* would embarrass her uncle. The old man and his disciple sat outside in the sunshine with Galton warmly wrapped in blankets. The two old friends talked animatedly about the work of the Eugenics Laboratory and the obvious shortcomings of certain members of the Council of the Eugenics Education Society. Galton reminisced on how much the Copley Medal meant to him and the numerous congratulatory notes he had received about it. Pearson departed the next day thinking Galton no worse physically than at any time in the past five years, and quite evidently with all his mental faculties still intact.

Three weeks later, on January 17, 1911, Francis Galton was dead. Eva wrote his friend Lady Pelly the next day that she had "the saddest news for you—dear Uncle Frank died last night—he had a sharp attack of bronchitis and died of heart failure, not having the strength to fight against it—he suffered much discomfort, but very little pain, and just at the last he was very peaceful."[190] Under the terms of Galton's will £45,000 was left to the University of London to endow the Chair of Eugenics, with the express wish that Pearson be its first occupant.[191] Pearson duly became the first in a series of a very distinguished Galton Professors who made outstanding contributions in biometry and human genetics. In 1963 the Francis Galton Laboratory of National Eugenics was renamed the Galton Laboratory of the Department of Human Genetics and Biometry. His only nephew, Bessie's son Edward Galton Wheler, received £15,000. His sole surviving niece, Adèle's daughter, the widowed Milly with whom he corresponded so frequently, received a lesser amount as did his great niece, Eva Biggs, who had taken such good care of her Uncle Frank for so many years. And his faithful servant of 40 years, Albert Gifi, received £200. Thus his will seemed to reflect his priorities in life—science first, male relatives second, women third, and servants last. By the middle of 1911 plans were well underway for the First International Congress of Eugenics, held in July 1912. The pedigree that Galton had done so much to popularize had triumphed, but it was not the pedigree of ancestral heredity, but the black-and-white pedigree of discontinuous inheritance of Mendel and Bateson.

EPILOGUE

Out of Pandora's Box
The First International Congress of Eugenices

> The tendency of heredity is to produce an environment which perpetuates that heredity: thus, the licentious parent makes an example which greatly aids in fixing habits of debauchery in the child.
>
> —R. L. Dugdale, *The Jukes*[1]

The First International Congress of Eugenics opened on July 24, 1912, at the majestic Hotel Cecil, sited on the Victoria Embankment with its sweeping views of the bustling Thames. The congress badge honored Galton, who had died a year earlier, with a likeness of his head in profile.[2] Major Leonard Darwin, the next to last of Charles Darwin's five surviving sons, was president of the Congress. He had served for 20 years in the Royal Engineers, retiring with his majority in 1890. Although he lacked the scientific creativity of his brothers, Horace, George, and Francis, all Royal Society members, he compensated with his enthusiasm for eugenics.

The International Congress of Eugenics was organized by the Eugenics Education Society and announced in the *Eugenics Review* in mid-1911 shortly after Galton's death.[3] It was evident that it would be a prestigious affair, as its vice-presidents included Lord Alverstone, the Lord Chief Justice; Sir Thomas Barlow, president of the College of Surgeons; Sir William Church, past president of the College of Physicians; the Bishops of Ripon and Birmingham; Sir William Collins, vice-chancellor of the University of London; and Principal Miers, the chief officer of the University of London. Alfred Ploetz, president of the Gesellschaft für Rassen Hygiene, the German equivalent of the Eugenics Education Society, led the German Consultative Committee, while David Starr Jordan, an outstanding biologist and chancellor of Stanford University, represented the American Consultative Committee. Several prominent doctors and biologists organized the French contributions, and consultative committees from Belgium, Italy, and Spain were formed.

By registration on Wednesday, July 24, 1912, the vice-presidential list had expanded to include notables like the Right Honourable Winston Churchill, First Lord of the Admiralty; Dr. Alexander Graham Bell; Dr. Charles W. Elliott, president emeritus of Harvard University; the Lord Mayor of London; and the German biologist August Weismann, father of the germline theory of biological continuity. Each consultative committee was adorned with the names of prominent biologists and eugenicists. The congress was to be divided into four sections: Biology and Eugenics, Practical Eugenics, Sociology and Eugenics, and Medicine and Eugenics. They met sequentially on separate days so concurrent sessions were avoided, with the English, French, German, and Italian languages being used on an equal footing. Membership rose to 750, which to *Nature* indicated "the widespread interest taken in the subject," but the reporter added tongue-in-cheek that "the series of brilliant entertainments organised by the hospitality committee, under the secretaryship of Mrs. Alec Tweedie, was a bait which attracted many."[4] These included the inaugural banquet at the Hotel Cecil, receptions hosted by Her Grace the Duchess of Marlborough, the Lord Mayor of London, the American Ambassador Whitelaw Reid, and the Congress president, Leonard Darwin. There was a special luncheon and garden party on the grounds of Coombe Park, Sevenoaks, to which congress members were conveyed by special train. A limited number of invitations to tea on the terrace of the House of Commons was available, with tickets being provided to view debates on the floor.

Between 400 and 500 attended the inaugural banquet presided over by Major Darwin. Arthur Balfour, the blue-blooded former prime minister, guided by what Austen Chamberlain called "the finest brain that has been applied to politics in our time,"[5] gave a speech to toast the "foreign friends and guests." His endorsement of eugenics was qualified. The Congress had "two great tasks allotted to it."[6] It must convince the public "that the study of eugenics is one of the greatest and most pressing necessities of our age."[7] But it also must also persuade the public that eugenics was "one of the most difficult and complex"[8] tasks science had ever undertaken. He continued in the same vein paying tribute to the eugenicists, but urging them to heed the flashing caution light by critically examining their premises, arguments, and data. Balfour speculated that "there were probably more differences of opinion" among scientists "with regard to certain fundamental principles lying at the root of heredity" than there were in the decades following the general acceptance of "the great Darwin's doctrines."[9] He warned that "every faddist"[10] would seize "hold of the eugenic problem as a machinery for furthering his own particular method of bringing the millenium upon the earth."[11] Balfour then voiced a concern as valid today as in 1912. Scientists writing and speaking about eugenics might "occasionally use language which is incorrect in itself and which is apt to produce a certain prejudice upon the impartial public."[12]

Then Balfour brilliantly exposed a paradox in eugenic thinking. "We say that the fit survive. But all that means is that those who survive are fit."[13] The perennial eugenic worry that "the biologically fit are diminishing in number through the diminution of the birth rate"[14] must be wrong by the "doctrine of natural selection" as he conceived it. If families of the professional class were "so small that it is impossible for them to keep up their numbers, they are biologically unfit for this very reason."[15] He admonished his audience that "the idea that you can get a society of the most perfect kind by merely considering certain questions about the strain and ancestry and the health and the physical vigour of various components of that society—that I believe is a most shallow view of a most difficult question."[16] Later on J. A. Lindsay, reviewing the Congress, caught the meaning of Balfour's ambiguous message perfectly. "Mr. Balfour, as the principal guest at the inaugural banquet of the Eugenics International Congress, inverted the part of the prophet Balaam. Invited to bless, he remained to curse."[17]

While Balfour had warned his audience to sip cautiously the heady mead of eugenics from its gilded chalice, Leonard Darwin the next day took a deep draft in his presidential address. He pointed to the almost universal acceptance of evolutionary theory, arguing that to be "practically useful" for human beings we needed to "know how and why succeeding generations of mankind have resembled or differed from each other."[18] Because "the environment of one generation" largely depended on that of preceding generations, much attention would always be "devoted to the factor of environment in the evolutionary process."[19] But nature was important too, as it was "increasingly evident that the inborn qualities of the child are derived from its ancestors in accordance with laws which, though now but imperfectly known, are gradually but surely being brought to light."[20] Darwin assembled the scaffolding of his case deftly. He acknowledged the importance of environment, but urged his audience to focus on heredity, as it would be unwise "to attempt to cover too much ground on one occasion."[21] He cleverly put nature in the driver's seat, arguing that not only were men's careers "largely influenced by their inborn qualities, but the surroundings into which each man steps" at birth depended on the inborn qualities of his "ancestors and predecessors who were instrumental in moulding that environment."[22] Darwin let his flock glimpse the promised land of eugenic utopia. Steps taken "to improve racial characteristics of the generations of the immediate future will undoubtedly benefit the countless millions of the more distant future as regards the heritage they will receive at birth in the form, not only of inborn qualities, but also of improved surroundings."[23] Eugenics would become not only a grail, a substitute for religion as Galton had hoped, but a "paramount duty" whose tenets would presumably become enforceable.

Darwin raised the lengthening shadow of inevitable racial degeneracy, wondering gloomily "whether the progress of Western civilisation is not now at a standstill, and, indeed, whether we are not in danger of an actual retrograde movement."[24] The overarching problem for humanity was that natural selection, "playing the part of the breeder of cattle in refusing to breed from inferior stocks," could no longer do its job as "we have been doing our best to prevent further progress being made by this same means. The unfit amongst men are now no longer necessarily killed off by hunger and disease, but are cherished with care, thus being enabled to reproduce their kind, however bad that kind may be."[25] Darwin's logic seemed impeccable. Red in tooth and claw no longer reigned supreme amongst humanity, so those who should die lived and propagated. But not wanting to appear mean he said "we cannot but glory in this saving of suffering" and this "spirit" of mankind "which leads to the protection of the weak."[26] Never again could we return "to the crude methods of natural selection," but even so "the effects likely to be produced by our charity on future generations is, to the say the least, but weakness and folly."[27]

Darwin had carefully placed each block in the eugenic argument and his message was simple. We can improve the race through selective breeding, but the race was currently declining. Civilizations might begin to fail, since we were preserving the weak, the genetically undesirable, and allowing them to breed absent the winnowing effects of natural selection. Something had to be done despite "the blanks in our knowledge of the laws of life."[28] Darwin reverently invoked the name of the late patron saint of eugenics. "Certainly, Sir Francis Galton, whose name we hope will ever in future be associated with the science of eugenics, a science to which he devoted the best years of his long life, declared with no uncertain voice that something should be attempted without further delay."[29] After all if cattle breeders were urged to delay stock improvement because the laws of heredity were incompletely understood, "they would simply laugh at us," as they knew which were the best and worst stocks. Galton had discovered the mystical box that Darwin now opened, out of which flapped eugenics accompanied by its courtiers: involuntary segregation, sterilization, and racial intolerance. They would spread a pestilence through Europe, America, and beyond that would rage in its most virulent and hideous form in the Nazi Germany of the 1930s and 40s.

Darwin's rousing speech was cheered loudly. Three themes characterized the stream of papers that followed: natural selection, pedigree analysis, and Mendelian genetics. The absence of natural selection in human populations meant that artificial selection for the fit and against the unfit had to serve as a substitute, an approach successfully used with domestic animals and cultivated plants. Galton had popularized pedigree analysis. He had used this tool to estimate the heritability of what he called "talent and character" in prominent families, but the pedigrees presented at the Congress featured the heritability of undesirable traits among the dregs of society. Mendelian genetics seemed perfectly suited to pedigrees. The problem was that, while Mendelian analysis was ideal for clear-cut discontinuously varying traits such as yellow versus green seed color in pea plants, it was misapplied grossly by the eugenicists to complex traits involving both nature and nurture, such as "feeblemindedness" and alcoholism.

On the day of Darwin's presidential speech, Dr. David Fairchild Weeks, medical superintendent of the New Jersey State Village for Epileptics, updated his earlier work on the heritability of epilepsy in a paper decorated with 19 figures depicting pedigrees.[30] They included grab-bags of presumed deleterious traits other than epilepsy such as "feeblemindedness," alcoholism, insanity, "criminalistic" tendencies, and tuberculosis. Weeks recognized that his complex pedigrees did not justify classifying this smorgasbord of detrimental characteristics as resulting from a single Mendelian trait. However, he concluded that while "epilepsy itself cannot be considered as a Mendelian factor, when considered by itself . . . epilepsy and feeble-mindedness are Mendelian factors of the recessive type."[31] In short, Weeks, like so many enthusiastic eugenicists, had forced his unwilling data into a Mendelian framework.

Another paper in similar vein on insanity was delivered by Dr. F. W. Mott, pathologist to the London County Hospital.[32] "There is a correlation," Mott stated, "between the wage-earning capacity of a population, pauperism, insanity, and tuberculosis."[33] He invoked natural selection, as it was "well-known that the feeble-minded are especially prone to tuberculosis, which is one of Nature's methods of eliminating the unfit."[34] Even though imbeciles and idiots were "often sterile, which is one mode by which a completely degenerate stock may die out,"[35] the frightening prospect was that "degenerate stocks generally contain feeble-minded of all grades, the majority of which will not die out, but propagate freely."[36] With the aid of lantern slides Mott paraded depressing pedigrees before his audience. One showed a family with one fairly normal branch and another branch peppered with drunks and lunatics. "The two stocks show a marked difference" said Mott, "one side, the maternal, is practically free from taint; almost every member of the paternal stock is unsound."[37] Some of Mott's pedigrees were obtained from Ernest Lidbetter, whose research Mott strongly supported.[38] Employed by the Poor Law Authority in London to investigate case histories of poor relief applicants, Lidbetter was an enthusiastic collector of pedigrees of the poor. They routinely showed that paupers begat paupers at an alarming rate, not to mention the tubercular, feebleminded, and insane (Fig. E-1). Even worse, the pauper pedigrees revealed the poor to be highly fecund, suggesting that the unfit would inherit the earth absent eugenic intervention. These pedigrees did not exhibit

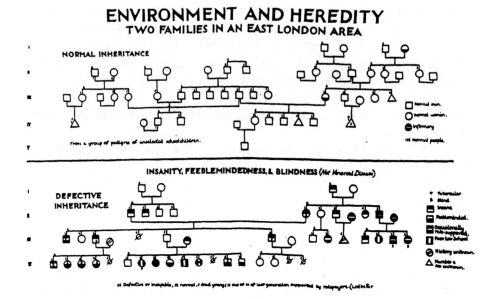

Fig. E-1 A Lidbetter pedigree showing two families living next door to one another, one filled with paupers, insane, and tubercular individuals and the other with none. The pedigree aimed to show the dominance of heredity over environment. From Pauline M. H. Mazumdar, *Eugenics, Human Genetics and Human Failings*. London: Routledge, 1992.

the "vulgar Mendelism" of those displayed by the American eugenicists who tried to trace families of the unfit to a single defective ancestor, "but a kind of non-quantitative population genetics, the interrelatedness of the entire pauper class as well as the transmission of its civic defects."[39]

Between lecture sessions, delegates and guests gathered in the handsome exhibition section to browse at their leisure among the elaborately illustrated pedigrees.[40] There were charts showing Mendelian inheritance in peas and Andalusian fowl and a pedigree highlighting the interrelatedness of the Darwin, Galton, and Wedgwood families. The American Breeders' Association put on an eye-catching display complete with impressive statistical tables on defectives and charts that illustrated the principles of heredity. Included were a group of 16 pedigrees collected by field workers at Davenport's Eugenics Record Office at Cold Spring Harbor. Exhibitions like this were prototypes for the Fitter Families competitions held in the "human stock" sections of American state fairs in the 1920s. Bleecker van Wagenen, chairing a committee of the American Breeders' Association studying "the best practical means for cutting off the defective germ-plasm in the human population," presented a preliminary progress report.[41] One of the committee members, Harry H. Laughlin, superintendent of Davenport's Eugenics Record Office, would soon become the most effective proponent of involuntary sterilization and immigration restriction for eugenic purposes in the history of the movement.

The Germans contributed the largest section to the exhibition, incorporating much material displayed in the *Internationale Hygiene-austellung* in Dresden the previous year.[42] Some charts emphasized the importance of environmental factors, but others showed the crushing genetic load that flattened "the unfortunate final recipient of the degeneration of several families, like a circus strong-man supporting too many acrobats on his shoulders."[43] Race hygiene, as eugenics was called in Germany, had an active following of well-known scientists, many of whom were represented on the German consultative committee.

One of the most prophetic papers was read by Vernon L. Kellog, a Stanford biologist.[44] Kellog reviewed the eugenic consequences of militarism. He concluded that war "results in the temporary or permanent removal from the general population of a special part of it, and the deliberate exposure of this part to death and disease, disease that may have a repercussion on the welfare of the whole population."[45] This "removal" not only disturbed "the sex equilibrium of the population" preventing "normal and advantageous sexual selection" of men "having the greatest life expectancy" and of the "greatest sexual vigor and fecundity," but it left behind men who fell "below this standard."[46] Kellog had voiced a concern, soon to be echoed during the Great War, that the lives of the best and brightest young men were being snuffed out in the trenches and barbed wire entanglements of France. Frederick Hoffman, statistician for the Prudential Insurance Company, addressed another classic eugenic concern.[47] Brandishing detailed maternity statistics for the State of Rhode Island based on the 1905 census, Hoffman worryingly recorded that while 17.5 percent of foreign-born married women in Rhode Island were childless, the frequency rose to 28.4 percent among native-born women. He concluded ominously that "the increase in the proportion of childless families among the native-born of native stock, is evidence of physical deterioration, and must have a lasting and injurious effect on national life and character."[48]

Professor Alfredo Niceforo of the University of Naples, speaking in French, probed the causes of inferiority of physical and mental characteristics of the lower social classes.[49] He concluded that the "lower classes present, in comparison with subjects of the higher classes, a lesser development of the figure of the cranial circumference, of the sensibility, of the resistance to mental fatigue, delay in the epoch when puberty manifests itself, a slowness in the growth, a larger number of anomalies and of cases of arrested development."[50] The lower classes also had greater birth and death rates, a "precocity in the age of marriage" and "a predilection for certain forms of crime." But M. Lucien March, directeur de la Statistique Générale de France and a vice-president of the French consultative committee to the Congress, struck a slightly discordant note.⁵¹ His detailed analysis suggested that "in France paupers, vagabonds, prisoners, etc., the descendants of whom form an undesirable addition to the population, are rather lower than the average in productiveness."⁵² But he added reassuringly that "among these are reckoned many who are in confinement, whose fecundity is therefore temporarily put a stop to."⁵³ The Whethams, authors of the 1909 book *Family and Nation*, tackled the problem of race.⁵⁴ They were the proud parents of six children, demonstrating through personal example that the "abler" classes could indeed be prolific. After a whirlwind cruise through history and geography, they divided the European races into three categories: Mediterranean, Alpine or Armenoid, and Northern. Of these three the Northern had the "acknowledged supremacy."⁵⁵ The worry for England was the incursion of the Southern races so that "the poorer parts of many towns" contained "the shorter, darker elements" and their rabbitlike tendency to reproduce meant a selective increase in "the racial elements of Southern origin" was posing a serious problem, as they were "the least productive of men of ability and genius in England."⁵⁶ Hence, "the British nation and perhaps the nations of Western Europe generally" were likely "to find themselves becoming darker, shorter, less able to take and keep an initiative, less steadfast and persistent, and possibly more emotional, whether in government, science or art."⁵⁷ So all the standard worries of eugenics were aired by one speaker or another and the bottom line was that race, class, and nation were under threat of being overrun by legions of loathsome feebleminded, poor, and racially suspect.

Reginald Punnett, professor of biology at Cambridge, and a protégé of Bateson's, was one of the bona fide geneticists present. His name adorns the simple square of Mendelian segregation patterns familiar to all who have taken introductory biology. His remarks also illustrated a level of scientific rigor absent from most of the other presentations. Punnett warned that precise knowledge was available on the inheritance of very few Mendelian characters in man.⁵⁸ He allowed that "speaking generally, the available evidence" suggested that feeblemindedness "is a case of simple Mendelian inheritance."⁵⁹ But, although phenomena of considerable genetic complexity could be unravelled in appropriate plant and animal systems, "the direct method is hardly feasible in man," although much could be learned "by collecting accurate pedigrees and comparing them with standard cases worked out in other animals."⁶⁰ Punnett cautioned that "the collection of such pedigrees is an arduous undertaking demanding high critical ability, and only to be carried out satisfactorily by those who have been trained in and are alive to the trend of genetic research."⁶¹ But later Punnett, convinced by the pedigree data collected by Henry Goddard and others, would conclude that feeblemindedness was inherited as a simple Mendelian recessive.⁶² In fact, a remarkable number of the most competent geneticists of the period subscribed to this notion, with T. H. Morgan being a notable exception.⁶³

On July 30 the final papers were read and Major Darwin delivered the President's Farewell Address.[64] It was short and to the point. He could not estimate when eugenicists would gain "ultimate victory," although "they should conquer in time."[65] He was interrupted by cheers. But the Congress just concluded might "have practical consequences in hastening on legislation such as that now being discussed in Parliament."[66] Although wishing to avoid controversial matters "all would place legislation tending to stamp out feeble-mindedness from future generations in a leading place in their programmes."[67] And so it went, with Darwin spinning out jubilant phrases and the audience interrupting. Afterwards the departing eugenicists, giddy with ideas, excitedly discussed corresponding about new developments and researching subjects of eugenic interest. Most of all they wanted to spread far and wide the great new creed with its glittering goal of race and class improvement through selective breeding.

And what were the consequences? In Great Britain, where eugenics began, they were rather benign.[68,69] The Eugenics Education Society kept track of legislation of interest. British eugenicists wrote letters to the *Times* expressing their opinions and dispatched deputations to Parliament to air their views on the poor laws, legislation to prevent propagation of the feebleminded by compulsory institutionalization, etc. Such was the enthusiasm for eugenics that a highly literate young London matron named Bolce in 1913 bore a daughter she christened "Eugenette," widely touted as the first eugenic baby. In the United States eugenics took a far more sinister turn and its targets included not only the feebleminded, but immigrants arriving in masses from southern and eastern Europe. Indiana passed the first involuntary sterilization law in 1907, and by 1913, 16 states had similar statutes.[70] Initially, only feebleminded men were sterilized using vasectomy, since tubal ligation was considered too dangerous for women, but by 1920 a relatively safe method for that operation had been perfected.

During World War I many state sterilization laws were struck down and involuntary sterilization programs became largely quiescent except in California. In 1922 Harry H. Laughlin, superintendent of the Eugenics Record Office, published an extraordinarily detailed study, *Eugenical Sterilization in the United States*, qualifying him as the foremost American expert on the subject. An upsurge in sterilization legislation occurred concomitantly, so by 1926, 23 states had enacted sterilization laws and 17 had active programs. In April 1927, the most important case in sterilization history was argued before the U.S. Supreme Court.[71] It involved a "feebleminded" Virginia woman named Carrie Buck whose mother and illegitimate daughter were also purported to be feebleminded. Laughlin was the expert witness and the Court voted 8 to 1 in favor of sterilization, with Justice Oliver Wendell Holmes writing the majority opinion containing this famous phrase: "Three generations of imbeciles are enough." This flawed decision provided a rationale at the federal level for a practice that had been legalized by many state legislatures.

Laughlin, acting as the "Expert Eugenical Agent" of the House Committee on Immigration and Naturalization, provided "scientific" advice to that committee supporting the opinions of the anti-immigration lobby. That committee wrote the Immigration Act of 1924, which through 1927 limited entry into the United States of immigrants from any European country to a small percentage of the foreign-born of the same national origin recorded in the census of 1890. Its main purpose was to restrict severely the swelling numbers of the "socially inadequate" from eastern and southern European countries like Poland and Italy and to favor the immigration of members of the desirable Nordic and Teutonic "races" from northwestern Europe. They had represented a much higher proportion of the immigrants in 1890.

The general unease concerning the quality of Eastern and Southern European immigrants was exacerbated by the finding that immigrants scored poorly on IQ tests, usually because their English was limited, and they had poor knowlege of American customs.[72] IQ had been coupled to heredity in Henry Goddard's famous study of the Kallikaks, and the hereditarian theory was promulgated by others, notably Lewis Terman of Stanford University, who actually coined the term "IQ." The extensive IQ testing of U.S. army recruits, beginning in 1917 under the auspices of the National Research Council, led to Robert Yerkes's enormous monograph in 1921, *Psychological Examining in the United States Army*, which drew three main conclusions: the average mental age of white American adults was just above that of morons; European immigrants from southern and eastern parts of the continent were less intelligent than those from northern and western parts; and blacks were at the bottom of the heap. In 1923 Yerkes's disciple Carl C. Brigham, an assistant professor of psychology at Princeton, published *A Study of American Intelligence*. Brigham's short book claimed that the "army data constitute the first really significant contribution to the study of race differences in mental traits."[73] It was available just in time to be considered during the debate on immigration-restriction legislation in Congress.

Now two insidious notions could be linked together to support prevailing prejudices. Intelligence as measured by an IQ test correlated with ability and had a strong heritable component. And African Americans and immigrants from eastern and southern Europe had lower than normal intelligence. Furthermore, as set forth in Madison Grant's 1916 book, *The Passing of the Great Race*, "anthropological" data argued that the Nordic race was deteriorating due to intermarriage with Alpines and Mediterraneans, a thesis that enjoyed a considerable vogue. The Immigration Act of 1924 was passed by large majorities in both houses and signed into law by President Calvin Coolidge. He had earlier remarked "America must be kept American, Biological laws show . . . that Nordics deteriorate when mixed with other races."[74]

German social Darwinists were deeply concerned about the future of the race, since medical care for "the weak" was perceived to have eliminated the struggle for existence. Furthermore, paupers and the socially troubled were thought to be multiplying more rapidly than the talented and fit.[75-77] The German race hygiene movement emerged in response to these fears late in the nineteenth century, with its principal proponents being Alfred Ploetz, Wilhelm Schallmeyer, Eugen Fischer, and Fritz Lenz. Fischer, Lenz, and Schallmeyer had all studied with August Weismann who was the first to enounce clearly the principle of the continuity of the germ plasm and its separation from the soma. This notion, combined with the resurgence of Mendelian genetics, caused race hygienists to argue that social change would be insufficient to alter the human condition. Ploetz founded *Archiv für Rassen- und Gesellschaftsbiologie* in 1904. In 1921 Fischer, Lenz, and the famous German geneticist Erwin Baur coauthored a two-volume work, *Grundriss der menschlichen Erblichkeitslehre und Rassenhygiene* (Outline of Human Genetics and Racial Hygiene). This was the classic text on the subject in Germany for the next 20 years. Its significance was such that two American eugenicists translated it into English in 1931. In the concluding chapter, Lenz presented an elaborate vision of race and race hygiene, making the case "that characteristics of the mind, no less than those of the body" were hereditarily determined, and environment could "do nothing more than help or hinder the flowering of hereditary potentialities."[78] Lenz's anthropological classification corresponded to conventional assumptions, with blacks at the bottom, followed up the scale by "Mongols," "Alpines," "Mediterraneans," and finally the pinnacle of humanity, the "Nordic."

Eugenic enactments in the United States were followed with interest by German race hygienists. Thus Géza von Hoffmann, for several years the Austrian vice-consul in California, regularly informed his colleagues and the German public about eugenic developments in America, publishing a book on the subject in 1913. Books by American racial anthropologists like Madison Grant and Lothrop Stoddard were translated into German, and Dugdale's study of the Jukes and Goddard's of the Kallikaks were widely cited. The first German proponent of eugenic sterilization was Gerhard Boeters, a district physician, who campaigned unsuccessfully for legislation in the early 1920s. In 1932, just before the collapse of the Weimar Republic, the Deutsche Gesellschaft für Rassenhygiene and the medical community drafted a law permitting the *voluntary* sterilization of certain classes of individuals perceived as being hereditarily defective, but the law required proof that the defective traits had a genetic basis. In the summer of 1933, shortly after the Nazis came to power, a restrictive version of the Weimar sterilization statute was passed. This authorized compulsory sterilization of individuals expressing supposed single-gene diseases like feeblemindedness, alcoholism, and schizophrenia. The new law was administered by genetic health courts and their appellate counterparts. These courts were usually attached to civil courts and presided over by a lawyer and two doctors, one of whom was an "expert" on genetic pathology. The number of persons sterilized under this law was around 400,000. Relatively few cases were appealed and most appeals failed. A flurry of other negative eugenic legislation followed rapidly, being capped by the three Nuremberg Laws of 1935 designed to "cleanse" the German population of unwanted elements. The first two were aimed primarily at the Jews, disenfranchising them as citizens, redesignating them as residents, and placing severe restrictions on the amount of Jewish blood permissible in interracial marriages with Aryan Germans. The third law required a premarital medical examination for the prospective husband and wife to see if "racial damage" might result from the marriage. It also prevented marriage of individuals having presumptive "genetic infirmities" like feeblemindedness. By 1938 the Nazis began to engage in the physical elimination of children and later adults who were deemed, for genetic reasons, to have "lives not worth living."

The Nazis also engaged in positive eugenic programs encouraging racially suitable women to have large families to increase the number of desirable, that is Nordic, individuals in the population. Special loans equivalent to a year's salary were given to men whose wives agreed to give up work outside the home. The amount to be paid back on each loan was reduced by a quarter for each child born. Severe restrictions were placed on abortions, such that operations could be conducted only when a woman's life was at risk. Unmarried women were treated as pariahs, being assigned to a subordinate category that included Jews. Nazi positive eugenic policies showed initial promise since the number of births per thousand increased from 14.7 in 1933 to 18 in 1934.

The barbaric nature of Nazi "eugenic" policies, culminating in the "final solution" and the holocaust, led to a total revulsion against eugenics following World War II when the extent of killing in the death camps became apparent. Nevertheless eugenic sterilizations continued to be carried out occasionally in the United States, in Scandinavia, Switzerland, and in the Canadian province of Alberta up until the 1970s.[79] Furthermore, China's Law on Maternal and Infant Health Care, enacted in 1995, makes premarital medical examination for serious genetic diseases compulsory for all.[80] If the disorder is serious enough, long-term contraception or tubal ligation is used to prevent conception. During pregnancy, prenatal testing is also compulsory, followed by termination if the fetus has a serious genetic defect, though it is unclear how widespread this practice will be among Chinese women.

Is all of this the malign legacy of Francis Galton? On the contrary, though Galton coined the term "eugenics" and campaigned vigorously in its behalf, it should be recognized that he was simply extrapolating Darwin's theory of evolution through natural selection to mankind. Since human beings were no longer subject to its force and since artificial selection had worked so successfully in establishing the many different breeds of domestic animals, a logical conclusion, given Galton's time and place, was that application of a benign selection regime to human beings ought eventually to yield a much fitter race. He would have been horrified had he known that within little more than 20 years of his death forcible sterilization and murder would be carried out in the name of eugenics, for Galton was not a mean or vindictive man. He was an ever-curious, inventive Victorian scientist whose contributions should be thought of in the context of the nineteenth-century world of British science in which he lived and worked. He is largely responsible for the development of fingerprinting as a forensic method and he made important contributions to psychology, especially in the case of mental imagery. He was an explorer, travel writer, and discoverer of the anticyclone. His legacy to modern human genetics includes pedigree analysis and twin studies. He discovered correlation and regression, and helped to found and nurture the statistical methods that today have extremely broad applications in many fields including human genetics. He was in the end a self-confident, optimistic man who was one of the first to bring quantitative methods into biology.

NOTES

Abbreviations

DNB, *Dictionary of National Biography*. Oxford: Oxford University Press.
DSB, *Dictionary of Scientific Biography*. New York: Charles Scribner's & Sons.
GA, Galton Archive, Manuscripts Room, University College, London.
Life, Karl Pearson, *The Life, Letters and Labours of Francis Galton*. I (1914), II (1924), vols. IIIA and IIIB (1930), Cambridge: Cambridge University Press.
Memories, Francis Galton, *Memories of my Life*, Third Edition. London: Methuen, 1909.
Victorian Genius, D. W. Forrest, *Francis Galton: The Life and Work of a Victorian Genius*. New York: Taplinger Publishing Co., 1974.

Preface

1. *Life*, vols. I, II, IIIA, and IIIB.
2. *Victorian Genius*.
3. Adrian Desmond, *Huxley: From Devil's Disciple to Evolution's High Priest* (Reading: Addison-Wesley, 1997), 617.

Prologue: Francis Galton in Perspective

1. Paul Gray, "Cursed by Eugenics," *Time* 153 (no. 1, 1999): 84–85.
2. Charles R. Darwin, *On the Origin of Species by Means of Natural Selection, or the Preservation of Favored Races in the Struggle for Life* (London: Murray, 1859).
3. Philip R. Reilly, *The Surgical Solution: A History of Involuntary Sterilization in the United States* (Baltimore: Johns Hopkins University Press, 1991).
4. Robert N. Proctor, *Racial Hygiene: Medicine under the Nazis* (Cambridge, Mass.: Harvard University Press, 1988).
5. Paul Weindling, *Health, Race and German Politics between National Unification and Nazism, 1870–1945* (Cambridge: Cambridge University Press, 1989), 539.
6. Francis Galton, *The Art of Travel; or Shifts and Contrivances Available in Wild Countries* (London: Murray, 1855; Phoenix Press, 2001).
7. Francis Galton, "Hereditary Talent and Character," *Macmillan's Magazine* 12 (1865): 157–66, 318–27.
8. *Life*, II: 86.
9. Francis Galton, *Hereditary Genius* (London: Macmillan, 1869).
10. Charles R. Darwin, *The Descent of Man, and Selection in relation to Sex*, 2 vols. (London: Murray, 1871).
11. Alphonse de Candolle, *Histoire des Sciences et des Savants depuis Deux Siècles* (Geneva: Georg, 1873).
12. Francis Galton, *English Men of Science: Their Nature and Nurture* (London: Macmillan, 1874).
13. Raymond Fancher, *Pioneers of Psychology* (New York: W. W. Norton & Co., 1979), 276.
14. Francis Galton, "The History of Twins, as a Criterion of the Relative Powers of Nature and Nurture," *Fraser's Magazine* 12 (1875): 566–76.
15. Lawrence Wright, "Double Mystery," *The New Yorker* (August 1995): 46–62.
16. Charles R. Darwin, *The Variation of Animals and Plants under Domestication*. 2 vols. (London: J. Murray, 1868).
17. *Life*, IIIA: 340–41.
18. Peter L. Bernstein, *Against the Gods: The Remarkable Story of Risk* (New York: John Wiley & Sons Inc., 1996), chap. 7.
19. Francis Galton, *Inquiries into Human Faculty and its Development* (London: Macmillan, 1883).
20. Web site http//ash.gene.ucl.ac.uk/face/
21. Robert H. Logie and Michel Denis, eds., "Mental Images in Human Cognition," *Advances in Psychology* 80 (Amsterdam: North Holland, 1991).
22. Graham Dean and Peter E. Morris, "Imagery and Spatial Ability: When Introspective Reports Predict Performance," *Advances in Psychology* 80 (1991): 331–47.
23. Francis Galton, *Finger Prints* (London: Macmillan, 1892).
24. Francis Galton, *Finger Print Directories* (London: Macmillan, 1895).
25. Francis Galton, "Co-relations and their Measurements, chiefly from Anthropometric Data," *Proc. of the Royal Society* 45 (1888): 135–45.
26. Francis Galton, *Natural Inheritance* (London: Macmillan, 1889).
27. William Bateson, *Materials for the Sudy of Variation treated with especial regard to Discontinuity in the Origin of Species* (London: Macmillan, 1894).

28. Francis Galton, "The Average Contribution of Each Several Ancestor to the Total Heritage of the Offspring," *Proc. of the Royal Society* 61 (1897): 401–13.
29. Ronald A. Fisher, "The Correlation Between Relatives on the Supposition of Mendelian Inheritance," *Trans. of the Royal Society of Edinburgh* 52 (1918): 399–433.

Chapter 1: An Enviable Pedigree

1. *Life*, I: 60.
2. Desmond King-Hele, *Doctor of Revolution: The Life and Genius of Erasmus Darwin* (London: Faber & Faber, 1977). The account of Erasmus Darwin's life given here is taken largely from this sympathetic and highly readable biography. Specific page citations are usually given only when passages are quoted from the book.
3. Desmond King-Hele, *Essential Writings of Erasmus Darwin* (London: MacGibbon & Kee, 1968), 46.
4. Robert E. Schofield, *The Lunar Society of Birmingham* (London: Oxford University Press, 1963). This work describes the formation of the Lunar Society from its predecessor the Lunar Circle and gives highly informative biographies of its talented membership. Sketches of Lunar Society members given in the present biography are drawn either from this book or from King-Hele's, *Doctor of Revolution*. Specific page citations are only given when passages are quoted from the book.
5. King-Hele, *Doctor of Revolution* 88–89.
6. This account of *Zoonomia* is based largely on Desmond King-Hele's *Essential Writings of Erasmus Darwin*.
7. Ibid., 87.
8. King-Hele, *Doctor of Revolution*, 245.
9. King-Hele, *Essential Writings*, 163–64.
10. *Life*, I: 13.
11. Ibid.
12. King-Hele, *Doctor of Revolution*, 314.
13. D. B. Wyndam Lewis and Charles Lee, *The Stuffed Owl: An Anthology of Bad Verse* (New York: Dutton, Everyman's Library, 1978).
14. *Life*, I: 22–25.
15. Francis Galton's ancestry on his paternal side is described in much interesting detail by Karl Pearson in *Life*, I: 26–54, which is the basis for the account presented in this chapter. Specific pages are only cited when quotations are given.
16. See the quotation in *Life*, I: 43.
17. Factors used to convert the value of the pound in the nineteenth century to its current value vary to some degree with the author, but are in the range of 38 to 50-fold so, in this book, I have usually given this spread in valuation.

18. See Schofield, *The Lunar Society of Birmingham* 194–204 on Priestley.
19. *Life*, I: 45.
20. Schofield, *The Lunar Society of Birmingham*, 221.
21. *Life*, I: 52–53.
22. Ibid., 52.
23. Ibid., 58.

Chapter 2: Metamorphosis: From Birth to Medical School

1. *Memories*, 22.
2. *Life*, I: 62.
3. Ibid., 63.
4. *Life*, I: 63.
5. Ibid., 63.
6. Ibid., 63.
7. Ibid., 63.
8. Ibid., 63.
9. Ibid., 63.
10. Violetta Galton's Diary, *GA*, List No. 53.
11. *Life*, I: 64.
12. Ibid., 65.
13. Ibid., 67.
14. *GA*, List No. 52.
15. *Memories*, 16–18.
16. *Life*, I: 71.
17. *GA*, List No. 108B.
18. *Memories*, 16–17.
19. *Life*, I: 77.
20. Ibid., 116.
21. Ibid., 81.
22. Sheldon Rothblatt, *The Revolution of the Dons* (Cambridge: Cambridge University Press, 1981), 53.
23. *Life*, I: 81–90 gives an account of Francis Galton at King Edward's School.
24. *DNB*, 10: 809–10.
25. *Memories*, 20.
26. Ibid., 22–23.
27. A. J. Youngson, *The Scientific Revolution in Victorian Medicine* (New York: Holmes & Meier Publishers, Inc., 1979), 12–13.
28. *Life*, I: 92–98 gives an account of the European tour.
29. *DNB*, 22: 242–43.
30. *Life*, I: 95.
31. Ibid., 95.
32. Ibid., 95.
33. Ibid., 95.
34. Ibid., 96.
35. Ibid., 96.
36. Ibid., 96.
37. Erwin C. Lessner and Ann M. Lingg Lessner, *The Danube: The Dramatic History of the Great River and the People Touched by Its Flow* (Garden City, New York: Doubleday, 1961), 418.

38. *Life*, I: 97.
39. *Memories*, 25.
40. *Life*, I: 97.
41. Youngson, *The Scientific Revolution in Victorian Medicine* 24–26.
42. Ibid., 28–30.
43. *Memories*, 34–35.
44. Youngson, *The Scientific Revolution in Victorian Medicine*, 32–34.
45. *Life*, I: 100.
46. Ibid., 102.
47. Ibid., 102.
48. *Memories*, 27–28.
49. Ibid., 34.
50. Ibid., 31.
51. Ibid., 39–42 describes Francis Galton's studies at King's College.
52. *DNB*, 15: 432–33.
53. *Memories*, 39.
54. *Life*, I: 106.
55. Ibid., 110–11.
56. Ibid., 116–18.
57. Ibid., 119.
58. *Memories*, 42–43.
59. *Life*, I: 107.
60. Janet Browne, *Charles Darwin: Voyaging* (New York: Alfred A. Knopf, 1995). Chapter 17 describes Darwin's life at Macaw cottage.
61. See Ibid., 42–64 for an excellent description of Charles Darwin's introduction to medicine.
62. Ibid., 55.
63. Ibid., Chapter 4.
64. Ibid., 117.
65. *Life*, I: 110.
66. Youngson, *The Scientific Revolution in Victorian Medicine*, 13.
67. *Life*, I: 110.
68. Ibid., 110.
69. Ibid., 110.
70. Ibid., 110.
71. *Life*, I: 113.
72. *DNB*, 13, 429–30.
73. See *Life* I: 126–39 for description and letters of Galton's summer trip to Europe and Asia Minor.
74. Ibid., 131.
75. Ibid., 130.
76. *DNB*, 13, 429–30.
77. Ibid., 132.
78. Lessner and Lingg Lessner, *The Danube*, 15.
79. *Life*, I: 134.
80. *GA*, List No. 66.
81. Ibid.
82. *Life*, I: 135.
83. *GA*, List No. 66.
84. *Memories*, 51.
85. *Life*, I: 136.
86. Ibid., 136.
87. *Memories*, 51.
88. *Life*, I: 136.
89. Taken from the account of Dr. A. Neale in 1806, which was accurate, in the main, as late as 1875, according to *A Handbook for Travellers in Turkey in Asia* (London: John Murray, 1875), 64–65.
90. Ibid., 64–65.
91. Ibid., 64–65.
92. Ibid., 114–15.
93. *Life*, I: 138.
94. *Memories*, 54–55.

Chapter 3: A Poll Degree from Cambridge

1. *Memories*, 60.
2. This account of the beginnings of Trinity College is taken from George M. Trevelyan, *Trinity College, An Historical Sketch* (Cambridge: Cambridge University Press, 1943), 3–12.
3. This account of the functions of coaches, fellows, professors, and tutors is based on the fine account in chapter 6 "Donnishness," in the book by Sheldon Rothblatt, *The Revolution of the Dons* (Cambridge: Cambridge University Press, 1981).
4. Ibid., chapter 5 for a description of Cambridge examinations and degrees.
5. Ibid., chapter 5.
6. *Memories*, 59.
7. Ibid.
8. Janet Browne, *Charles Darwin: Voyaging* (New York: Alfred A. Knopf, 1995), 117.
9. See *Life*, I: 149–50 and plate 51 for a description of Galton's rooms.
10. Trevelyan, *Trinity College*, 91–92.
11. *Victorian Genius*, 20.
12. *Life*, I: 142.
13. Ibid., 77.
14. *GA*, List No. 68.
15. *DNB*, 14: 764.
16. *Life*, I: 144–45.
17. The relationship between mathematics and athletics in Victorian Cambridge is discussed in an interesting article by Andrew Warwick, "Exercising the Student Body," in *Science Incarnate: Historical Embodiments of Natural Knowledge*, ed. Christopher Lawrence and Steven Shapin (Chicago: University of Chicago Press, 1998), chap. 8.
18. *Life*, I: 144.
19. *Life*, I: 145.
20. Ibid., 149–510.
21. Ibid., 151.
22. Ibid., 152–53.
23. Ibid., 153.
24. Ibid., 154.
25. Rothblatt, *The Revolution of the Dons*, 198–204; *DNB* 9: 1233–34.

26. See *Life*, I: 155–61 for an account of the summer at Browtop.
27. Ibid., 157.
28. *DNB*, 20, 1365–74.
29. See Allison Winter, "A Calculus of Suffering: Ada Lovelace and the Bodily Constraints on Women's Knowledge in Early Victorian England," in Lawrence and Shapin, ed. *Science Incarnate*, chap. 6.
30. *DNB*, 20: 1371.
31. Trevelyan, *Trinity College*, 91–95 for an account of Christopher Wordsworth's years as Master of Trinity College.
32. *Life*, I: 158.
33. *Memories*, 60.
34. *Life*, I: 160–62.
35. Ibid., 163.
36. Rothblatt, *The Revolution of the Dons*, 190–92.
37. Ibid., 191.
38. Trevelyan, *Trinity College* 96–102 for an account of Whewell's years as Master of Trinity College.
39. *DNB*, 8: 982–83.
40. *DNB*, 12: 787–90.
41. *DNB*, 22: 487–88.
42. *DNB*, 19: 472–74.
43. *Life*, I: 164.
44. *GA*, List No. 68.
45. *DNB*, 10: 1137–38.
46. *DNB*, 3: 557–58.
47. *Life*, I: 165.
48. Ibid., 166.
49. *DNB*, 22: 401–2.
50. *DNB*, 20: 202–3.
51. *Life*, I: 168.
52. *GA*, List No. 67.
53. Ibid.
54. *Memories*, 73.
55. Ibid., 73.
56. *Life*, I: 169.
57. Ibid., 170.
58. *GA*, List No. 56.
59. *Life*, I: 171–72.
60. *Life*, I: 176.
61. Rothblatt, *The Revolution of the Dons*, 182–83.
62. *GA*, List No. 66.
63. *Life*, I: 192.
64. Ibid., 192.

Chapter 4: Drifting

1. *Life*, I: 210.
2. Ibid., 196.
3. *Memories*, 74–75.
4. *Life*, I: 51–52.
5. Ibid., 193.
6. *Memories*, 84–85.
7. Ibid., 85.
8. *Life*, I: 198–99.
9. *Memories*, 85.
10. Ibid., I: 200.
11. Ibid., 199–203; *Memories*, 86–100. Unfortunately, Galton failed to describe his impressions of the city in his two accounts of the journey up the Nile, both written long after the trip.
12. Amelia A. B. Edwards, *A Thousand Miles up the Nile* (London: Century, reissued 1989), 4–5. Edwards gives a vivid account of an adventure up the Nile made 28 years after Galton's and so I have paraphrased her descriptions to infer what Galton and his friends saw and encountered in Cairo and on the first part of their Nile journey.
13. Robert O. Collins and Robert L. Tignor, *Egypt and the Sudan* (Englewood Cliffs, N. J.: Prentice Hall, Inc., 1967), 53–59.
14. Karl Baedeker, *Egypt and Sˊdˏn*, eighth revised edition, (Leipzig: Karl Baedeker, 1929), 73.
15. Edwards, *A Thousand Miles up the Nile*, 23.
16. Ibid., 39.
17. Ibid., 11–12.
18. *Memories*, 86.
19. *GA*, List No. 94.
20. *Life*, I: 200.
21. *GA*, List No. 95.
22. Edwards, *A Thousand Miles up the Nile* 194–97.
23. *Life*, I: 200.
24. Paul Santi and Richard Hill, *The Europeans in the Sudan 1834-1878* (Oxford: Oxford at the Clarendon Press, 1980), 52–53. By 1835 Muhammad Ali Pasha had become convinced that he must employ European engineers and technologists if he was going to carry out prospecting and extraction of the gold in a systematic fashion. In 1836 a Piedmontese metallugist named Boreani in the service of the Egyptian government was instructed to accompany a group of Austrian mining experts led by Josef Russegger to survey the Blue Nile gold deposits. Russegger sent back an optimistic report, which was contradicted by Boreani whose instructions were to check on Russegger's findings. A frustrated Muhammad Ali Pasha decided to visit the gold workings himself, so he travelled down the Blue Nile to the village of Kiri near the gold deposits and close to the Ethiopian border.
25. *Life*, I: 200.
26. Ibid., 201.
27. Ibid., 201.
28. Alan Moorehead, *The Blue Nile* (New York: Harper and Row, 1962), 156.
29. *Memories*, 91.
30. Santi and Hill, *The Europeans in the Sudan 1834-1878*, 75.
31. Ibid., 81–82.
32. Ibid., 81–82.
33. Ibid., 3; *Memories*, 93–94.

34. Santi and Hill, *The Europeans in the Sudan 1834-1878*, 79.
35. Ibid., 79.
36. *Memories*, 92-95; *Life*, I: 201-2.
37. *Memories*, 93.
38. Ibid., 95-97; *Life*, I: 202-03.
39. *GA*, List No. 94.
40. *Memories*, 101-9; *Life*, I: 203-5.
41. *Life*, IIIB: 454, *GA*, List No. 93.
42. Ibid., 454-55, Ibid.
43. *Cooks' Tourist's Handbook for Palestine and Syria* (London: Thos. Cook and Co., 1911), 328-29.
44. Ibid., 67-70.
45. *Memories*, 110-20.
46. *DNB*, 14, 15-16.
47. *Memories*, 114.
48. *Life*, I: 156-57.
49. Ibid., 157.
50. Ibid., 157.
51. Ibid., 180.
52. *GA*, List No. 81.
53. *Victorian Genius*, 37.
54. *Memories*, 119.
55. *Life*, I: 209.
56. Ibid., 212.

Chapter 5: South Africa

1. *Memories*, 122.
2. Ibid., 122.
3. Tim Jeal, *Livingstone* (New York: G. P. Putnam's & Sons, 1973), 93.
4. *DNB*, 22 (supplement), 691-94.
5. *Life*, I: 214-15.
6. Robert A. Stafford, *Scientist of Empire* (Cambridge: Cambridge University Press, 1989), 153.
7. Ibid., 4-32.
8. *Memories*, 123.
9. Charles G. Andersson, *Lake Ngami; or Four Years in Africa* (Philadelphia: John E. Potter, 1885). Charles Andersson's account of Galton's expedition to what is now Namibia is one of the two describing the trip the other being Galton's own work, *Travels in South Africa*, these works have been used as primary sources for chapter 5-7. Page references are only cited in the case of quoted passages. The two books are complementary in emphasis, Andersson being aware of the plants and animals around him while Galton concentrates on leading the expedition and making quantitative observations when possible.
10. *DNB*, 25: 245.
11. Francis Galton, *Narrative of an Explorer in Tropical South Africa*, second edition (London: Ward, Lock & Co., 1889), 2. See comment under 9 above.
12. *GA*, List No. 100.

13. *Memories*, 125.
14. *Life*, I: 218-19.
15. *Memories*, 124.
16. Ibid.
17. Andersson, *Lake Ngami; or Four Years in Africa*.
18. Galton, *Narrative of an Explorer in Tropical South Africa*, 3.
19. Andersson, *Lake Ngami; or Four Years in Africa*.
20. *Life*, I: 218-219.
21. Galton, *Narrative of an Explorer in Tropical South Africa*, 3.
22. Andersson, *Lake Ngami; or Four Years in Africa*, 5.
23. Ibid., 8.
24. Joseph H. Lehmann, *Remember you are an Englishman: A Biography of Sir Harry Smith, 1787-1860* (London: Jonathan Cape, 1977). My brief account of Sir Harry Smith's career is based on this book.
25. Harry Smith, *The Autobiography of Lieutenant General Sir Harry Smith* (London: John Murray, 1901), I: 200.
26. Ibid., 200.
27. Ibid., 200.
28. *Life*, I: 219-20.
29. Galton, *Narrative of an Explorer in South Africa*.
30. Andersson, *Lake Ngami; or Four Years in Africa*, 53.

Chapter 6: Making Peace with Jonker Afrikaner

1. Francis Galton, *Narrative of an Explorer in Tropical South Africa* (London: Ward, Lock and Co., 1889), 7. See comment under 9, chapter 5.
2. I. Goldblatt, *History of Southwest Africa from the Beginning of the Nineteenth Century* (Cape Town: Juta & Co. Ltd., 1971), 12.
3. Galton, *Narrative of an Explorer in Tropical South Africa*, 10.
4. *GA*, List No. 97.
5. *GA*, List No. 99B.
6. Charles J. Andersson, *Lake Ngami; or Four Years in Africa* (Philadelphia: John E. Potter, 1885), 18. See comment 9, chap. 5.
7. Galton, *Narrative of an Explorer in Tropical South Africa*, 29-30.
8. Andersson, *Lake Ngami, or Four Years in Africa*, 39.
9. Ibid., 39.
10. Ibid., 39.
11. Ibid., 40.
12. Olga Levinson, *Story of Namibia* (Capetown: Tafelberg Publishers, Ltd., 1978), 21-32.

13. Ibid., 21–32.
14. Ibid., 33–35.
15. Galton, *Narrative of an Explorer in Tropical South Africa*, 43.
16. *Victorian Genius*, 44.
17. Andersson, *Lake Ngami; or Four Years in Africa*, 58.
18. Ibid., 59–60.
19. Ibid., 61.
20. Galton, *Narrative of an Explorer in Tropical South Africa*, 53.
21. Ibid., 54.
22. Ibid., 54.
23. Ibid., 54.
24. Ibid., 57.
25. *Life*, I: 227–28.
26. Galton, *Narrative of an Explorer in Tropical South Africa*, 70.
27. Levinson, *Story of Namibia*, 36–37.

Chapter 7: Expedition to Ovampoland

1. Olga Levinson, *Story of Namibia* (Capetown: Tafelberg Publishers, Ltd., 1978), 15. Note that there are several different spellings for Ovampo including Ovambo and Owambo. I have chosen the version used by Galton.
2. Francis Galton, *Narrative of an Explorer in Tropical South Africa* (London: Ward, Lock and Co., 1889); and Charles J. Andersson, *Lake Ngami: or Four Years in Africa* (Philadelphia: John E. Potter, 1885?), 39–40. See comment 9, chap. 5.
3. Andersson, *Lake Ngami; or Four Years in Africa*, 104.
4. Galton, *Narrative of an Explorer in Tropical South Africa*, 80.
5. Ibid., 81.
6. Ibid., 81.
7. Ibid., 82.
8. Francis Galton, *Hereditary Genius*, Second American Edition (New York: Appleton, 1879), 339.
9. Galton, *Narrative of an Explorer in Tropical South Africa*, 89.
10. Galton, *Hereditary Genius*, 339.
11. Ibid., 339.
12. Galton, *Narrative of an Explorer in Tropical South Africa*, 85.
13. Andersson, *Lake Ngami; or Four Years in Africa*, 117.
14. Galton, *Narrative of an Explorer in Tropical South Africa*, 93.
15. Ibid., 94.
16. Andersson, *Lake Ngami; or Four Years in Africa*, 119.
17. *GA*, List No. 97.
18. Galton, *Narrative of an Explorer in Tropical South Africa*, 110.
19. *Life*, IIIB, plate 59.
20. Andersson, *Lake Ngami; or Four Years in Africa*, 140.
21. Levinson, *Story of Namibia*, 118.
22. Galton, *Narrative of an Explorer in Tropical South Africa*, 125.
23. *GA*, List No. 99A.
24. Galton, *Narrative on an Explorer in Tropical South Africa*, 127.
25. This description of the organization of the Ovampo kingdom and Nangoro's history is taken from Frieda-Nela Williams, *Precolonial Communities of Southwestern Africa: A History of Owambo Kingdoms 1600–1920*, Archeia No. 16 (Windhoek: National Archives of Namibia, 1991). Note that there is one important discrepancy between F.-N. Williams history and the accounts given by Galton and Andersson. According to Williams, Nangoro, Nangolo as she calls him, spared his stepbrother Shipanga who succeeded him on the throne of the Ndonga kingdom. The account given by Andersson and Galton, in contrast mentions Chipanga, the daughter of Nangoro's sister-in-law, as the heiress to the throne. Chipanga does not seem to be mentioned by F.-N. Williams. Since I have not been able to resolve these discrepant accounts, I have chosen to follow the narratives of Galton and Andersson, although I strongly suspect they may be wrong, as it is difficult to believe that Nangoro would have offered his potential successor as a temporary wife to Galton.
26. Galton, *Narrative of an Explorer in Tropical South Africa*, 130.
27. Ibid., 132.
28. Ibid., 132.
29. Ibid.,
30. *GA*, List No. 99B.
31. *Memories*, 143. For some reason Galton is coy about discussing this event in his *Narrative of an Explorer in Tropical South Africa*, but has no such hesitation in *Memories*.
32. Galton, *Narrative of an Explorer in Tropical South Africa*, 139.
33. Ibid., 140.
34. Ibid., 141.
35. Andersson, *Lake Ngami; or Four Years in Africa*, 167.
36. Galton, *Narrative of an Explorer in Tropical South Africa*, 166.
37. Ibid., 177.
38. Levinson, *Story of Namibia*, 15–16.
39. Ibid., 15–16.
40. Ibid., 36.
41. Galton, *Narrative of an Explorer in Tropical South Africa*, 195–99. This appendix to the book is extracted from the *Proc. of the Royal Geographical Society* (1858) and describes the journey of the two

missionaries and Green to Ovampoland including Green's letter and Galton's public comments following the reading of Green's letter.
42. Levinson, *Story of Namibia*, 17.
43. Galton, *Narrative of an Explorer in Tropical South Africa*, 198.

Chapter 8: Fame and Marriage

1. *Life*, I: 241.
2. Francis Galton, "Recent expedition into the Interior of South-western Africa," *J. of the Royal Geographical Society* 22 (1852): 140–63.
3. Ibid., 140–63.
4. *Memories*, 152.
5. Edward Graham, *The Harrow Life of Henry Montagu Butler* (London: Longmans, Green, and Co., 1920), chap. 1 and 2.
6. *Victorian Genius*, 55–56.
7. *GA*, List No. 56. Contains his appointment books for winter and spring, 1853.
8. Patrick Beaver, *The Crystal Palace* (London: Hugh Evelyn, 1970).
9. *Illustrated London News* 21 (August 7, 1852): 94–95; 23 (November 5, 1853): 383.
10. *Life*, I: 240.
11. Hugh R. Mill, *The Record of the Royal Geographical Society, 1830–1930* (London: Royal Geographical Society, 1930), 19–21, 48–49.
12. *J. of the Royal Geographical Society* 23 (1853):ix and lviii–lix.
13. Ibid., ix and lviii–lix.
14. Ibid., ix and lviii–lix.
15. Ibid., ix and lviii–lix.
16. *Life*, I: 241.
17. *Westminster Review* 4 (1853: 270.
18. *Life*, I: 240.
19. *Victorian Genius*, 58.
20. *GA*, List No. 53 contains first part of Louisa Galton's yearly diary.
21. There were "scarcely two dozen books in Galton's library as we now have it which we can assert he must have purchased to forward his own work. There are masses of measurements and observations of Galton's own, but unlike Darwin he did not start by analysing published material." *Life*, II: 12. Pages 11–12 also describe Galton's house.
22. *Memories*, 158.
23. Biographies of all individuals described herein are in the *DNB* as follows: Sir Rutherford Alcock, 22: 29–30; Thomas Atkinson, 22: 84–85; George Bentham, 2: 263–67; Sir Richard Burton, 22: 349–56; James Fergusson, 6: 1224–26; Russell Gurney, 8: 808–9; Sir Joseph Hooker, Twentieth Century (1901–1911): 294–99; Sir John Lubbock, 12: 227–28; Laurence Oliphant, 14: 1027–31; Sir Lewis Pelly, 15: 720–22; Herbert Spencer, Twentieth Century (1901–1911): 360–69; William Spottiswoode, 28: 826.
24. *Memories*, 169.
25. Mill, *The Record of the Royal Geographical Society, 1830–1930*, 244.
26. Francis Galton, *The Art of Travel; or, Shifts and Contrivances Available in Wild Countries*, reprint of the Fifth Edition (London: Phoenix Press, 2001).
27. Dorothy Middleton, Ibid., "Introduction."
28. *GA*, List No. 56.
29. *Victorian Genius*, 59.
30. *GA*, List No. 190. Letter from Sir Samuel White Baker.
31. Galton, *The Art of Travel*, 53–65.
32. Ibid., 36.
33. Ibid., 295.
34. Ibid., 296.
35. Ibid., 308.
36. Ibid.
37. Ibid., 309.
38. Ibid., 314.
39. *Blackwood's Magazine* 79 (1856): 597.
40. *Westminster Review* 7 (1855): 587.
41. *Chamber's Journal* 23 (1855): 374.
42. *Memories*, 163.
43. This outline of Crimean War events is drawn from A. J. Barker, The Vainglorious War 1854–56 (London: Weidenfeld & Nicolson, 1970).
44. *Life*, II: 14–18.
45. *Times*, 25 September, 1855.
46. Ibid.
47. Francis Galton, "A Visit to Northern Spain at the Time of the Eclipse in 1860," *Vacation Tourists and Notes of Travel in 1860*, ed. Francis Galton (London: Macmillan, 1861), 422–454.
48. *North American Review* 92 (1861): 271.
49. *North American Review* 95 (1862): 558.
50. *Victorian Genius*, 75–76.
51. Francis Galton, "Sun Signals for the Use of Travellers," *Proc. of the Royal Geographical Society* 4 (1859): 14–19.
52. *Memories*, 165.
53. Francis Galton, "The Exploration of Arid Countries," *Proc. of the Royal Geographical Society* 2 (1857–58): 60–77.
54. Unsigned article by Francis Galton, "Recent Discoveries in Australia," *The Cornhill Magazine* 5 (1862): 354–64.
55. *Life*, II: 67–69.
56. Edward C. Mack and W. H. G. Armytage, *Thomas Hughes: The Life of the Author of Tom Brown's Schooldays* (London: Ernest Benn Ltd., 1952), 124–26.
57. *GA*, List No. 190, letter from Tom Hughes to Francis Galton, January 24, 1863.
58. *Life*, II: 67–69.

59. A. D. Orange, "The Beginnings of the British Association, 1831–185," in *The Parliament of Science: The British Association for the Advancement of Science 1831–1981*, ed. Roy MacLeod and Peter Collins (Northwood: Science Reviews Ltd., 1981), 43–64.

60. Roy MacLeod, "Retrospect: The British Association and its Historians," Ibid., 17–42.

61. Ibid., 24.

62. Ibid., 24.

63. Edward N. da Costa Andrade, *A Brief History of the Royal Society* (London: Royal Society, 1960).

64. Marie B. Hall, *All Scientists Now: The Royal Society in the Nineteenth Century* (Cambridge: Cambridge University Press, 1984).

65. *Memories*, 151.

66. Frank R. Cowell, *The Athenaeum: Club and Social Life in London 1824-74* (London: Heinemann, 1975).

67. Ibid., 94.

Chapter 9: Riding High with the Royal Geographical Society. I. The Great Lakes of Africa

1. *Memories*, 198.

2. Unpublished account of the Royal Geographical Society by Clements Markham in the Royal Geographical Society Archives, 354.

3. Ibid., 352.

4. Francis Galton to Norton Shaw, March 3, 1856; and Francis Galton to N. Shaw letters of November 28, 1856 and December 3, 1856. Royal Geographical Society Archives.

5. Ibid., October 22, 1856.

6. Norton Shaw to Francis Galton, October 24, 1856. Royal Geographical Society Archives.

7. Francis Galton to Norton Shaw, November 28, 1856, Royal Geographical Society Archives.

8. Francis Galton, Nicolay and Clark to President, Royal Geographical Society, December 3, 1856.

9. Francis Galton to Norton Shaw, February 22, 1858. Royal Geographical Society Archives. Note that the correspondence for 1857 is missing.

10. Ibid., July 10, 1858.

11. Ibid., April 4, 1862.

12. *Victorian Genius*, 71.

13. *Memories*, 210.

14. Alan Moorehead, *The White Nile* (New York: Harper & Brothers, 1960). The account of the search for the source of the Nile given in this chapter relies to a great extent on Alan Moorehead's marvelous synthesis published many years ago plus the individual biographies listed below.

15. Ibid., 4.

16. *Proc. of the Royal Geographical Society* 1 (1855–56, 1856–57).

17. Byron Farwell, *Burton: A Biography of Sir Richard Francis Burton* (New York: Holt, Rinehart and Winston, 1963). This is still one of the best biographies of Burton available and has been used as the basis for this account.

18. *Proc. of the Royal Geographical Society* 1 (1855–56, 1856–57).

19. Ibid., 58–60.

20. Ibid., 93.

21. Francis Galton. "The Exploration of Arid Countries," *Proc. of the Royal Geographical Society* 2 (1857–1858): 60–77.

22. Ibid., 75–76.

23. Ibid., 76.

24. Alexander Maitland, *Speke* (London: Constable, 1971), 98. This full-length portrait of John Hanning Speke has been used as the basis for this account.

25. Ibid., 98.

26. *Proc. of the Royal Geographical Society* 3 (1858–1859): 208–10.

27. Ibid., 212.

28. Maitland, *Speke*, 99.

29. *Memories*, 199.

30. Ibid., 199.

31. *Proc. of the Royal Geographical Society* 3 (1858–1859): 217.

32. Ibid., 218.

33. Ibid., 219.

34. Ibid., 305.

35. Ibid., 356.

36. *Victorian Genius*, 72.

37. Ibid., 72.

38. *Proc. of the Royal Geographical Society* 5 (1860–1861): 96–97.

39. Ibid., (1860–1861): 139–40.

40. Fawn M. Brodie, *The Devil Drives: A Life of Sir Richard Burton* (New York: W. W. Norton & Co. Inc., 1967), 220.

41. *GA*, List No. 320, John Hanning Speke letter to Francis Galton, February 26, 1863.

42. Ibid.

43. Ibid.

44. *Proc. of the Royal Geographical Society* 7 (1862–1863): 108–10.

45. Ibid., 110.

46. *Times*, 26 May, 1863.

47. *Proc. of the Royal Geographical Society* 7 (1862–1863): 213–25.

48. Brodie, *The Devil Drives*, 222.

49. Ibid., 222.

50. Ibid., 222.

51. Ibid., 223–24.

52. Farwell, *Burton*.

53. Isabel Burton, *The Life of Captain Sir Richard Francis Burton* (London: Duckworth & Co., 1898), 241.

54. Ibid., 241.
55. *Memories*, 201.
56. *Illustrated London News* 45 (Sept. 24, 1864): 301–11.
57. Ibid., 301–11.
58. Ibid., 301–11.
59. James Tunstall, *Rambles about Bath and its Neighborhood* (London: Simpken & Marshall & Co., seventh edition, 1876).
60. Burton, *The Life of Captain Richard Francis Burton*, 241.
61. *Times*, 16 September, 1864.
62. Ibid.
63. Burton, *The Life of Captain Richard Francis Burton*, 241.
64. *Memories*, 202.
65. *Times*, 17 September, 1864.
66. Ibid.
67. *Illustrated London News* 45 (Sept. 24, 1864): 301–11.
68. *Proc. of the Royal Geographical Society* 9 (1864–1865): 2.
69. Ibid., 6.
70. Ibid., 8.
71. Ibid., 10.
72. Ibid., 8.
73. Ibid., 13.
74. *Memories*, 204; Maitland, *Speke*, 225–28.
75. *Proc. of the Royal Geographical Society* 10 (1865–66): 6–22.
76. Ibid., 22–23
77. Ibid., 22–23.
78. Tim Jeal, *Livingstone* (New York: G. P. Putnam's Sons, 1973). This exhaustive biography of Livingstone has been used as the basis for this account.
79. *Proc. of the Royal Geographical Society* 11 (1866–1867): 111–12.
80. Ibid., 142–43.
81. Ibid., 175–76.
82. Jeal, *Livingstone*, 315.
83. *Proc. of the Royal Geographical Society* 14 (1869–70): 8–16.

Chapter 10: Riding High with the Royal Geographical Society. II. Stanley Faces Off with the Geographers

1. *Memories*, 206.
2. *DNB*, 16: 771–74
3. *Proc. of the Royal Geographical Society* 16 (1871–1872): 81.
4. Ibid., 88.
5. Ibid., 125.
6. Ibid., 185.
7. Frank J. McLynn, *Stanley: The Making of an African Explorer* (London: Constable, 1989). This is an excellent account of Stanley's life from birth through his second great expedition across Africa from Zanzibar to the mouth of the Congo River. It has been relied on extensively, particularly for details of Stanley's life.
8. Ibid., 226.
9. *New York Herald*, 21 May 1872, contains these and other excerpts from British papers.
10. Ibid.
11. *Proc. of the Royal Geographical Society* 16 (1871–1872): 241.
12. *New York Herald*, 10 May 1872.
13. *Proc. of the Royal Geographical Society* 16 (1871–1872): 291–375.
14. Ibid., 370
15. Ibid., 376.
16. Ibid., 419–21.
17. Ibid., 412.
18. Ibid., 419–21.
19. Ibid., 419–21.
20. Ibid., 419–21.
21. Excerpt in the *New York Herald*, 6 July, 1872.
22. Ibid., July 2, 1872.
23. Ibid., July 4, 1872, contains this excerpt.
24. Ibid., July 26, 1872.
25. Ibid., July 27, 1872.
26. Ibid., July 4, 1872.
27. *Daily News*, 2 August, 1872.
28. *Times* 27 July, 1872.
29. McLynn, *Stanley: The Making of an African Explorer*, 205–8.
30. *Proc. of the Royal Geographical Society* 16 (1871–1872): 429–30.
31. *Illustrated London News* 61 (August 17, 1872): 152–54, 162–63.
32. *Daily News*, 16 August 1872.
33. Ibid.
34. Ibid., 17 August 1872.
35. Ibid.
36. Ibid.
37. Ibid.
38. Ibid.
39. Ibid.
40. Ibid.
41. Ibid.
42. Ibid.
43. Ibid.
44. Ibid.
45. Ibid.
46. Ibid.
47. Ibid.
48. Ibid.
49. Ibid.
50. Henry M. Stanley, *How I Found Livingstone: Travels, Adventures, and Discoveries in Central Africa* (New York: Charles Scribner's & Sons, 1899), 687.

51. *Daily News*, 19 August 1872.
52. Ibid.
53. Ibid.
54. Ibid., 22 August 1872.
55. Ibid.
56. Ibid.
57. Ibid.
58. Ibid.
59. *Memories*, 207.
60. McLynn, *Stanley: The Making of an African Explorer*, 214.
61. *Manchester Examiner*, 2 September 1872.
62. Ibid.
63. Ibid.
64. At 18, Stanley set sail as a cabin boy on the packet *Windermere* bound for New Orleans, where he debarked to seek his fortune in America. The next day he offered his services to a distinguished-looking passerby saying "Do you want a boy, sir?" Stanley was given a quick literacy test and a trial period of a week. His benefactor, named Henry Stanley, took him on as a junior clerk and John Rowlands changed his name to Henry Stanley; the Morton came later. See McLynn, *Stanley: The Making of an African Explorer*.
65. *GA*, List No. 76. William B. Carpenter to Galton, September 2, 1872.
66. Ibid.
67. Ibid.
68. Ibid.
69. Ibid.
70. *Proc. of the Royal Geographical Society* 16 (1871–1872): 430.
71. *Victorian Genius*, 119–20.
72. Ibid., 119–120.
73. *Daily News*, 30 August 1872.
74. Ibid.
75. Ibid.
76. Ibid.
77. Ibid., September 2, 1872.
78. Ibid.
79. Ibid.
80. Ibid.
81. Ibid.
82. Ibid., September 4, 1872.
83. Ibid.
84. Ibid.
85. Ibid.
86. McLynn, *Stanley: The Making of an African Explorer*, 371.
87. *GA*, List No. 76. Letter from Sir Clements R. Markham to Francis Galton, which includes Stanley's letter to Markham.
88. Ibid.
89. Ibid.
90. McLynn, *Stanley: The Making of an African Explorer*, 216.
91. *GA*, List No. 76.
92. *Daily News*, 5 September, 1872.
93. Ibid., 6 September 1872.
94. Ibid., 7 September 1872.
95. Ibid.
96. McLynn, *Stanley: The Making of an African Explorer*, 216.
97. Ibid., 216.
98. *Daily News*, 9 September 1872.
99. Ibid.
100. McLynn, *Stanley: The Making of an African Explorer*, 219–20.
101. Ibid., 219–20.
102. *GA*, List No. 76. William B. Carpenter to Francis Galton, September 12, 1872.
103. McLynn, *Stanley: The Making of an African Explorer*, 220.
104. Ibid., 220.
105. Ibid., 220.
106. *Daily News*, 22 October 1872.
107. Ibid.
108. McLynn, *Stanley: The Making of an African Explorer*, 222.
109. Stanley, *How I Found Livingstone*, 687.
110. Ibid., 687.
111. Ibid., 687.
112. *Proc. of the Royal Geographical Society* 16 (1871–1872): 430–33.
113. Ibid.
114. *Daily News*, 17 September 1872.
115. Ibid., 18 September 1872.
116. Ibid.
117. Ibid.
118. Ibid.
119. *Life*, II: 30–31.
120. Ibid., 30–31.
121. Ibid., 30–31.
122. Ibid., 30–31.
123. Ibid., 30–31.
124. *GA*, List No. 76. Note made by Francis Galton to himself on Stanley's origin.
125. *GA*, List No. 76.
126. Using a factor of 50 (see chap. 1, note 17) to convert these numbers to their late twentieth century equivalents, this means that around £25,000 ($41,750) was set aside for research and £2,500 ($4,175) for each lecture. This is not a great deal of money for research by current standards, but an enormous amount for individual scientific lectures where $200 is a generous honorarium.
127. Clements Markham, unpublished history of the Royal Geographical Society, 173–75.
128. Ibid., 449.
129. Ibid., 190.
130. Ibid., 190.
131. Ibid., 449.
132. Hugh R. Mill, *The Record of the Royal Geographical Society, 1830-1930* (London: The Royal Geographical Society, 1930), 249–50.
133. Ibid., 107–12.

134. *DNB*, 23 (1901–1911): 166–68.
135. Mill, *The Record of the Royal Geographical Society, 1830–1930*, 111.

Chapter 11: Weather Maps and the Anticyclone

1. Gisela Kutzbach, *The Thermal Theory of Cyclones: A History of Meteorological Thought in the Nineteenth Century* (Boston: American Meteorological Society, 1979), 58.
2. *Illustrated London News* 22 (October 29, 1853): 362.
3. *Life*, II: 21–22.
4. Ibid., 22.
5. Francis Galton, "On Stereoscopic Maps, taken from Models of Mountainous Countries," *J. of the Royal Geographical Society* 35 (1865): 99–104.
6. *Life*, II: 36–40.
7. Kutzbach, *The Thermal Theory of Cyclones*. The account of the evolution meteorological theory and the making of weather maps is taken largely from this source.
8. H. Landsberg, "Storm of Balaklava and the Daily Weather Forecast," *The Scientific Monthly* 7 (1954): 347–52.
9. A. J. Barker, *The Vainglorious War 1854–56* (London: Weidenfeld & Nicolson, 1970), 196.
10. *Illustrated London News* 25 (Dec. 12, 1854): 576.
11. Ibid.
12. Ibid.
13. Ibid.
14. Ibid., (Dec. 16, 1854): 606.
15. Roger Warren Prouty, *The Transformation of the Board of Trade 1830–1855* (London: William Heinemann Ltd., 1957). Chapter 3 discusses the role of the Board of Trade in regulating merchant shipping and the meteorological office.
16. This sketch of FitzRoy's life is drawn from the book by H. E. I. Mellersh, *FitzRoy of the Beagle* (London: Rupert Hart-Davis, 1968) and from the *DNB*, 7, 207–9.
17. *Illustrated London News* 35 (Nov. 5, 1859): 448.
18. Ibid., 441.
19. Ibid. (Nov. 12, 1859): 467.
20. *Memories*, 230–231.
21. H. Mellersh, *FitzRoy of the Beagle*.
22. *Times*, 18 June 1864.
23. Ibid.
24. Ibid.
25. *Daily News*, 2 May 1865.
26. *GA*, List No. 118/1.
27. Ibid.
28. Ibid.
29. Ibid.
30. Ibid.
31. Ibid.
32. Ibid.
33. Ibid.
34. Ibid.
35. *Life*, II: 53–58.
36. Francis Galton, "Barometric Predictions of Weather," *British Association Report* 40 (1870):31–33. Also in *Nature* 2 (1870): 501–3.
37. *Life*, II:55.
38. Ibid., 58.
39. *DNB*, 17:563–65; *DSB*, 12:49–53.
40. *Memories*, 227.
41. *Life*, II: 80.
42. *Memories*, 154–55.
43. *Victorian Genius*, 86.
44. Ibid.
45. Ibid.
46. *GA*, List No. 53, Louisa Galton's diary.
47. Ibid.
48. *Memories*, 155.

Chapter 12: Hereditary Talent and Character

1. *Life*, I: 1
2. *Memories*, 287–88.
3. Ibid., 287–88.
4. Ibid., 287–88.
5. Francis Galton, "Hereditary Talent and Character," *Macmillan's Magazine* 12 (1865): 157–66, 318–27.
6. See *Macmillan's Magazine* vols. 1, 2, 7, 10, and 12 for the articles mentioned. The discussion of the founding of *Macmillan's Magazine* is taken from Charles Morgan, *The House of Macmillan (1843–1943)* (New York: The Macmillan Company, 1944), chap. 4 and 5.
7. Charles R. Darwin, *On The Origin of Species by Means of Natural Selection, or the Preservation of Favored Races in the Struggle for Life* (J. Murray: London, 1859), 29.
8. *Memories*, 288.
9. Galton, *Macmillan's Magazine*, 12: 161.
10. Ibid., 163.
11. Ibid., 322.
12. Ruth Schwartz Cowan, *Sir Francis Galton and the Study of Heredity in the Nineteenth Century* (Ann Arbor, MI, 1969), 30–31.
13. Ibid., 30.
14. Galton, *Macmillan's Magazine*, 12: 326.
15. *Memories*:289.
16. *Life*, II: 86.
17. Ibid., 87.
18. Charles R. Darwin, *The Variation of Animals and Plants under Domestication*, Second Edition (London: John Murray & Sons, 1875), reprinted in the *Works of Charles Darwin*, ed. Paul H. Barrett and R. B. Freeman (London: William Pickering, 1988), 20.

19. *Memories*, 304.
20. William Spottiswoode, "On Typical Mountain Ranges: An Application of the Calculus of Probabilities to Physical Geography," *J. of the Royal Geographical Society*, 31 (1861): 149–54.
21. Ibid.
22. *Memories*, 304.
23. Stephen M. Stigler, *The History of Statistics: The Measurement of Uncertainty before 1900* (Cambridge: Belknap Press, Harvard University, 1986), 169.
24. Ibid., 206–9.
25. Francis Galton, *Hereditary Genius: An Inquiry into its Laws and Consequences* Revised edition (Appleton: New York: 1879), 10.
26. Ibid., 37.
27. Ibid., 40.
28. Ibid., 42.
29. Ibid., 62.
30. Ibid., 82.
31. Ibid., 82.
32. Ibid., 83.
33. Ibid., 84.
34. Ibid., 84.
35. Ibid., 84.
36. Ibid., chapter 8.
37. Ibid., 131.
38. Ibid., 132.
39. Ibid., 132.
40. Desmond King-Hele, *Doctor of Revolution: The Life and Genius of Erasmus Darwin* (London: Faber & Faber, 1977), 296.
41. Galton, *Hereditary Genius*, 274.
42. Ibid., 282.
43. Ibid., 274.
44. Galton, "Statistical Inquiries into the Efficacy of Prayer," *Fortnightly Review* 12 (1872): 125–35.
45. Edwin Mallard Everett, *The Party of Humanity: The Fortnightly Review and its Contributors 1865-1874* (Chapel Hill: The University of North Carolina Press, 1939). The descriptive material included in the text is abstracted from the first four chapters of this interesting book. The quote from Anthony Trollope is on p. 18.
46. *GA*, List No. 120/4.
47. Galton, *Fortnightly Review* 12: 126.
48. Ibid., 126.
49. Ibid., 127.
50. Ibid., 127.
51. Ibid., 128.
52. Ibid., 134.
53. Ibid., 135.
54. *Spectator*, 3 August 1872, 974–75.
55. Ibid.
56. Ibid.
57. Ibid.
58. Ibid., 17 August 1872, 1038–39.

59. Ibid.
60. *Life*, II: 175.
61. Galton, *Hereditary Genius*, 328.
62. Ibid., 328.
63. Ibid., 338.
64. Ibid., 338.
65. Richard Herrnstein and Charles Murray, *The Bell Curve* (New York: Free Press, 1994).
66. Galton, *Hereditary Genius*, 339.
67. Ibid., 352–53.
68. Ibid., 357.
69. *GA*, List No. 53, Louisa Galton's diary.
70. *Memories*, 290.
71. *Life*, II: 132–33; *DNB*, 18: 144–45.
72. *Life*, II: 132.
73. *The Daily News*, 16 December 1869; *GA*, List No. 120/5.
74. *Times*, 7 January 1870; *GA*, List No. 120/5.
75. Ibid.
76. *Chambers's Journal* 7 (February 10, 1870): 118–22; *GA*, List No. 120/5.
77. Ibid.
78. *Morning Post*, *GA*, List No., 120/5. Date is missing from clipping.
79. *Saturday Review*, December 25, 1869, 832–833; *GA*, List No. 120/5.
80. Ibid.
81. Emel Aileen Gökyigit, "The Reception of Francis Galton's *Hereditary Genius* in the Victorian Periodical Press," *J. History of Biology* 27 (1994): 215–240.
82. Ibid., 231.
83. Ibid., 215–40.
84. Ibid., 221.
85. Ibid., 223–26.
86. Ibid., 229.
87. Francis Galton, *Hereditary Genius: An Inquiry into its Laws and Consequences* (London: Macmillan & Co., reissue of 1892), viii.
88. *Nation*, 6 April 1893; *GA*, List No. 192.
89. Ibid.
90. *Blackburn Standard*, 3 December 1892; *GA*, List No. 192.
91. *National Observer*, 4 February 1893; *GA*, List No. 192.
92. *Daily Chronicle*, 28 October 1892; *GA*, List No. 192.

Chapter 13: Gemmules, Rabbits, Germs, and Stirps

1. *Memories*, 288.
2. Ruth Schwartz Cowan, *Sir Francis Galton and the Study of Heredity in the Nineteenth Century* (Ann Arbor, Michigan, 1969), 96–98; Galton's copy of Charles R. Darwin, *The Variations of Ani-*

mals and Plants under Domestication (London: John Murray & Sons, first edition, 1868), Galton Laboratory, University College, London.

3. See Peter J. Vorzimmer, "Charles Darwin and Blending Inheritance," *Isis* 54 (1963): 371–90 for a discussion of the evolution of Charles Darwin's ideas on inheritance.

4. Charles R. Darwin, *Variation in Animals and Plants under Domestication*, Second edition (London: J. Murray, 1875), reprinted in *The Works of Charles Darwin*, ed. Paul H. Barrett and R. B. Freeman (London: William Pickering, 1988), 20: 321.

5. Ibid., 321.
6. Ibid., 321.
7. Ibid., 328.
8. Ibid., 328.
9. Francis Galton, *Hereditary Genius: An Inquiry into Its Laws and Consequences*, Revised edition (Appleton: New York, 1879), 370.
10. Ibid., 370.
11. Ibid., 371.
12. Adrian Desmond and James Moore, *Darwin* (London: Michael Joseph, 1991), 532.
13. George H. Lewes, "Mr. Darwin's Hypothesis," *Fortnightly Review* 4 n.s. (1868): 492–509.
14. Ibid., 503.
15. Ibid., 503.
16. Ibid., 503.
17. Ibid., 503.
18. Ibid., 507.
19. *Life*, II: 157.
20. Galton, *Hereditary Genius*, 363.
21. *Memories*, 297–98.
22. *Life*, II: 157. All Galton's letters to Darwin about the pangenesis experiments from December 1869 through May 1871 are reproduced on pages 157–163 and the quotations given come from them. According to Pearson, *Life*, II: 156, Darwin's replies to Galton's letters have not been found.
23. Ibid., 158.
24. Ibid., 161.
25. Francis Galton, "Experiments in Pangenesis, by breeding from Rabbits of a pure variety, into whose Circulation Blood taken from other Varieties had previously been Largely Transfused," *Proc. of the Royal Society* 19 (1871): 397.
26. *Memories*, 297–98.
27. Galton, *Proc. of the Royal Society* 19: 393–410.
28. Ibid., 394.
29. Ibid., 394.
30. Ibid., 394.
31. Ibid., 395.
32. Ibid., 404.
33. *Life*, II: 162.
34. Darwin, *Variation of Animals and Plants under Domestication*, 303. There is a long and rueful footnote of Darwin's that manfully summarizes the critiques of pangenesis that came out following publication of the first edition of *Variation* in 1868.

35. *Life*, II: 162.
36. Ibid., 162.
37. Ibid., 163.
38. Ibid., 163.
39. Ibid., 164.
40. Ibid., 164.
41. Ibid., 165.
42. Ibid., 168–70.
43. Ibid., 175.
44. Ibid., 169.
45. Francis Galton, "On Blood-relationship," *Proc. of the Royal Society* 20 (1872): 392–402.
46. Cowan, *Sir Francis Galton and the Study of Heredity in the Nineteenth Century*. Fortunately both Dr. Schwartz Cowan and Pearson, *Life*, II: 170–74, have made valiant attempts to decipher the meaning of "On Blood-relationship," for which this author is most grateful.
47. Darwin, *Variation of Animals and Plants under Domestication*, 338.
48. Ibid., 391–92.
49. Today we would explain reversion in terms of recessive mutations carried in stocks that sometimes segregate to yield occasional individuals that are pure (homozygous) for the recessive.
50. Galton, *Proc. of the Royal Society* 20: 396.
51. *Life*, II: 181.
52. The first definition of germ in a recent edition of *The New Shorter Oxford English Dictionary* (1993) is a "part of a living organism which is capable of developing into a similar organism or part of one." The second definition is even more explicit. "A thing from which something may spring or develop; an elementary principle; a rudiment."
53. *Life*, II: 182.
54. Ibid., 183.
55. Ibid., 183.
56. Ibid., 184.
57. Patricia Thomas Srebrnik, *Alexander Strahan, Victorian Publisher* (Ann Arbor, Michigan: The University of Michigan Press, 1986). This book has an excellent account of the history of the *Contemporary Review*.
58. Francis Galton, "A Theory of Heredity," *Contemporary Review* 27 (1875): 80–95.
59. Ibid., 81.
60. Ibid., 82.
61. Ibid., 83.
62. Ibid., 84.
63. Ibid., 84.
64. Ibid., 94.
65. *Life*, IIIA, 340–41.
66. *Life*, II: 187.
67. Ibid., 187.
68. Ibid., 187.

69. Ibid., 189.
70. Ibid., 189.
71. Ibid., 189.
72. Ibid., 190.
73. Francis Galton, "A Theory of Heredity," *J. of the Anthropological Institute* 5 (1875): 329–48.
74. *Life*, II: 187.
75. *Life*, II: 192.
76. Francis Galton, *Hereditary Genius: An Inquiry into Its Laws and Consequences* (London: Macmillan and Co., reissue of 1892), xiv–xv.
77. This account of Mendel's life is drawn from Robert Olby, *Origins of Mendelism* (Chicago: The University of Chicago Press, 2nd ed., 1985), chapter 5.
78. Gregor Mendel, "Versuch über Pflanzen-Hybriden," *Verhandlungen des naturforschenden Vereines in Brünn* 4 (1865): 3–47.
79. Olby, *Origins of Mendelism*, 216–19.
80. Ibid., 103.

Chapter 14: Nature and Nurture

1. Francis Galton, "The History of Twins, as a Criterion of the Relative Powers of Nature and Nurture," *Fraser's Magazine* 12 (1875): 566–76.
2. Francis Galton, *Hereditary Genius: An Inquiry into its Laws and Consequences* (New York: Appleton, New and Revised American Edition, 1879), 210.
3. Raymond Fancher, "Alphonse de Candolle, Francis Galton, and the Early History of the Nature-nurture Controversy," *J. History of Behavioral Sciences* 19 (1983): 341–51.
4. See Ibid., 345. Translation of specific passages of de Candolle's are quoted here and below from Fancher's paper. These have been amplified in a couple of instances. Also the quotes from Galton's private notebook and the Darwin quote are found in this article.
5. Ibid., 345.
6. Ibid., 345.
7. Ibid., 345.
8. Ibid., 345–46.
9. Ibid., 347.
10. Ibid., 346.
11. Ibid., 345.
12. Ibid., 347.
13. *Life*, II: 135.
14. Ibid., 135.
15. Ibid., 135.
16. Fancher, *J. History of Behavioral Sciences* 19: 346.
17. *Life*, II: 136.
18. Fancher, *J. History of Behavioral Sciences* 19: 348.
19. *Life*, II: 136.
20. Fancher, *J. History of Behavioral Sciences* 19: 348.
21. Francis Galton, "On the Causes which Operate to Create Scientific Men," *Fortnightly Review* 13 (1873): 345–51.
22. Ibid., 346.
23. Ibid., 346.
24. Ibid., 346.
25. Ibid., 348.
26. Ibid., 351.
27. Ibid., 351.
28. Fancher, *J. History of Behavioral Sciences* 19: 349.
29. Francis Galton, *English Men of Science: Their Nature and Nurture* (New York: Appleton, 1875).
30. *Life*, II: 178.
31. Raymond Fancher, "A Note on the Origin of the Term 'Nature and Nurture,'" *J. History of Behavioral Sciences* 15 (1979): 321–22.
32. Galton, *English Men of Science*, 9.
33. Ibid., 9.
34. Ibid., 9.
35. Ibid., 88.
36. Ibid., 169.
37. Ibid., 189–90.
38. Raymond Fancher, *Pioneers of Psychology* (New York: W.W. Norton, 1979), 276.
39. Galton, *Fraser's Magazine* 12: 566.
40. Ibid., 567.
41. *GA*, List No. 122/1.
42. *GA*, List No. 122/1c–j.
43. See reference 1.
44. Francis Galton, "Short Notes on Heredity etc., in Twins," *J. of the Anthropological Institute* 5 (1875): 324–29; Francis Galton, "The History of Twins, as a Criterion of the Relative Powers of Nature and Nurture," *J. of the Anthropological Institute* 5 (1876): 391–406.
45. Waldo Hilary Dunn, *James Anthony Froude: A Biography, 1857-1894* (London: Oxford University Press, 1963), chap. 23.
46. Galton, *Fraser's Magazine* 12: 566.
47. Ibid., 566.
48. Ibid., 568.
49. Ibid., 570.
50. Ibid., 571.
51. Ibid., 573.
52. Ibid., 575.
53. Ibid., 576.

Chapter 15: Sweet Peas and Anthropometrics

1. *Memories*, 244.
2. Francis Galton, "Hereditary Improvement," *Fraser's Magazine* 7 (1873): 116–30.
3. Ibid., 116.
4. Ibid., 117.
5. Ibid., 117.
6. Ibid., 117.
7. Ibid., 119.
8. Ibid., 121.
9. Ibid., 123.
10. Ibid., 123.

11. Ibid., 124.
12. Ibid., 125.
13. Ibid., 125.
14. Ibid., 126.
15. Ibid., 127.
16. Ibid., 127.
17. Ibid., 129.
18. Ibid., 129.
19. Francis Galton, "Proposal to Apply for Anthropological Statistics from Schools," *J. of the Anthropological Institute* 3 (1874): 308–11.
20. J. M. Tanner, "Galton on Human Growth and Form," in *Sir Francis Galton, FRS*, Milo Keynes, ed. (London: The Macmillan Press, Ltd., 1993), 108–18.
21. Walter Fergus and G. F. Rodwell, "On a Series of Measurements for Statistical Purposes Recently made at Marlborough College," *J. of the Anthropological Institute* 4 (1874): 126–29.
22. Francis Galton, "Notes on the Marlborough School Statistics," *J. of the Anthropological Institute* 4 (1874): 130–37.
23. Francis Galton, "Men of Science, their Nature and their Nature," *Nature* 9 (1874): 344–45.
24. Francis Galton, "On a Proposed Statistical Scale," *Nature* 9 (1874): 342–43.
25. Francis Galton, "Statistics of Intercomparison, with Remarks on the Law of Frequency of Error," *Philosophical Magazine* 49 (1875): 33–46.
26. Stephen M. Stigler, *The History of Statistics: The Measurement of Uncertainty before 1900* (Cambridge: Belknap Press of Harvard University Press, 1986), 271.
27. Francis Galton, "On the Height and Weight of Boys Aged 14, in Town and Country Public Schools," *J. of the Anthropological Institute* 6 (1876): 174–81.
28. Ibid., 174.
29. Ibid., 174.
30. This account is taken from Tanner, "Galton on Human Growth and Form."
31. Ibid., 112.
32. Francis Galton, "Typical Laws of Heredity," *Proc. of the Royal Institution* 8 (1877): 282–301. Also in *Nature* 15 (1877): 492–95, 512–14, 532–33.
33. See Thomas Martin, *The Royal Institution* (London: Longmans, Green & Co., 1949); *Royal Institution of Great Britain: The Record of the Royal Institution of Great Britain* (London: Wm. Clowes, 1941).
34. Galton, *Nature* 15: 492.
35. Ibid., 492.
36. Ibid., 493.
37. Ibid., 493.
38. Stigler points out in his history of statistics that the "device was an analog for a binomial experiment; each row of pins subjecting a shot to an independent disturbance, equally likely to be left or right." Stigler, *The History of Statistics*, 276.
39. Galton, *Nature* 15: 495.
40. *Memories*, 300.
41. Galton, *Nature* 15: 512.
42. Ruth Schwartz Cowan, "Francis Galton's Statistical ideas: The Influence of Eugenics," *Isis* 63 (1972): 509–28.
43. Galton, *Nature* 15: 512.
44. *Life*, II: 181.
45. Galton, *Nature* 15: 513.
46. Ibid., 513.
47. Ibid., 513.
48. Ibid., 513.
49. Ibid., 513.
50. *Life*, IIIa: 3
51. Cowan, "Francis Galton's Statistical Ideas," 522.
52. Galton, *Nature* 15: 513.
53. Ibid., 513.
54. Ibid., 514.
55. Ibid., 514.
56. Charles R. Darwin, *The Descent of Man in Relation to Sex* (New York: Modern Library edition, Random House), 916
57. *Life*, IIIa: 6.
58. *Victorian Genius*, 188.
59. Francis Galton, "The Measurement of Fidget," *Nature* 32 (1885): 174–75.
60. *Life*, II: 340.
61. *GA*, List No. 53, Louisa Galton's diary.
62. Ibid.
63. Ibid.
64. Ibid.
65. Ibid.
66. Ibid.
67. Ibid.
68. Ibid.
69. Ibid.
70. Francis Galton, "The Anthropometric Laboratory," *Fortnightly Review* 31 (1882): 332–38.
71. Ibid., 334.
72. Francis Galton, *Inquiries into Human Faculty and Its Development* (London: J. M Dent, 1928).
73. Ibid., 24.
74. Ibid., 24.
75. *Saturday Review*, 26 May 1883, 668–69.
76. Ibid.
77. Ibid.
78. Ibid.
79. Ibid.
80. *Spectator*, 11 August 1883, 1029–30.
81. Ibid.
82. Ibid.
83. *Guardian*, 4 April 1883: 1001, *GA*, List No. 124
84. Ibid.

85. Ibid.
86. George J. Romanes, *Nature* 28 (1883): 97–98.
87. Ibid.
88. Ibid.
89. *Victorian Genius*, 172–73.
90. Francis Galton, "Medical Family Registers," *Fortnightly Review* 34 (1883): 244–50.
91. Ibid., 244.
92. Ibid., 246.
93. Ibid., 246.
94. Francis Galton, *Record of Family Faculties* (London: Macmillan, 1884).
95. *Life*, II: 366–370.
96. *GA* List No, 53, Louisa Galton's diary.
97. Ibid.
98. *Illustrated London News* 84 (May 10, 1884): 108.
99. *Illustrated London News* 85 (Aug. 2, 1884): 90–95.
100. Ibid., 106.
101. *Life*, II: 213.
102. Ibid., 213.
103. Ibid., 370–78.
104. Williams and Randall, his successor at South Kensington, were both referred to as Sargeant or Serjeant. Francis Galton appears to be using the term in the now obsolete form that refers to a servant or attendant (see *Shorter Oxford English Dictionary* (Oxford: Clarendon Press, 1993).
105. Francis Galton, "On the Anthropometric Laboratory at the Late International Health Exhibition," *J. of the Anthropological Institute* 14 (1884): 205–21.
106. *Memories*, 245.
107. Galton, *J. of the Anthropological Institute* 14: 210.
108. *Memories*, 246.
109. Ibid., 246.
110. Galton, *J. of the Anthropological Institute* 14: 206.
111. *Memories*, 249.
112. *Life*, II: 379.
113. Ibid., 379.
114. Francis Galton, "Some Results of the Anthropometric Laboratory," *J. of the Anthropological Institute* 14 (1884): 275–87.
115. Tanner, "Galton on Human Growth and Form."
116. Francis Galton, "Some Results of the Anthropometric Laboratory," *J. of the Anthropological Institute* 14 (1884): 205–21. See pg. 278.
117. *Life*, II: 375.
118. *Life*, II: 380.

Chapter 16: Probing the Mind

1. Francis Galton, *Inquiries into Human Faculty and Its Development* (New York: Everyman's Library edition, E. P. Dutton, 1928), 1.
2. *Memories*, 80.
3. An account of Mesmer's career can be found in Raymond Fancher, *Pioneers of Psychology* (New York: W. W. Norton & Co., 1979), 170–78.
4. Alison Winter, *Mesmerized: Powers of the Mind in Victorian Britain* (Chicago: University of Chicago Press, 1998).
5. An account of George Combe's life is found in Paul Alfred Erickson, "Phrenology and Physical Anthropology: The George Combe Connection," *Occasional Papers in Anthropology*, No. 6, (Halifax, Nova Scotia: Department of Anthropology, Saint Mary's University, 1979).
6. Daniel J. Kevles, *In the Name of Eugenics* (Cambridge: Harvard University Press, 1995), 6.
7. *Life*, II: 62–63.
8. *DNB*, Twentieth Century (1912–1921), 136–37.
9. *DNB*, 9, 1119–21.
10. *Life*, II: 62–67.
11. Francis Galton, Address to Section D—Biology, British Association, *Nature* 16 (1877): 344–47.
12. *DNB*, Twentieth Century (1901–1911): 528–29.
13. Galton, *Nature* 16: 346.
14. Ibid., 346.
15. Ibid., 346.
16. Francis Galton, "Composite Portraits," *J. of the Anthropological Institute* 8 (1878): 132–44.
17. Ibid., 135.
18. Ibid., 135.
19. Ibid., 140.
20. Ibid., 143.
21. Ibid., 143.
22. Galton, *Inquiries into Human Faculty and Its Development*, 11.
23. *Life*, II: 201.
24. *Memories*, 262–63.
25. Ibid., 262–63.
26. Galton, *Inquiries into Human Faculty and Its Development*, 10.
27. Ibid., 10.
28. Francis Galton, "Photographic Composites," *The Photographic News* 29 (April 17 and 24, 1885): 234–45.
29. *Life*, II: 293.
30. http://ash.gene.ucl.ac.uk/face/
31. Ibid.
32. Galton, *Inquiries into Human Faculty and Its Development*, 3.
33. Francis Galton, "Psychometric Facts," *Nineteenth Century* 5 (1879): 425–33.
34. Ibid., 425.
35. Ibid., 425.
36. Ibid., 426.
37. Ibid., 429.
38. Ibid., 430.
39. Ibid., 430.
40. Ibid., 430.
41. Ibid., 432.

42. Ibid., 433.
43. Ibid.
44. Francis Galton, "Psychometric Experiments," *Brain* 2 (1879): 149–62.
45. *Victorian Genius*, 148.
46. David Burbridge, "Galton's 100: an Exploration of Francis Galton's Imagery Studies," *British Journal of the History of Science* 27 (1994): 443–63. See p. 445. Based on his own research, Burbridge disputes the claim dating back to Karl Pearson *Life*, II: 12 that Francis Galton failed, with the exception of the works of Charles Darwin and, perhaps Adolphe Quetelet, to read widely in the fields that interested him.
47. Edwin Garrigues Boring, *A History of Experimental Psychology* (New York: Appleton-Century-Crofts, second edition, 1950), 634, 639.
48. *Life*, IIIB: 464.
49. Boring, *A History of Experimental Psychology*, 281.
50. This account of Galton's questionnaire and the responses he received is taken from Burbridge's "Galton's 100" article *British Journal of the History of Science* 27: 443–63.
51. Ibid., 448.
52. Francis Galton, "Statistics of Mental Imagery," *Mind* 5 (1880): 301–18.
53. Ibid., 302.
54. Ibid., 302.
55. Ibid., 304.
56. Ibid., 318.
57. Ibid., 318.
58. Burbridge, *British Journal of the History of Science* 27: 461.
59. Ibid., 454.
60. Ibid., 461.
61. Ibid., 461.
62. Ibid., 461.
63. Ibid., 461.
64. Ibid., 446.
65. *DNB*, 2, 474–475.
66. Francis Galton, "Visualized Numerals," *J. of the Anthropological Institute* 10 (1880): 85–102.
67. Ibid., 92.
68. Ibid., 98.
69. Ibid., 102.
70. Francis Galton, "Mental Imagery," *Fortnightly Review* 28 (1880): 312–24.
71. Ibid., 313.
72. Ibid., 314.
73. Ibid., 315.
74. Ibid., 316.
75. *Times*, 3 September 1880.
76. Ibid.
77. Ibid.
78. Ibid.
79. Ibid.
80. Ibid.
81. Ibid.
82. Ibid.
83. Ibid.
84. Burbridge, *British Journal of the History of Science* 27: 446.
85. What the *Times* review and Galton's unpublished rebuttal highlight is the question of Galton's awareness of current literature and theory in this as well as other fields he sought to investigate. It is a striking characteristic of Galton's articles and books that the contributions of others, with the exception of Darwin and Quetelet, are rarely cited. Pearson (*Life*, II: 12) believed that Galton simply did not bother to read the literature. "Galton took up his problems one after another and worked on them largely disregarding their past history, when indeed they had one." But David Burbridge, in his analysis of Galton's imagery studies, makes a convincing case from Galton's own notes that he was probably much better read, at least in psychology, than Pearson's statement or the *Times* reviewer's comments imply. Please see n. 46 above. Galton's failure in this regard may have been largely one of citation.
86. Fancher, *Pioneers of Psychology*, chapter 7.
87. Richard J. Herrnstein and Charles Murray, *The Bell Curve* (New York: The Free Press, 1994).
88. Allan Paivio, *Images in the Mind* (New York: Harvester-Wheatsheaf, 1991), 185, 188, 309, 311, 373.
89. Robert H. Logie and Michel Denis, eds., "Mental Images in Human Cognition," *Advances in Psychology* 80 (1991).
90. Norman E. Wetherick, "What Goes on in the Mind When we Solve Syllogisms?" Ibid., 255–67.
91. John T. E. Richardson, "Gender Differences in Imagery, Cognition, and Memory," Ibid., 271–303.
92. Ibid.
93. Graham Dean and Peter E. Morris, "Imagery and Spatial Ability: When Introspective Reports Predict Performance," Ibid., 331–47.
94. Boring, *A History of Experimental Psychology*, 461–62.
95. Michael M. Sokal, ed., *An Education in Psychology: James McKeen Cattell's Journal and Letters from Germany and England, 1880–1888* (Cambridge, Mass.: MIT Press, 1981), 1–18.
96. Raymond Fancher, *The Intelligence Men: Makers of the IQ Controversy* (New York: W. W. Norton & Co., 1985), 44–49.
97. Sokal, ed., *An Education in Psychology*, 47–49.
98. Fancher, *The Intelligence Men*, 44–49.
99. Sokal, ed., *An Education in Psychology* 90.
100. Ibid., 113.
101. Ibid., 191.

102. Ibid., 215.
103. *J. of the Anthropological Institute* 16 (1887): 154–74.
104. Sokal, ed., *An Education in Psychology*, 215.
105. Ibid., 221–23.
106. Francis Galton, "On Recent Designs for Anthropometric Instruments," *J. of the Anthropological Institute* 16 (1887): 2–9.
107. Sokal, ed., *An Education in Psychology*, 297–98
108. Ibid., 300.
109. Fancher, *The Intelligence Men*, 44–49.
110. Sokal, ed., *An Education in Psychology*, 339.
111. Fancher, *The Intelligence Men*, 5–18. Provides an excellent account of J. S. Mill's psychological theories.
112. Ibid., 49–83. Provides an excellent account of Alfred Binet's psychological contributions.
113. Ibid., 52.
114. Theta Holmes Wolf, *Alfred Binet* (Chicago: University of Chicago Press, 1973), 85–86.
115. Francis Galton, "Psychology of Mental Arithmaticians and Blindfold Chess-players," *Nature* 51 (1894): 73–74.
116. Ibid., 73.
117. Fancher, *The Intelligence Men*, 49–83.

Chapter 17: Fingerprints

1. Douglas G. Browne and Alan Brock, *Fingerprints: Fifty Years of Scientific Crime Detection* (London: George G. Harrap & Co. Ltd, 1953). This highly entertaining history of fingerprinting has been a particularly useful source in writing this chapter.
2. Francis Galton, "Personal Identification and Description," *Nature* 38 (1888): 173–77, 201–02.
3. Ibid., 173.
4. Ibid., 174.
5. Henry T. F. Rhodes, *Alphonse Bertillon* (New York: Abelard-Schuman, 1956). The account of Bertillon's life given in this chapter is based largely on this interesting book.
6. Galton, *Nature* 38: 175.
7. Ibid., 201.
8. Ibid., 201.
9. Henry Faulds, "On Skin Furrows in the Hands," *Nature* 22 (1880): 605.
10. *GA*, List No. 190. Faulds' letter to Darwin and Darwin's letter to Galton.
11. Ibid.
12. Ibid.
13. Ibid.
14. *Life*, II: 195.
15. *GA*, List No. 190 indicates that both the Faulds and Darwin letters were found at the Anthropological Institute in 1894 by William Peck who then returned them to Galton.

16. William J. Herschel, "Skin Furrows of the Hand," *Nature* 23 (1880): 76.
17. Faulds, *Nature* 22: 605.
18. Ibid.
19. Ibid.
20. Ibid.
21. Ibid.
22. Herschel, *Nature* 23: 76.
23. Ibid.
24. Douglas Woodruff, *The Tichborne Claimant: A Victorian Mystery* (New York: Farrar, Straus and Cudahy, 1957). The account of the Tichborne case given here is taken from Woodruff's engrossing account.
25. Ibid.
26. Herschel, *Nature* 23: 76.
27. Francis Galton. Presidential Address, January 1888. *J. of the Anthropological Institute* 17 (1888): 35.
28. Francis Galton. "I. The Patterns in Thumb and Finger Marks—On their Arrangement into Naturally Distinct Classes, the Permanence of the Papillary ridges that make them, and the Resemblance of their Classes to Ordinary Genera," *Philosophical Trans. of the Royal Society of London, Series B* 182 (1891): 1–23.
29. *Life*, IIIA: 168.
30. Francis Galton, "Method of Indexing Fingerprints," *Proc. of the Royal Society* 54 (1891): 540–45.
31. Francis Galton, "Identification by Fingerprints," *Nineteenth Century* 30 (1891): 303–11.
32. Ibid., 303.
33. Ibid., 303.
34. Ibid., 304.
35. Ibid., 304.
36. Ibid., 304.
37. Ibid., 305.
38. Ibid., 311.
39. Francis Galton, *Finger Prints* (London: MacMillan and Co., 1892).
40. Stephen M. Stigler, "Galton and Identification by Fingerprints," *Genetics* 140 (1995): 857–60. Stigler's short article gives a nice succinct summary of Galton's contributions to fingerprinting and his explanation of Galton's calculation of independence is especially well done.
41. Galton, *Finger Prints*, 109.
42. Ibid., 109.
43. Ibid., 110–11.
44. Stigler (*Genetics* 140: 857–60) seems satisfied with Galton's analysis. "To be accepted today, Galton's modelling would require more detail, but with minor qualifications (and acceptance of Galton's personal experience with fingerprint patterns as an adequate basis upon which to form estimates) it can be rigorously defended as correct and conservative." But Pearson quarrels with Galton's assumption of independence of his six-ridge squares. "While convinced that the chance of two individu-

als actually possessing the same finger-print in all its *minutiae* is infinitesimally small—as small as the chance that two woodcutters given the same topic would produce two blocks identical in every line and dot—yet one recognizes that Galton's treatment, however ingenious, lacks the power of compelling conviction. Nature probably works more definitely to form a whole pattern than can be mimicked by Galton's 24 'independent variable' squares." *Life*, IIIA: 183.

45. Galton, *Finger Prints*, 189.
46. Ibid., 186.
47. As Stigler (Genetics 140: 857–60) points out, "present research shows that even monozygotic twins are not identical in fingerprints."
48. Galton, *Finger Prints* 192–93.
49. Ibid., 196.
50. Ibid., 197.
51. *Times*, 10 November 1892.
52. *St. James's Gazette*, 1 December 1892; *GA*, List No. 192.
53. Ibid.
54. Ibid.
55. *Perthshire Advertiser*, 12 December 1892. *GA*, List No. 192.
56. *National Observer*, 3 February 1893; *GA*, List No. 192.
57. *The Scotsman*, 7 November 1892; *GA*, List No. 192.
58. *British Medical Journal*, 24 February 1893; *GA*, List No. 192.
59. *Saturday Review*, 14 January 1893.
60. *Life*, IIIA: 207.
61. Ibid., 214.
62. *Life*, IIIA: 148–54.
63. Ibid., 149.
64. Ibid., 149.
65. Ibid., 149.
66. Ibid., 150.
67. *Standard*, 22 March 1894; *GA*, List No. 192.
68. *Yorkshire Post*, 14 March 1894; *GA*, List No. 192.
69. *Daily Chronicle*, 19 March 1894; *GA*, List No. 192.
70. *Evening Standard*, 21 March 1894; *GA*, List No. 192.
71. Ibid.
72. *Morning Post*, 27 March 1894; *GA*, List No. 192.
73. Henry T. F. Rhodes, *Alphonse Bertillon*, 139–41.
74. *Life*, IIIA: 199.
75. Henry Faulds, "On the identification of habitual criminals by finger-prints," *Nature* 50 (1894): 548.
76. Ibid.
77. Ibid.
78. William J. Herschel, "Finger-Prints," *Nature* 51 (1894): 77.
79. Ibid.
80. Ibid.
81. Ibid.
82. *Life*, IIIA: 145.
83. Ibid., 145.
84. Ibid., 150.
85. Francis Galton, "Review of Guide to Finger-print Identification by Henry Faulds," *Nature* Supplement to 72 (1905): iv–v.
86. Ibid., iv.
87. Ibid., iv.
88. Ibid., iv.
89. *Victorian Genius*, 213. Forrest reasons that because "of Galton's unbounded admiration for those of high scientific and social status who came from gifted families," he was bound to favor Herschel. "Herschel was the perfect example of the kind of person Galton was liable to overestimate. If a choice had to be made between the relative achievements of Faulds and Herschel there could be little doubt that in Galton's mind to whom the credit should be given."
90. Rhodes, *Alphonse Bertillon*, 139–41.
91. Browne and Brock, *Fingerprints: Fifty Years of Scientific Crime Detection*, 47–48.
92. *Life*, IIIA: 151.
93. Ibid., 151.
94. *DNB* (1931–1940): 421–22.
95. Browne and Brock, *Fingerprints: Fifty Years of Scientific Crime Detection* 45–47.

Chapter 18: The Birth of Biometrics

1. Peter L. Bernstein, *Against the Gods: The Remarkable Story of Risk* (New York: John Wiley & Sons, Inc., 1996), 168–69.
2. Francis Galton, "Co-relations and their Measurement, chiefly from Anthropometric Data," *Proc. of the Royal Society* 45 (1888): 135–45.
3. *GA*, List No. 53, Louisa Galton's diary for 1888.
4. "The British Association," *Nature* 32 (1885), 507–510.
5. Stephen M. Stigler, *The History of Statistics: The Measurement of Uncertainty Before 1900* (Cambridge: Belknap Press, Harvard University, 1986), 283–93. The discussion of regression and reversion toward the mean in this chapter relies heavily on Stephen M. Stigler's excellent account.
6. Francis Galton, *Natural Inheritance* (London: Macmillan and Co., 1889), 90.
7. Francis Galton, "Regression towards Mediocrity in Hereditary Stature," *J. of the Anthropological Institute* 15 (1886): 246–63.
8. To obtain the median Galton multiplied $69.2 \times 18 = 1245.6$ and $70.2 \times 14 = 982.8$, added the two = 2228.4 and divided by 32 total adult children to give 69.6 so his calculated median of 69.5 is a little low.

9. Later a mildly embarrassed Galton republished this table in *Natural Inheritance* with a footnote recording "a small blunder since discovered in sorting the entries between the first and second lines. It is obvious that 4 children cannot have 5 midparents." *Natural Inheritance*, 208.

10. Galton, *J. of the Anthropological Institute* 15: 252

11. Ibid., 249.

12. Ibid., 254.

13. Ibid., 254.

14. Despite Galton's scrupulously honest attempt to describe the "smoothing" process, it is difficult to reconstruct exactly what he did, since he now created a diagram that summarized the height data for only about one–third of the progeny (Fig. 18-2).

15. Galton, *Natural Inheritance*, chapter 9.

16. Ibid., chapter 10.

17. *Memories*, 302.

18. Ibid., 302.

19. Galton, *J. of the Anthropological Institute* 15: 255.

20. Galton recognized that he was accounting in some way for the variability of his data and continued as follows: "They were all clearly dependent on three elementary data, supposing the law of frequency of error to be applicable throughout; these data being (1) the measure of racial variability, whence that of the mid-parentages may be inferred as has already been explained, (2) that of co-family variability (counting the offspring of like midparentages as members of the same co-family), and (3) the average ratio of regression." Stigler (*The History of Statistics*, 288) explains Galton's "three elementary data" in current terms as follows: "In modern terminology we would say that he assumed normal distributions and required three numbers. These were (1) the population variance, from which the variance of midparent height P followed, (2) the conditional variance of child's height p given midparent's height P, and (3) the slope of the regression line, r."

21. *Memories*, 302.

22. Ibid., 303.

23. Galton, *J. of the Anthropological Institute* 15: 255.

24. Ibid., 255.

25. As Stigler (*The History of Statistics*, 289) says the "first key point that Galton had hit upon was the strict interdependence of the quantities involved. The regression coefficient, conditional variance, and the population were bound together." But the second key point "the consequences of the symmetry of the situation regarding child and midparent" that was so evident in his table "did not register immediately."

26. Bernstein, *Against the Gods: The Remarkable Story of Risk*, 170.

27. Maurice G. Kendall, Edgeworth's biographer (see note 28 below), believes that two elements are to be found that characterize much of Edgeworth's later work. First, Edgeworth "assumes that pleasure can be quantified, and that the amounts of pleasure experienced by different sentient beings can be added together. Consequently, at least in theory, a hedonic calculus can be set up in mathematical terms. Secondly, he asserts that most of the problems of physical ethics relate to optimization: for example, given a certain quantity of stimulus to be distributed among a given set of sentients, we may require to find the law of distribution productive of the greatest quantity of pleasure." This is not as weird as it sounds, for as Stephen M. Stigler (*The History of Statistics*, 306) points out, Edgeworth's treatment of the subject "showed a confident and creative mastery of the calculus of variations, not to mention some knowledge of mathematical physics, and a thorough familiarity with the mathematical psychophysics of Fechner, Delboeuf, Helmholtz, and Wundt." Soon Edgeworth would begin to apply his ever-increasing armamentarium of mathematical tools to economic and social statistics.

28. Maurice G. Kendall, "Francis Ysidro Edgeworth, 1860–1906," *Biometrika* 55 (1968): 269–75.

29. Stigler, *The History of Statistics: The Measurement of Uncertainty Before 1900*, 305–25 discuss Edgeworth's life and work.

30. Ibid., quoted from letter on 306–7.

31. Francis Galton, "The Statistics of Intercomparison," *Philosophical Magazine* 49 (1875): 33–46.

32. As Stigler (*The History of Statistics*, 313) puts it the "classical theory provided a rationale for the appearance of probability distributions (in particular, normal curves) where the goal was determining a fixed, objectively defined quantity from observations that were subject to measurement errors. Now these distributions could be rationalized for social data at arbitrary levels of classification, just as a quincunx could be cut at arbitrary levels. And the distributions, once rationalized, themselves served as objects for comparison."

33. Stigler (Ibid., 315) summarizes by saying that "much of Edgeworth's work through 1885 can be interpreted as using Galton's 1875 insight as a rationale for bringing the techniques of classical theory to bear on problems in the social sciences. His appreciation of the potential of Galton's further development of regression and correlation came, for the most part, much later."

34. Francis Galton, "Family Likeness in Stature," *Proc. of the Royal Society*, 40 (1886): 42–72.

35. After wearily plodding through "Family Likeness in Stature" with its numerous ogives, Pearson concluded that these curves were "by no means helpful for elucidating correlation, as the reader of the first ten pages of the Royal Society paper will find." *Life*, IIIa: 12.

36. As Stigler puts it (*The History of Statistics* 290), "the phenomenon of regression was freed from its status as temporal or intergenerational in origin and, as we shall see later, this symmetry eventually led to the notion of 'correlation."

37. *GA*, List No. 53, Louisa Galton's diary for 1889.

38. Fiona MacCarthy, *William Morris: A Life for Our Time* (New York: Alfred A. Knopf, 1995), 336–40.

39. *Memories*, 300.

40. *Life*, II: 393.

41. Stigler, *The History of Statistics* 298.

42. Galton probably used "co-relate" rather than "Correlate" in his title on purpose. Stigler's analysis (Ibid., 297–98) is as follows. The physicist W. R. Grove had published *The Correlation of Physical Forces* in 1846, a work that went through many editions. Correlated ideas, he wrote, are "inseparable even in mental conception," but he also emphasized that "one cannot take place without the other." In 1874 Jevons published *Principles of Science* writing that things "are correlated (*con, relata*) when they are so related or bound to each other that *where one is the other is*, and *where one is not the other is not*." Galton scribbled a marginal entry next to this passage in Jevons's book noting that correlation was a "Nice wd. never so with common meaning (Groves's). Thus where motion is there heat may *not* be, but when motion is not then heat will [not] be." Both definitions imply some degree of causation. But there may be another reason why Galton chose "co-relation" rather than "correlation" (see *The Oxford Universal Dictionary* [Oxford: Clarendon Press, Third Edition, 1955] 394, 399). "Co-relate," a word of fairly recent vintage first used in 1839, meant a joint or mutual relation. In this sense it could be taken as meaning kinship, a subject in which as we have seen, Galton was intensely interested. The older word "correlation," whose usage dated from at least 1551, refers to a mutual relationship of two or more things. Thus, Galton may have felt that co-relate expressed his intent to tie his concept to kinship as opposed to "things."

43. Galton, *Natural Inheritance* 1.

44. John G. McKendrick, "On the Modern Cell Theory and Phenomenon of Fecundation," *Proc. Philosophical Society of Glasgow* 19 (1887–88): 71–125. McKendrick's review is filled with precise diagrams showing chromosomes and mitotic divisions, but it is clear from the text that the regular movements of chromosomes had not yet been associated with the behavior of genetic elements.

45. Galton, *Natural Inheritance* 4.

46. Ibid., 4–5.

47. Ibid., 9.

48. Ibid., 10.

49. Ibid., 11.

50. Ibid., 12.

51. The inheritance of eye color in human beings is still a subject of debate. There is reason to believe that it is polygenic in nature although several major genes influencing eye color have also been reported. See *Online Mendelian Inheritance in Man* http//:www3.ncbi.nlm.nih.gov/Omim/

52. Galton, *Natural Inheritance* 28.

53. Ibid., 28.

54. Ibid., 187.

55. Ibid., 34.

56. Ibid., 83.

57.–59. Ibid.

60. While Galton's discussion of regression is sometimes hard to follow, William B. Provine (*The Origins of Theoretical Population Genetics* (Chicago: The University of Chicago Press, 1971, 20–21) has generalized the concept and made Galton's quantitative measurement of regression readily understandable. "Suppose groups I and II are chosen from a population with a median measure M of some character. Then the median measure of the character in group I may be expressed as $M \pm D$, and in group II $M \pm kD$. The quantity k Galton defined as the regression of group II on group I with respect to the chosen character."

61. Galton again reported that the average regression of mid-filial stature upon mid-parental height was about [bu88]2[cm3[cm. He admitted that he had "smoothed" his data for the real number he obtained was 3/5 and explained the reason why. By making the assumption that both parents contributed equally to stature, Galton could use the mid-parental value to argue that "the average Regression from the Parental to the Mid-Filial Stature must be one half of two-thirds, or one-third." But when he tried to plot regression of "uniparental" on filial data they seemed "to show a Regression of about two-fifths, which differs from that of one-third in the ratio of 6 to 5." Galton dismissed the observation because his data points showed so much scatter. [See Galton, *Natural Inheritance*, 98–99.] As Provine points out, (see note 60 above) Galton's reason for wanting to use 2/3 rather than 3/5 was that he was desirous of calculating mid-filial regression on a single parent, but he felt his data were insufficient to allow him to do this directly.

62. Ibid., 132.

63. Karl Pearson, "Mathematical Contributions to the Theory of Evolution. On the Law of Ancestral Heredity," *Proc. of the Royal Society* 62 (1898): 386–412.

64. Pearson (*Life*, IIIA: 24) has pointed out that Galton's original reasoning was defective since these regression coefficients are not independent, saying that "I do not think Galton's method of

deducing the degree of resemblance between kinsmen of various degrees of blood relationship from the single datum of the regression of a filial array on its midparent, will pass muster; it is extraordinarily suggestive—no one had thought before of giving a quantitative measurement to the various types of kinship." And Provine (*The Origins of Theoretical Population Genetics*, 22–23) adds that "if Galton's derivations of his regression coefficients and his law of ancestral heredity were questionable, he nevertheless opened the door to a statistical analysis of correlations of characters, an analysis which was to have immense influence upon evolutionary thought. The biometricians were later to point to *Natural Inheritance* as the starting point of biometry."

65. Francis Galton, "Pedigree Moth-Breeding, as a Means of Verifying Certain Important Constants in the General Theory of Heredity," *Trans. of the Entomological Society of London*, Part I (1887): 19–28.

66. *Life*, IIIA: 49.

67. This chapter was, essentially, a reprint of "Family Likeness in Eye-Colour," *Proc. of the Royal Society* 40 (1886): 402–17.

68. Galton, *Natural Inheritance*, 139.
69. Ibid., 152.
70. *Life*, IIIA: 39.
71. Galton, *Natural Inheritance*, 155.
72. Ibid., 155.
73. Ibid., 157.
74. Ibid., 158.
75. *Life*, IIIA: 67.
76. Ibid., 67.
77. Galton, *Natural Inheritance*, 162.
78. *Life*, IIIA: 69.
79. Galton, *Natural Inheritance*, 175.
80. Ibid., 185–86.
81. *GA*, List No. 53 Louisa Galton's diary for 1889.
82. *Scottish Leader*, 14 March 1889; *GA*, List No. 192.
83. Ibid.
84. Ibid.
85. Ibid.
86. Ibid.
87. *The Spectator*, 2 July 1889.
88. Ibid.
89. Ibid.
90. Stigler, *The History of Statistics* (5), 301.
91. *GA*, List No. 190. Letters of Florence Nightingale and Francis Galton.
92. *DNB*, 22 (supplement), 69–94.
93. *GA*, List No. 190.
94. Ibid.
95. Ibid.
96. Ibid.
97. Ibid.
98. Ibid.

Chapter 19: Galton's Disciples

1. *Life*, IIIa: 57.
2. This history of Karl Pearson's early life is taken largely from Churchill Eisenhart, Karl Pearson, *DSB*, 10: 447–73; Daniel J. Kevles, *In the Name of Eugenics* (Cambridge: Harvard University Press, 1995), chapter 2; and Egon S. Pearson, *Karl Pearson: An Appreciation of Some Aspects of His Life and Work* (Cambridge: Cambridge University Press, 1938), 1–50.
3. Pearson, *Karl Pearson*, 3.
4. Ibid., 3.
5. *DNB*, 22 (1901–1911): 233–35.
6. Pearson, *Karl Pearson*, 3.
7. Kevles, *In the Name of Eugenics*, 22.
8. Pearson, *Karl Pearson*, 4.
9. Ibid., 4.
10. Ibid., 4.
11. J. B. S. Haldane, "Karl Pearson," (1857–1957). A centenary lecture delivered at University College London," originally in *Biometrika* 44 (1957): 303–313; reprinted in Egon S. Pearson and MG Kendall, eds., *Studies in the History of Statistics and Probability* (London: Griffin, 1970), I: 427–37.
12. Pearson, *Karl Pearson* 8.
13. Ibid., 7.
14. Ibid., 7.
15. Fiona MacCarthy, *William Morris: A Life for Our Time* (New York: Alfred A. Knopf, 1995), 462–503.
16. Ruth First and Ann Scott, *Olive Schreiner* (New York: Schocken Books, 1980), 140.
17. Ibid. This account of the relationship between Pearson and Schreiner is taken from 145–170 of this interesting biography.
18. Pearson, *Karl Pearson*, 18.
19. Ibid., 14.
20. Ibid., 14.
21. Ibid., 14.
22. Ibid., 14.
23. First and Scott, *Olive Schreiner*, 145–170.
24. Ibid., 161.
25. Ibid., 162.
26. Ibid., 165.
27. Ibid., 166.
28. Ibid., 169.
29. Kevles, *In the Name of Eugenics* 26.
30. Pearson, *Karl Pearson*, 16–17.
31. Pearson, *Karl Pearson*, 20–21 gives the history of Gresham College.
32. According to Stephen M. Stigler it was "the view of a man wearing blinders. He saw only the application at hand, and if he explored the book's implications at all, it was as a contribution to questions related to sexual selection and to social policy (a topic on which Galton was notably silent)."

Stephen M. Stigler, *The History of Statistics; The Measurement of Uncertainty Before 1900* (Cambridge: Belknap Press, Harvard University, 1986), 303–4.
33. Karl Pearson, *The Grammar of Science* (London: Dent, 1937 republication of 1892 edition).
34. Pearson, *Karl Pearson*, Appendices I: 132–41 and II: 142–53.
35. Ibid., 18.
36. Ibid., 19.
37. Stigler, *The History of Statistics*, 327.
38. Maurice G. Kendall, "Francis Ysidro Edgeworth, 1845–1926," originally in *Biometrika* 55 1968: 269–275. Reprinted in *Studies in the History of Statistics and Probability*, ed. Egon S. Pearson and Maurice G. Kendall (London: Griffin, 1970), 1: 257–263, see p. 261.
39. Ibid., 261.
40. *Life*, IIIB: 486.
41. Galton enlarged the possible applications of the normal curve when he suggested applying logarithms in analyzing the observations. Stigler, *The History of Statistics*, 329–42.
42. Ibid., 331.
43. Ibid., 333.
44. Ibid., 339.
45. Ibid., 341.
46. Kendall, "Francis Ysidro Edgeworth, 1845–1926," 262.
47. Karl Pearson, "Notes on the History of Correlation," *Biometrika* 13 (1920): 25–45, reprinted in *Studies in the History of Statistics and Probability*, eds. Egon S. Pearson and Maurice G. Kendall (London: Griffin, 1970), 1: 185–205, see p. 189.
48. *GA*, List No. 237. Edgeworth's letters to Galton. The quotation is from his letter of March 21, 1894.
49. Karl Pearson, "Walter Frank Raphael Weldon, 1860–1906," *Biometrika* 5 (1906): 1–52, reprinted in *Studies in the History of Statistics and Probability*, eds. Egon S. Pearson and Maurice G. Kendall, 1:265–321. The description of Weldon's life relies heavily on this excellent biography published shortly after Weldon's death.
50. W. F. R. Weldon, "The Variations Occurring in certain Decapod Crustaceans. I. *Crangon vulgaris*," *Proc. of the Royal Society* 47 (1890): 445–53.
51. Ibid., 445.
52. Ibid., 447.
53. Ibid., 451.
54. Ibid., 453.
55. Ibid., 45[?? 68].
56. *Life*, IIIb, 483.
57. W. F. R. Weldon, "Certain Correlated Variations in *Crangon vulgaris*," *Proc. of the Royal Society* 51 (1892): 2–21.
58. Ibid., 2.
59. Ibid., 11.
60. Pearson, "Notes on the History of Correlation," 201.
61. W. F. R. Weldon, "On certain Correlated Variations in *Carcinus moenas*," *Proc. of the Royal Society* 54 (1893): 318–33.
62. Ibid., 324.
63. Ibid., 324.
64. Ibid., 328.
65. Ibid., 329.
66. Ibid., 329.
67. Kevles, *In the Name of Eugenics*, 29.
68. Pearson, "Walter Frank Raphael Weldon, 1860–1906", 285–87. Pearson gives an interesting account of the revolt of the professors at University College on which this discussion is largely based.
69. Ibid., 286.
70. Ibid., 286.
71. Ibid., 287.
72. Ibid., 288.
73. Ibid., 287.
74. Ibid., 284.

Chapter 20: Evolution by Jumps

1. John Maynard Smith, "Galton and Evolutionary Theory," in *Sir Francis Galton, FRS: The Legacy of His Ideas*, ed. Milo Keynes (London: Macmillan Press, Ltd., 1993), 166.
2. Francis Galton, *Natural Inheritance* (London: Macmillan, 1889), 32–33.
3. The account of the first part of William Bateson's life in this chapter is drawn from the first 60 pages of his wife's memoir (Beatrice Bateson, *William Bateson, F.R.S.: Naturalist* [Cambridge: Cambridge University Press, 1928]). Specific page citations are only given in the case of quotations.
4. Ibid., 10.
5. *Balanoglossus* is a hemichordate, or acorn worm, in the line of evolution of the phylum Chordata, which contains the vertebrates.
6. *DSB*, 501–2.
7. Beatrice Bateson, *William Bateson, F.R.S.: Naturalist*, 388.
8. William K. Brooks, *The Law of Heredity: A Study of the Cause of Variation and the Origin of Living Organisms* (Baltimore: John Murphy, 1883). Also see William B. Provine, *The Origins of Theoretical Population Genetics* (Chicago: The University of Chicago Press, 1971), 37–39.
9. Brooks, Ibid., 297.
10. Beatrice Bateson, *William Bateson, F.R.S.: Naturalist*, 58–59.
11. Ibid., 58–59.
12. Ibid., 21–24.
13. Ibid., 17.

14. William Bateson, "On some Variations of *Cardium edule* apparently Correlated to the Conditions of Life," *Phil. Trans. Royal Society of London, Series B* (1889). Reprinted in *Scientific Papers of William Bateson*, ed. Reginald C. Punnett (Cambridge: Cambridge University Press, 1928), 1: 33–71.
15. *GA*, List No. 198 Bateson letters.
16. *Life*, IIIA: 288.
17. Beatrice Bateson, *William Bateson, F.R.S.: Naturalist*, 42.
18. Ibid., 51.
19. Ibid., 27.
20. William Bateson and A. Bateson, "On the Variations in Floral Symmetry of Certain Parts having Irregular Corollas," *J. of the Linnean Society (Botany)* 28 (1891). Reprinted in *Scientific Papers of William Bateson*, ed. Reginald C. Punnett (Cambridge: Cambridge University Press, 1928), 1: 126–61.
21. Ibid., 159.
22. William Bateson and H. H. Brindley, "Some cases of Variation in Secondary Sexual Characteristics Statistically Examined," *Proceedings of the Zoological Society* (1893). Reprinted in *Scientific Papers of William Bateson*, ed. Reginald C. Punnett (Cambridge: Cambridge University Press, 1928), 1: 193–201.
23. *GA*, List No. 198. Bateson's letters.
24. Beatrice Bateson, *William Bateson, F.R.S.: Naturalist*, 28.
25. William Bateson, *Materials for the Study of Variation treated with especial regard to Discontinuity in the Origin of Species* (London: Macmillan, 1894).
26. Ibid., vi.
27. Ibid., 1.
28. Ibid., 2.
29. Ibid., 4. Note the use of genetically in the sense of origin prior to Bateson's coining the term to describe the science of genetics (*Nature* 146 (1906): 1). See *Oxford English Dictionary*, 6 (1989) for uses of the term genetic.
30. Ibid., 5.
31. Ibid., 6.
32. Ibid., 15.
33. Ibid., 17.
34. Ibid., 36.
35. Ibid., 42.
36. Francis Galton, "Discontinuity in Evolution," *Mind* (n.s.) 3 (1894): 362–72.
37. Ibid., 368.
38. Ibid., 368.
39. Ibid., 369.
40. Ibid., 369.
41. Provine, *The Origins of Theoretical Population Genetics*, 44.
42. W. F. R. Weldon, "The Study of Animal Variation," *Nature* 50 (1894): 25–26.
43. Ibid., 25.
44. Ibid., 25.
45. Ibid., 26.
46. Alfred Russel Wallace, "The Method of Organic Evolution," parts I and II. *Fortnightly Review* 63 (1895): 211–24, 435–45.
47. H. Lewis McKinney, Alfred Russel Wallace, *DSB*, 14, 133–40.
48. Wallace, *Fortnightly Review* 63: 212.
49. Ibid., 212.
50. Ibid., 212.
51. Ibid., 212.
52. Ibid., 212.
53. Ibid., 213.
54. Ibid., 214–215.
55. Ibid., 216.
56. William Bateson, *Materials for the Study of Variation*, 568.
57. Wallace, *Fortnightly Review* 63: 218.
58. Ibid., 220.
59. Ibid., 223.
60. Ibid., 435.
61. Ibid., 438.
62. Ibid., 439.
63. Francis Galton, "The Patterns in Thumb and Finger Marks—On their Arrangement into Naturally Distinct Classes, the Permanence of the Papillary Ridges that make them, and the Resemblance of their Classes to Ordinary Groups," *Phil. Trans. Royal Society of London, Series B* 181 (1891): 1–23.
64. Ibid., 22.
65. Ibid., 22.
66. Ibid., 23.
67. Wallace, *Fortnightly Review* 63: 441.
68. Ibid., 444.
69. Ibid., 445.
70. Beatrice Bateson, *William Bateson, F.R.S.: Naturalist*, 57.
71. W. T. Thiselton-Dyer, "Variation and Specific Stability," *Nature* 51 (1895): 459–61.
72. William Bateson, "The Origin of the Cultivated *Cineraria*," *Nature* 51 (1895): 605–7.
73. Ibid., 607.
74. W. T. Thiselton-Dyer, "The Origin of the Cultivated *Cineraria*," *Nature* 52 (1895): 3–4.
75. Ibid., 3.
76. Ibid., 3.
77. Bateson, "The Origin of the Cultivated *Cineraria*," *Nature* 52 (1895): 29.
78. Ibid., 29.
79. W. F. R. Weldon, "The Origin of the Cultivated *Cineraria*," *Nature* 52 (1895): 54.
80. Ibid., 54.
81. Ibid., 54.
82. Provine, *The Origins of Theoretical Population Genetics*, 47.
83. Ibid., 47.
84. Ibid., 47.
85. Ibid., 48.

86. W. T. Thiselton-Dyer, "The Origin of the Cultivated *Cineraria,*" *Nature* 52 (1895): 128–29. Weldon, "The Origin of the Cultivated Cineraria," *Nature* 52 (1895): 129.
87. Ibid., 129.
88. Provine, *The Origins of Theoretical Population Genetics*, 48.
89. *Life*, IIIA: 128–29.
90. Ibid., 130–31.
91. Karl Pearson, "Walter Frank Raphael Weldon. 1860–1906," *Biometrika* 5 (1906): 1–52, reprinted in *Studies in the History of Statistics and Probability*, eds. Egon S. Pearson and Maurice G. Kendall, 1, see 286–91.
92. W. F. R. Weldon, "I. Remarks on Variation in Animals and Plants," *Proc. of the Royal Society* 57 (1895): 361–79.
93. Weldon, "II. Remarks on Variations in Animals and Plants," *Proc. of the Royal Society* 57 (1895): 379–82.
94. Ibid., 380.
95. Ibid., 379.
96. Provine, *The Origins of Theoretical Population Genetics*, 49.
97. *Life*, IIIA: 127.
98. Ibid., 127.
99. Provine, *The Origins of Theoretical Population Genetics*, 50.
100. Ibid., p. 50.
101. *Life*, IIIB: 127–28.
102. *Life*, IIIB: 128.
103. *GA*, List No. 53 Louisa Galton's diary for 1896.
104. Francis Galton, "Intelligible Signals from Neighboring Stars," *Fortnightly Review* 60 (1896): 657–64.
105. *GA*, List No. 53 Louisa Galton's diary for 1896.
106. *GA*, List No. 53 Francis Galton's entry for 1897.
107. *Life*, IIIB: 502.
108. Ibid., 502.
109. Ibid., 503.

Chapter 21: The Mendelians Trump the Biometricians

1. Robert De Marrais, "The Double-edged Effect of Sir Francis Galton: A Search for the Motives in the Biometrician-Mendelian Debate," *J. History of Biology* 7 (1974): 141–74, see p. 142.
2. He attempted to address the problem in a paper from the Anthropometric Laboratory. Francis Galton, "Family Likeness in Stature," *Proc. of the Royal Society* 40 (1886): 42–72.
3. He returned to the same problem in Francis Galton, "Family Likeness in Eye-colour," *Proc. of the Royal Society* 40 (1886): 402–17.
4. Francis Galton, "The Average Contribution of each Several Ancestor to the Total Heritage of the Offspring," *Proc. of the Royal Society* 61 (1897): 401–13.
5. Galton also wrote an equation that tried to take into account parental means for a given character and the deviation of the progeny from these mean. William B. Provine in *The Origins of Theoretical Population Genetics* (Chicago: The University of Chicago Press, 1971), discusses Galton's ancestral law on 51–54, together with Pearson's modifications of the law, and then devotes an entire appendix (pp. 179–87) to summarizing the technical considerations that led to considerable confusion in applying the law. Michael Bulmer has made a useful clarification of the rather confused arguments Galton used to derive the ancestral law ("Galton's Law of Ancestral Heredity," *Heredity* 81 [1998]: 579–85).
6. Galton, *Proc. of the Royal Society* 61: 408.
7. Karl Pearson, "Mathematical Contributions to the Theory of Evolution: On the Law of Ancestral Heredity," *Proc. of the Royal Society, Series B* 62 (1898): 386–412. See 386.
8. Ibid., 386.
9. *Life*, IIIB: 504.
10. Pearson, *Proc. of the Royal Society, Ser. B* 62: 412.
11. William Bateson, "Problems of Heredity as a subject for Horticultural Investigation," *J. Royal Horticultural Society* 25 (1900): parts I and II. Reprinted in Beatrice Bateson, *William Bateson, F. R. S. Naturalist, his Essays and Addresses* (Cambridge: Cambridge University Press, 1928), 171–80. The quotation is on 173–74.
12. Provine, in *The Origins of Theoretical Population Genetics*, discusses the relationship between Bateson and de Vries on 66–69.
13. Beatrice Bateson, *William Bateson*, 73.
14. *GA*, List No. 198, Bateson letters to Galton.
15. William Bateson, *Mendel's Principles of Heredity: A Defence* (Cambridge: Cambridge University Press, 1902), 21–22.
16. Karl Pearson, "Mathematical Contributions to the Theory of Evolution.—IX. On the Principle of Homotyposis and its Relation to Heredity, to the Variability of the Individual, and to that of the Race. Part I.—Homotyposis in the Vegetable Kingdom," *Philosophical Trans. Royal Society, Series A* 197 (1901): 285–379.
17. Ibid., 287.
18. Ibid., 288.
19. Ibid., 288.
20. *Life*, IIIA: 241.
21. William Bateson, "Heredity, Differentiation, and other Conceptions of Biology: A Consideration of Professor Karl Pearson's Paper 'On the Principle of Homotyposis,'" *Proc. of the Royal Society* 69

(1901). Reprinted in *Scientific Papers of William Bateson*, ed. Reginald C. Punnett (Cambridge: Cambridge University Press, 1928), 1: 404–18.
22. Ibid., 411.
23. *Life*, IIIA: 241.
24. Editorial, *Biometrika* 1: 1 (1902).
25. Ibid., 2.
26. Ibid., 5.
27. Ibid., 6.
28. *Life*, IIIA: 127.
29. Francis Galton, "Biometry," *Biometrika* 1 (1902): 9.
30. *Life*, IIIB: 507.
31. Ibid., 510.
32. Ibid., 512.
33. *Victorian Genius*, 246–47.
34. Margaret S. Drower, *Flinders-Petrie* (London: Victor Gollancz Ltd., 1985), 476–77.
35. Ibid., 303.
36. *Life*, IIIB: 520.
37. Hesketh Pearson, *Modern Men and Mummers* (London: George Allen & Unwin, Ltd., 1921), 72.
38. Ibid., 63–79. Pearson's treatment of Louisa (p. 71), long dead before he met Galton, was harsh. But he failed to recognize that Galton's total preoccupation with his work; his frequent absences for scientific reasons; vacations cut short by the British Association meeting; and the absence of any offspring that she could mother made a lonely life for Louisa.
39. Ibid., 66.
40. Ibid., 65.
41. Ibid., 74–75.
42. *Memories*, 315–316.
43. W. F. R. Weldon, "Mendel's Laws of Alternative Inheritance in Peas," *Biometrika* 1 (1902): 228–54.
44. Ibid., 232.
45. Ibid., 233.
46. Ibid., 235.
47. Ibid., 236.
48. Provine, *The Origins of Theoretical Population Genetics*, 63.
49. Ibid., 63.
50. Bateson, *Mendel's Principles of Heredity: A Defence*, vi.
51. Ibid., vi.
52. Ibid., 108.
53. Ibid., 117–18.
54. Ibid., 208.
55. Karl Pearson, "Walter Frank Raphael Weldon, 1860–1906," *Biometrika* 5 (1906): 1–52, reprinted in *Studies in the History of Statistics and Probability*, eds. Egon S. Pearson and Maurice G. Kendall (London: Griffin, 1970), 1: 309.
56. Karl Pearson, "On the Fundamental Conceptions of Biology," *Biometrika* 1 (1902): 320–44.
57. Ibid., 320.
58. Ibid., 320.
59. Ibid., 321.

60. Ibid., 321.
61. Ibid., 322.
62. Ibid., 323.
63. Ibid., 329–30.
64. Ibid., 344.
65. William Bateson and E. R. Saunders, "Experimental Studies in the Physiology of Heredity," *Reports to the Evolution Committee of the Royal Society*, Report I (1902), reprinted in part in *Scientific Papers of William Bateson*, ed. Reginald C. Punnett (Cambridge: Cambridge University Press, 1928), 1: 29–68.
66. Ibid., 60.
67. Ibid., 65.
68. Ibid., 66.
69. W. F. R. Weldon, "On the Ambiguities of Mendel's Characters," *Biometrika* 2 (1903): 44–55.
70. Ibid., 53.
71. Ibid.
72. W. F. R. Weldon, "Professor de Vries on the Origin of Species," *Biometrika* 1 (1902): 365–374
73. Provine, *The Origins of Theoretical Population Genetics*, 64–70.
74. Today we know these differences arise because *O. lamarckiana* is genetically balanced in such a way that each individual differs by many genes that can segregate as blocks in subsequent generations.
75. Provine, *The Origins of Theoretical Population Genetics*, 67.
76. Ibid., 67.
77. Weldon, *Biometrika* 1: 369.
78. Ibid., 369.
79. Ibid., 374.
80. Weldon, *Biometrika* 1: 244.
81. Ibid., 244.
82. A. D. Darbishire, "Note on the Results of Crossing Japanese Waltzing Mice with European Albino Races," *Biometrika* 2 (1903): 101–04.
83. Provine, *Origins of Theoretical Population Genetics*, 74.
84. Darbishire, "Second Report on the Result of Crossing Japanese Waltzing Mice with European Albino Races," *Biometrika* 2 (1903): 165–73.
85. William Bateson, "Mendel's Principles of Heredity in Mice," *Nature* 67 (1903): 462–63.
86. Ibid., 462.
87. W. F. R. Weldon, "Mendel's Principles of Heredity in Mice," *Nature* 67 (1903): 512.
88. Bateson, "Mendel's Principles of Heredity in Mice," *Nature* 67 (1903): 585–86.
89. Ibid.
90. Ibid.
91. Ibid.
92. Ibid.
93. Weldon, "Mendel's Principles of Heredity in Mice," *Nature* 67 (1903): 610.
94. Ibid., 610.
95. Ibid., 610.

96. Bateson, "Mendel's Principles of Heredity in Mice," *Nature* 67 (1903): 33–34. Weldon, "Mendel's Principles of Heredity in Mice," *Nature* 67 (1903): 34.
97. Ibid., 34.
98. Ibid., 34.
99. Ibid., 34.
100. William Bateson, "The Present State of Knowledge of Colour-heredity in Mice and Rats," *Proc. Zoological Society of London* 2 (1903): 71–99, reprinted in *Scientific Papers of William Bateson*, ed. Reginald C. Punnett (Cambridge: Cambridge University Press, 1928) 2: 76–108.
101. Ibid., 81.
102. Ibid., 81.
103. Ibid., 103.
104. A. D. Darbishire, "Third Report on Hybrids between Waltzing Mice and Albino Races," *Biometrika* 2 (1903): 282–85.
105. W. F. R. Weldon, "Mr. Bateson's Revisions of Mendel's Theory of Heredity," *Biometrika* 2 (1903): 286–98.
106. Darbishire, "On the Result of Crossing Japanese Waltzing Mice with Albino Mice," *Biometrika* 3 (1904): 1–51.
107. Ibid., 25.
108. Provine, *Origins of Theoretical Population Genetics*, 77–80.
109. Ibid., 78.
110. In cross (i) the grandparents are both hybrid for albino (a) and its wild type allele (+). The progeny of this cross will consist of phenotypically wild type (+/+, +/a) and albino (a/a) progeny in a 3:1 ratio and a genotypic ratio of 1 +/+ : 2 +/a : 1 a/a. Darbishire pooled the phenotypically wild type F_1 progeny mice in his crosses incorrectly, assuming that they were all hybrid when 0.33 were in fact homozygous +/+ and 0.67 were +/a. These mice were then mated to yield the progeny Darbishire actually scored and, assuming mating occurred at random, the expectation is that matings would have occurred in the following frequencies: +/+ x +/+ (0.33)(0.33) = 0.11; +/a x +/a (0.67)(0.67) = 0.45; +/+ x +/a 2(0.33)(0.67) = 0.44. Since only the +/a x +/a cross will yield homozygous a/a progeny in a 3:1 ratio, the expected frequency of albino progeny, assuming random mating, is (0.45)(0.25) = 0.113 compared to an observed frequency of 0.11. Cross (ii) adds an albino grandparent, which means that one of the two crosses yields only +/a F_1 progeny. Therefore, when these rabbits are crossed matings will be made in the following frequencies : +/+ x +/a (0.33)(1.0) = 0.33 and +/a x +/a (0.67)(1.0) = 0.67. Therefore, the frequency of albino F_2 progeny should be (0.67)(0.25) = 0.167 expected versus 0.187 observed. Adding a second albino grandparent in cross (iii) means, by definition, that all F_1 progeny are +/a so that a/a F_2 progeny are expected in a 3:1 ratio, that is a frequency of 0.25 compared to an observed frequency of 0.25. In short Darbishire's results fit Mendel's predictions with uncanny accuracy.
111. Provine, *Origins of Theoretical Population Genetics*, 79.
112. Ibid., 79.
113. W. E. Castle and Glover M. Allen, "The Heredity of Albinism," *Proc. American Academy of Arts & Sciences* 38 (1903): 602–21.
114. Ibid., 612.
115. Ibid., 612.
116. Maurice G. Kendall, "George Udny Yule, 1871–1951," originally in *J. Royal Statistical Society* 115A (1952): 156–61, reprinted in *Studies in the History of Statistics and Probability*, ed. Egon S. Pearson and Maurice G. Kendall (London: Griffin, 1970), 418–25.
117. George Udny Yule, "Mendel's Laws and their Probable Relations to Intra-racial Heredity," *New Phytologist* 1 (1902): 193–207, 222–38.
118. Ibid., 224.
119. Ibid., 235.
120. "Zoology at the British Association," *Nature* 70 (1904): 538–41.
121. William Bateson, Presidential Address to the Zoological Section, British Association: Cambridge Meeting, 1904, reprinted in Beatrice Bateson, *William Bateson* (8), 233–59.
122. Ibid., 234.
123. Ibid., 243.
124. William Bateson, "A Text-book of Genetics" (Review of Lotsy's *Vorlesungen über Descendenz-theiorien*, I. Theil), *Nature* 74 (1906): 146.
125. *Times*, Saturday, 20 August 1904.
126. Ibid.
127. Ibid.
128. Ibid.
129. Ibid.
130. Ibid.
131. *Life*, IIIB: 528.
132. *Times*, Saturday, 20 August, 1904.
133. Ibid.
134. Ibid.
135. Ibid.
136. Ibid.
137. Ibid.
138. Pearson, "Walter Frank Raphael Weldon, 1860–1906," 308.
139. A. H. Sturtevant, *A History of Genetics* (New York: Harper & Row, 1965), 37–38.
140. Ibid., 39–41.

Chapter 22: The Triumph of the Pedigree: Eugenics

1. *Memories*, 322.
2. For an excellent discussion of the concern that developed in Britain over deterioration and

decline in the quality of the British population, see the first few chapters in Richard A. Soloway's book *Demography and Degeneration* (Chapel Hill: University of North Carolina Press, 1995).

3. Robert A. Nye, "The Rise and Fall of the Eugenics Empire: Recent Perspectives on the Impact of Biomedical Thought in Modern Society," *The Historical Journal* 36 (1993): 687–700.

4. *DNB*, (1912–1921): 48–50.

5. David Englander and Rosemary O'Day, eds., *Retrieved Riches: Social Investigation in Britain 1840-1914* (Aldershot: Scolar Press, 1995). Many references to Beatrice Potter Webb's work.

6. Charles Booth, *Life and Labour in London* (London: MacMillan, 1902), 1 (Poverty): 156.

7. Francis Galton, "The Possible Improvement of the Human Breed under the Existing Conditions of Law and Sentiment," *Essays in Eugenics* (London: Eugenics Education Society, 1909), 1–34. See p. 11.

8. Ibid., 17–18.
9. Ibid., 19.
10. Ibid., 19.
11. Ibid., 20.
12. Ibid., 19.
13. Ibid., 22.
14. Ibid., 22.
15. Ibid., 22.
16. Ibid., 23.
17. Ibid., 24.
18. Ibid., 24.
19. Ibid., 25.
20. Ibid., 26.
21. Ibid., 27.
22. Ibid., 33.
23. Ibid., 34.

24. Francis Galton, "The Possible Improvement of the Human Breed under the Existing Conditions of Law and Sentiment," *Nature* 64 (1901): 659–65. *Nature* published Galton's lecture in full.

25. *Life*, IIIA: 236.
26. *Life*, IIIA: 238.
27. Ibid., IIIB: 521.
28. Ibid., 521.
29. *Sociological Papers* 1 (1904): 284.
30. Ibid., 289.

31. Francis Galton, "Eugenics: Its Definition, Scope and Aims," Ibid., 45–50.

32. Ibid., 47.
33. Ibid., 48.
34. Ibid., 48.
35. Ibid., 49.
36. Ibid., 50.
37. Ibid., 52–53.
38. Ibid., 52.
39. Ibid., 53.
40. Ibid, 53.
41. Ibid., 53.
42. Ibid., 53.
43. Ibid., 54.
44. Ibid., 54.
45. Ibid., 54.
46. Ibid., 54.
47. Ibid., 55.
48. Ibid., 56–58.
49. Ibid., 56–57.
50. Ibid., 58–60.
51. Ibid., 59.
52. Ibid., 59.
53. Ibid., 74.
54. Ibid., 74.
55. *Life*, IIIA: 222.
56. *Sociological Papers* 1 (1904): 80.

57. Francis Galton, "Restrictions in Marriage," Ibid., 2 (1905): 2–13.

58. Ibid., 5.

59. Francis Galton, "Studies in National Eugenics," Ibid. 14–17.

60. Ibid., 19
61. Ibid., 15.
62. Ibid., 15.
63. Ibid., 17.
64. Ibid., 18–19.
65. Ibid., 18.
66. Ibid., 21–22.
67. Ibid., 21.
68. Ibid., 21.
69. Ibid., 21.
70. Ibid., 21.
71. Ibid., 21.
72. Ibid., 27.
73. Ibid., 30–33.
74. Ibid., 31.

75. See discussion following Galton's first speech to the Sociological Society in 1904 (24), p. 62; and *Life*, IIIB: 536–39.

76. *Sociological Papers* 1 (1904): 62.
77. Ibid., 62.
78. *Life*, IIIB: 539.
79. Ibid., IIIA: 276.
80. Ibid., 276.
81. Ibid., 276.

82. Francis Galton, *Noteworthy Families* (London: John Murray, 1909), ix–xlii.

83. Ibid., xxxix.
84. Ibid., xli.
85. *Life*, IIIB: 555–56.

86. Galton had put much time and effort into thinking about the presentation of pedigrees and proper pedigree notation as discussed in the following papers: Francis Galton, "Pedigrees," *Nature* 67 (1903): 586–87; Francis Galton, "Nomenclature and Tables of Kinship," *Nature* 69 (1904): 294–95; Francis Galton, "Average Number of Kinsfolk in

Each Degree," *Nature* 70 (1904): 529; Francis Galton, "Average Number of Kinsfolk in Each Degree," *Nature* 70 (1904): 626 and 71 (1905): 248; Francis Galton, "Nomenclature of Kinship—Its Extension," *Nature* 73 (1905): 150–51.
87. *Life*, IIIB: 559.
88. Ibid., 563.
89. Ibid., 564–65.
90. Ibid., 565.
91. Francis Galton, "Measurement of Resemblance," *Nature* 74 (1906): 562–63.
92. *Life*, IIIB: 565–66.
93. Ibid., IIIA: 281.
94. Ibid., 292–96.
95. Ibid., 292.
96. Ibid.
97. Ibid.
98. Ibid.
99. Ibid., 291.
100. Ibid., 300.
101. Ibid., 301, 303.
102. Ibid., 304.
103. Ibid., 305–07.
104. Ibid., 309.
105. Ibid., 311.
106. Ibid., 313.
107. Ibid., 314.
108. Ibid.
109. Ibid., 315–16.
110. Ibid., 316.
111. Ibid., 317.
112. Ibid., 329.
113. Ibid., IIIA: 330.
114. *Memories*, 310–23.
115. Ibid., 373.
116. Ibid., 323.
117. *Life*, IIIB: 587.
118. *Life*, IIIB: 588.
119. Ibid., IIIA: 323.
120. Ibid., 335.
121. Ibid.
122. Pauline M. H. Mazumdar, *Eugenics, Human Genetics and Human Failings: The Eugenics Society, its Sources and its Critics in Britain* (New York: Routledge, Chapman and Hall, Inc. 1992), 28–30. An excellent account of the early days of the Eugenics Education Society.
123. *Life*, IIIA: 339.
124. Mazumdar, *Eugenics, Human Genetics and Human Failings*, 29.
125. *Life*, IIIA: 347–49.
126. Sheila Faith Weiss, *Race Hygiene and National Efficiency: The Eugenics of Wilhelm Schallmayer* (Berkeley: University of California Press, 1987), chap. 3.
127. William H. Schneider, *Quality and Quantity: The Quest for Biological Regeneration in Twentieth-Century France* (Cambridge: Cambridge University Press, 1990), chap. 3.
128. Ibid., 61.
129. Ibid., 62.
130. Daniel J. Kevles, *In the Name of Eugenics: Genetics and the Uses of Human Heredity* (Cambridge: Harvard University Press, 1995), chap. 3.
131. *Life*, IIIA: 340–42.
132. Ibid., 340.
133. Ibid., 347.
134. Ibid., 342.
135. Francis Galton, "Local Associations for Promoting Eugenics," *Essays in Eugenics* (London: Eugenics Education Society, 1909), 101–9. Also see *Nature* 78 (1908): 645–47.
136. Ibid., 105.
137. Ibid., 105.
138. Ibid., 108.
139. *Life*, IIIA: 362.
140. Ibid., 371.
141. Ibid.
142. Ibid., 372.
143. Ibid.
144. Ibid.
145. Ibid., 374.
146. Ibid., 375.
147. Ibid., 377.
148. Ibid., 379.
149. Ibid.
150. Ibid., 380–81.
151. Ibid., IIIB: 597.
152. Ibid., 597.
153. Ibid., IIIA: 386.
154. Egon S. Pearson, *Karl Pearson: An Appreciation of Some Aspects of his Life and Work* (Cambridge: Cambridge University Press, 1938), 46.
155. Ibid., 61.
156. Ibid.
157. Ibid.
158. Ibid.
159. *Life*, IIIA, 405–07.
160. Ibid., 406.
161. Ibid.
162. Ibid.
163. Ibid.
164. Ibid., 396–400.
165. Ibid., 408.
166. Ibid., 408–9.
167. Ibid., 409.
168. Ibid., IIIA: 431.
169. Ibid., 400–401.
170. Ibid., 413–25. Pearson's detailed description of *Kantsaywhere* with much of the remaining text interpolated.
171. Ibid., 414.
172. Ibid.
173. Ibid.

174. Ibid.
175. Ibid., 416.
176. Ibid., 415.
177. Ibid., 416.
178. Ibid.
179. Ibid., 418.
180. Ibid., 419.
181. Ibid.
182. Ibid., 420.
183. Ibid., 422.
184. Ibid.
185. Ibid.
186. Ibid.
187. Ibid., IIIB: 615–16.
188. Galton's executors, presumably at his behest, planned to destroy the manuscript of *Kantsaywhere*, but the niece charged with the job had second thoughts and got in touch with a cousin saying that while she had "destroyed *all* the story" there were a good many pages she felt she could not judge. She returned the mutilated manuscript to her cousin, suggesting that it be sent to Pearson, if she and Eva Biggs agreed, so the fragmentary novel was forwarded to Pearson minus, among other things, the dubious love scenes. Ibid., 411–13.
189. Ibid., 432.
190. Ibid., 433.
191. *Victorian Genius*, 288.

Epilogue: Out of Pandora's Box: The First International Congress of Eugenics

1. Richard I. Dugdale, *The Jukes: A Study in Crime, Pauperism, Disease, and Heredity* (New York: G. P. Putnam's Sons, 1877), 65.
2. *Eugenics Review* 4 (1912–13): 217.
3. "International Congress," *Eugenics Review* 3 (1911–12): 182.
4. "The First International Congress of Eugenics," *Nature* 89 (1912): 558–61.
5. Robert K. Massie, *Dreadnought: Britain, Germany, and the Coming of the Great War* (New York: Random House, 1991), 312.
6. *The Times*, 25 July 1912, reprinted Balfour's speech.
7. Ibid.
8. Ibid.
9. Ibid.
10. Ibid.
11. Ibid.
12. Ibid.
13. Ibid.
14. Ibid.
15. Ibid.
16. Ibid.

17. J. A. Lindsay, "The Case for and against Eugenics," *The Nineteenth Century and After* 72 (1912): 546–57.
18. *The Times*, 26 July 1912, reprinted Darwin's speech.
19. Ibid.
20. Ibid.
21. Ibid.
22. Ibid.
23. Ibid.
24. Ibid.
25. Ibid.
26. Ibid.
27. Ibid.
28. Ibid.
29. Ibid.
30. David Fairchild Weeks, "The Inheritance of Epilepsy," *Problems in Eugenics: Papers communicated to the first International Eugenics Congress held at the University of London, July 24 to 30th, 1912* (London: Eugenics Education Society, 1912), 62–99.
31. Ibid., 78.
32. F. W. Mott, "Heredity and Eugenics in Relation to Insanity," Ibid., 400–428.
33. Ibid., 408.
34. Ibid., 408.
35. Ibid., 408.
36. Ibid., 408.
37. Ibid., 417.
38. Pauline M. H. Mazumdar, *Eugenics, Human Genetics and Human Failings: The Eugenics Society, its Sources and its Critics in Britain* (New York: Routledge, Chapman and Hall, Inc., 1992), 73.
39. Ibid., 87.
40. Ibid., 89–90.
41. Bleecker Van Wagenen, "Preliminary Report of the Committee of the Eugenic Section of the American Breeder's Association to Study and Report on the Best Practical Means for Cutting Off the Defective Germ-Plasm in the Human Population," *Problems in Eugenics*, 460–79.
42. Mazumdar, *Eugenics, Human Genetics and Human Failings*, 90–95.
43. Ibid., 92.
44. Vernon L. Kellog, "Eugenics and Militarism," *Problems in Eugenics*, 220–31.
45. Ibid., 224.
46. Ibid.
47. Frederick L. Hoffman, "Maternity Statistics of the State of Rhode Island, State Census of 1905," Ibid., 334–40.
48. Ibid., 337.
49. Alfredo Niceforo, "The Cause of the Inferiority of Physical and Mental Characters in the Lower Social Classes," Ibid., 189–94.
50. Ibid., 192.

51. M. Lucien March, "The Fertility of Marriages according to Profession and Social Position," Ibid., 208–20.
52. Ibid., 218.
53. Ibid., 218.
54. W. C. D. and C. D. Whetham, "The Influence of Race on History," Ibid., 237–46.
55. Ibid., 241.
56. Ibid., 246.
57. Ibid., 246.
58. Reginald C. Punnett, "Genetics and Eugenics," Ibid., 137–38.
59. Ibid., 137.
60. Ibid., 138.
61. Ibid., 138.
62. Diane B. Paul and Hamish G. Spencer, "The hidden science of eugenics," *Nature* 374 (1995): 302–4.
63. David. Barker, "The Biology of Stupidity: Genetics, Eugenics and Mental Deficiency in the Inter-War Years," *British Journal for the History of Science* 23 (1989): 347–75.
64. *Times*, 31 July 1912.
65. Ibid.
66. Ibid.
67. Ibid.
68. Daniel J. Kevles, *In the Name of Eugenics* (Cambridge: Harvard University Press, 1995).
69. Richard A. Soloway, *Demography and Degeneration: Eugenics and the Declining Birthrate in Twentieth-Century Britain* (Chapel Hill: University of North Carolina Press, 1995).
70. Philip R. Reilly, *The Surgical Solution: A History of Involuntary Sterilization in the United States* (Baltimore: Johns Hopkins University Press, 1991). Reilly details the whole story of involuntary sterilization in the United States and its relationship to eugenics.
71. Robert J. Cynkar, "Buck v. Bell: Felt Necessities v. Fundamental Values," *Columbia Law Review* 81 (1981): 1418–61; Stephen J. Gould, "Carrie Buck's Daughter," *Natural History* (July, 1984), 14–18.
72. Stephen J. Gould, *The Mismeasure of Man* (New York: W. W. Norton, 1981). Gould documents the many fallacies inherent in IQ testing carried out during the early decades of the twentieth century.
73. Kevles, *In the Name of Eugenics*, 82.
74. Ibid., 97.
75. Robert N. Proctor, *Racial Hygiene: Medicine under the Nazis* (Cambridge: Harvard University Press, 1988). This superb book relates the whole story of the progression from involuntary sterilization, to racial laws, to euthanasia, and finally to the Holocaust in Nazi Germany.
76. Paul Weindling, *Health, Race, and German Politics between National Unification and Nazism, 1870-1945* (Cambridge: Cambridge University Press, 1989). This important book highlights important aspects of German health policy in the context of social change and politics.
77. Stefan Kühl, *The Nazi Connection* (New York: Oxford University Press, 1994). This book documents the influence of the eugenics movement in the United States on German eugenics.
78. Proctor, *Radical Hygiene: Medicine under the Nazis* 50.
79. Frank Dikötter, "Race Culture: Recent Perspective on the History of Eugenics," *American Historical Review* 103 (1998): 467–78.
80. "Western eyes on China's eugenics law," *The Lancet* 346: 131 (1995).

BIBLIOGRAPHY

Note that the list of articles and books by Francis Galton is incomplete and only represents those that I actually consulted. For a complete list of papers and books the reader can consult D. W. Forest, *Francis Galton: The Life and Work of a Victorian Genius*. New York: Taplinger, 1974; or Milo Keynes, ed., *Sir Francis Galton, FRS: The Legacy of His Ideas*. London: Macmillan, 1993.

Articles

Barker, David. The Biology of Stupidity: Genetics, Eugenics and Mental Deficiency in the Inter-war Years. *British Journal for the History of Science* 23 (1989): 347–75.
Bateson, William. On some Variations of *Cardium edule* apparently Correlated to the Conditions of Life. *Phil. Trans. Royal Soc. of London, Series B* 153 (1889). Reprinted in *The Scientific Papers of Bateson, William*, ed. Reginald Punnett, 1: 33–71. Cambridge: Cambridge University Press, 1928.
———. The Origin of Cultivated *Cineraria*. *Nature* 51 (1895): 605–07.
———. The Origin of the Cultivated *Cineraria*. *Nature* 52 (1895): 29.
———. Problems of Heredity as a Subject for Horticultural Investigation. *J. Royal Horticultural Society* 25 (1900): parts I and II. Reprinted in Beatrice Bateson, *Bateson, William, F.R.S. Naturalist, his Essays and Addresses*, 171–80. Cambridge: Cambridge University Press, 1928.
———. Heredity, Differentiation, and other Conceptions of Biology: A Consideration of Professor Karl Pearson's paper 'On the Principle of Homotyposis.', *Proc. of the Royal Society* 69, 1901. Reprinted in *The Scientific Papers of Bateson, William*, ed. Reginald C. Punnett, 1: 404–18. Cambridge: Cambridge University Press, 1928.
———. Mendel's Principles of Heredity in Mice. *Nature* 67 (1903): 462–63.
———. Mendel's Principles of Heredity in Mice. *Nature* 67 (1903): 585–86.
———. Mendel's Principles of Heredity in Mice. *Nature* 67 (1903): 33–34.
———. The Present State of Knowledge of Colour-heredity in Mice and Rats. *Proc. of the Zoological Society of London* 2 (1903): 71–99. Reprinted in *The Scientific Papers of Bateson, William*, ed. Reginald C. Punnett, 2: 76–108. Cambridge: Cambridge University Press, 1928.
———. Presidential address to the Zoological Section, British Association: Cambridge Meeting (1904) reprinted in Beatrice Bateson, *Bateson, William, F.R.S. Naturalist, His Essays and Addresses*, 233–59. Cambridge: Cambridge University Press, 1928.
———. A Text-book of Genetics (Review of Lotsy's *Vorlesungen über Descendenztheiorien*, I. Theil). *Nature* 74 (1906): 146.
Bateson, William, and A. Bateson. On the Variations in Floral Symmetry of Certain Parts having Irregular Corollas. *J. of the Linnean Society (Botany)* 28 (1891). Reprinted in *The Scientific Papers of Bateson, William*, ed. Reginald C. Punnett, 1: 126–61. Cambridge: Cambridge University Press, 1928.
Bateson, William, and H. H. Brindley. Some Cases of Variation in Secondary Sexual Characteristics Statistically Examined. *Proc. of the Zoological Society* (1893). In *The Scientific Papers of Bateson, William*, ed. Reginald C. Punnett, 1: 193–201. Cambridge: Cambridge University Press, 1928.
Bateson, William, and E. R. Saunders. Experimental Studies in the Physiology of Heredity. *Reports to the Evolution Committee of the Royal Society*, Report I (1902). Reprinted in part in *Scientific Papers of Bateson, William*, ed. Reginald C. Punnett, 1: 29–68. Cambridge: Cambridge University Press, 1928.
Englander, David, and Rosemary O'Day, eds. *Retrieved Riches: Social Investigation in Britain 1840–1914*. Aldershot: Scolar Press, 1995.
Bulmer, Michael. Galton's Law of Ancestral Heredity. *Heredity* 81 (1998): 579–85.
Burbridge, David. Galton's 100: An Exploration of Francis Galton's Imagery Studies. *British J. of the History of Science* 27 (1994): 443–63.
Castle, W. E., and Glover M. Allen. The Heredity of Albinism. *Proc. of the American Academy of Arts & Sciences* 38 (1903): 602–21.
Cowan, Ruth Schwartz. Francis Galton's Statistical Ideas: The Influence of Eugenics. *Isis* 63 (1972): 509–28.

Darbishire, A. D. Note on the Results of Crossing Japanese Waltzing Mice with European Albino Races. *Biometrika* 2 (1903): 101–4.
———. Second Report on the Result of Crossing Japanese Waltzing Mice with European Albino Races. *Biometrika* 2 (1903): 165–73.
———. Third Report on Hybrids between Waltzing Mice and Albino Races. *Biometrika* 2 (1903): 282–85.
———. On the Result of Crossing Japanese Waltzing Mice with Albino Mice. *Biometrika* 3 (1904): 1–51.
Dean, Graham, and Peter E. Morris. Imagery and Spatial Ability: When Introspective Reports Predict Performance. *Advances in Psychology* 80 (1991): 331–47.
De Marrais, Robert. The Double-edged Effect of Sir Francis Galton: A Search for the Motives in the Biometrician-Mendelian Debate. *J. History of Biology* 7 (1974): 141–74.
Dikötter, Frank. Race Culture: Recent Perspectives on the History of Eugenics. *American Historical Review* 103 (1998): 467–78.
Erickson, Paul Alfred. Phrenology and Physical Anthropology: The George Combe Connection. *Occasional Papers in Anthropology*, No. 6. Halifax, Nova Scotia: Department of Anthropology, Saint Mary's University, 1979.
Fancher, Raymond. A Note on the Origin of the Term 'Nature and Nurture.' *J. History of Behavioral Sciences* 15 (1979): 321–22.
———. Alphonse de Candolle, Francis Galton, and the Early History of the Nature-Nurture Controversy. *J. History of Behavioral Sciences* 19 (1983): 341–51.
Faulds, Henry. On Skin Furrows in the Hands. *Nature* 22 (1880): 605.
Faulds, Henry. On the Identification of Habitual Criminals by Finger-prints. *Nature* 50 (1894): 548.
Fergus, Walter, and G. F. Rodwell. On a Series of Measurements for Statistical Purposes Recently made at Marlborough College. *J. of the Anthropological Institute* 4 (1874): 126–29.
Fisher, Ronald A. The Correlation Between Relatives on the Supposition of Mendelian Inheritance. *Trans. of the Royal Society of Edinburgh* 52 (1918): 399–433.
Galton, Francis. The Exploration of Arid Countries. *Proc. of the Royal Geographical Society* 2 (1857–58): 60–77.
———. Sun Signals for the Use of Travellers. *Proc. of the Royal Geographical Society* 4 (1859): 14–19.
——— (unsigned article). Recent Discoveries in Australia. *The Cornhill Magazine* 5 (1862): 354–64.
———. Hereditary Talent and Character. *Macmillan's Magazine* 12 (1865): 157–66, 318–27.
———. Barometric Predictions of Weather. *British Association Report* 40 (1870): 31–33.
———. Experiments in Pangenesis, by Breeding from Rabbits of a Pure Variety, into whose Circulation Blood taken from Other Varieties had previously been Largely Transfused. *Proc. of the Royal Society* 19 (1871): 393–410.
———. Statistical Inquiries into the Efficacy of Prayer. *Fortnightly Review* 12 (1872): 125–35.
———. Hereditary Improvement. *Fraser's Magazine* 7 (1873): 116–30.
———. On the Causes which Operate to Create Scientific Men. *Fortnightly Review* 13 (1873): 345–51.
———. Notes on the Marlborough School Statistics. *J. of the Anthropological Institute* 4 (1874): 130–37.
———. Proposal to apply for Anthropological Statistics from Schools. *J. of the Anthropological Institute* 3 (1874): 308–11.
———. Men of Science, their Nature and their Nurture. *Nature* 9 (1874): 344–45.
———. On a Proposed Statistical Scale. *Nature* 9 (1874): 342–43.
———. Statistics of Intercomparison, with Remarks on the Law of Frequency of Error. *Philosophical Magazine* 49 (1875): 33–46.
———. The History of Twins, as a Criterion of the Relative Powers of Nature and Nurture. *Fraser's Magazine* 12 (1875): 566–76.
———. Short Notes on Heredity etc., in Twins. *J. of the Anthropological Institute* 5 (1875): 324–29.
———. The History of Twins, as a Criterion of the Relative Powers of Nature and Nurture. *J. of the Anthropological Institute* 5 (1875): 391–406.
———. A Theory of Heredity. *Contemporary Review* 27 (1875): 80–95.
———. A Theory of Heredity. *J. Anthropological Institute* 5 (1875): 329–48.
———. On the Height and Weight of Boys aged 14, in Town and Country Public Schools. *J. of the Anthropological Institute* 6 (1876): 174–81.
———. Typical Laws of Heredity. *Proc. of the Royal Institution* 8 (1877): 282–301. Also in *Nature* 15 (1877): 492–95, 512–14, 532–33.
———. Address to Section D—Biology, British Association, *Nature* 16 (1877): 344–47.
———. Composite Portraits. *J. of the Anthropological Institute* 8 (1878): 132–44.

Galton, Francis. Psychometric Facts. *Nineteenth Century* 5 (1879): 425–33.
———. Psychometric Experiments. *Brain* 2 (1879): 149–62.
———. Statistics of Mental Imagery. *Mind* 5 (1880): 301–18.
———. Visualized Numerals. *J. of the Anthropological Institute* 10 (1880): 85–102.
———. Mental Imagery. *Fortnightly Review* 28 (1880): 312–24.
———. The Anthropometric Laboratory. *Fortnightly Review* 31 (1882): 332–38.
———. Medical Family Registers. *Fortnightly Review* 34 (1883): 244–50.
———. On the Anthropometric Laboratory at the Late International Health Exhibition. *J. of the Anthropological Institute* 14 (1884): 205–21.
———. Some Results of the Anthropometric Laboratory. *J. of the Anthropological Institute* 14 (1884): 275–87.
———. The Measurement of Fidget. *Nature* 32 (1885): 174–75.
———. Family Likeness in Stature. *Proc. of the Royal Society* 40 (1886): 42–72.
———. Family Likeness in Eye-colour. *Proc. of the Royal Society* 40 (1886): 402–17.
———. Regression Towards Mediocrity in Hereditary Stature. *J. of the Anthropological Institute* 15 (1886): 246–63.
———. Pedigree Moth-Breeding, as a Means of verifying Certain Important Constants in the General Theory of Heredity. *Trans. of the Entomological Society of London* Part I (1887): 19–28.
———. On Recent Designs for Anthropometric Instruments. *J. of the Anthropological Institute* 16 (1887): 2–9.
———. Personal Identification and Description. *Nature* 38 (1888): 173–77, 201–2.
———. Co-relations and their Measurements, chiefly from Anthropometric Data. *Proc. of the Royal Society* 45 (1888): 135–45.
———. A Visit to Northern Spain at the Time of the Eclipse in 1860. Reprinted from *Vacation Tourists*, 1860 and 1861, together with Galton, Francis. *Narrative of an Explorer in Tropical South Africa*. London: Ward, Lock & Co., 1889.
———. I. The Patterns in Thumb and Finger Marks—On their Arrangement into Naturally Distinct Classes, the Permanence of the Papillary ridges that make them, and the Resemblance of their Classes to Ordinary Genera. *Phil. Soc. Transactions of the Royal Soc. of London*, Series B 182 (1891): 1–23.
———. Method of Indexing Fingerprints. *Proc. of the Royal Society* 54 (1891): 540–45.
———. Identification by Fingerprints. *Nineteenth Century* 30 (1891): 303–11.
———. Psychology of Mental Arithmeticians and Blindfold Chess-players. *Nature* 51 (1894): 73–74.
———. Discontinuity in Evolution. *Mind* (n.s.), 3 (1894): 362–72.
———. Intelligible Signals from Neighboring Stars. *Fortnightly Review* 60 (1896): 657–64.
———. The Average Contribution of Each Several Ancestor to the Total Heritage of the Offspring. *Proc. of the Royal Society* 61 (1897): 401–13.
———. The Possible Improvement of the Human Breed under the Existing Conditions of Law and Sentiment. *Nature* 64 (1901): 659–65.
———. Biometry. *Biometrika* 1 (1902): 9
———. Pedigrees. *Nature* 67 (1903): 586–87.
———. Nomenclature and Tables of Kinship. *Nature* 69 (1904): 294–95.
———. Average Number of Kinsfolk in Each Degree. *Nature* 70 (1904): 529, 626; 71 (1905): 248.
———. Eugenics: Its Definition, Scope and Aims. *Sociological Papers* 1, 45–50.
———. Nomenclature of kinship—its extension. *Nature* 73 (1905): 150–51.
———. Restrictions in Marriage. *Sociological Papers* 2 (1905): 2–13.
———. Studies in National Eugenics. *Sociological Papers* 2 (1905): 14–17.
———. Measurement of Resemblance. *Nature* 74 (1906): 562–63.
———. The Possible Improvement of the Human Breed under the Existing Conditions of Law and Sentiment. In *Essays in Eugenics*, 1–34. London: Eugenics Education Society, 1909. Also see *Nature* 64 (1901): 659–65.
———. Local Associations for Promoting Eugenics. In *Essays in Eugenics*, 101–9. London: Eugenics Education Society, 1909. Also see *Nature* 78 (1908): 645–47.
Gökyigit, Emel Aileen. The Reception of Francis Galton's *Hereditary Genius* in the Victorian Periodical Press. *J. History of Biology* 27 (1994): 215–40.
Gray, Paul. Cursed by Eugenics. *Time* 153 (no. 1, 1999): 84–85.
Haldane, Karl Pearson, J.B.S. (1857–1957). A centenary lecture delivered at University College London. Originally in *Biometrika* 44 (1957): 303–13. Reprinted in *Studies in the History of Statistics and Probability*, ed. Egon S. Pearson and M. G. Kendall, 1: 427–47. (London: Griffin, 1970).

Herschel, William J. Skin Furrows of the Hand. *Nature* 23 (1880): 76.
———. Finger-Prints. *Nature* 51 (1894): 77.
Hoffman Frederick L. Maternity Statistics of the State of Rhode Island, State Census of 1905, The Inheritance of Epilepsy. In *Problems in Eugenics: Papers communicated to the first International Eugenics Congress held at the University of London, July 24 to 30th, 1912*, 334–40. London: Eugenics Education Society, 1912.
Kellog, Vernon L. Eugenics and militarism, The inheritance of epilepsy. In *Problems in Eugenics: Papers communicated to the first International Eugenics Congress held at the University of London, July 24 to 30th, 1912*, 220–31. London: Eugenics Education Society, 1912.
Kendall, Maurice G. George Udny Yule, 1871–1951. *J. Royal Statistical Society* 115A (1952): 156–61. Reprinted in *Studies in the History of Statistics and Probability*, ed. Egon S. Pearson and Maurice G. Kendall, 1: 418–25. London: Griffin, 1970.
———. Francis Ysidro Edgeworth, 1845–1926. *Biometrika* 55 (1968): 269–75. Reprinted in *Studies in the History of Statistics and Probability*, ed. Egon S. Pearson and Maurice G. Kendall, 1: 257–63. London: Griffin, 1970.
Lewes, George H. Mr. Darwin's Hypothesis. *Fortnightly Review*, 4 n.s. (1868): 492–509.
Lindsay, J. A. The Case For and Against Eugenics. *The Nineteenth Century and After* 72 (1912): 546–57.
McKendrick, John G. On the Modern Cell Theory and Phenomenon of Fecundation. *Proc. Philosophical Society of Glasgow* 19 (1887–88): 71–125.
MacLeod, Roy. Retrospect: The British Association and its Historians. In *The Parliament of Science: The British Association for the Advancement of Science 1831–1981*, ed. Roy MacLeod and Peter Collins, 17–42. Northwood: Science Reviews Ltd., 1981.
March, M. Lucien. The Fertility of Marriages According to Profession and Social Position, The Inheritance of Epilepsy. In *Problems in Eugenics: Papers communicated to the first International Eugenics Congress held at the University of London, July 24 to 30th, 1912*, 208–20. London: Eugenics Education Society, 1912.
Maynard Smith, John. Galton and Evolutionary Theory. In *Sir Francis Galton, FRS: The Legacy of His Ideas*, 158–169, ed. Milo Keynes. London: Macmillan, 1993.
Mendel, Gregor. Versuch über Pflanzen-Hybriden. *Verhandlungen des naturforschenden Vereines in Brünn* 4 (1865): 3–47.
Mott F. W. Heredity and Eugenics in Relation to Insanity. In *Problems in Eugenics: Papers communicated to the first International Eugenics Congress held at the University of London, July 24 to 30th, 1912*, 400–28. London: Eugenics Education Society, 1912.
Neale, A. Description of Constantinople. In *A Handbook for Travellers in Turkey in Asia*, 64–65. London: John Murray, 1875.
Niceforo, Alfredo. The Cause of the Inferiority of Physical and Mental Characters in the Lower Social Classes, The Inheritance of Epilepsy. In *Problems in Eugenics: Papers communicated to the first International Eugenics Congress held at the University of London, July 24 to 30th, 1912*, 189–194. London: Eugenics Education Society, 1912.
Nye, Robert A. The Rise and Fall of the Eugenics Empire: Recent Perspectives on the Impact of Biomedical Thought in Modern Society. *The Historical Journal* 36 (1993): 687–700.
Orange, A. D. The Beginnings of the British Association, 1831–1851. In *The Parliament of Science: The British Association for the Advancement of Science 1831–1981*, 43–64. Ed. Roy MacLeod and Peter Collins, Norwood: Science Reviews Ltd., 1981.
Paul, Diane B. and Hamish G. Spencer. The hidden science of eugenics. *Nature* 374 (1995): 302–4.
Pearson, Karl. Mathematical Contributions to the Theory of Evolution. On the Law of Ancestral Heredity. *Proc. of the Royal Society* 62 (1898): 386–412.
———. Mathematical Contributions to the Theory of Evolution.—IX. On the Principle of Homotyposis and its Relation to Heredity, to the Variability of the Individual, and to that of the Race. Part I.—Homotyposis in the Vegetable Kingdom. *Philosophical Trans. of the Royal Society, Series A* 197 (1901): 285–379.
———. On the Fundamental Conceptions of Biology. *Biometrika* 1 (1902): 320–44.
———. Walter Frank Raphael Weldon, 1860–1906. *Biometrika* 5 (1906): 1–52. Reprinted in *Studies in the History of Statistics and Probability*, ed. Egon S. Pearson and Maurice G. Kendall, 1: 265–321. London: Griffin, 1970.
———. Notes on the History of Correlation. *Biometrika* 13 (1920): 25–45. Reprinted in *Studies in the History of Statistics and Probability*, ed. Egon S. Pearson and Maurice G. Kendall, 1: 185–205. London: Griffin, 1970.
Punnett, Reginald C. Genetics and Eugenics, The Inheritance of Epilepsy. *Problems in Eugenics: Papers communicated to the first International Eugenics Congress held at the University of London, July 24 to 30th, 1912*, 137–38. London: Eugenics Education Society, 1912.

Richardson, John T. E. Gender Differences in Imagery, Cognition, and Memory. *Advances in Psychology* 80 (1991): 271–303.
Spottiswoode, William. On Typical Mountain Ranges: An Application of the Calculus of Probabilities to Physical Geography. *J. of the Royal Geographical Society* 31 (1861): 149–54.
Stigler, Stephen M. Galton and Identification by Fingerprints. *Genetics* 140 (1995): 857–60.
Tanner, J. M. Galton on Human Growth and Form. In *Sir Francis Galton, FRS*, ed. Milo Keynes, 108–18. London: The Macmillan Press, Ltd., 1993.
Thiselton-Dyer, W. T. Variation and Specific Stability. *Nature* 51 (1895); 459–61.
———. Origin of Cultivated *Cineraria*. *Nature* 52 (1895). 3–4.
———. The Origin of the Cultivated *Cineraria*. *Nature* 52 (1895): 128–29.
Van Wagenen, Bleecker. Preliminary Report of the Committee of the Eugenic Section of the American Breeder's Association to Study and Report on the Best Practical Means for Cutting Off the Defective Germ-Plasm in the Human Population. *Problems in Eugenics: Papers communicated to the first International Eugenics Congress held at the University of London, July 24 to 30th, 1912*, 460–79. London: Eugenics Education Society, 1912.
Vorzimmer, Peter J. Charles Darwin and Blending Inheritance. *Isis* 54 (1963): 371–90.
Wallace, Alfred Russel. The Method of Organic Evolution. Parts I and II, *Fortnightly Review* 63 (1895): 211–24, 435–45.
Warwick, Andrew. Exercising the Student Body. In *Science Incarnate: Historical Embodiments of Natural Knowledge*, ed. Christopher Lawrence and Steven Shapin, chapter 8. Chicago: University of Chicago Press, 1998.
Weeks, David Fairchild. The Inheritance of Epilepsy. *Problems in Eugenics: Papers communicated to the first International Eugenics Congress held at the University of London, July 24 to 30th, 1912*, 62–99. London: Eugenics Education Society, 1912.
Weldon, W. F. R. The Variations Occurring in Certain Decapod Crustaceans. I. *Crangon vulgaris. Proc. of the Royal Society* 47 (1890): 445–53.
———. Certain Correlated Variations in *Crangon vulgaris. Proc. of the Royal Society* 51 (1892): 2–21.
Weldon, W. F. R. On Certain Correlated Variations in *Carcinus moenas. Proc. of the Royal Society* 54 (1893): 318–33.
———. The Study of Animal Variation. *Nature* 50 (1894): 25–26.
———. The Origin of the Cultivated *Cineraria*. *Nature* 52 (1895): 54.
———. The Origin of the Cultivated *Cineraria*. *Nature* 52 (1895): 29.
———. I. Remarks on Variation in Animals and Plants. *Proc. of the Royal Society* 57 (1895): 361–79.
———. II. Remarks on Variations in Animals and Plants. *Proc. of the Royal Society* 57 (1895). 379–82.
———. Mendel's Laws of Alternative Inheritance in Peas. *Biometrika* 1 (1902): 228–54.
———. Professor de Vries on the Origin of Species. *Biometrika* 1 (1902): 365–74.
———. On the Ambiguities of Mendel's Characters. *Biometrika* 2 (1903): 44–55.
———. Mendel's Principles of Heredity in Mice. *Nature* 67 (1903): 34.
———. Mendel's Principles of Heredity in Mice. *Nature* 67 (1903): 512.
———. Mendel's Principles of Heredity in Mice. *Nature* 67 (1903): 610.
———. Mr Bateson's Revisions of Mendel's Theory of Heredity. *Biometrika* 2 (1903): 286–98.
Wetherick, Norman E. What Goes on in the Mind When we Solve Syllogisms? *Advances in Psychology* 80 (1991): 255–67.
Whetham, W. C. D. and C. D. The Influence of Race on History, The Inheritance of Epilepsy. *Problems in Eugenics: Papers communicated to the first International Eugenics Congress held at the University of London, July 24 to 30th, 1912*, 237–46. London: Eugenics Education Society, 1912.
Winter, Alison. A Calculus of Suffering: Ada Lovelace and the Bodily Constraints on Women's Knowledge in Early Victorian England. In *Science Incarnate: Historical Embodiments of Natural Knowledge*, ed. Christopher Lawrence and Steven Shapin, chapter 6. Chicago: University of Chicago Press, 1998.
Wright, Lawrence. Double Mystery. *The New Yorker* (August, 1995): 46–62.
Yule, G. Udny. Mendel's Laws and Their Probable Relations to Intra-racial Heredity. *New Phytologist* 1 (1902): 193–207, 222–38.

Books

Andersson, Charles G. *Lake Ngami; or Four Years in Africa*. Philadelphia: John E. Potter., Co., 1885?.
Andrade, E. N. da Costa. *A Brief History of the Royal Society*. London, 1960.

Baedeker, Karl. *Egypt and Sudan*. Eighth revised edition. Leipzig: Karl Baedeker, 1929.
Barker, A. J. *The Vainglorious War 1854-56*. London: Weidenfeld & Nicolson, 1970.
Bateson, Beatrice. *Bateson, William, F.R.S.: Naturalist*. Cambridge: Cambridge University Press, 1928.
Bateson, William. *Materials for the Study of Variation treated with especial regard to Discontinuity in the Origin of Species*. London: Macmillan, 1894.
Beaver, Patrick. *The Crystal Palace*. London: Hugh Evelyn, 1970.
Bernstein, Peter L. *Against the Gods: The Remarkable Story of Risk*. New York: John Wiley & Sons, Inc., 1996.
Booth, Charles. *Life and Labour in London*, 1 (Poverty). London: Macmillan, 1902.
Boring, Edwin Garrigues. *A History of Experimental Psychology*. Second edition. New York: Appleton-Century-Crofts, 1950.
Brodie, Fawn M. *The Devil Drives: A Life of Sir Richard Burton*. New York: W. W. Norton, 1967.
Brooks, William K. *The Law of Heredity: A Study of the Cause of Variation and the Origin of Living Organisms*. Baltimore: John Murphy, 1883.
Browne, Douglas G., and Alan Brock. *Fingerprints: Fifty Years of Scientific Crime Detection*. London: George G. Harrap, 1953.
Browne, Janet. *Charles Darwin: Voyaging*. New York: Alfred A. Knopf, 1995.
Burton, Isabel. *The Life of Captain Sir Richard Francis Burton*. London: Duckworth, 1898.
Collins, Robert O., and Robert L. Tignor. *Egypt and the Sudan*. Englewood Cliffs: Prentice Hall, 1967.
Cooks' Tourist's Handbook for Palestine and Syria. London: Thos. Cook, 1911.
Cowan, Ruth Schwartz. *Sir Francis Galton and the Study of Heredity in the Nineteenth Century*. Ann Arbor, Michigan, 1969.
Cowell, Frank R. *The Athenaeum: Club and Social Life in London 1824-74*. London: Heinemann, 1975.
Darwin, Charles R. *The Origin of Species by Means of Natural Selection, or the Preservation of Favored Races in the Struggle for Life*. Murray: London, 1859.
———. *The Variation of Animals and Plants under Domestication*. 2 vols. London: J. Murray, 1868.
———. *The Variation of Animals and Plants under Domestication*. Second edition. London: J. Murray, 1875. Reprinted in the *Works of Charles Darwin*, ed. Paul H. Barrett and R. B. Freeman. London: William Pickering, 1988.
———. *The Descent of Man, and Selection in relation to Sex*. 2 Vols. London: Murray, 1871.
———. *The Descent of Man in Relation to Sex*. Modern Library Edition. New York: Random House.
de Candolle, Alphonse. *Histoire des Sciences et des Savants depuis Deux Siécles*. Geneva: Georg, 1873.
Desmond, Adrian. *Huxley: From Devil's Disciple to Evolution's High Priest*. Reading: Addison-Wesley, 1997
Desmond, Adrian, and James Moore. *Darwin*. London: Michael Joseph, 1991.
Dugdale, Richard L. *The Jukes: A Study in Crime, Pauperism, Disease, and Heredity*. New York: G. P. Putnam's Sons, 1877.
Dunn, Waldo Hilary. *James Anthony Froude: A Biography, 1857-1894*. London: Oxford University Press, 1963.
Edwards, Amelia A. B. *A Thousand Miles up the Nile*. London: Century, reissued 1989.
Everett, Edwin Mallard. *The Party of Humanity: The Fortnightly Review and its Contributors 1865-1874*. Chapel Hill: University of North Carolina Press, 1939.
Fancher, Raymond. *Pioneers of Psychology*. New York: W. W. Norton, 1979.
———. *The Intelligence Men: Makers of the IQ Controversy*. New York: W. W. Norton, 1985.
Farwell, Byron. *Burton: A Biography of Sir Richard Francis Burton*. New York: Holt, Rinehart and Winston, 1963.
First, Ruth, and Ann Scott. *Olive Schreiner*. New York: Schocken Books, 1980.
Forrest, D. W. *Francis Galton: The Life and Work of a Victorian Genius*. New York: Taplinger, 1974.
Galton, Francis. *Hereditary Genius: An Inquiry into its Laws and Consequences*. London: Macmillan, 1869.
———. *Hereditary Genius: An Inquiry into its Laws and Consequences*. Second American edition. New York: Appleton, 1879.
———. *Hereditary Genius: An Inquiry into Its Laws and Consequences*. London: Macmillan, 1892.
———. *English Men of Science: their Nature and Nurture*. New York: Appleton, 1875.
———. *Inquiries into Human Faculty and its Development*. London: Macmillan, 1883.
———. *Record of Family Faculties*. London: Macmillan, 1884.
———. *Narrative of an Explorer in Tropical South Africa*. Second edition. London: Ward, Lock, 1889.
———. *Natural Inheritance*. London: Macmillan, 1889.
———. *Finger Prints*. London: Macmillan, 1892.
———. *Finger Print Directories*. London: Macmillan, 1895.
———. *Noteworthy Families*. London: J. Murray, 1909.
———. *Memories of my Life*. Third edition. London: Methuen, 1909.

Galton, Francis. *Inquiries into Human Faculty and its Development.* Everyman's Library edition. New York: E. P. Dutton, 1928.
———. *The Art of Travel; or Shifts and Contrivances Available in Wild Countries.* London: Murray, 1855.
———. *The Art of Travel; or, Shifts and Contrivances Available in Wild Countries.* Reprint of the fifth edition. London: Phoenix Press, 2001.
Goldblatt, I. *History of Southwest Africa from the Beginning of the Nineteenth Century.* Cape Town: Juta, 1971.
Gould, Stephen J. *The Mismeasure of Man.* New York: W. W. Norton, 1981.
Graham, Edward. *The Harrow Life of Henry Montagu Butler.* London: Longmans Green, 1920.
Hall, Marie B. *All Scientists Now: The Royal Society in the Nineteenth Century.* Cambridge: Cambridge University Press, 1984.
Herrnstein, Richard and Charles Murray. *The Bell Curve.* New York: Free Press, 1994.
Jeal, Tim. *Livingstone.* New York: G. P. Putnam's Sons, 1973.
Kevles, Daniel J. *In the Name of Eugenics.* Cambridge: Harvard University Press, 1995.
King-Hele, Desmond. *Doctor of Revolution: The Life and Genius of Erasmus Darwin.* London: Faber & Faber, 1977.
———. *Essential Writings of Erasmus Darwin.* London: MacGibbon & Kee, 1968.
Kühl, Stefan. *The Nazi Connection.* New York: Oxford University Press, 1994.
Kutzbach, Gisela. *The Thermal Theory of Cyclones: A History of Meteorological Thought in the Nineteenth Century.* Boston: American Meteorological Society, 1979.
Lawrence, Christopher, and Steven Shapin, eds. *Science Incarnate: Historical Embodiments of Natural Knowledge.* Chicago: University of Chicago Press, 1998.
Lehmann, Joseph H. *Remember you are an Englishman: A Biography of Sir Harry Smith, 1787-1860.* London: Jonathan Cape, 1977.
Lessner, Erwin C., and Ann M. Lingg Lessner. *The Danube: The Dramatic History of the Great River and the People Touched by Its Flow.* Garden City, N.Y.: Doubleday, 1961.
Levinson, Olga. *Story of Namibia.* Capetown: Tafelberg Publishers, 1978.
Logie, Robert H. and Michel Denis, eds. *Mental Images in Human Cognition. Advances in Psychology* 80: 1991.
MacCarthy, Fiona. *William Morris: A Life For Our Time.* New York: Alfred A. Knopf, 1995.
Mack, Edward C. and W. H. G. Armytage. *Thomas Hughes: The Life of the Author of Tom Brown's Schooldays.* London: Ernest Benn, 1952.
McLynn, Frank J. *Stanley: The Making of an African Explorer.* London: Constable, 1989.
Maitland, Alexander. *Speke.* London: Constable, 1971.
Martin, Thomas. *The Royal Institution.* London: Longmans Green, 1949.
Massie, Robert K. *Dreadnought: Britain, Germany, and the Coming of the Great War.* New York: Random House, 1991.
Mazumdar, Pauline M. H. *Eugenics, Human Genetics and Human Failings: The Eugenics Societies, its Sources and its Critics in Britain.* New York: Routledge, Chapman and Hall, 1992.
Mellersh, H. E. I. *FitzRoy of the Beagle.* London: Rupert Hart-Davis, 1968.
Mill, Hugh R. *The Record of the Royal Geographical Society, 1830-1930.* London: Royal Geographical Society, 1930.
Moorehead, Alan. *The Blue Nile.* New York: Harper and Row, 1962.
———. *The White Nile.* New York: Harper & Brothers, 1960.
Morgan, Charles. *The House of Macmillan (1843-1943).* New York: Macmillan, 1944.
Olby, Robert. *Origins of Mendelism.* Second edition. Chicago: The University of Chicago Press, 1985.
Paivio, Allan. *Images in the Mind.* New York: Harvester-Wheatsheaf, 1991.
Pearson, Egon S. *Karl Pearson: An Appreciation of Some Aspects of His Life and Work.* Cambridge: Cambridge University Press, 1938.
Pearson, Hesketh. *Modern Men and Mummers.* London: George Allen & Unwin, Ltd., 1921.
Pearson, Karl. *The Grammar of Science.* London: Dent, 1937 republication of 1892 edition.
———. *The Life, Letters and Labours of Francis Galton.* Cambridge: Cambridge University Press, I, 1914; II, 1924; IIIA and IIIB, 1930.
Proctor, Robert N. *Racial Hygiene: Medicine under the Nazis.* Cambridge: Harvard University Press, 1988.
Prouty, Roger Warren. *The Transformation of the Board of Trade 1830-1855.* London: William Heinemann, 1957.
Provine, William B. *The Origins of Theoretical Population Genetics.* Chicago: The University of Chicago Press, 1971.

Reilly, Philip R. *The Surgical Solution: A History of Involuntary Sterilization in the United States*. Baltimore: Johns Hopkins University Press, 1991.
Rhodes, Henry T. F. *Alphonse Bertillon*. New York: Abelard-Schuman, 1956.
Rothblatt, Sheldon. *The Revolution of the Dons*. Cambridge: Cambridge University Press, 1981.
Royal Institution of Great Britain: The Record of the Royal Institution of Great Britain. London: Wm. Clowes & Sons, 1941.
Santi, Paul, and Richard Hill. *The Europeans in the Sudan 1834-1878*. Oxford: Oxford at the Clarendon Press, 1980.
Schneider, William H. *Quality and Quantity: The Quest for Biological Regeneration in Twentieth-Century France*. Cambridge: Cambridge University Press, 1990.
Schofield, Robert E. *The Lunar Society of Birmingham*. London: Oxford University Press, 1963.
Smith, Harry. *The Autobiography of Lieutenant General Sir Harry Smith*, I. London: J. Murray, 1901.
Sokal, Michael M., ed. *An Education in Psychology: James McKeen Cattell's Journal and Letters from Germany and England, 1880-1888*. Cambridge, Mass.: MIT Press, 1981.
Soloway, Richard A. *Demography and Degeneration*. Chapel Hill: The University of North Carolina Press, 1995.
Srebrnik, Patricia Thomas. *Alexander Strahan, Victorian Publisher*. Ann Arbor: The University of Michigan Press, 1986.
Stafford, Robert A. *Scientist of Empire*. Cambridge: Cambridge University Press, 1989.
Stanley, Henry M. *How I found Livingstone: Travels, Adventures, and Discoveries in Central Africa*. New York: Charles Scribner's & Sons, 1899.
Stigler, Stephen M. *The History of Statistics: The Measurement of Uncertainty before 1900*. Cambridge: Belknap Press of Harvard University Press, 1986.
Sturtevant, A. H. *A History of Genetics*. New York: Harper & Row, 1965.
Trevelyan, George M. *Trinity College, An Historical Sketch*. Cambridge: Cambridge University Press, 1943.
Tunstall, James. *Rambles about Bath and its Neighborhood*. Seventh edition. London: Simpken & Marshall, 1876.
Weindling, Paul. *Health, Race and German Politics between National Unification and Nazism, 1870-1945*. Cambridge: Cambridge University Press, 1989.
Weiss, Sheila Faith. *Race Hygiene and National Efficiency: The Eugenics of Wilhelm Schallmayer*. Berkeley: University of California Press, 1987.
Williams, Frieda-Nela. *Precolonial Communities of Southwestern Africa: A History of Owambo Kingdoms 1600-1920*. Archeia No. 16. Windhoek: National Archives of Namibia, 1991.
Winter, Alison. *Mesmerized: Powers of the mind in Victorian Britain*. Chicago: University of Chicago Press, 1998.
Wolf, Theta Holmes. *Alfred Binet*. Chicago: University of Chicago Press, 1973.
Woodruff, Douglas. *The Tichborne Claimant: A Victorian Mystery*. New York: Farrar, Straus and Cudahy, 1957.
Youngson, A. J. *The Scientific Revolution in Victorian Medicine*. New York: Holmes & Meier, 1979.

INDEX

A Study of American Intelligence (Brigham), 354
A Swan and Her Friends (Lucas), 18
"A Theory of Heredity" Galton (1875), 181
Aberfeldy, Scotland, 44, 45
Aberystwyth, Wales, 26
Abu Hamad, Sudan, 51–52
Acacias, 69
Acre (Akko), Israel, 54
Aden, Lebanon, 55
Aden (now Yemen), 112
Adult elements, 180
Advances in Psychology, 8, 226
Against the Gods: The Remarkable Story of Risk (Bernstein), 7, 250, 256
Airy, Sir George, 103
Alcock, Sir Rutherford, 98, 132
Aldershot, Hampshire, 102
Alexander, Sir James, 119
Alexandria, Egypt, 48
Algoa Bay, South Africa, 64, 68
Ali, Dragoman, 48–49, 51, 54, 56
Aliwal (India), 64
Aliwal North (South Africa), 64
Aliwal South (South Africa), 64
Alkaptonuria, 312
Allelomorph, compound 312–314
Allfancy, A., 342–343
Allfancy, T., 343
al-Matamma, Sudan, 52, 54
Alma River, Crimea, 101
Alpine Journal, 103
Alverstone, Lord, 345
Ambiguity of Mendel's Categories, Weldon (1903), 313
Amboise, France, 99
American Breeders' Association, 350
Amiral, 90
Ana, 69
Ancestral law, 263
Ancestral Law of Heredity, 303–06, 310, 314–19
Andersson, Charles J., 62–63, 65, 68–71, 74–75, 77, 79, 82–86, 88–91, 99, 136
  Brush with heatstroke, 71
  Encounter with a rhinoceros, 75
  Observations on Damaras, 71–72
  Observations on Hill Damaras, 75
Anthropometric Laboratory, 7, 210–14, 227, 250–51, 264, 266, 269
Anticyclone, 140, 147
Archdall-Reid, Dr. F.W., 336
Archiv für Rassen- und Gesellschaftsbiologie, 336, 340, 355
Arnold, Matthew, 156
Arrowsmith, John, 62, 122
Art of Travel (Galton), 4, 98–101
Artistic ability, inheritance, 264, 266
Asquith, H.H., 244, 340
Assagai, 72–74, 82, 84–85
Athenaeum, 273
Athenaeum Club, 106
Atkinson, Thomas, 98
Atwood, Rev., 25, 39
Aveling, E., 274

Baboons, 85
Back, Admiral Sir George, 126
Badajoz, Spain, 64
Baden, Germany, 97
Bagamayo (Bagamoyo), Tanzania, 125
Bahr-el-Ghazal, White Nile tributary, 125, 128–29
Baker, Sir Samuel, 99, 107, 115, 119–22, 137
Balaklava, Crimea, 101, 144
Balaklava Bay, Crimea, 144
Balanoglossus, 286–288
Balfour, Arthur
  Address at First International Congress of Eugenics, 346–47
Balfour, Francis, 279–80, 286
Ball, J., 224
Ballot, A.C. Buys, 144
Bam, Rev. Mr., 68–69
Barclay, Hedworth, 48, 49, 52–54, 63
Barclay, Robert, 21
Barlow, Sir T., 345
Barmen, Namibia, 72–73, 75, 77, 79
Baroness Burdett-Coutts, 127
Barrington, A., 334
Barth, Dr., 112, 119
Basset hounds, 303–05, 311

Baster, 73
Bateson, Beatrice
   Wife of William Bateson, 288, 296, 305
Bateson, William Henry
   Father of William Bateson, 286
Bateson, William, 9, 186, 286, 288, 294, 303–05, 307, 310–11, 313–14, 317, 320–22, 329, 338, 340, 344, 352
   Books
      Materials for Study of Variation (1894), 290, 292, 294, 311–312
      Mendel's Principles of Heredity: A Defence (1902), 310, 311, 314, 320
   Cambridge meeting of the BAAS (1904), 320–23
   Criticizes Weldon's shore crab results, 300
   Critique of Darbishire's experiments, 314–19
   Defense of Mendel, 310–12
   Discontinuous evolution, 294
   Discontinuous variation in beetles and earwigs, 289–90
   Discontinuous variation in *Streptocarpus*, 289
   Evolution Committee of the Royal Society, 300–01
   Family and Education, 286–87
   Fellow of St. John's College, 288
   Galton's influence, 288–90
   Genetics of coat color in mice, 317
   Inadequacy of ancestral law, 305
   Influence of Francis Balfour, 286
   Influence of W. K. Brooks and saltatory evolution, 287–88
   Pearson and Bateson, 300, 305–07, 310–12, 322
   Personality, 288
   Rediscovery of Mendel's laws, 305
   Research on *Balanoglossus*, 286–88
   Stewardship of St. John's College, 289
   Variation in cockles *Cardium edule*, 288
   Wallace's critique of discontinuous evolution, 293–94
   Weldon and Bateson, 186, 286, 293, 296–98, 300, 310–11, 313–19, 321–23
   "The Facts of Heredity in the Light of Mendel's Discovery" (1902), 312–13
Battle of the Pyramids, 48
Baur, Erwin, 355
Bayuda Desert, Sudan, 54
Beachy Head, East Sussex, 141
Beagle Journal (Darwin), 31
Beagle, HMS, 145
Beaufort, Rear Admiral Francis, 107

Beauty-map of the British Isles, 309
Beirut, Lebanon, 54
Beke, Charles, 129
Bell, Dr. A. G., 346
Bella, ship, 237
Belper Committee, 248–49
Benguela, Angola, 65
Bennett, James Gordon Jr., 125
Bentham, George, 98, 104
Berber, Sudan, 52
Bergdamas. *See* Hill Damara
Bernstein, Peter, 7, 250, 256
Bertillon, Alphonse, 8, 232, 234, 240, 245, 248
   Classification system, 232
Bertillonage, 234, 237, 245, 247–48
Besant, Annie, 274
Bidder, George Jr., 224–25, 229
Bidder, George Sr., 224–25, 229
Biggs, Evelyne (Eva)
   Great niece of Francis Galton, 308, 332–33, 340, 344
Binet, Alfred, 229–30
Biometrics Laboratory, 340
Biometrika, 9, 307, 309, 313–14, 318, 323, 333
Biometrika, founding of, 307
Birmingham, Bishop of, 345
Birmingham, England, 13, 18, 20–22
Bishari (Nubian) desert, Sudan, 51, 54
Bishop of Chichester, 127
Bishop, Isabella, 138–39
Blackburn Standard, 171
Blackmore, Richard, 156
Blackwood, John, 115
Blackwood's Magazine, 100, 115
Blake, C., 317
Blois, France, 99
Blood River, battle of, 65
Blue Nile, 51, 53, 108
Board of Trade, 145–47
Bob, Arab boy, 49, 52, 54
Boers, 65–66
Boeters, G., 355
Bolce, Eugenette, 353
Boomplaats, battle of, 65
Booth, Alfred, 324
Booth, Charles, 3, 324–25
Bordeaux, France, 103
Boring, Edwin, 226
Boulogne, France, 25–26
Boulton, Hugh William, 26
Boulton, Mathew, 16, 26
Boulton, Mathew P.W., 26, 39, 48
Boulton, Montagu, 48, 53–55
Bowen, 104

Bowman, Sir William, 27–28, 30, 32
Brachydactyly, 337
Bradshaw, Henry, 272–73, 276
Brain, 221
Brandes, H.W., 142
Breadalbane, Lord, 45
Breakfast-table questionnaire, 221
Brigham, C. C., 354
Brindley, James, 16
British Association for the Advancement of Science
  (BAAS), 105, 206, 225; Aberdeen, (1885), 250;
  Cambridge (1904), 320–23; Dover (1899), 308;
  Plymouth (1877), 216
British Journal of Inebriety, 341
British Quarterly Review, 171
Brock, A., 231
Brooke, Sir James, 99
Brooks, W.K., 287
Browne, Sir J.C., 224
Browning, O., 272
Browtop, Keswick, Cumbria, 40–42
Bruce, James, 108
Buck, Carrie, 353
Buckland, Frank, 157
Bulletin International de l'Observatoire de Paris, 145
Bunbury, Mrs. Robert
  Sister of Francis Galton, 47, 151, 210, 328
  See also Galton, Adelé
Bunbury, Millicent (Milly)
  Niece of Francis Galton, 47
  See also Millicent Lethbridge
Bunbury, Robert Shirley, 47
Burrows, Cordy, 126
Burslem-Trent canal, 14
Burton, Isabel
  Wife of Sir Richard Burton, 118
Burton, Sir Richard, 98, 107, 110–121, 135, 137
Bury, Mr., 25
Bushmen, 83, 85–87
Bushwoman, 76
Butler, Emily
  Sister of Louisa Galton, 95
Butler, George
  Father-in-law of Francis Galton, 94–95
Butler, Henry Montagu, 94
  Brother of Louisa Butler Galton, 94, 209, 327
Butterworth, Capt., 63, 140–41
Buxton, Charles, 44–45

Cairns, J.E., 104
Cairo, Egypt, 48, 54
Calcutta, India 236
Cambrian System, 62
Cambridge Review, 272
Cambridge University
  Coaches (private tutors), 38–40
  Donnishness, 42
  Fellows, 37–38
  Long Vacation, 38
  Pollmen (students with ordinary degrees), 38, 45
  Professors, 38
  Reading parties, 38, 46
  Tripos (honors examination), 38–40, 45
  Tutors, 37–38, 40–41, 44
  Wranglers (students with highest scores on mathematics tripos), 39, 44
Camelthorn, 69–70, 77, 83
  See also thorn trees and thorn bushes
Campbell, Frederick, 43
Canning, George, 17–18, 48
Cape of Good Hope, South Africa, 61–62
Cape Town, South Africa, 63–64, 66–67, 72, 94
Carcinus moenas, shore crab, 281, 283, 299–300
Cardium edule, cockle, 288
Carlyle, Thomas, 106, 193
Caron, A., 337
Carpenter, William .B., 127, 130–31, 134–35, 226
Castle, W.E., 319, 323, 337
Castro, Thomas, 237
Cataract Arab, Shellalae, 49
Catholic World, 171
Cattell, James McKean, 227–30
Cayley, Arthur, 44, 99, 272
Chamberlain, Austin, 346
Chamber's Magazine, 101
Chambord, France, 99
Channell, Mr. Justice, 249
Chapupa, 84–85
Charcot, J.M., 229
Chenonceaux, France, 99
Chepmell, Dr., 301
Chesapeake Zoological Laboratory, 287
Chik (Chikorongo-onkompè), 85–86, 88
China's Law on Maternal and Infant Health Care, 1995, 356
Chipanga, 88
Christo, courier, 48
Chromosome theory of heredity, 323

Chromosomes, 259
Church, Sir W., 345
Churchill, Winston, 346
Cineraria controversy, 296–99, 315
Clark, W.G., 103
Classification and Uses of Fingerprints (Henry), 248
Claverdon, Warwickshire, 47, 94
Clerk-Maxwell, J., 272
Clifford, W., 276
Cobb, H., 274
Cobwebs to Catch Flies, 24
Cold Spring Harbor experimental station, 337
Colenso, J. W., 118
Coleridge, M., 308
Collins, Detective-Sargeant, 249
Collins, Sir William, 345
Composite photography, 7, 215–20
Comte, August, 167–68
Congo River, Congo, 122, 137
Constantinople (Istanbul), Turkey, 53
Contemporary Review, 181
Contréxéville, France, 97
Conway, J., 270
Cooley, 110
Coolidge, President Calvin, 354
Coombe, George, 58, 215
"Co-relations and their Measurements, chiefly from Anthropometric Data, Galton" (1888), 258
Cornhill Magazine, 104
Correlation, 9, 250, 257–258, 298
Correlation coefficient, 9, 258, 281
Cory, William Johnson, 43
Crackanthorpe, Montague, 336, 341
Craddock Nowell: A Tale of the New Forest (Blackmore), 156
Craik, G. L., 18
Crangon vulgaris, common shrimp, 280–81, 285, 289
Crichton-Browne, Sir J., 336
Crimean War, 101–02, 144, 267
Crookes, Sir William, 216
Croom-Robertson, George, 211
Crystal Palace, 94
Curzon, Marquis George, 139
Cyclonic storms, 144–147

Dabby bushes, Namibia, 68
Dacka, 75
Dahabeyah, 49–50, 52, 54
Dahomey (Benin), 118
Daily Chronicle, 171, 245
Daily News, 126, 129, 130, 132–34, 136, 170

Daily Telegraph, 124, 126, 131, 137
Dalhousie, sailing ship, 63, 140–41
Dalyell, Robert, 62
Damaras, 71–91
Damascus, Syria, 54–55
Dar Fur, Sudan, 52
Darbishire, A.D., 314–15, 317–19, 321, 323
d'Arnaud Bey, Joseph Pons, 51
Darwin, Charles Robert, 1, 3, 5–6, 9, 13, 17–18, 31, 62, 96–98, 101, 104, 145, 155–57, 164–65, 167–69, 190–91, 202, 206, 216, 235–36, 245, 260, 286, 288–90, 293–94, 296, 313, 338, 345–46, 356
  Correspondence with Galton, 96, 101, 167, 169, 175–81, 183–85, 191, 202, 216, 235
  Hereditary mechanisms and Pangenesis, 173–79, 183
Darwin, Charles
  Son of Erasmus and Mary Darwin, 18
Darwin, Emma Wedgwood
  Wife of Charles Darwin, 31, 176
Darwin, Erasmus, 3, 13–19, 40, 48, 57, 155, 165, 259
  Affair with Miss Parker, 18
  Appearance, 13–14
  Contributions to natural sciences, 16–17
  Friendship with Josiah Wedgwood, 14, 16
  Invention of speaking machine, 16
  Lunar Society of Birmingham, 16
  Marriage to Mary Howard,
  Paternal grandmother of Charles Darwin, 18
  Maternal grandfather of Francis Galton, 13–14, 17–18, 22, 32
  Paternal grandfather of Charles Darwin, 13, 32
  Marriage to Elizabeth Collier Sacheveral-Pole, Maternal grandmother of Francis Galton, 18–19
Darwin, Erasmus Jr.
  Son of Erasmus and Mary Darwin, 18–19, 31
Darwin, Francis Sacheverall
  Godfather of Francis Galton, 19
Darwin, Sir George
  Son of Charles Darwin, 169, 177, 278, 183, 332, 342, 345
Darwin, Henrietta later Litchfield
  Daughter of Charles Darwin, 17, 176
Darwin, Major Leonard
  Son of Charles Darwin, 219, 345–49, 353
Darwin, Mary (Polly) Howard
  First wife of Erasmus Darwin, 18

Darwin, Robert Waring
  Father of Charles Darwin, 18–19
Darwin, Sir Francis
  Son of Charles Darwin, 345
Darwin, Sir Horace
  Son of Charles Darwin, 345
Darwin, Susannah Wedgwood
  Mother of Charles Darwin, 18
Darwin-Wallace Celebration of the Linnean Society of London, 1908, 338
Davenport, C., 307, 323, 337, 350
Davy, Sir Humphry, 199
Dawson, Lieutenant. Lewellyn, 124–25, 131–33, 136–37
de Candolle, Alphonse, 5, 187–92
de Forest, E., 278
de Marrais, R., 303
de Poitiers, Diane, 99
de Vries, Prof. Hugo, 305, 312–14
Dead Sea, 56
Deal, Kent, 140
Dean, G., 226
Decipherment of Blurred Finger Prints (Galton), 243
Delagoà Bay (Baia de Maputo), Mozambique, 63
Denbigh, North Wales, 125, 131
Deptford Murders, 249
Dewar, Sir James, 256
Dewey, John, 267
Dial and Morning Star, 279
Diamandi, 229–30
Dickson, J. H., 256
Dictionary of Men of the Time (1865), 158
Die Fronica (Pearson), 276
Die Mutationstheorie (de Vries), 313
Diogenes, 109–10
Directional selection, 263–64
Discovery of the Source of the Nile (Speke), 117
Disease, inheritance, 266–67
Disraeli, Benjamin, 106
Disraeli, Isaac, 106
Dongola, Sudan, 54
Donkin, Bryan, 274–75
Donoghue, I., 342, 343
Donovan, phrenologist who examined Galton, 58
"Double Mystery," Wright (1995), 6
Doughty, Sir E., 237
Dover, Kent, 94
Drummond Castle, steamship, 247–48
Du Cane, Sir Edmund, 216–17
Duddeston House, Birmingham
  Home of Samuel Galton, 19, 23
Duff, Sir Mountstuart Grant, 139
Dugdale, Richard L., 345
Dungeness, Kent, 141
Dutch East India Company, 63
Dyer, W. Thiselton, 296–300

Earwigs, 290–92
Ebner, Rev. Mr. Johann Leonhard, 74
Economic Journal, 277
Eddis, 40
Edgeworth, F.B., 256
Edgeworth, Francis Ysidro, 256–57, 269, 276–78
Edinburgh Review, 137, 170
Edward II, King, 37
Eikhams (Windhoek), Namibia, 74, 76–77, 89
Elderton, Ethel, Francis Galton Scholar, 334, 339–40
Elderton, W. Palin, 332, 334, 340
Elephant Fountain, Namibia, 89–90
Elephantine Island, Egypt, 49
Elephants, 84, 90
Eliot, George, 175
Elliotson, John, 216
Elliott, C. W., 346
Ellis, George, 48
Ellis, Guy, 308
Ellis, Havelock, 274–75, 331, 339
Embryonic elements, 180
Emperor Napoleon III, 127
Empress Eugenie, 127
Encyclopedia of Physiology (Bowman and Todd), 30
English Men of Science: Their Nature and Nurture (Galton), 5, 191–93, 251
Epilepsy, heritability of, 349
Erasmus Darwin (Krause), 17
Ercoles, Senorita R.F., 256
Erhardt, James, 110–11
Erongo, Namibia, 77
Eschbach, 308
Espy, J.P., 142
Ethiopia, 52–53
Etosha salt pan, 86
Eugenic sterilization, Germany, 355
Eugenic sterilization, United States, 348, 351, 353, 357
Eugenic sterilization, various countries, 2, 356
Eugenical Sterilization in the United States (Laughlin), 353
Eugenics, 1, 2, 10, 308, 309, 323–57
  Definition, 207
  Negative, 1
  Positive, 1

Eugenics Education Society, 336, 338–42, 344–45, 353
Eugenics Laboratory, 334–35, 339–42, 344
Eugenics Record Office, Cold Spring Harbor, 350
Eugenics Record Office, University of London, 330, 334, 337
Eugenics Review, 339–40, 345
Eupatoria, Crimea, 101, 144
Euphorbia, 77
Evans, Frederick, 147
Evard, Courier, 48, 49, 51, 54
Evening Standard, 245
Everest, Lt. Col George, 107
Evolution Committee of the Royal Society, 298–302, 307, 312
Eye color, inheritamce, 264

Family and Nation (Whetham and Whetham), 352
"Family Likeness in Stature," Galton (1886), 257
Fancher, Raymond, 5, 193, 226
Faraday, Michael, 199
Farmer, Benjamin, 19
Farmer, James
  Brother-in-law of Samuel Galton, 19
Farmer, Mary
  Wife of Samuel Galton, 19
Farrar, Rev. F., 197
Farrar, T. H., 147
Farrow, 249
Faulds, Henry, 235–36, 240–41, 245–47
Fechner, Gustav, 221
Fergus, Walter, 197
Fergusson, James, 98
Fernando Po, Equitorial Guinea 118
Findlay, 115
Finger Print Directories (Galton), 8, 243, 246
Finger Prints (Galton), 8, 241–43, 246–47
Fingerprints, 8, 213, 295
  Characterization, 235, 237, 239
  Classification, 243
  Heritability, 242
  Making imprints, 235–36, 241
  Permanence, 239
  Race, 242–43
  Uniqueness, 241–42
  Uses, 235–37, 240–41
Fingerprints: Fifty Years of Scientific Crime Detection (Browne and Brock), 231
Firman, 48
First International Congress of Eugenics (1912), 10, 323, 344–46

Fischer, Eugen, 355
Fisher, R.A., 10, 312
FitzRoy, Captain Robert, 145–48
FitzRoy's cones, 146
Florence, Italy, 96
Flycatcher, 70
Foam, schooner, 66–67
Fortnightly Review, 165–66, 175, 181, 190, 207, 209, 225, 293
Founder's Medal, Royal Geographical Society, 95
Francolin, redbilled, 70
Fraser's Magazine, 193, 195
Frederick, native leader, 68
French, Mrs., 25
Frere, Hookham, 48
Frere, Robert, 47
Freshfield, D., 138–39
Freud, S., 221
Fuller, George, 119

Gabriel, 66, 76, 79
Galton Laboratory of Eugenics, 2, 334–35, 339–42
  Renamed in 1963, 344
Galton Professorship, 2–3, 344
Galton, Adèle (Delly)
  Sister of Francis Galton, 23–25, 31
  See also Mrs. Robert Bunbury
"Galton and Evolutionary Theory," (1993), Maynard-Smith, 286
Galton, Arthur
  Nephew of Francis Galton, 335
Galton, Captain Douglas
  Cousin of Francis Galton, 61–62, 267
Galton, Darwin
  Brother of Francis Galton, 23, 47, 56, 57, 328
Galton, Edward Wheler
  Nephew of Francis Galton, 333
Galton, Elizabeth Anne (Bessy)
  Sister of Francis Galton, 23, 25, 27, 30
  See also Mrs. Edward Wheler
Galton, Emma (Pemmy)
  Sister of Francis Galton, 23, 25, 31, 39, 46–47, 58, 94, 102, 206, 208, 215, 302, 308, 327, 328, 330
Galton, Erasmus
  Brother of Francis Galton, 23, 47, 327, 333, 339
Galton, Francis
  Appearance, 39, 260–61, 309
  Anthropometrics, 197–99, 205, 207, 209, 211–14, 250–56

Awards and honors
  Congress badge, First International Congress of Eugenics (1912), 345
  Copley Medal of the Royal Society (1910), 342
  Darwin Medal of the Royal Society (1902), 327
  Darwin-Wallace Medal, Linnean Society of London (1908), 338
  Elected to Athenaeum Club (1856), 106
  Elected to Royal Society (1856), 105
  Founder's Medal, Royal Geographical Society (1853), 95
  Honorary Fellow of Trinity College (1902), 327
  Honorary President of the Eugenics Education Society, 1908, 336
  Knighthood, 1909, 340
  Presidential invitation for the BAAS (1905), 332
  Silver medal of the French Geographical Society (1854), 105
Beauty map of the British Isles, 309
Biographical sketch, 3–10
Biometrika, founding of, 307
Birth, 23
Boat accident, 30, 31
Books
  Art of Travel (1855), 4, 98–101
  Decipherment of Blurred Finger Prints (1893), 243
  English Men of Science: Their Nature and Nurture (1874), 5, 191–93, 251
  Finger Prints (1892) 8, 241–42, 246–47
  Finger Print Directories (1895), 8, 243, 246
  Hereditary Genius (1869), 4–6, 81, 157–65, 167–75, 185, 187, 189, 192, 194, 267, 303, 327, 329
  Hints for Travellers, (1865), 98
  Inquiries into Human Faculty and its Development (1883), 7, 207–09, 215, 217
  Kantsaywhere, unpublished (1910), 2, 10, 342–44
  Knapsack Guide for Travellers in Switzerland (1904), 103
  Memories of my Life (1908), 23, 56, 61, 107, 123, 131, 173, 195, 324, 335
  Meteorographica (1863), 142
  Natural Inheritance (1889), 9, 214, 250–51, 253, 258–67, 269, 276, 280, 285–86, 289, 292, 327
  Physical Index to 100 Persons Based on their Measures and Finger Prints (1894), 243
  Noteworthy Families (1906), 332
  Tropical South Africa (1853), 67, 91, 94–96, 98
  Vacation Tourists and Notes of Travel (1864) 102–03
Correspondence with Darwin, 96, 101, 167, 169, 175–81, 183–85, 191, 202, 216, 235
Education
  Cambridge University, 37–46
  See also Cambridge University,
  Early education, 23–26
  Medical studies, 26–31
Efficacy of Prayer, 165, 166, 167
Eugenics
  Address to Sociological Society (1904), 328–30
  Address to Sociological Society (1905), 330–32
  Address to Eugenics Education Society (1908), 338–39
  Defined, 207
  Endowment of Galton Research Fellow, 330
  Eugenic certificates, 333, 334
  Eugenics Education Society, 336, 338–42
  Galton Professorship at the University of London, 234, 344
  Herbert Spencer lecture, Oxford (1907), 334–35
  Hereditary Improvement of the human race, 195–97
  Huxley Lecture (1901), 324–27
  Mediates disputes between the Eugenics Education Society and the Eugenics Laboratory, 339–42
Evolution
  Directional selection, 263–64
  Evolution by discontinuous steps, 286, 288–90, 292
  Evolution Committee of the Royal Society, 298–301
  Hypothesis of organic stability, 260–61, 292–93, 295, 303
  Wallace's critique of discontinuous evolution, 293–96
Fidgets, 205–06
Heredity
  Acquired characteristics, 174, 182–83, 259–60
  Address to Royal Institution (1877), 199–205
  Ancestral Law, 9, 263, 303–05
  Artistic ability, 264, 266
  Disease, heritability of, 266–67, 332

Eye color, 264–65
Galton's theory of inheritance, 179–85
Hereditary elements, 164, 173, 180–85, 261–62
Hybrization of gemmules, 183–84
Mendel, failure to appreciate, 305
Nature versus nurture, 187–93
Pangenesis experiments, 175–79
Pangenesis, critique of, 177
Pangenesis, Galton's apology to Darwin, 178
Predicts reduction division (meiosis), 262
Pedigrees of famous men, 161–65, 167–68
Stature, 250–56, 262–63
Sweetpea experiments, 202–05
Twin studies, 187, 193–94
Hunting and fishing (1847–49), 56, 57
Inventions
  Balance, 40
  Galton's toys, 40
  Quincunx, 200–03, 262
  Registrator, 205
  Rotatory steam engine, 40
  Sun signal, 103
Telegraph machine, 58
Lectures on art of survival (1855–56), 102
Management Committee, Kew Observatory, 149–50
Meteorology
Anticyclone, discovered, 147
Chairs board of inquiry, 147–48
Meteorological Committee member, 148
Weather maps, 142–43, 147
Wind and sailing ships, 149
Namibian expedition
  Aftermath, 90–92
  Andersson joins expedition, 62
  Cape Town, South Africa, 63, 64
  Encounters Chapupa, 84–85
Encounters Chik, 85–87
  Encounters Chipanga, 89
  Encounters Jonker Afrikaner, 67, 72–78, 89–91
  Encounters Kahikene, 81–82
  Encounters lion, 71
  Encounters Nangoro, 87–89
  Ovampoland to Schmelen's Hope, 89
  Schmelen's Hope to Ovampoland, 81–87
  Schmelen's Hope to Tounobis, 89–90
  Walfisch Bay to Schemelen's Hope, 67–77
  Hires Hans Larsen, 72

Hires personnel for expedition, 66
  Observations on Damaras, 80–81
  Observations on Ovampo, 89
  Outfits expedition, 62
  Personnel problems, 79
  Proposes expedition to Lake Ngami, 62
  Rhenish missions and missionaries, 68–77
  Sails to South Africa, 63
  Sir Harry Smith changes Galton's plans, 64–65
Papers
  "Co-relations and their Measurements, chiefly from Anthropometric Data" (1888), 258
  "Discontinuity in evolution" (1894), 292
  "Family Likeness in Stature" (1886), 257
  "Hereditary Talent and Character" (1865), 4, 155–57
  "Hereditary Improvement" (1873), 195–97
  "Identification by Finger-Tips" (1891), 240–41
  "Medical Family Registers " (1883), 209
  "Mental Imagery" (1880), 225
  "On Blood Relationship" (1872), 179–181
  "Psychometric facts" (1879), 220
  "The Median Estimate" (1899), 308
  "Typical laws of heredity" (1877), 199–205
Personal identification
  Address to Royal Institution (1888), 231–35
  Composite photography, 215–20
  Fingerprinting, 213, 237–48
  Fingerprint characterization and classification, 237–40
  Fingerprint heritability, 242
  Fingerprint permanence, 239
  Fingerprint uniqueness, 24–42
  Fingerprints, racial variation, 242, 243
  Fingerprints, relations with Henry Faulds, 235–36, 245–47
  Fingerprints, relations with William Herschel, 235, 239, 240, 247
  Measuring facial resemblance, 333
  Tichborne trial, 237
  Physical activity and mental ability, 39
Personal
  Courtship and marriage, 94–96
  Death of Francis Galton, 344
  Death of Louisa Galton, 301, 302
  Evelyne 'Eva' Biggs, 307–09
  Failure to have children, 96
  Family pedigrees, Table 1-1, 13; Table 2-1, 24

Homes, Victoria Street flat (1854), 96
Homes, 42 Rutland Gate (1857), 97
Nervous breakdowns, 45, 150–51
Personality, 108, 309
Personal difficulties, 206, 209–10
Social life, 97, 98
Strain of his commitments on his wife, 206–07
Psychology
  Breakfast table questionnaire, 221
  Cattell, James, influence on, 227–28
  Explores his own mind, 220
  Impact of field of psychology, 226–27
  Mental imagery, 8, 220–26
  Mesmerism, 215
  Phrenology, 57–58, 215
  Qualities of different races, 168
  Reviews Binet's *Psychologie des Grands Calculaterus et Joueurs d'Échecs*, 229–30
  Spiritualism, 216
Reader, The 104–05
Royal Geographical Society
  Admission of women, 138–39
  Aridity and dry climates, 110–11
  Competitive examinations in geography, 138
  Council service, 107–09, 138
  Dawson relief expedition, 132, 136
  Elected a member, 62
  Founding the Proceedings, 107
  Friction with Stanley, 123, 126–138
  Geographical instruction at universities, 138
  Lake Victoria's climate, 117
  Livingstone's remarks about Galton, 122
  Nile's source, 113, 116, 120
  Opinion of Burton and Speke, 113
  Scientific Purposes Committee, 138
  Shaw affair, 107–08
  Speke 115–16, 119–21
Statistics
  Address to BAAS, Aberdeen (1885) 250, 256–57
  Advises Weldon, 280–81, 283
  Correlation, 250, 257–58, 280–83
  Edgworth correspondence, 257, 278
  Family Variability, 203, 205
  Mid-parent, 251–54
  Nightingale, correspondence, 267–68
  Normal distribution, 157–61, 168, 200–05, 251–56, 280–83, 325–27
  Ogive, 197–98, 262–63
  Quincunx, 200–03, 257, 262
  Percentiling, 214, 262–63

Regression, 199, 203, 205, 214
Regression analysis, 262, 266
Regression coefficient, 7, 203
Regression line, 7, 203
Regression (reversion) to the mean, 203–05, 251–56, 258, 261–63, 325–27
Travels
  Egypt and the Middle East (1845–46), 47–56
  Europe with Bowman and Russell (1838), 27–28
  Germany (1843), 46
  Germany to Constantinople (1840), 33–36
  Eva Biggs, Egypt (1899–1900), 308; Europe (1899), 308; (1902), 327; (1905), 332–33
  Louisa Galton, Europe (1853), 96; (1867–68), 151; (1875–76), 206; (1882), 206; (1897), 301–02
Galton, Hubert, 21
Galton, Louisa Butler
  Wife of Francis Galton, 94–96, 98, 102–03, 150–51, 200, 206, 209–10, 250, 301–02, 328
Galton, Lucy Barclay
  Paternal grandmother of Francis Galton, 21
Galton, Lucy Harriot
  Sister of Francis Galton, 23, 26
  See also Mrs. James Moillet
Galton, Samuel
  Great grandfather of Francis Galton, 14, 19, 20
Galton, Samuel Jr.
  Paternal grandfather of Francis Galton, 20–21
Galton, Samuel Tertius
  Father of Francis Galton, 13, 21–23, 25–27, 29–35, 37, 39–40, 42, 44–47
Galton, Theodore Howard, 39
Galton, Violetta Darwin
  Mother of Francis Galton, 19, 21, 23, 150, 206
Gammage, Mr., 213
Garibaldi, 103
Garrod, Sir Archibald, 312
Garson, J.G., 248
Gassiott, Henry, 93
Gattaca, 10
Gauss, 149
Geddes, Prof. P., 267
Gemmules (germs), 6, 173–181, 183–85
Genera Plantarum (Linnaeus), 16

General Hospital, Birmingham, 26–28, 30, 33
Genetics of human facial features, 220
Gentleman's Magazine, 20
Genty, Madame Eugénie, 210
George III, King, 14
Germ line theory, 6, 157, 183, 276
Ghou Damup also Xou-Daman. *See* Hill Damara
Gibbs, W.F., 47
Giessen, Germany, 33
Gifi, A., 344
Giraffe, 71
Gladstone, W.E., Prime Minister, 213
Glasgow Herald, 274
Gnu, 90
Goddard, Henry, 352, 354–55
Godwin, W., 17
Gondokoro (Juba), Sudan, 115–16, 121
Gonville and Caius College, Cambridge, 39
Gothenberg, Sweden, 62
Gotto, Sybil, 335–36, 338–40
Grant Duff, Sir Mountstuart E., 121
Grant, Captain James A., 106, 115–17, 119, 121, 128–29, 135, 139
Grant, Madison, 354–55
Grant, Ulysses S., 125
Grauwacke, 62
Gray, Lord, 62
Gray. Paul, 1
Great Exhibition, 210
Greeley, Horace, 125
Green, 91
Gresham College, 276
Grey, Maria, 169
Griffiths, Maj. A., 244
Grove, George, 102
Grove, Sir William, 301
Grundriss der menschlichen Erblichkeitslehre und Rassenhygiene (Baur, Fischer, Lenz), 355
Guide to Finger-Print Identification (Faulds), 247
Guinea fowl, 71
Guinea pigs, 321
Gumption-Reviver, 39
Gurney, Emily, 98, 301
Gurney, Russell, 98

Haeckel, Ernst, 338
Hahn, Rev. Mr. Hugo, 73, 76
Haldane, J.B.S., 272
Hallam, Arthur, 43
Hallam, Henry, 43, 47
Hamilton, L., 308

Han Houri, 55
Harriman, M., 337
Harris, Henry
  Galton Professor, 3
Harrismith, South Africa, 64
Hastings, East Sussex, 151
Helmholtz, H., 211
Helvellyn, Cumbria, 41
Hemophilia, 337
Henn, Lt. William, 124
Henri II, King, 99
Henri IV, battleship, 144
Henry VIII, King, 37
Henry, Joseph, 144
Henry, Sir Edward, 8, 235, 245, 248, 249
Henslow, Prof.John S., 32, 38
Herdman, W.A., 332
Hereditary Genius (Galton) 4–6, 81, 157–65, 167–75, 185, 187, 189, 192, 194, 267, 303, 327, 329
"Hereditary Improvement," Galton (1873), 195–97
"Hereditary Talent and Character," Galton (1865), 4, 155–57
Herero. *See* Damara
Heritability of alcoholism, 340
Heron, D., Francis Galton Fellow, 334, 339–40
Herrnstein, Richard, 226
Herschel, Sir John, 62, 103, 142, 149
Herschel, Sir William, 8, 235–37, 239–41, 245–48
Hickson, Prof., 322
Hill Damara, 71, 73, 75, 77
Hints to Travellers (Galton), 98
Hipperley, E.P., 131
Hippopotamuses, 83
Histoire des Sciences et des Savants depuis Deux Siecles (de Candolle), 5, 187–191
History of English Literature (Craik), 18
History of Experimental Psychology (Boring), 226–27
History of Statistics (Stigler), 197
History of the Theory of Elasticity (Todhunter), 272, 276
Hodgkin, Thomas, 108
Hodgson, Joseph, 23, 26–27, 32, 46
Hoffman, Frederick, 351
Hoffmann, G. von, 355
Holmes, Oliver Wendell, 353
Home, David D., 216
Homotyposis, 306–07, 311–12
Hooker, J. D., 97, 104, 192, 202, 260, 338
Hope, A. J. Beresford, 42

Hopkins, T., 109
Hopkins, William, 40, 42–44, 270
Hopkinson, David, 7
Horner, Leonard, 24
Hospitals, Victorian, 28–29
Hottentots, 69, 76
　See also Khoikhoi
Houghton, Lord, 121
How I Found Livingstone (Stanley), 130, 136
Hoy, Orkney Islands, 57
Hudibras (Butler), 24
Huggins, Sir W., 327
Hughes, Tom, 104
Hugo Hahn, Rev. Mr., 72, 91
Hull, Humberside, 62
Huntington's chorea, 337
Huxley, Thomas H., 3, 104, 156, 175, 284, 287, 292, 338
Hyde Park, London, 47
Hyena, 76

Ichabo Island, Namibia, 67
"Identification by Finger-Tips," Galton (1891), 240–41
Iliad (Homer), 24
Images in Mind (Pavio), 226
Immigration Act of 1924, United States, 354
Inaudi, 229–30
Inkerman, Crimea, 101, 144
Inquiries into Human Faculty and its Development (Galton), 7, 207–09, 215, 217
International Health Exhibition (1884), 210–13
Internationale Hygiene-austellung, Dresden (1911), 351
Intracellular Pangenesis (de Vries), 313
IQ test, 230, 354

Jaffa (Tel-Aviv), Israel, 55
Jager Afrikaner, 73
James, Henry, 106
James, Paul Moon, 21
Jebel Kasyun, Syria, 55
Jephson, Dr., 151
Jerusalem, Israel, 55
Jeune, Francis
　Headmaster, King Edward's School, 26, 57
Jevons, W.S., 257
John o'Groat's, Scotland, 57
Johnson, George, 31
Johnson, H. Vaughan, 47
Jones, Mrs.
　Mother of H.M. Stanley, 131
Jonker Afrikaner, 67, 72–78, 89–91

Jonker, Jan, 78
Jordan River, 56
Jordan, David Starr, 345
Journal of Heredity, 338
Journal of Linnean Society, 289
Journal of the Anthropological Institute, 181, 185, 193, 198, 214, 255
Journal of the Royal Geographical Society, 93, 98, 107, 115–17
Jukes, 355

Kadîsha (Abu Ali) river, Lebanon, 55
Kaffir corn, 85
Kahikenè, 73, 81–82
Kalahari Desert, 61, 64
Kallikaks, 354, 355
Kalulu, 130
Kantsaywhere (Galton), 2, 10, 342–44
Kaondoka, 85
Kaptein, 73
Karuma Falls, 115, 121
Katcha Bay, Crimea, 144
Kay, Eben, 44, 47
Kay, Joseph, 43–45
Keir, James, 16
Kellog, Vernon L., 351
Kelvin, Lord, 37, 40
Kenilworth, 25–26, 39
Keswick, Cumbria, 40, 57
Kew, geophysical observatory, 149, 150
Khamseen, desert wind, 54
Khartoum, Sudan, 51–54, 106, 115–16
Khawr al-Adi, Blue Nile tributary, 51
Khoikhoi, 69, 73, 74
　See also Hottentots
Kierie, a Damara club, 72, 82
King Edward's School, Birmingham, 26, 57
King's College, Cambridge, 37, 43, 46
King's College, London, 27, 30–31
Kingsley, Charles, 104
Kingsley, Henry, 156
Kirk, John, 122–24
Kirkwall, Orkney Island, 57
Kleinschmidt, Rev. Mr. F.H. Heinrich, 72–73
Knapsack Guide for Travellers to Switzerland (Galton), 103
Knowles, General, 102
Kolbe, Rev. Mr., 72–73, 78
Konai, R., 236
Kordofan, Sudan, 52
Korosko, Egypt, 50, 52
Krapf, Johann Ludwig, 109
Kudu, 75
Kuisip (Kuiseb) River, Namibia, 68–69

Kunene (Cunene) River, Angola, 87, 89, 90
Kurumen, South Africa, 63

Ladismith (South Africa), 64
Ladysmith (South Africa), 64
Ladywood near Birmingham
 First home of Samuel Tertius Galton, 21
Lake Albert, 99, 115, 119, 121–22, 126
 See also Luta Nzigé
Lake Bangweolo (Bangweulu), Zambia, 122, 125
Lake District, Cumbria, 38, 40
Lake Kivu, 128
Lake Kyoga, 121
Lake Lincoln, 125
Lake Moero (Mweru), Zambia, 122
Lake Ngami (Andersson), 90
Lake Ngami, Botswana, 61–63, 65, 74, 79, 81, 89, 90, 93, 99
Lake Nyasa, 109, 117, 120, 122
Lake of Tiberias (Sea of Galilee), Israel, 56
Lake Tanganyika, 112–14, 117, 120, 122–23, 125–26, 128
Lake Victoria, 109, 111–17, 120–22, 137
 See also Victoria Nyanza or Nyanza
Lamarck, J.-B., 293
Langtry, Lillie, 211
Lankester, Sir E. Ray, 279, 283–84, 289, 300, 338
Lapouge, Georges V. de, 337
Larsen, Hans, 72, 74–77, 80, 82, 84, 89
Lassalle, Ferdinand, 273
Latent elements, 179–82
Laughlin, Harry H., 350, 353–54
Lavater, J.C., 216
Lazarette (quarantine station), 56
Le Verrier, U.J.J., 145, 148
Leamington, Warwickshire, 21–22, 25, 47, 56–57
Lebanon, 54–55
Lebensborn (Well of Life) program, 2
Lenz, Fritz, 355
Leopard, 77
Lethbridge, Millicent Bunbury
 Niece of Francis Galton, 209, 322, 328, 332–33, 335, 344
Lewes, George, 104, 165
 Critique of pangenesis, 175
Lichfield, Staffordshire, 18
Lidbetter, E., 349
Liebig, Justus, 33
Life and Labour of the People of London (Booth), 3, 325
Life-History Album (Galton), 209

The Million of Facts (Phillips), 156
Lions, 69–71
Lisbon, Portugal, 19
Lives of the Chancellors (Lord Campbell), 164
Lives of the Judges (Foss), 161
Livingstone Search and Relief Expedition, 123–25, 131–32, 136–37
Livingstone, David, 4, 61, 63–65, 93, 106, 110, 117, 120, 122–130, 132–33, 135–37
Livingstone, Oswell, 124–25
Locke, John, 228
Lockyer, Norman, 104–05
London Phrenological Institute, 58
London Standard, 124
Longfellow, Henry Wadsworth, 156
Loomis, Elias, 142, 144
Lord Granville, 131, 135
Lord Mayor of London, 346
Lorna Doone (Blackmore), 156
Lourenço Marques (Maputo), Mozambique, 65
Louries, Grey, 71
Loxton, Somersetshire, 47
Lualaba River, Congo, 122, 125, 129
Lubbock, Sir John, 98
Lucas, L.V., 18
Ludlow, J.M., 104
Lunar Circle, 16
Lunar Society of Birmingham, 16, 20–22
Luta Nzigé. *See* Lake Albert
Lutyens, Lady E., 336
Lutyens, Sir Edwin, 336
Lychnis, Campion, 313, 321
Lyell, Sir Charles, 104, 118, 156

Macmillan, Alexander, 102, 156
Macmillan's Magazine, 4, 155–57
Macnaghten, M.L., 244
Madeira, 63
Madison, President James, 64
Maginn, William, 193
Maherero, 90
Mahomed, Dr. F.A., 219
Maine, Henry, 43
Malta, 47
Mameluke, 48
Manchester Examiner, 131
March, Lucien, 337, 351
Markham, Sir Clements, 107, 128, 133–35, 138–39
Marlborough School, 197, 199
Marlborough, Duchess of, 346
Marseilles, France, 56
Marshall, Alfred, 257

Marshall, Cordelia
  Wife of William Whewell, 41
Marshall, John, 224
Marx, Eleanor, 274
Marx, Karl, 273
Maskelyne, Sir Neville, 93
Materials for Study of Variation (Bateson), 9, 290, 92–4, 311–12
Mathematical Psychics: An Essay on the Application of Mathematics to the Moral Sciences (Edgworth), 256
Mathison, 40
Maudsley, Henry, 221, 329
Maxwell, James Clerk, 40
Maynard-Smith, John, 286
McCullough, Hugh, 135
McKendrick, John, 259
McQueen, J., 110, 111, 112, 117
"Medical Family Registers," Galton (1883), 209
Memories of My Life (Galton), 23, 56, 61, 107, 123, 131, 173, 195, 324, 335
Mendel, Gregor, 3, 10,185–86, 259, 303, 305–06, 309–18, 320–22, 329, 341
Mendelian recessive, 312
Mendelians triumph over Ancestrians, BAAS, Cambridge (1904), 320–23
Mendel's Principles of Heredity: A Defence (Bateson), 310, 311, 314, 320
Mental imagery, 8, 220–26
"Mental Imagery," Galton (1880), 225
Menzies, Sir Neil and Lady, 44
Mercier, Dr., 329
Merivale, Herman, 170
Meroë, Sudan, 52. 54
Merrifield, F., 263
Mesmerism, 215–16
Metamorphoses (Ovid), 25
Meteorographica (Galton), 142
Meteorological Committee, 139, 148–49
Meteorological Department
  Board of Trade, 145–47
Mid-parent, 251–254
Miers, Principal, 345
Mill, James, 228
Mill, John Stuart, 228–29
Millais, Sir Everett, 303
Millais, Sir John Everett, 303
Miller, William, 33
Mind, 215, 222, 292, 293
Minot, Prof., 321
Mitchel Grove, ship, 141
Mivart, St. G., 288
Moilliet, James, 47

Moilliet, Mrs. James
  Sister of Francis Galton, 47, 308
  See also Galton, Lucy Harriot
Monro, Alexander, 31
Mont Dore les Bains, France, 97
Moral Education League, 336
Morgan, Thomas Hunt, 352
Morley, John, 166
Morning Advertiser, 117
Morning Post, 170, 245
Morris, P.E., 226
Morta, John, 66, 71, 74, 77, 80, 83
Mott, Dr. F. W., 331, 336, 349
Mouse genetics, 314–19
"Mr Bateson's Revision of Mendel's Theory of Heredity," Weldon (1903), 317
Mt. Kenya, Kenya 109
Mt. Eshuameno, Namibia, 82
Mt. Ja Kabaka, 82
Mt. Kilima-Njaro (Kilimanjaro), Tanzania 109
Mt. Omatako, Namibia, 82
Mt. Omuvereoom, Namibia, 82, 83
Muir, R., 249
Multan, Pakistan, 55
Murchison, Sir Roderick Impey, 62, 95, 104, 106, 108–09, 111, 113, 115–124, 132–33
Murie, Dr., 176
Murray, Charles, 226
Murray, John, publisher, 94
Mytton, Jack, 56
Mytton, Jack Sr., 56

Naarip plain, Namibia, 70, 75
Nablus, Jordan, 103
Naegeli, Karl, 186
Nama *See* Namaqua
Namaqua, 72, 73–74, 76, 78–79, 90
Namaqualand, 73
Namibian tribes described, 73–74
Nangoro, 81, 84–88, 91
'nara vine, Namibia, 68
Nasheer, Sheikh Nair Abu, 56
Natal, South Africa, 65
National Reformer, 171
Natural Inheritance (Galton), 9, 214, 250–51, 253, 258–67, 269, 276, 280, 285–86, 292, 327
Nature, 171, 178, 208, 235–37, 240, 245, 24–47, 273, 278, 293, 297–98, 307–08, 315–16, 321, 333, 346
Nembungu, 87
Netjo, 85–86

Neverwas, Mr., 342
New Phytologist, 320
New York Herald, 124–26, 134, 137
New, Rev. Charles, 126
Newton, Isaac, 38, 293
Niceforo, A., 351
Nightingale, Florence, 26–68
Nile, 99
Nineteenth Century, 220–21, 227, 240
Nordau, Max, 331
North American Review, 103
Noteworthy Families (Galton), 332
Nuremberg Laws of 1935, 356
Nyanza, *See* Lake Victoria.

O'Brien, Matthew, 39–40
Odyssey (Homer), 24
Oenothera lamarckiana, evening primrose, 313
Ogive, 197–98 262–63
Okahandja, Namibia, 73
Okamabuti, Namibia, 84
Oliphant, Laurence, 98, 115, 118
Oliver, D., 279
Omagundè, 81–82
Omagundè's son, 81
Omanbondè, Namibia, 83
Omobondè, Namibia, 81
Omoramba River, Namibia, 83, 84
Omutchamatunda, Namibia, 86
"On Blood Relationship" Galton (1872), 179–181
Onanis, Namibia, 75
Ondonga, Namibia, 86
Onesimus, 79–80
Orange River, South Africa, 63–65, 73
Organic Stability, 260–61, 292–93, 295, 303
Origin of Species (Darwin), 1, 32, 155, 335
Orlam, 73, 78
Orton, Arthur, 237
Oswell, W.C., 93
Otchikongo, Namibia, 85
Otchikoto, Namibia, 86
Otjikango, Namibia, 73
Otjimbinguè, Namibia, 70, 77, 90
Otjironjuba, Namibia, 83
Ovampo, 81, 84–89, 91–92
Oxpeckers, yellowbilled, 89

Paivio, Alan, 226
Pall Mall Gazette, 274
Palmerston, Lord, 102
Pangenes (de Vries), 313
Pangenesis, Provisional Hypothesis, 6, 164, 168, 173–75, 177–78, 180–81, 185

Parker, Miss
   Mistress of Erasmus Darwin, 18
Parker, R., 274
Parker, Sir Hyde, 62–63, 94
Parkyns, Mansfield, 53, 54, 107
Parry, Maj., 33
Partridge, Richard, 30–31, 39
Pasha, Muhammad Ali, 48
Patent elements, 180–81
Patron's Medal, Royal Geographical Society, 95
Pearson, Amy
   Sister of Karl Pearson, 270
Pearson, Arthur
   Brother of Karl Pearson, 270
Pearson, Egon
   Son of Karl Pearson, 273
Pearson, Fanny
   Mother of Karl Pearson, 270
Pearson, Hesketh, 308–09
Pearson, Karl, 4, 9–10, 22, 58, 149, 157, 203, 205, 214, 219, 239, 243, 258, 263–64, 266, 269–79, 281, 283–85, 300, 303, 307, 309–12, 314, 318–19, 321, 328, 330, 333–35, 338–42, 344
   Ancestral Law, 305
   Bateson and Pearson, 300, 305–07, 310–12, 322
   Books
     Die Fronica (1887), 276
     The Ethic of Freethought (1888), 273–74
     The Grammar of Science (1892), 273, 276
     The New Werther (1880), 273
   Boyle lecture, Oxford (1907), 334
   Cambridge meeting of the BAAS (1904), 322
   Changes Carl to Karl, 272
   Director, Galton Laboratory, University of London, 334
   Edgeworth and Pearson, 277–78
   Education, 270–72
   Family, 269–70
   Fellowship, King's College, 272
   Galton Professor of Eugenics, University College, London, 2, 276, 323, 344
   Galton's influence on, 276–77
   Germany, passion for, 273
   Goldsmid Professor at University College, 275–76
   Gresham Lecturer, Gresham College, 276
   Homotyposis, 306–07, 311–12
   Men and Women's Club, 274, 276
   Professor of Applied Mathematics, University of

London, 320
  Proposes Francis Galton Laboratory for the Study of National Eugenics, 334
  Schreiner, Olive and Pearson, 274–75
  Socialism, 273, 274
  University of London politics, 283–85
  Weldon and Pearson, 283–85, 300, 307, 310
Pearson, William
  Father of Karl Pearson, 269–70
Peas (*Pisum*), 185–86, 309–310
Peel, Robert, Prime Minister, 41
Pelly, Sir Lewis, 98
Pembroke College, 44
Peninsular War, 64
Penrose, Lionel
  Galton Professor, 3
Pentland Firth, 57
Percentiling, 214, 262–63
Pereira, I., 65
Petherick, John, 115–16, 120, 129
Petrie, W. M. Flinders, 308
Pharoah's Rat (ichneumon), 54
Phillips, Sir Thomas, 156
Phillipus, 79, 80
Philosophical Magazine, 278
Philosophical Transactions of the Royal Society, 278, 288
Phrenology, 57–58, 215–16
Physical Index to 100 Persons Based on their Measures and Finger Prints (Galton), 243
Physigonomy, 216
Pinard, A., 337
Pinkerton's Travels, 98
Pioneers of Psychology (Fancher), 5, 193, 226
Pitt, William Prime Minister, 17
Ploetz, Alfred, 336, 345, 355
Plymouth, Devon, 140
Political Justice (Godwin), 17
Polygenic, 313
Poole, Mr., 222
Pope, William, 105
Port Elizabeth, South Africa, 64
Pretorius, Andries, 65
Priestley, Joseph, 16, 20–22
Prince, HMS, 144
Principia (Newton), 293
Proceedings of the Royal Geographical Society, 108–10, 117, 125, 133, 139
Proceedings of the Royal Society, 9, 177, 181, 250, 257–58, 280, 281
Psychological Examining in the United States Army (Yerkes), 354
Psychologie des Grands Calculaterus et Joueurs d'Échecs (Binet), 229

"Psychometric Facts," Galton (1879), 220
Ptolemy, 108
Puericulture, 337
Punch, 214
Punnett, Reginald, 321, 352
Purkinje, Johann E., 234
Purple Thorn Moth, 263
Pyrenees, 103

Quarterly Journal of Science, 216
Queen Victoria, 131, 134–35, 138, 146
Quetelet, Adolph, 158, 200, 257, 277
Quillimane (Quelimane), Mozambique, 65
Quincunx, 200–03, 257, 262

Rabbits, 173, 175–77, 179
Radburn Hall, near Derby, Derbyshire, 18
Randall, Sgt., 269
Rath, Rev. Mr., 71
Rawlinson, Gen. Sir Henry, 123–25, 127, 129, 134–36
Reader, The, 104–05
Rebmann, Johann, 110
Red Nation, 73–74
Redfield, W.C., 142
Regression, 199, 203, 205, 214
Regression analysis, 262, 266
Regression coefficient, 7, 203
Regression line, 7, 203
Regression (reversion) to the mean, 9, 203–05, 251–56, 258, 260, 261–63, 314, 325–27
Rehoboth, Namibia, 73
Reid, W., 346
Rëis (captain), 49
Rhinoceros, 74–75, 90
Rhyl Journal, 131
Richardson, J. T. E., 226
Richterfeldt, Namibia, 70–72, 74–77
Riebeeck, Jan van, 63
Ripon, Bishop of, 345
Ripon, The Earl of, 114
Robben Island, South Africa, 63
Roberts, Charles, 199
Robertson, L., 216
Robertson, George C., 222
Rodwell, G.F., 197
Rollet, M. Brun, 114
Romanes, George, 208
Rome, Italy, 96
Rosebery, Lord, 335
Rossetti, Christina, 156
Routh, Edward J., 40, 270
Rovuma River, Tanzania, 122

Rowlands, John, 125, 131
  See also Stanley, Henry Morton
Royal Anthropological Institute, 324
Royal Charter, ship, 146
Royal Geographical Society, 4, 62, 95, 107–17, 120–22, 123–25, 126–29, 131–39
Royal George, ship, 62
Royal Institution, 199–200, 231
Royat, France, 97, 301
Ruabah, Emir, 56
Rücker, Sir Arthur, 330
Rusizi River, 112, 125, 127–28
Russell, 27–28
Russell, G.W.E., 106

Sabine, Gen.l Sir Edward, 149–50, 177
Sacheveral-Pole, Colonel Edward
  First husband of Elizabeth Darwin, 18
Sacheveral-Pole, Elizabeth Collier
  Second wife of Erasmus Darwin, 18
St. Asaph, North Wales, 125, 131
St. Helena, John, 76, 80
St. Helena, Atlantic Ocean, 67, 90
St. John's College, Cambridge, 280, 286, 288–89
St. George's Hospital Reports, 199
St. Giles citadel, Tripoli, Lebanon, 55
St. Magnus cathedral, Orkney Island, 57
Salahieh, Syria, 55
Saleeby, C.W., 339, 341
Saltations, 254, 288, 294
Samson, HMS, 144
Saturday Review, 117, 170, 207
Saunders, Miss C.B., 313
Saunders, Miss E.R., 312, 321
Schallmeyer, Wilhelm, 336, 355
Scheppmann, Rev. Mr., 69
Scheppmansdorf, Namibia, 68–69, 74–75
Schimmelpenninck, Mary Anne, 19, 21
Schmelen's Hope, Namibia, 72–74, 77, 89
Schmidt, German phrenologist, 57
Schöneberg, Rhenish missionary, 68
Schreiner, Olive, 274–75
Schuster, Edgar
  Galton Research Fellow, 330, 332–34
Schweinfurth, 129
Scotland, 54
Scott, Sir Walter, 24
Scottish Leader, 267
Sea of Uniamesi, 109
Sebastopol, Crimea, 101–02, 144
Second Sikh War, 55
Sedgwick, Adam, 62, 289
Seward, G.E., 122

Shambok, 74
Sharpe, Elizabeth, 274
Sharpe, Maria
  Wife of Karl Pearson, 275
Shaw, Dr. H. Norton, 109–11
Shaw, George Bernard, 274, 329–30
Sheik of the Bishari Desert, 51
Sheikh of Aden, 55
Sheikh of the Cataract, 49
Shendi, Sudan, 52
Sheppard, W.F., 277
Shirreff, Emily, 169–170
Shrike, fiscal, 74
Shrubsole, W.H., 280
Silurian System, 62
Simon, T., 230
Singh, Ranjur, 64
Skewed normal distributions, 277–78
Skiddaw Mountains, Cumbria, 40, 41
Slaughter, Dr. J.W., 339
Small, Dr. William, 16
Smith, Lady Juana
  Wife of Sir Harry Smith, 64
Smith, Sir Harry, 64–65, 74
Smithfield, South Africa, 64
Socialist Song Book, 273
Society for the Study of Inebriety, 336, 341
Sociological Society, 328, 330, 332, 336
Somalia, 109
Somerville, Mary, 41
South Ronaldsay, Orkney Islands, 57
Spa, Belgium, 97
Solar eclipse, 102
Spectator, 104, 167, 208, 267
Speke, Captain John H., 107, 110–22, 128, 137
Spencer, Herbert, 97, 104–05, 156, 260, 269
Sperm banks, 10
Spiritualism, 216
Spottiswoode, William, 3, 97, 109, 142, 158, 209, 260
Springbok, 86
Standard, 245
Stanley, Henry Morton, 4, 107, 121, 123–38
Stanton, H. de, 37
Statistical Methods (Davenport), 307
Stature, inheritance, 250–56, 262–63
Steatopygia, 73
Steinbok, 77
Stewart, General J. Shaw, 121
Stewartson, 69–71
Stigler, Stephen, 197
Stink-wood, 84
Stirp, 173, 181–83
Stoddard, Lothrop, 355

Stokes, G., 272
Story of Namibia (Levinson), 79
Strachey, Sir Richard, 138–39, 150
Strachey, Lytton, 106
Strangford, Viscount, 117
Strasburger, E., 338
Stratford-on-Avon, Warwickshire, 47
Stratton brothers, 249
Streptocarpus rexii, 289
Structureless elements, 180
Stubbing, Rev. T.R., 322
"Studies on Plant Hybridization" Mendel (1865), 185
Sudd, Sudan, 53
Sun Signal, 103
Sutton, W.S., 323
Swakop River, Namibia, 69–73, 77, 80
Swartboy, 79
Sweet peas, 202–03, 205, 251, 253, 269
Switzerland, 96
Sykes, Colonel, 110, 112
Synoptic weather charts, 142
Syria, 54–55

Table Bay, South Africa, 63
Table Mountain, South Africa, 63
Tabora, Tanzania, 111
Tanner, J.M., 197
Taylor, Tom, 43, 104
Temple of Osiris at Philae, Egypt, 49
Tennyson, Alfred Lord, 43, 156
Terman, Lewis, 354
Thackeray, W.M., 193
The Anti-Jacobin, 17
The Basset Hound Club Rules and Studbook (Millais), 303
The Bell Curve (Herrnstein and Murray), 226
The Botanic Garden (E. Darwin) Parts I and II, 16, 48
The Common Sense of the Exact Sciences (Clifford), 276
The Connexion of the Physical Sciences (Somerville), 41
The Descent of Man and Selection in Relation to Sex (Darwin), 5, 205
"The Double-Edged Effect of Sir Francis Galton: A Search for the Motives in the Biometrician-Mendelian Debate," de Marrais (1974), 303
The Ethic of Freethought (Pearson), 273–74
The Grammar of Science (Pearson), 273, 276
The Hillyars and the Burtons: A Story of Two Families (Henry Kingsley), 156

"The History of Twins, As a Criterion of the Relative Powers of Nature and Nurture," Galton (1875), 193
The Jukes (Dugdale), 345
The Lake Regions of Central Africa (Burton), 115
The Larches, Second home of Samuel Tertius Galton, 21–22, 25
The Law of Heredity: A Study of the Cause of Variation and the Origin of Living Organisms (Brooks), 287
The Loves of The Triangles (Canning), 17, 48
The New Werther (Pearson), 273
The Passing of the Great Race (Grant), 354
The Record of Family Faculties (Galton), 209
The Silurian System (Murchison), 62
"The Statistics of Intercomparison" Galton (1875), 257
The Stuffed Owl, 18
The Temple of Nature (E. Darwin), 17
The Trinity; a Nineteenth Century Passion Play (Pearson), 272
Thicknesses, R., 274
Thompson, Dr., 55
Thorn trees and thorn bushes, 80, 83, 86, 90
See also camelthorn
Tichborne, Roger, 237, 240, 243
Tidman, A., 61
Timboo, 66, 70, 74, 80, 84
Time magazine, 1
Times, 102, 124, 126, 139, 146–48, 158, 170, 225, 226, 283–84, 322, 332, 341–42
Times Literary Supplement, 335
Tjobis Fountain, Namibia, 71
Tjobis River, Namibia, 71
Todd, Prof. Richard, 30
Todhunter, I., 272
Tom Brown's Schooldays (Hughes), 104
Tounobis, Botswana, 89–90
Transiliencies, 292
Trans-Orangia (South Africa), 65
Traveller's Club, 106
Treasury of Human Inheritance (Pearson), 339
Trinity College, Cambridge, 37–45, 327
Tripoli (Tarabulus), Lebanon, 55
Trollope, Anthony, 165
Tropical South Africa (Galton), 67, 91, 94–95, 98
Troup Committee, 244–45
Troup, C.E., 244
Tuke, Dr. D. H., 224
Tunbridge, J.B., 245
Twass, Botswana, 90
Tweedie, A., 346

Twins, 5, 187, 193–94
Tyndall, John, 104, 199
"Typical Laws of Heredity," Galton (1877), 199–205

Ujiji. *See also* Lake Tanganyika, 110, 112, 123–124, 127
University of London, reform, 270, 283–84

Vacation Tourists and Notes of Travel, (Galton), 102–03
Vaillant, Marshall J. B. P., 145
van Wagenen, Bleecker, 350
Variation of Animals and Plants under Domestication (Darwin), 6, 157, 173, 175, 179
Venable, 44
Venn, John, 278
Vererbung und Auslese im Lebenslauf der Völker (Schallmeyer), 336
Vichy, France 97
Vickery, Dr. Alice D., 331
Victoria Nile, 115

Wadi Halfa, Sudan, 52, 54, 308
Waggoner, John, 66, 79
Wales, Prince of, 117
Walfisch (Walvis) Bay, Namibia, 66–68, 89–90, 93
Wallace, Alfred Russel, 185, 293–96, 298–99, 338
Walpole, Horace., 311
Washburn, Ambassador, 126
Washington, D.C., 64
Waterloo, Belgium, 64
Watt, James, 16
Webb, Beatrice, 274, 325
Webb, Sidney, 274
Weber, 149
Wedgwood, Josiah
    Grandfather of Charles Darwin, 14
Weeks, Dr. D. F., 349
Weismann, August, 6, 157, 183, 276, 338, 346, 355
Welch's Hotel, Cape Town, South Africa, 63
Weldon, Anne
    Mother of W.F.R. Weldon, 279
Weldon, Dante
    Brother of W.F.R. Weldon, 279
Weldon, Florence Tebb
    Wife of W.F.R. Weldon, 280
Weldon, W.
    Father of W.F.R. Weldon, 279
Weldon, W.F.R., 9, 269, 279–86, 280, 289, 293, 298–300, 303, 307, 314, 317, 319, 320, 321, 322, 323, 329, 330, 333

"Ambiguity of Mendel's Categories" (1903), 313
Ancestral Law experiments, 314–18
Bateson and Weldon, 186, 286, 293, 296–98, 300, 310–11, 313–19, 321–23
Cambridge meeting of the BAAS (1904), 321–22
Criticizes Mendel's theory, 309–11, 313–14
Death of, 323, 333
Demonstrator at Cambridge, 280
Evolution Committee of the Royal Society, 299
Family and Education, 279–80
Jodrell Professor at University College, London, 289
Lecturer at St. John,s College, Cambridge, 280
Marriage, 280
Pearson and Weldon, 283–85, 300, 307, 310
University of London politics, 283–85
Variation studies 280–83
Weldon Medal, 333
Weldon Memorial, 333
Weldon's Register of Facts and Occurrences relating to Literature, the Sciences and the Arts (W. Weldon), 279
Wellington, Duke of, 64, 102
Wells, H.G., 329
Wenzel, Abraham, 66, 79
Westminster Gazette, 328, 336
Westminster Review, 95, 101
Wetherick, N.E., 226
Wheler, Edward, 47
Wheler, Mrs. Edward
    Sister of Francis Galton, 22, 47, 150, 308, 327, 333
    See also Galton, Elizabeth Anne
Whetham, C. D., 352
Whetham, W. C. D., 352
Whewell, William
    Master of Trinity College, Cambridge, 3, 37, 41–42, 106
White Nile, 53, 108, 111, 113, 116, 121, 125
Wildbad, Germany, 97, 301
Williams, John, 66
Williams, Serjeant, 211
Wissler, Clark, 228
Withering, William, 22
Wordsworth, Christopher
    Master of Trinity College, Cambridge, 38, 41–42
Worshipful Company of Drapers, 340
Wright, Lawrence, 6
Wundt, Wilhelm, 211, 226–27

Xhosa, 64, 65
Xylotrupes gideon, Javanese beetle, 290, 292

Yates, 137
Yeats, W.B., 106
Yerkes, Robert, 354
Yorkshire Post, 245
Yr Herald Cymraeg, 125

Yule, Col., 224
Yule, George U., 278, 319–20

Zambezi River, 65, 108
Zanzibar, 109, 115, 125, 137
Zebras, 86, 90
Zoonomia, or the Laws of Organic Life (E. Darwin), 16–17
Zulu, 65

Lightning Source UK Ltd.
Milton Keynes UK
UKOW042255130213

206265UK00004B/23/A